T0221116

Space Flight Dynamics

Aerospace Series

Space Flight Dynamics

Craig A. Kluever
University of Missouri-Columbia, USA

This edition first published 2018
© 2018 John Wiley & Sons Ltd

The right of Craig A. Kluever to be identified as the author of this work has been asserted in accordance with law.

Registered Offices
John Wiley & Sons, Inc., 111 River Street, Hoboken, NJ 07030, USA
John Wiley & Sons Ltd, The Atrium, Southern Gate, Chichester, West Sussex, PO19 8SQ, UK

Editorial Office
The Atrium, Southern Gate, Chichester, West Sussex, PO19 8SQ, UK

For details of our global editorial offices, customer services, and more information about Wiley products visit us at www.wiley.com.

Wiley also publishes its books in a variety of electronic formats and by print-on-demand. Some content that appears in standard print versions of this book may not be available in other formats.

Library of Congress Cataloging-in-Publication Data

Names: Kluever, Craig A. (Craig Allan), author.
Title: Space flight dynamics / by Craig A. Kluever.
Description: First edition. | Hoboken, NJ : John Wiley & Sons, 2018. |
   Includes bibliographical references and index. |
Identifiers: LCCN 2017042818 (print) | LCCN 2017054455 (ebook) | ISBN
   9781119157908 (pdf) | ISBN 9781119157847 (epub) | ISBN 9781119157823 (cloth)
Subjects: LCSH: Astrodynamics. | Space flight.
Classification: LCC TL1050 (ebook) | LCC TL1050 .K555 2018 (print) | DDC
   629.4/1–dc23
LC record available at https://lccn.loc.gov/2017042818

Cover design by Wiley
Cover image: An Atlas V rocket with NASA's Juno spacecraft lifts off from Space Launch Complex 41 of the Cape Canaveral Air Force Station in Florida. Photo credit: Pat Corkery, United Launch Alliance

Set in 10/12pt Warnock by SPi Global, Pondicherry, India

Printed in the UK

Contents

Series Preface

The field of aerospace is multi-disciplinary and wide-ranging, covering a large variety of products, disciplines and domains, not solely in engineering but also in many related supporting activities. These combine to enable the aerospace industry to produce innovative and technologically advanced vehicles. The wealth of knowledge and experience that has been gained by expert practitioners in the various aerospace fields needs to be passed onto others working in the industry and also researchers, teachers and the student body in universities.

The *Aerospace Series* aims to provide a practical, topical and relevant series of books aimed at people working within the aerospace industry, including engineering professionals and operators, engineers in academia, and allied professions such commercial and legal executives. The range of topics is intended to be wide-ranging, covering design and development, manufacture, operation and support of aircraft, as well as topics such as infrastructure operations and current advances in research and technology.

There is currently a renewed interest world-wide in space, both in terms of interplanetary exploration, and its commercialisation via a range of different opportunities including: communications, asteroid mining, space research and space tourism. Several new companies have been set up with the aim of exploiting the commercial opportunities. A fundamental issue for any space mission is how to get the system into space and then how to control its trajectory and attitude to complete the mission objectives.

This book, *Space Flight Dynamics*, provides a comprehensive coverage of the topics required to enable space vehicles to achieve their design goals whilst maintaining the desired performance, stability and control. It is a very welcome addition to the Wiley Aerospace Series.

Peter Belobaba, Jonathan Cooper and Allan Seabridge

Preface

This textbook is intended for an introductory course in space flight dynamics. Such a course is typically required for undergraduates majoring in aerospace engineering. It is also frequently offered as an elective in mechanical and aerospace engineering curricula. Whether taken for required or elective credit, this course is usually taken in the junior or senior year, after the student has completed work in university physics, rigid-body dynamics, and differential equations. A brief survey of university catalogs shows that titles for these courses include *Orbital Mechanics, Astrodynamics, Astronautics,* and *Space Flight Dynamics.* The principal topic covered in essentially all courses is two-body orbital motion, which involves orbit determination, orbital flight time, and orbital maneuvers. A secondary topic that appears in many of these courses is spacecraft attitude dynamics and attitude control, which involves analyzing and controlling a satellite's rotational motion about its center of mass. A number of space flight courses also cover topics such as orbital rendezvous, launch trajectories, rocket propulsion, low-thrust transfers, and atmospheric entry flight mechanics. The primary goal of this textbook is to provide a comprehensive yet concise treatment of all of the topics that can comprise a space flight dynamics course. To my knowledge, a single space flight textbook that covers all the topics mentioned above does not exist.

A secondary goal of this textbook is to demonstrate concepts using real engineering examples derived from actual space missions. It has been my experience that undergraduate students remain engaged in a course when they solve "real-world" problems instead of academic "textbook" examples. A third goal is to produce a readable textbook with a conversational style inspired by my textbook-author role model, John D. Anderson, Jr. *Space Flight Dynamics* is a distillation of 20 years of course notes and strategies for teaching space flight in the Mechanical and Aerospace Engineering Department at the University of Missouri-Columbia.

Chapter 1 is a brief historical overview of the important figures and events that have shaped space flight. Chapter 2 provides the foundation of this textbook with a treatment of orbital mechanics. Here we are able to obtain analytical expressions for the orbital motion of a small body (such as a satellite) relative to a large gravitational body (such as a planet). Chapter 3 extends these concepts with a discussion of orbit determination, that is, the process of completely characterizing a satellite's orbit. In Chapter 4 we present Kepler's time-of-flight equations which allow us to predict a satellite's orbital position at a future (or past) time. We also discuss Lambert's problem: the process of determining an orbit that passes through two points in space separated by a particular flight time.

Chapter 5 introduces orbital perturbations that arise from the non-spherical shape of the attracting body, third-body gravity forces, and atmospheric drag. Perturbations cause the satellite's motion to deviate from the analytical solutions we obtained for the two-body motion studied in Chapters 2–4. We also introduce the restricted three-body problem where gravitational forces from two primary bodies (such as the Earth and moon) simultaneously influence the satellite's motion.

Chapter 6 presents fundamentals of rocket propulsion and launch trajectories. This chapter serves as a key transitional link to subsequent chapters that involve orbital maneuvers. Chapter 6 shows that burning a given quantity of rocket propellant corresponds to a change in orbital velocity, or Δv. The next four chapters involve orbital maneuvers, where the performance metric is typically the Δv increment. Chapter 7 discusses orbital changes achieved by so-called impulsive maneuvers where a rocket thrust force produces a velocity change in a relatively short time. Chapter 8 treats relative motion and orbital rendezvous, where a satellite moves in proximity to a desired orbital location or another orbiting satellite. In Chapter 9, we discuss low-thrust orbit transfers where an electric propulsion system provides a continuous but small perturbing thrust force that slowly changes the orbit over time. Interplanetary trajectories are treated in Chapter 10. Here we analyze a space mission by piecing together three flight segments: a planetary departure phase, an interplanetary cruise phase between planets, and a planetary arrival phase.

Chapter 11 introduces atmospheric entry or the flight mechanics of a spacecraft as it moves from orbital motion to flight through a planetary atmosphere. Here we develop analytical solutions for entry flight both with and without an aerodynamic lift force.

Chapters 2–11 involve particle dynamics, where we treat the satellite as a point mass. The last two chapters involve analyzing and controlling the rotational motion of a satellite about its center of mass. Chapter 12 presents attitude dynamics, or the analysis of a satellite's rotational motion. Topics in Chapter 12 include rotational motion in the absence of external torques, spin stability, and the effect of disturbance torques on rotational motion. Chapter 13 presents an introduction to attitude control. Here we primarily focus on controlling a satellite's angular orientation by using feedback and attitude control mechanisms such as reaction wheels and thruster jets.

Numerous examples are provided at key locations throughout Chapters 2–13 in order to illustrate the topic discussed by the particular section. Chapters 2–13 also contain end-of-chapter problems that are grouped into three categories: (1) conceptual problems; (2) MATLAB problems; and (3) mission applications. Many of the example and end-of-chapter problems illustrate concepts in space flight by presenting scenarios involving contemporary and historical space missions.

Appendix A presents the physical constants for celestial bodies. Appendix B provides a brief review of vectors and their operations and Appendix C is a review of particle kinematics with respect to inertial and rotating coordinate frames.

My intent was to write a comprehensive yet concise textbook on space flight dynamics. A survey of 35 space flight courses offered by US aerospace engineering programs shows that nearly half (17/35) are "orbits only" courses that focus on orbital mechanics, orbit determination, and orbital transfers. The remaining (18/35) courses include a mix of orbital motion and attitude dynamics and control. In addition, more than one-third (13/35) of the surveyed courses cover rocket performance and atmospheric entry. Few existing space flight textbooks adequately cover all of these topics. I believe that this textbook has the breadth and depth so that it can serve all of these diverse space flight courses.

Several people have contributed to the production of this textbook. Many reviewers provided valuable suggestions for improving this textbook and they are listed here:

Jonathan Black, Virginia Polytechnic Institute and State University
Craig McLaughlin, University of Kansas
Eric Monda, United Launch Alliance
Erwin Mooij, Delft University of Technology
Henry Pernicka, Missouri University of Science and Technology
David Spencer, The Pennsylvania State University
Srinivas Rao Vadali, Texas A&M University
Ming Xin, University of Missouri-Columbia

I am grateful for Jonathan Jennings' help with figures and illustrations. Finally, I would like to thank my wife Nancy M. West for her patience, encouragement, and skilled editorial work throughout this project. This book is dedicated to her.

University of Missouri-Columbia, May 2017   *Craig A. Kluever*

About the Companion Website

Don't forget to visit the companion website for this book:

www.wiley.com/go/Kluever/spaceflightmechanics

There you will find valuable material designed to enhance your learning, including:

- Solutions
- M files

Scan this QR code to visit the companion website

1

Historical Overview

1.1 Introduction

Before we begin our technical discussion of space flight dynamics, this first chapter will provide a condensed historical overview of the principle contributors and events associated with the development of what we now commonly refer to as *space flight*. We may define space flight as sending a human-made satellite or spacecraft to an Earth orbit or to another celestial body such as the moon, an asteroid, or a planet. Of course, our present ability to launch and operate satellites in orbit depends on knowledge of the physical laws that govern orbital motion. This brief chapter presents the major developments in astronomy, celestial mechanics, and space flight in chronological order so that we can gain some historical perspective.

1.2 Early Modern Period

The fields of astronomy and celestial mechanics (the study of the motion of planets and their moons) have attracted the attention of the great scientific and mathematical minds. We may define the *early modern period* by the years spanning roughly 1500–1800. This time frame begins with the late Middle Ages and includes the Renaissance and Age of Discovery. Figure 1.1 shows a timeline of the important figures in the development of celestial mechanics during the early modern period. The astute reader will, of course, recognize these illuminous figures for their contributions to mathematics (Newton, Euler, Lagrange, Laplace, Gauss), physics (Newton, Galileo), dynamics (Kepler, Newton, Euler, Lagrange), and statistics (Gauss). We will briefly describe each figure's contribution to astronomy and celestial mechanics.

The first major figure is Nicolaus Copernicus (1473–1543), a Polish astronomer and mathematician who developed a solar-system model with the sun as the central body. Galileo Galilei (1546–1642) was an Italian astronomer and mathematician who defended Copernicus' sun-centered (or "heliocentric") solar system. Because of his heliocentric view, Galileo was put on trial by the Roman Inquisition for heresy and spent the remainder of his life under house arrest.

Space Flight Dynamics, First Edition. Craig A. Kluever.
© 2018 John Wiley & Sons Ltd. Published 2018 by John Wiley & Sons Ltd.
Companion website: www.wiley.com/go/Kluever/spaceflightmechanics

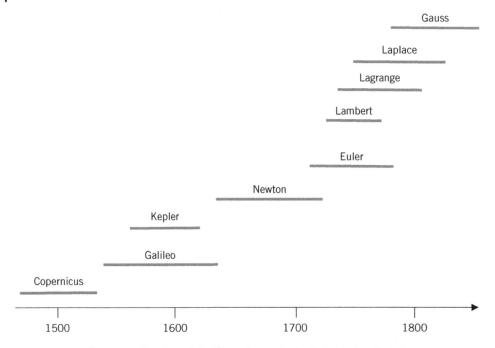

Figure 1.1 Timeline of significant figures in the Early Modern Period.

Johann Kepler (1571–1630) developed the fundamental laws for planetary motion based on astronomical observations of the planet Mars compiled by the Danish noble-man Tycho Brahe (1546–1601). Kepler's three laws are:

1) The orbit of a planet is an ellipse, with the sun located at a focus.
2) The radial line from the sun to the planet sweeps out equal areas during equal time intervals.
3) The square of a planet's orbital period for one revolution is proportional to the cube of the planet's "mean distance" from the sun.

The third law notes the planet's "mean distance" from the sun. In Chapter 2 we will define this "mean distance" as one-half of the length of the major axis of an ellipse. Kepler published his first two laws of planetary motion in 1609 and his third law in 1619. Kepler developed an expression for the time-of-flight between two points in an orbit; this expression is now known as *Kepler's equation*.

Isaac Newton (1642–1727) was an English astronomer, mathematician, and physicist who developed calculus and formulated the laws of motion and universal gravitation. Newton's three laws of motion are:

1) A body remains at rest or moves with a constant velocity unless acted upon by a force.
2) The vector sum of the forces acting on a body is equal to the mass of the body multiplied by its absolute acceleration vector (i.e., $\sum \mathbf{F} = m\mathbf{a}$).
3) When a body exerts a force on a second body, the second body exerts an equal-and-opposite force on the first body.

The first and second laws hold relative to a fixed or inertial reference frame. Newton published the three laws of motion in *Principia* in 1687. Newton's universal law of gravitation states that any two bodies attract one another with a force that is proportional to the product of their masses and inversely proportional to the square of their separation distance. Newton's laws of motion and gravitation explain the planetary motion that Kepler described by geometrical means.

Leonhard Euler (1707–1783), a Swiss mathematician, made many mathematical and scientific contributions to the fields of calculus, mathematical analysis, analytical mechanics, fluid dynamics, and optics. Euler also developed equations that govern the motion of a rotating body; these equations serve as the foundation for analyzing the rotational motion of satellites in orbit. Johann Heinrich Lambert (1728–1777), also a Swiss mathematician, formulated and solved the problem of determining the orbit that passes through two known position vectors with a prescribed transit time. Known today as *Lambert's problem*, its solution provides a method for the orbit-determination process as well as planning orbital maneuvers. Joseph-Louis Lagrange (1736–1813) was an Italian-born mathematician who made significant contributions in analytical mechanics and celestial mechanics, including the determination of equilibrium orbits for a problem with three bodies and the formulation of *Lagrange's planetary equations* for orbital motion. Pierre-Simon Laplace (1749–1827) was a French mathematician who, among his many mathematical contributions, formulated the first orbit-determination method based solely on angular measurements. Carl Friedrich Gauss (1777–1855), a German mathematician of great influence, made significant contributions to the field of orbit determination. In mid-1801 he predicted the orbit of the dwarf planet Ceres using a limited amount of observational data taken before Ceres became obscured by the sun. In late 1801, astronomers rediscovered Ceres just as predicted by Gauss.

1.3 Early Twentieth Century

Let us next briefly describe the important figures in the early twentieth century. It is during this period when mathematical theories are augmented by experimentation, most notably in the field of rocket propulsion. It is interesting to note that the important figures of this period were inspired by the nineteenth century science fiction literature of H.G. Wells and Jules Verne and consequently were tantalized by the prospect of interplanetary space travel.

Konstantin Tsiolkovsky (1857–1935) was a Russian mathematician and village school teacher who worked in relative obscurity. He theorized the use of oxygen and hydrogen as the optimal combination for a liquid-propellant rocket in 1903 (the same year as the Wright brothers' first powered airplane flight). Tsiolkovsky also developed theories regarding rocket propulsion and a vehicle's velocity change – the so-called "rocket equation."

Robert H. Goddard (1882–1945), a US physicist, greatly advanced rocket technology by combining theory and experimentation. On March 16, 1926, Goddard successfully launched the first liquid-propellant rocket. In 1930, Goddard moved his laboratory to New Mexico and continued to develop larger and more powerful rocket engines.

Hermann J. Oberth (1894–1989) was born in Transylvania and later became a German citizen. A physicist by training, he independently developed theories regarding human space flight through rocket propulsion. Oberth was a key figure in the German Society for Space Travel, which was formed in 1927, and whose membership included the young student Wernher von Braun. Von Braun (1912–1977) led the Nazi rocket program at Peenemünde during World War II. Von Braun's team developed the V-2 rocket, the first long-range rocket and the first vehicle to achieve space flight above the sensible atmosphere.

At the end of World War II, von Braun and members of his team immigrated to the US and began a rocket program at the US Army's Redstone Arsenal at Huntsville, Alabama. It was during this time that the US and the Soviet Union were rapidly developing long-range intercontinental ballistic missiles (ICBMs) for delivering nuclear weapons.

1.4 Space Age

On October 4, 1957, the Soviet Union successfully launched the first artificial satellite (Sputnik 1) into an Earth orbit and thus ushered in the *space age*. Sputnik 1 was a polished 84 kg metal sphere and it completed an orbital revolution every 96 min. The US successfully launched its first satellite (Explorer 1) almost 4 months after Sputnik on January 31, 1958. Unlike Sputnik 1, Explorer 1 was a long, tube-shaped satellite, and because of its shape, it unexpectedly entered into an end-over-end tumbling spin after achieving orbit.

Our abridged historical overview of the first half of the twentieth century illustrates the very rapid progress achieved in rocket propulsion and space flight. For example, in less than 20 years after Goddard's 184 ft flight of the first liquid-propellant rocket, Nazi Germany was bombarding London with long-range V-2 missiles. Twelve years after the end of World War II, the USSR successfully launched a satellite into orbit. Another point of interest is that in this short period, rocket propulsion and space flight transitioned from the realm of the singular individual figure to large team structures funded by governments. For example, the US established the National Aeronautics and Space Administration (NASA) on July 29, 1958.

The US and USSR space programs launched and operated many successful missions after the space age began in late 1957. Table 1.1 summarizes notable robotic space missions (i.e., no human crew). A complete list of successful space missions would be quite long; Table 1.1 is not an exhaustive list and instead presents a list of mission "firsts." It is truly astounding that 15 months after Sputnik 1, the USSR sent a space probe (Luna 1) to the vicinity of the moon. Equally impressive is the first successful interplanetary mission (Mariner 2), which NASA launched less than 5 years after Explorer 1. Table 1.1 shows that spacecraft have visited all planets in our solar system and other celestial bodies such as comets and asteroids.

On April 12, 1961, the USSR successfully sent the first human into space when Yuri Gagarin orbited the Earth in the Vostok 1 spacecraft. Less than 1 month later, the US launched its first human into space when Alan Shepard flew a suborbital mission in a Mercury spacecraft. Table 1.2 presents notable space missions with human crews (as with Table 1.1, Table 1.2 focuses on first-time achievements). Tables 1.1 and 1.2 clearly illustrate the accelerated pace of accomplishments in space flight. Table 1.2 shows

Table 1.1 Notable robotic space missions.

Mission	Date	Achievement	Country
Sputnik 1	October 4, 1957	First artificial satellite to achieve Earth orbit	USSR
Luna 1	January 2, 1959	First satellite to reach the vicinity of the moon	USSR
Mariner 2	December 14, 1962	First spacecraft to encounter (fly by) another planet (Venus)	US
Mariner 4	July 14, 1965	First spacecraft to fly by Mars	US
Luna 9	February 3, 1966	First spacecraft to land on another body (moon)	USSR
Luna 10	April 3, 1966	First spacecraft to orbit the moon	USSR
Venera 7	December 15, 1970	First spacecraft to land on another planet (Venus)	USSR
Mariner 9	November 14, 1971	First spacecraft to orbit another planet (Mars)	US
Pioneer 10	December 3, 1973	First spacecraft to fly by Jupiter	US
Mariner 10	March 29, 1974	First spacecraft to fly by Mercury	US
Viking 1	July 20, 1976	First spacecraft to land on Mars	US
Voyager 1	March 1979, November 1980	Fly by encounters with Jupiter, Saturn, and Saturn's moon Titan	US
Voyager 2	January 1986, August 1989	First spacecraft to fly by Uranus and Neptune	US
Galileo	December 8, 1995	First spacecraft to orbit Jupiter	US
Mars Pathfinder	July 4, 1997	First rover on the planet Mars	US
NEAR Shoemaker	February 12, 2001	First spacecraft to land on an asteroid (433 Eros)	US
Cassini-Huygens	July 2004, January 2005	First spacecraft to orbit Saturn (Cassini) and first spacecraft to land on the moon Titan (Huygens)	US and Europe
Stardust	January 16, 2006	First spacecraft to return samples from a comet	US
MESSENGER	March 18, 2011	First spacecraft to orbit Mercury	US
New Horizons	July 14, 2015	First spacecraft to fly by Pluto	US

Table 1.2 Notable space missions with human crews.

Mission	Date	Achievement	Country
Vostok 1	April 12, 1961	First human to reach space and orbit the Earth	USSR
Vostok 6	June 16, 1963	First woman in space	USSR
Voskhod 2	March 18, 1965	First human "spacewalk" outside of orbiting spacecraft	USSR
Gemini 6A	December 15, 1965	First orbital rendezvous	US
Apollo 8	December 24, 1968	First humans to orbit the moon	US
Apollo 11	July 20, 1969	First humans to land and walk on the moon	US
Salyut 1	April 19, 1971	First orbiting space station with crew	USSR
STS-1	April 12, 1981	First flight of a reusable spacecraft (Space Shuttle)	US
International Space Station	November 20, 1998	First multinational space station and largest satellite placed in Earth orbit	Russia, US, Europe, Japan, Canada

Table 1.3 Significant advances in space flight dynamics in the twentieth century.

Researcher(s)	Achievement
Dirk Brouwer Yoshihide Kozai	Developed pioneering work in the field of analytical satellite theory, including the perturbing effects of a non-spherical Earth
Theodore Edelbaum	Obtained analytical optimal trajectory solutions for spacecraft propelled by low-thrust electric propulsion engines
Richard Battin	Developed guidance and navigation theories for lunar and interplanetary spacecraft
Rudolf Kalman	Developed an optimal recursive estimation method (the *Kalman filter*) that has been applied to orbit determination and satellite navigation
W.H. Clohessy and R.S. Wiltshire	Developed closed-form solutions for the motion of a satellite relative to an orbiting target satellite (i.e., orbital rendezvous)
Derek Lawden	Developed theories for optimal rocket trajectories
A.J. Eggers and H.J. Allen Dean Chapman	Obtained analytical solutions for the entry flight phase of a ballistic capsule or lifting spacecraft returning to Earth from space
Robert Farquhar	Conceived of and managed space missions that targeted orbits where the satellite is balanced by the gravitational attracting of two celestial bodies
Ronald Bracewell Vernon Landon	Developed theories regarding the stability of a spinning satellite in orbit
Paul Cefola	Developed the Draper Semianalytical Satellite Theory (DSST) for rapid orbital calculations over a long time period

the very rapid progress in space missions with human crews in the 1960s, culminating with the first Apollo lunar landing on July 20, 1969. To date, three countries have developed human space flight programs: USSR/Russia (1961); US (1961); and China (2003).

We end this chapter with a brief summary of the significant twentieth century figures in the field of space flight dynamics. Table 1.3 presents these figures and their accomplishments. This list is certainly not exhaustive; furthermore, it is difficult to identify single individuals when the tremendous achievements in space flight over the past 60 years involve a large team effort.

2

Two-Body Orbital Mechanics

2.1 Introduction

In this chapter, we will develop the fundamental relationships that govern the orbital motion of a satellite relative to a gravitational body. These relationships will be derived from principles that should be already familiar to a reader who has completed a course in university physics or particle dynamics. It should be no surprise that we will use Newton's laws to develop the basic differential equation relating the satellite's acceleration to the attracting gravitational force from a celestial body. We will obtain analytical (or closed-form) solutions through the conservation of energy and angular momentum, which lead to "constants of motion." By the end of the chapter the reader should be able to analyze a satellite's orbital motion by considering characteristics such as energy and angular momentum and the associated geometric dimensions that define the size and shape of its orbital path. Understanding the concepts presented in this chapter is paramount to successfully grasping the subsequent chapter topics in orbit determination, orbital maneuvers, and interplanetary trajectories.

2.2 Two-Body Problem

At any given instant, the gravitational forces from celestial bodies such as the Earth, sun, moon, and the planets simultaneously influence the motion of a space vehicle. The magnitude of the gravitational force of *any* celestial body acting on a satellite with mass m can be computed using Newton's law of universal gravitation

$$F_{grav} = \frac{GMm}{r^2} \tag{2.1}$$

where M is the mass of the celestial body (Earth, sun, moon, etc.), G is the universal constant of gravitation, and r is the separation distance between the gravitational body and the satellite. It is not difficult to see that Eq. (2.1) is an *inverse-square* gravity law. The gravitational force acts along the line connecting the centers of the two masses. Figure 2.1 illustrates Newton's gravitational law with a two-body system comprising the Earth and a satellite. The Earth attracts the satellite with gravitational force vector

Space Flight Dynamics, First Edition. Craig A. Kluever.
© 2018 John Wiley & Sons Ltd. Published 2018 by John Wiley & Sons Ltd.
Companion website: www.wiley.com/go/Kluever/spaceflightmechanics

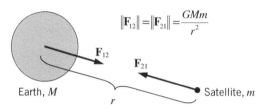

Figure 2.1 Newton's law of universal gravitation.

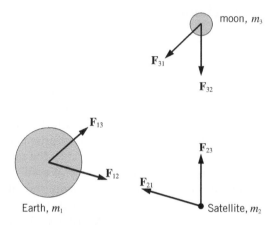

Figure 2.2 Gravitational forces for a three-body system.

$\mathbf{F}_{21}$ and the satellite attracts Earth with force $\mathbf{F}_{12}$. The reader should note that Eq. (2.1) presents the magnitude of the mutually attractive gravitational forces.

Figure 2.2 shows a schematic diagram of a three-body system (Earth, satellite, moon) with mutual gravitational forces among all three bodies. It should be clear from Figure 2.2 that $\mathbf{F}_{ij} = -\mathbf{F}_{ji}$. Equation (2.1) shows that the magnitudes are equal, or $\|\mathbf{F}_{ij}\| = \|\mathbf{F}_{ji}\|$. It is not difficult to imagine a diagram similar to Figure 2.2 with several (or N) gravitational bodies (however, an N-body diagram is very cluttered). The goal of this chapter (and the objective of this textbook) is to determine the motion of the satellite. Hence, a reasonable approach (similar to methods used in a basic dynamics course) would be to apply Newton's second law to a free-body diagram of the satellite. Applying Newton's second law to satellite mass m_2 for the three-body problem illustrated in Figure 2.2 yields

$$m_2\ddot{\mathbf{r}}_2 = \mathbf{F}_{21} + \mathbf{F}_{23} \tag{2.2}$$

where $\ddot{\mathbf{r}}_2$ is the satellite's acceleration vector relative to an *inertial* frame of reference or a frame that does not accelerate or rotate (we will use the over-dot notation to indicate a time derivative, e.g., $\dot{\mathbf{r}} = d\mathbf{r}/dt$ and $\ddot{\mathbf{r}} = d^2\mathbf{r}/dt^2$). We can extend Eq. (2.2) to an N-body system

$$m_2\ddot{\mathbf{r}}_2 = \sum_{\substack{j=1 \\ j \neq 2}}^{N} \mathbf{F}_{2j} \tag{2.3}$$

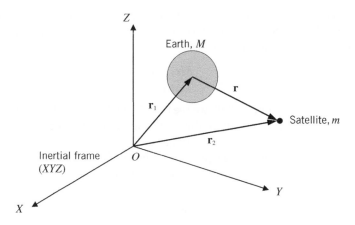

Figure 2.3 Two-body system.

Clearly, Eq. (2.3) is reduced to Eq. (2.2) when $N = 3$ as in Figure 2.2. Integrating Eq. (2.3) allows us to obtain the satellite's motion [velocity $\dot{\mathbf{r}}_2(t)$ and position $\mathbf{r}_2(t)$] in an N-body gravitational field. However, we cannot obtain analytical solutions of the general N-body problem [note that the inverse-square gravity (2.1) is a nonlinear function]. We must employ numerical integration schemes (such as Runge–Kutta methods) to obtain solutions to the N-body problem.

It is possible, however, to obtain analytical solutions for the satellite's motion if we only consider two bodies. These closed-form solutions will provide the basis for our analysis of space vehicle motion throughout this textbook. Figure 2.3 shows a two-body system comprising the Earth (mass M) and satellite (mass m). Coordinate system XYZ is an inertial Cartesian frame that does not rotate or accelerate. Vectors $\mathbf{r}_1$ and $\mathbf{r}_2$ are the inertial (absolute) positions of the Earth and satellite relative to the XYZ frame. The position of the satellite *relative* to the Earth is easily determined from vector addition:

$$\mathbf{r} = \mathbf{r}_2 - \mathbf{r}_1 \tag{2.4}$$

If the mutual gravitational forces are the only forces in the two-body system, then applying Newton's second law to each mass particle yields

$$\text{Earth:}\ M\ddot{\mathbf{r}}_1 = \frac{GMm}{r^2}\left(\frac{\mathbf{r}}{r}\right) \tag{2.5}$$

$$\text{Satellite:}\ m\ddot{\mathbf{r}}_2 = \frac{GMm}{r^2}\left(\frac{-\mathbf{r}}{r}\right) \tag{2.6}$$

Note that $\mathbf{r}/r$ is a unit vector pointing from the Earth's center to the satellite (hence $-\mathbf{r}/r$ is the direction of the Earth's attractive gravitational force on the satellite). Adding Eqs. (2.5) and (2.6) yields

$$M\ddot{\mathbf{r}}_1 + m\ddot{\mathbf{r}}_2 = \mathbf{0} \tag{2.7}$$

Integrating Eq. (2.7), we obtain

$$M\dot{\mathbf{r}}_1 + m\dot{\mathbf{r}}_2 = \mathbf{c}_1 \tag{2.8}$$

where c_1 is a vector of integration constants. Equation (2.8) is related to the velocity of the center of mass of the two-body system. To show this, let us express the inertial position of the two-body system's center of mass:

$$\mathbf{r}_{cm} = \frac{M\mathbf{r}_1 + m\mathbf{r}_2}{M + m} \tag{2.9}$$

Taking the time derivative of Eq. (2.9), we see that Eq. (2.8) is equal to the product of the total mass $(M + m)$ and the velocity of the center of mass. Therefore, we can conclude that the center of mass $\mathbf{r}_{cm}$ is not accelerating.

Our goal is to develop a governing equation for the satellite's motion relative to a single gravitational body M. Let us take the second time derivative of the relative position vector, Eq. (2.4):

$$\ddot{\mathbf{r}} = \ddot{\mathbf{r}}_2 - \ddot{\mathbf{r}}_1 \tag{2.10}$$

Next, we use Eqs. (2.5) and (2.6) to substitute for the absolute acceleration vectors of the Earth and satellite:

$$\ddot{\mathbf{r}} = \frac{GM}{r^2}\left(\frac{-\mathbf{r}}{r}\right) - \frac{Gm}{r^2}\left(\frac{\mathbf{r}}{r}\right)$$

or

$$\ddot{\mathbf{r}} = -\frac{G(M+m)}{r^3}\mathbf{r} \tag{2.11}$$

Note that although the denominator is r^3, Eq. (2.11) is still an inverse-square law because $\mathbf{r}/r$ is a unit vector. Equation (2.11) is a vector acceleration equation of the relative motion for the two-body problem.

Let us complete the two-body equation of motion by making use of the previous results and the assumption that the satellite's mass m is negligible compared with the mass of the gravitational body M. This assumption is very reasonable; for example, the mass ratio of a 1,000 kg satellite and the Earth is less than $2(10^{-22})$. Hence, we may assume that the two-body system center of mass and the center of the Earth are coincident. Furthermore, because the center of mass is not accelerating we can place an inertial frame at the center of the gravitational mass M. Figure 2.4 shows this scenario where the origin O of the

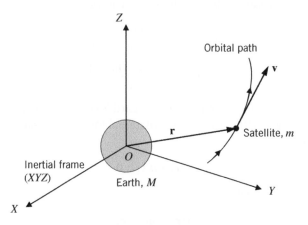

Figure 2.4 Two-body system with a body-centered inertial frame *XYZ*.

inertial frame *XYZ* is at the Earth's center. With this definition, vector **r** becomes the absolute or inertial position of the satellite. Finally, because mass *m* is negligible we have $G(M + m) \approx GM$. We define the *gravitational parameter* $\mu \equiv GM$ so that Eq. (2.11) may be rewritten as

$$\ddot{\mathbf{r}} = -\frac{\mu}{r^3}\mathbf{r} \tag{2.12}$$

Equation (2.12) is the two-body equation of motion. Solving Eq. (2.12) will yield the position and velocity vectors [$\mathbf{r}(t)$ and $\mathbf{v}(t) = \dot{\mathbf{r}}(t)$] of the satellite mass *m* relative to the central gravitational body *M*. Equation (2.12) is the fundamental equation for two-body motion that we will use for the remainder of the textbook. It is useful to summarize the assumptions that lead to Eq. (2.12):

1) The two bodies are spherically symmetric so that they may be considered as particles or point masses.
2) The mutually attractive gravitational forces are the only forces acting in the two-body system.
3) The mass of the satellite is negligible compared with the mass of the celestial body.

A final note is in order. The motion of an Earth-orbiting satellite is governed by Eq. (2.12) where the Earth's gravitational parameter is $\mu = 3.986(10^5)$ km^3/s^2. For a satellite orbiting the moon we may still use Eq. (2.12) but with the moon's gravitational parameter ($\mu_{moon} = 4{,}903$ km^3/s^2). We must remember to use the gravitational parameter μ that corresponds to the appropriate central attracting body. Table A.1 in Appendix A presents the gravitational parameters of several celestial bodies.

2.3 Constants of Motion

We begin to develop the analytical solution for two-body motion by determining constants associated with the two-body problem. The concepts presented in this section (momentum and energy) should be familiar to students with a background in basic mechanics. Many of the derivations that follow rely on vector operations such as the cross product and vector triple product; we summarize these operations in Appendix B.

2.3.1 Conservation of Angular Momentum

Linear momentum of a satellite is simply the product of its mass *m* and velocity vector **v**. Angular momentum **H** (or "moment of momentum") is defined by the cross product of position vector **r** and linear momentum *m***v**:

$$\mathbf{H} = \mathbf{r} \times m\mathbf{v} \tag{2.13}$$

The time derivative of angular momentum (for a satellite with constant mass) is

$$\dot{\mathbf{H}} = \dot{\mathbf{r}} \times m\mathbf{v} + \mathbf{r} \times m\dot{\mathbf{v}} \tag{2.14}$$

Because $\dot{\mathbf{r}} = \mathbf{v}$, the first cross product in Eq. (2.14) is zero. The term $m\dot{\mathbf{v}}$ is equal to the force **F** acting on the satellite. Hence, Eq. (2.14) becomes the familiar relationship between the time-rate of angular momentum and the external torque produced by force **F**

$$\dot{\mathbf{H}} = \mathbf{r} \times \mathbf{F} \tag{2.15}$$

For the two-body problem, the central-body gravitational force is the only force acting on the satellite. Furthermore, this attractive force is aligned with the position vector **r** and hence the cross product in Eq. (2.15) is zero. Consequently, the satellite's angular momentum **H** vector is constant for two-body motion.

We can arrive at the same result by performing vector operations on the governing two-body equation of motion (2.12). First, take the cross product of position **r** with each side of Eq. (2.12):

$$\mathbf{r} \times \ddot{\mathbf{r}} = \mathbf{r} \times \frac{-\mu}{r^3} \mathbf{r} \tag{2.16}$$

Clearly, the right-hand side of Eq. (2.16) is zero because we are crossing two parallel vectors. Hence, Eq. (2.16) becomes $\mathbf{r} \times \ddot{\mathbf{r}} = \mathbf{0}$. Next, we can carry out the following time derivative

$$\frac{d}{dt}(\mathbf{r} \times \mathbf{v}) = \dot{\mathbf{r}} \times \mathbf{v} + \mathbf{r} \times \dot{\mathbf{v}}$$

$$= \mathbf{v} \times \mathbf{v} + \mathbf{r} \times \ddot{\mathbf{r}} \tag{2.17}$$

$$= \mathbf{r} \times \ddot{\mathbf{r}}$$

Because Eq. (2.16) shows that $\mathbf{r} \times \ddot{\mathbf{r}} = \mathbf{0}$, the cross product $\mathbf{r} \times \mathbf{v}$ must be a constant vector. Referring back to Eq. (2.13), we see that $\mathbf{r} \times \mathbf{v}$ is angular momentum **H** divided by mass m. The "specific angular momentum" or angular momentum per unit mass of a satellite in a two-body orbit is

$$\mathbf{h} = \mathbf{r} \times \mathbf{v} = \text{constant vector} \tag{2.18}$$

Position and velocity vectors (**r** and **v**) will change as a satellite moves along its orbit but the angular momentum **h** remains a constant vector. Figure 2.5 shows an arc of a

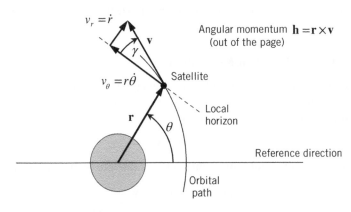

Figure 2.5 Angular momentum and flight-path angle, γ.

satellite's orbit where its current position and velocity vectors are denoted by $\mathbf{r}$ and $\mathbf{v}$. Because angular momentum $\mathbf{h}$ is the cross product $\mathbf{r} \times \mathbf{v}$, it is perpendicular to the plane containing vectors $\mathbf{r}$ and $\mathbf{v}$. Figure 2.5 shows counterclockwise satellite motion where vectors $\mathbf{r}$ and $\mathbf{v}$ are in the plane of the page and hence $\mathbf{h}$ is pointing out of the page. Because vector $\mathbf{h}$ is constant, the plane containing the motion of the satellite (known as the *orbital plane*) is also fixed in space for two-body motion. The orbital plane passes through the center of the gravitational body because it contains position vector $\mathbf{r}$. The angle γ in Figure 2.5 is called the *flight-path angle* and it is measured from the local horizon (perpendicular to $\mathbf{r}$) to the velocity vector $\mathbf{v}$. Flight-path angle is positive when the satellite's radial velocity component is positive, or $\dot{r} > 0$ (as shown in Figure 2.5). Conversely, $\gamma < 0$ if the length of the radius vector is decreasing, or $\dot{r} < 0$. If the satellite moves in a circular orbit where the radius is constant ($\dot{r} = 0$), then the flight-path angle is zero at all times and the velocity vector $\mathbf{v}$ remains perpendicular to position vector $\mathbf{r}$.

We can express the *magnitude* of the angular momentum vector in terms of radius r and the components of velocity vector $\mathbf{v}$. Let us define vectors $\mathbf{r}$ and $\mathbf{v}$ in terms of polar coordinates

$$\mathbf{r} = r\mathbf{u}_r \tag{2.19}$$

$$\mathbf{v} = v_r\mathbf{u}_r + v_\theta\mathbf{u}_\theta \tag{2.20}$$

where unit vector $\mathbf{u}_r$ points in the radial direction and unit vector $\mathbf{u}_\theta$ points in the transverse direction (perpendicular to $\mathbf{r}$ or along the local horizon in the direction of motion; see Section C.3 in Appendix C for additional details). The radial and transverse velocity components are $v_r = \dot{r} = v\sin\gamma$ and $v_\theta = r\dot{\theta} = v\cos\gamma$, respectively (see Figure 2.5). Using Eqs. (2.19) and (2.20), the angular momentum is

$$\mathbf{h} = \mathbf{r} \times \mathbf{v} = \begin{vmatrix} \mathbf{u}_r & \mathbf{u}_\theta & \mathbf{u}_k \\ r & 0 & 0 \\ v_r & v_\theta & 0 \end{vmatrix} = rv_\theta\mathbf{u}_k \tag{2.21}$$

where the unit vector $\mathbf{u}_k$ points normal to the orbital plane according to the right-hand rule. Equation (2.21) shows that the magnitude of the angular momentum vector is the product of the radius r and the transverse velocity component v_θ. Therefore, a satellite with purely *radial* velocity will have zero angular momentum – it has no angular motion! The following expression may be used to determine the angular momentum magnitude:

$$h = rv_\theta = r^2\dot{\theta} = rv\cos\gamma \tag{2.22}$$

From the equivalence of the terms above, it is easy to reconcile that the satellite's transverse velocity component is $v_\theta = r\dot{\theta} = v\cos\gamma$ (see Figure 2.5).

2.3.2 Conservation of Energy

We demonstrated the conservation of angular momentum by taking the vector (or cross) product of the governing two-body equation and position $\mathbf{r}$. Next, we will obtain a scalar result by taking the scalar (or dot) product of the velocity vector $\dot{\mathbf{r}}$ and both sides of the governing two-body equation of motion (2.12):

$$\dot{\mathbf{r}} \cdot \ddot{\mathbf{r}} = \dot{\mathbf{r}} \cdot \frac{-\mu}{r^3}\mathbf{r} \tag{2.23}$$

The left-hand side of Eq. (2.23) is the dot product $\mathbf{v} \cdot \dot{\mathbf{v}}$ while the right-hand side involves the dot product $\mathbf{r} \cdot \dot{\mathbf{r}}$. Therefore both sides contain a dot product between a vector and its time derivative. Figure 2.5 shows that the dot product $\mathbf{r} \cdot \mathbf{v} = \mathbf{r} \cdot \dot{\mathbf{r}}$ involves the projection of velocity vector $\mathbf{v}$ in the direction of position $\mathbf{r}$, that is

$$\mathbf{r} \cdot \dot{\mathbf{r}} = r v_r = r \dot{r} \tag{2.24}$$

Using this result, we obtain $\mathbf{v} \cdot \dot{\mathbf{v}} = v \dot{v}$, or the product of the velocity magnitude and the rate of change of the length of vector $\mathbf{v}$. Equation (2.23) becomes

$$v \dot{v} = \frac{-\mu}{r^3} r \dot{r} = \frac{-\mu}{r^2} \dot{r} \tag{2.25}$$

Note that each side of Eq. (2.25) can be written as a time derivative:

$$\frac{d}{dt}\left(\frac{v^2}{2}\right) = v \dot{v} \text{ and } \frac{d}{dt}\left(\frac{\mu}{r}\right) = \frac{-\mu}{r^2} \dot{r} \tag{2.26}$$

Therefore, Eq. (2.25) becomes

$$\frac{d}{dt}\left(\frac{v^2}{2}\right) = \frac{d}{dt}\left(\frac{\mu}{r}\right) \tag{2.27}$$

or

$$\frac{d}{dt}\left(\frac{v^2}{2} - \frac{\mu}{r}\right) = 0 \tag{2.28}$$

The bracketed term in Eq. (2.28) must be a constant. Integrating Eq. (2.28), we obtain

$$\xi = \frac{v^2}{2} - \frac{\mu}{r} = \text{constant} \tag{2.29}$$

where ξ is the specific energy (total energy per unit mass) of the satellite in its orbit. The reader should be able to identify the first term on the right-hand side ($v^2/2$) as kinetic energy per unit mass. The second term ($-\mu/r$) is the potential energy of the satellite per unit mass. A satellite's potential energy increases as its distance from the attracting body increases (similar to the "*mgh*" potential energy discussed in a university physics course). However, a satellite's minimum potential energy (occurring when r is equal to the radius of the attracting body) is negative and its maximum potential energy approaches zero as $r \to \infty$. We shall soon see that adopting this convention means that a satellite in a closed (or repeating) orbit has *negative* total energy while a satellite following an unbounded open-ended trajectory has *positive* energy. In either case, Eq. (2.29) tells us that the satellite's total energy ξ remains constant along its orbital path. The satellite may speed up during its orbit and gain kinetic energy but in doing so it loses potential energy so that total energy ξ remains constant.

We can also demonstrate the conservation of energy by using a gravitational potential function. From the gradient of a scalar potential function U, we can determine the gravitational force (or gravitational acceleration)

$$\ddot{\mathbf{r}} = \nabla U \tag{2.30}$$

where the "del" operator is a vector differential operation of partial derivatives with respect to position coordinates (such as *XYZ* Cartesian coordinates). The two-body potential function is

$$U = \frac{\mu}{r} \tag{2.31}$$

Note that the potential function U is the *negative* of the potential energy. Computing the gradient of potential function $U = \mu/r$ leads to the right-hand side of Eq. (2.12), the governing equation of motion for the two-body problem (we will present the details of the gradient operation in Chapter 5). Next, we may use the chain rule to write the time derivative of the scalar potential function:

$$\frac{dU}{dt} = \frac{\partial U}{\partial \mathbf{r}} \frac{d\mathbf{r}}{dt}$$

The first term on the right-hand side is $\nabla U = \ddot{\mathbf{r}}$. Therefore, the time derivative is

$$\frac{dU}{dt} = \ddot{\mathbf{r}} \cdot \dot{\mathbf{r}} = \dot{\mathbf{v}} \cdot \mathbf{v} = v\dot{v} \tag{2.32}$$

Equation (2.26) shows that $v\dot{v}$ is the time derivative of kinetic energy. Defining specific kinetic energy as $T = v^2/2$, we can write Eq. (2.32) as

$$\frac{dU}{dt} = \frac{dT}{dt}$$

or

$$\frac{d}{dt}(T - U) = 0 \tag{2.33}$$

Equation (2.33) shows that $T - U$ is constant. Because potential energy V is the negative of the potential function (i.e., $V = -\mu/r = -U$), Eq. (2.33) shows that the sum of kinetic energy and potential energy ($\xi = T + V$) is constant.

2.4 Conic Sections

So far we have determined that a satellite's angular momentum $\mathbf{h}$ is a constant vector (i.e., the orbital plane remains fixed in space) and total energy ξ is constant. However, we have not completely determined the orbital solution of the governing two-body equation of motion (2.12). One more vector manipulation of Eq. (2.12) will lead to an expression for the satellite's position in its orbit. The derivation of an orbital position solution follows.

2.4.1 Trajectory Equation

To begin, we take the cross product of the two-body equation (2.12) with angular momentum vector $\mathbf{h}$

$$\ddot{\mathbf{r}} \times \mathbf{h} = \frac{\mu}{r^3}(\mathbf{h} \times \mathbf{r}) \tag{2.34}$$

Note that the minus sign of the right-hand side term is cancelled by reversing the order of the cross product (i.e., $\mathbf{h} \times \mathbf{r} = -\mathbf{r} \times \mathbf{h}$). The left-hand side of Eq. (2.34) is the time derivative of the cross product $\dot{\mathbf{r}} \times \mathbf{h}$

$$\frac{d}{dt}(\dot{\mathbf{r}} \times \mathbf{h}) = \ddot{\mathbf{r}} \times \mathbf{h} + \dot{\mathbf{r}} \times \overset{0}{\cancel{\dot{\mathbf{h}}}} \tag{2.35}$$

The right-hand side of Eq. (2.34) can be expanded using $\mathbf{h} = \mathbf{r} \times \mathbf{v}$:

$$\frac{\mu}{r^3}(\mathbf{h} \times \mathbf{r}) = \frac{\mu}{r^3}(\mathbf{r} \times \mathbf{v}) \times \mathbf{r} \tag{2.36}$$

Using the vector triple product, Eq. (2.36) becomes

$$\frac{\mu}{r^3}(\mathbf{r} \times \mathbf{v}) \times \mathbf{r} = \frac{\mu}{r^3}[\mathbf{v}(\mathbf{r} \cdot \mathbf{r}) - \mathbf{r}(\mathbf{r} \cdot \mathbf{v})]$$

$$= \frac{\mu}{r^3}[r^2 \mathbf{v} - r\dot{r}\mathbf{r}] \tag{2.37}$$

$$= \frac{\mu}{r}\mathbf{v} - \frac{\mu\dot{r}}{r^2}\mathbf{r}$$

Note that the intermediate step in Eq. (2.37) involves the dot products $\mathbf{r} \cdot \mathbf{r} = r^2$ and $\mathbf{r} \cdot \dot{\mathbf{r}} = r\dot{r}$. The right-hand side of Eq. (2.37) can also be expressed as a time derivative:

$$\frac{d}{dt}\left(\frac{\mu}{r}\mathbf{r}\right) = \frac{\mu}{r}\dot{\mathbf{r}} - \frac{\mu}{r^2}\dot{r}\mathbf{r} \tag{2.38}$$

Therefore, the original cross-product, Eq. (2.34), may be expressed in terms of these two time derivatives:

$$\frac{d}{dt}(\dot{\mathbf{r}} \times \mathbf{h}) = \frac{d}{dt}\left(\frac{\mu}{r}\mathbf{r}\right) \tag{2.39}$$

Integrating Eq. (2.39) yields

$$\dot{\mathbf{r}} \times \mathbf{h} = \frac{\mu}{r}\mathbf{r} + \mathbf{C} \tag{2.40}$$

where $\mathbf{C}$ is a constant vector. Next, let us take the dot product of Eq. (2.40) with position vector $\mathbf{r}$ so that we can obtain a scalar equation:

$$\mathbf{r} \cdot (\dot{\mathbf{r}} \times \mathbf{h}) = \frac{\mu}{r}\mathbf{r} \cdot \mathbf{r} + \mathbf{r} \cdot \mathbf{C} \tag{2.41}$$

For a scalar triple product, we have $\mathbf{r} \cdot (\dot{\mathbf{r}} \times \mathbf{h}) = (\mathbf{r} \times \dot{\mathbf{r}}) \cdot \mathbf{h}$, or $\mathbf{h} \cdot \mathbf{h} = h^2$. Hence, Eq. (2.41) becomes

$$h^2 = \mu r + rC\cos\theta \tag{2.42}$$

where the dot product $\mathbf{r} \cdot \mathbf{C}$ is replaced by the product of the two magnitudes and the cosine of the angle θ between vectors $\mathbf{r}$ and $\mathbf{C}$. Dividing Eq. (2.42) by the gravitational parameter μ and factoring out r from the right-hand side yields

$$\frac{h^2}{\mu} = r\left(1 + \frac{C}{\mu}\cos\theta\right) \tag{2.43}$$

Solving for radial position we obtain

$$r = \frac{h^2/\mu}{1 + \dfrac{C}{\mu}\cos\theta} \tag{2.44}$$

Equation (2.44) is the equation of a *conic section* written in polar coordinates with the origin at a focus. Consulting a standard textbook in analytical geometry shows that the constant numerator term h^2/μ in Eq. (2.44) is the *parameter p* (or *semilatus rectum*) and the constant C/μ is the *eccentricity e*. Using these constants, we may express Eq. (2.44) as

$$r = \frac{p}{1 + e\cos\theta} \tag{2.45}$$

Equation (2.45) is known as the *trajectory equation* and it relates radial position r to polar angle θ. Because an ellipse is a conic section Eqs. (2.44) and (2.45) prove Kepler's first law. We will discuss the trajectory equation in more detail after presenting the geometry of the possible conic sections.

A conic section is the curve that results from the intersection of a right circular cone and a plane. Figure 2.6 shows two cones placed tip-to-tip and the three possible conic sections: (a) ellipse; (b) parabola; and (c) hyperbola. The ellipse is a closed curve that results from a cutting plane that intersects only one cone. Note that the circle (Figure 2.6a) is a special case of an ellipse where the cutting plane is parallel to the base of the cone (or perpendicular to the cone's line of symmetry). A parabola is an open curve that is produced when the cutting plane is parallel to the edge of the cone (Figure 2.6b). A hyperbola is also an open curve that is produced when the cutting plane intersects both cones; hence it consists of two branches (Figure 2.6c).

Figure 2.7 presents the geometrical characteristics of the three conic sections. All conics have two foci (labeled F_1 and F_2) where the gravitational body is located at the primary focus F_1 and F_2 is the secondary or "vacant focus." The foci lie on the *major axis* (the

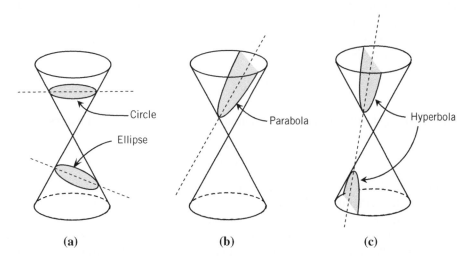

(a)   **(b)**   **(c)**

Figure 2.6 Conic sections: (a) ellipse and circle; (b) parabola; and (c) hyperbola.

"long" axis) and are separated by distance $2c$ for an ellipse (Figure 2.7a) and distance $-2c$ for a hyperbola (Figure 2.7c). The length of the major axis connecting the extreme ends of the conic section is $2a$ for an ellipse (Figure 2.7a) and $-2a$ for a hyperbola (Figure 2.7c). For a hyperbola, the distances $-2c$ and $-2a$ shown in Figure 2.7c are feasible because by convention the dimensions a and c are both taken as negative (the second branch of a hyperbola about focus F_2 is shown in Figure 2.7c as a dashed path so that the dimensions can be defined; a satellite on a hyperbolic trajectory follows the first branch). For a parabola (Figure 2.7b), the secondary focus F_2 is an infinite distance from the primary focus F_1 and therefore both dimensions c and a are infinite. The minor axis of an ellipse spans its narrow width and is perpendicular to the major axis. The dimension a is called the *semimajor axis*. For an ellipse a is half of the length of the major axis and b is half the length of the minor axis. The parameter p is the perpendicular distance from the gravitational body to the conic section. Parameter p is a positive, finite distance for all three conics sections shown in Figure 2.7. Because *all* conic sections obey the polar equation (2.44) or (2.45), the parameter is related to the angular momentum of the orbit:

$$p = \frac{h^2}{\mu} \tag{2.46}$$

It is useful to present some basic relationships for conic sections. The eccentricity e is defined as

$$e = \frac{c}{a} \tag{2.47}$$

Eccentricity increases as the two foci move farther apart and the ellipse becomes "long and skinny." For a closed conic section, the dimension c is always less than a and hence $e < 1$ for an ellipse. Eccentricity becomes smaller as the two foci move closer together and c decreases. When the two foci coincide, $c = 0$ and we have a *circular orbit* with $e = 0$. Eccentricity for a parabola is exactly equal to unity. Figure 2.7c shows that for a hyperbola the distance $-2c$ is greater than distance $-2a$ and hence $e > 1$.

The ratio of the minor and major axes will be used to derive an expression for the flight time on an ellipse in Chapter 4. Because the radial distance from a focus to the minor-axis crossing is equal to a (Figure 2.7a) we have a right triangle where $b^2 + c^2 = a^2$. Substituting $c = ae$, we arrive at the relationship

$$\frac{b}{a} = \sqrt{1 - e^2} \tag{2.48}$$

We may use Eq. (2.48) along with the definition $p = b^2/a$ to obtain an expression for parameter

$$p = a(1 - e^2) \tag{2.49}$$

Equation (2.49) holds for an ellipse and hyperbola but not for a parabola.

Figure 2.7 shows that the point of closest approach to the gravitational body is the *periapsis* (or "near apse"). The *apoapsis* is the farthest point from the focus and it only exists for an ellipse. The *apse line* connects the apoapsis and periapsis and it coincides with the major axis. Polar angle θ is the angle between a vector pointing from the primary focus to the periapsis direction and the position vector $\mathbf{r}$. In other words, polar angle θ

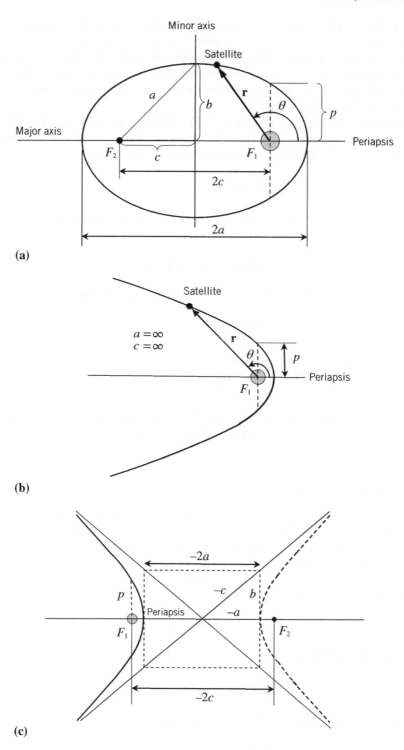

Figure 2.7 Geometrical characteristics of conic sections: (a) ellipse; (b) parabola; and (c) hyperbola.

is measured from the major axis (in the periapsis direction) to the satellite's current position vector **r** in the direction of motion (the satellite in Figure 2.7a is moving counter-clockwise in the elliptical orbit). We call polar angle θ the *true anomaly*. When $\theta = 0$, the satellite is at periapsis and when $\theta = 180°$, the satellite is at apoapsis. Of course, a parabola and hyperbola do not have an apoapsis because they are open-ended curves with branches that extend to infinity.

We may use the trajectory equation (2.45) and the parameter equation (2.49) to develop concise expressions for the radial distances for periapsis and apoapsis. At periapsis, we have $\theta = 0$ and the trajectory equation (2.45) yields

$$\text{Periapsis:} \quad r_p = \frac{p}{1+e} \tag{2.50}$$

Substituting Eq. (2.49) for parameter p, we obtain another expression for periapsis position r_p

$$\text{Periapsis:} \quad r_p = a(1-e) \tag{2.51}$$

At apoapsis, true anomaly $\theta = 180°$ and the trajectory equation yields

$$\text{Apoapsis:} \quad r_a = \frac{p}{1-e} \tag{2.52}$$

We may express Eq. (2.52) in terms of semimajor axis and eccentricity

$$\text{Apoapsis:} \quad r_a = a(1+e) \tag{2.53}$$

Equations (2.50) and (2.51) are valid for all conic sections, while the apoapsis equations, Eqs. (2.52) and (2.53), only apply to elliptical orbits.

2.4.2 Eccentricity Vector

Recall that in our derivation of the trajectory equation (2.45) we defined the polar angle θ (i.e., the true anomaly) as the angle between constant vector **C** and position vector **r**. Hence, vector **C** points in the periapsis direction. Comparing Eqs. (2.44) and (2.45) shows that the eccentricity e is related to the magnitude of vector **C**; that is, $e = C/\mu$. Therefore, the *eccentricity vector* $\mathbf{e} = \mathbf{C}/\mu$ also points in the direction of periapsis. We can solve Eq. (2.40) for the constant vector **C**

$$\mathbf{C} = \dot{\mathbf{r}} \times \mathbf{h} - \frac{\mu}{r}\mathbf{r} \tag{2.54}$$

Substituting $\dot{\mathbf{r}} = \mathbf{v}$ and $\mathbf{h} = \mathbf{r} \times \mathbf{v}$ into Eq. (2.54) yields

$$\mathbf{C} = \mathbf{v} \times (\mathbf{r} \times \mathbf{v}) - \frac{\mu}{r}\mathbf{r} \tag{2.55}$$

Using the vector triple product, Eq. (2.55) becomes

$$\mathbf{C} = [\mathbf{r}(\mathbf{v} \cdot \mathbf{v}) - \mathbf{v}(\mathbf{v} \cdot \mathbf{r})] - \frac{\mu}{r}\mathbf{r} \tag{2.56}$$

Using $\mathbf{v} \cdot \mathbf{v} = v^2$ and dividing Eq. (2.56) by μ we obtain the eccentricity vector $\mathbf{e} = \mathbf{C}/\mu$

$$\mathbf{e} = \frac{1}{\mu}\left[\left(v^2 - \frac{\mu}{r}\right)\mathbf{r} - (\mathbf{r} \cdot \mathbf{v})\mathbf{v}\right] \tag{2.57}$$

The magnitude of the eccentricity vector is the eccentricity of the orbit, or $e = \|\mathbf{e}\|$.

The eccentricity vector $\mathbf{e}$ and the angular momentum vector $\mathbf{h}$ define the orbit's orientation in three-dimensional space. Both $\mathbf{e}$ and $\mathbf{h}$ are computed from the satellite's position and velocity vectors $\mathbf{r}$ and $\mathbf{v}$. Referring to Figure 2.4, we see that position and velocity vectors $(\mathbf{r},\mathbf{v})$ may be expressed in a body-centered Cartesian coordinate frame. Furthermore, even though $\mathbf{r}$ and $\mathbf{v}$ coordinates change as the satellite moves along its orbital path, the angular momentum and eccentricity vectors $\mathbf{h}$ and $\mathbf{e}$ remain constant and fixed in inertial space. We will use this information for the three-dimensional orbit determination problem discussed in Chapter 3.

2.4.3   Energy and Semimajor Axis

We have already shown that total energy ξ consists of kinetic energy T and potential energy V and that this sum $(T + V)$ remains constant along the orbital path. It is useful to develop an expression that relates total energy to a geometric property of the conic section. To show this, let us use Eq. (2.29) to determine the total energy using the satellite's position and velocity at periapsis, r_p and v_p

$$\xi = \frac{v_p^2}{2} - \frac{\mu}{r_p} \tag{2.58}$$

Equation (2.22) shows that (constant) angular momentum can be computed using the periapsis position and velocity

$$h = r_p v_p \tag{2.59}$$

Note that flight-path angle γ is zero at periapsis (and at apoapsis) because the radial velocity component $\dot{r}$ is zero as the satellite passes through its minimum (or maximum) radial position. Using Eq. (2.59), periapsis velocity squared is

$$v_p^2 = \frac{h^2}{r_p^2} = \frac{\mu p}{r_p^2} = \frac{\mu a(1-e^2)}{r_p^2} \tag{2.60}$$

Equation (2.60) has made use of $p = h^2/\mu$ [Eq. (2.46)] and the relationship $p = a(1-e^2)$ [Eq. (2.49)]. Using Eq. (2.60) in the energy equation (2.58), we obtain

$$\xi = \frac{\mu a(1-e^2)}{2\,r_p^2} - \frac{\mu}{r_p} = \frac{\mu a(1-e^2) - 2\mu r_p}{2\,r_p^2} \tag{2.61}$$

We may use Eq. (2.51) and substitute $r_p = a(1-e)$ into Eq. (2.61) to yield

$$\xi = \frac{\mu a(1-e^2) - 2\mu a(1-e)}{2a^2(1-e)^2} \tag{2.62}$$

Table 2.1 Orbital characteristics.

Orbit	Semimajor axis	Eccentricity	Energy
Circle	$a > 0$	$e = 0$	$\xi < 0$
Ellipse	$a > 0$	$0 \leq e < 1$	$\xi < 0$
Parabola	$a = \infty$	$e = 1$	$\xi = 0$
Hyperbola	$a < 0$	$e > 1$	$\xi > 0$

Equation (2.62) can be reduced to the very simple relationship

$$\xi = -\frac{\mu}{2a} \tag{2.63}$$

Equation (2.63) shows that total energy is solely a function of the semimajor axis a. Some very important conclusions may be drawn from Eq. (2.63): (1) because $a > 0$ for an elliptical orbit, its total energy is *negative*; (2) because $a = \infty$ for a parabolic orbit, its total energy is *zero*; and (3) because $a < 0$ for a hyperbolic orbit, its total energy is *positive*.

Table 2.1 summarizes the important characteristics of conic sections and two-body orbits. Note that although a circular orbit is a special case of an ellipse it is included in Table 2.1 for completeness. For closed (repeating) orbits (circles and ellipses) the semimajor axis is positive, eccentricity is less than unity, and the total energy is negative. Hyperbolas are open-ended trajectories where semimajor axis is negative, eccentricity is greater than unity, and energy is positive. The parabola has infinite semimajor axis, eccentricity of exactly one, and zero energy. In this textbook we will tend to refer to circles and ellipses as *orbits* (closed paths) and parabolas and hyperbolas as *trajectories*. As a final summary of this section, we should note that the expressions for the conservation of momentum [Eqs. (2.18) and (2.22)], conservation of energy [Eqs. (2.29) and (2.63)], and the trajectory equation (2.45) are valid for all conic sections.

Example 2.1 A tracking station determines that an Earth-orbiting satellite is at an altitude of 2,124 km with an inertial velocity of 7.58 km/s and flight-path angle of 20° (Figure 2.8; not to scale). Determine (a) total specific energy, (b) angular momentum, (c) eccentricity, and also the type of orbit (conic section) (Example 2.1).

a) Total energy can be computed using Eq. (2.29)

$$\xi = \frac{v^2}{2} - \frac{\mu}{r}$$

with the given inertial velocity $v = 7.58$ km/s and radius $r = 2,124$ km $+ R_E$

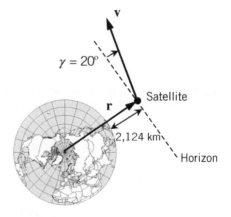

Figure 2.8 Earth-orbiting satellite (Example 2.1).

where R_E is Earth's radius. It is extremely important for the reader to note that r is the radius from the center of the gravitational body to the satellite, and therefore when an altitude is given (as in this case), we must add the radius of the celestial body. For problems involving Earth-orbiting satellites in this textbook, we will use $R_E = 6{,}378$ km. The reader should also note that two-body orbital calculations require the appropriate gravitational parameter; for the example problems in this textbook we will use $\mu = 3.986(10^5)$ km^3/s^2 for Earth. The reader may consult Appendix A to obtain more precise numerical values for the physical constants.

Using $R_E = 6{,}378$, km we find that $r = 2{,}124$ km $+ R_E = 8{,}502$ km. The total energy is

$$\xi = (7.58)^2/2 - 3.986\left(10^5\right)/8{,}502 = \boxed{-18.1549 \text{ km}^2/\text{s}^2}$$

Because energy is negative, we know that the satellite is following a closed orbit.

b) Angular momentum is computed using Eq. (2.22)

$$h = rv\cos\gamma = (8{,}502)(7.58)\ \cos\left(20°\right) = \boxed{60{,}558.64 \text{ km}^2/\text{s}}$$

c) Eccentricity can be computed from Eq. (2.49)

$$p = a\left(1 - e^2\right) \ \text{ or } \ e = \sqrt{1 - \frac{p}{a}}$$

We obviously need semimajor axis a and parameter p. Semimajor axis can be computed from total energy using Eq. (2.63)

$$\xi = -\frac{\mu}{2a} \ \text{ or } \ a = -\frac{\mu}{2\xi} = 10{,}977.76\,\text{km}$$

and parameter can be computed directly from angular momentum using Eq. (2.46)

$$p = \frac{h^2}{\mu} = 9{,}200.57\,\text{km}$$

Using these dimensions, the eccentricity is $e = \sqrt{1 - \dfrac{p}{a}} = \boxed{0.4024}$.

Because energy $\xi < 0$ and eccentricity is in the range $0 < e < 1$, the satellite is in an elliptical orbit.

This example demonstrates how we will use basic orbital relationships for the remainder of this textbook, namely that semimajor axis a can be computed from total energy ξ and parameter p is solely a function of angular momentum h. Eccentricity e is a function of a and p or energy and angular momentum.

2.5 Elliptical Orbit

We developed several relationships between conic-section geometry (i.e., a, p, and e) and orbital characteristics (energy ξ and angular momentum h) in the previous section. This section will continue to develop relationships for elliptical orbits. The reader should note that a circular orbit is a special case of an ellipse ($e = 0$) and therefore we will treat it in this section. Furthermore, we will present a few "standard orbits" (circles and ellipses) that are frequently used for Earth-orbiting satellites.

2.5.1 Ellipse Geometry

Figure 2.9 shows a satellite in an elliptical orbit about the Earth. If we compare Figure 2.7a and Figure 2.9, we see that the length of the major axis ($2a$) is equal to the sum of the perigee and apogee radii ($r_p + r_a$); therefore, the semimajor axis a can be computed using

$$a = \frac{r_p + r_a}{2} \tag{2.64}$$

Of course, Eq. (2.64) is valid for an elliptical orbit about any celestial body where r_p is the periapsis radius and r_a is the apoapsis radius. We can also determine eccentricity e from perigee and apogee radii. Comparing Figure 2.7a and Figure 2.9, we see that the distance between the foci is $2c = r_a - r_p$. Because eccentricity is the ratio of c and a [see Eq. (2.47)], we can determine e by dividing $2c = r_a - r_p$ by $2a = r_p + r_a$ to yield

$$e = \frac{r_a - r_p}{r_p + r_a} \tag{2.65}$$

Equation (2.65) is a useful formula for determining the eccentricity for an ellipse. Clearly, when perigee radius r_p is nearly equal to apogee radius r_a the eccentricity is very small and the orbit is nearly circular. The reader must remember that r_p and r_a are radial distances from the center of the gravitational body and *not* altitudes above the surface of the celestial body.

2.5.2 Flight-Path Angle and Velocity Components

Figure 2.9 also shows the satellite's flight-path angle, γ. Recall that we measure flight-path angle from the local horizon (perpendicular to the radius vector **r**) to the velocity vector **v**. Flight-path angle γ is *always* between $-90°$ and $+90°$ and it is zero when the satellite is at perigee or apogee. When a satellite is traveling from periapsis to apoapsis (as shown in Figure 2.9), its true anomaly is between zero and $180°$ and its flight-path angle is positive.

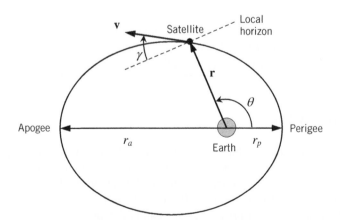

Figure 2.9 Elliptical orbit about the Earth.

Conversely, flight-path angle is negative when a satellite approaches periapsis (i.e., $180° < \theta < 360°$).

The reader should note that Eq. (2.22) is the only formula we have developed thus far for computing flight-path angle. Using Eq. (2.22) to find γ requires taking the inverse cosine of the positive quantity $h/(rv)$. A calculator or computer inverse-cosine operation of a positive argument will always place the angle in the first quadrant (i.e., $0° < \gamma < 90°$). From the previous discussion, we know that $\gamma < 0$ for half of an elliptical orbit when the satellite approaches periapsis. We can resolve this quadrant ambiguity by expressing the tangent of the flight-path angle as the ratio of the radial velocity component v_r and the transverse velocity component v_θ

$$\tan\gamma = \frac{v_r}{v_\theta} \tag{2.66}$$

Figure 2.5 shows this simple geometric relationship between the velocity components. We obtain the radial velocity by taking the time derivative of the trajectory equation (2.45) via the chain rule:

$$v_r = \dot{r} = \frac{dr}{d\theta}\frac{d\theta}{dt} = \frac{pe\sin\theta}{(1+e\cos\theta)^2}\frac{d\theta}{dt} \tag{2.67}$$

Equation (2.22) shows that the time rate of true anomaly is

$$\frac{d\theta}{dt} = \frac{h}{r^2} = \frac{h(1+e\cos\theta)^2}{p^2} \tag{2.68}$$

Note that the trajectory equation (2.45) has been squared and substituted for r^2 in Eq. (2.68). Finally, substituting Eq. (2.68) for $d\theta/dt$ in Eq. (2.67) and using $p = h^2/\mu$ we obtain a simplified expression for radial velocity

$$v_r = \frac{\mu}{h}e\sin\theta \tag{2.69}$$

Equation (2.69) shows that $\dot{r} = 0$ at periapsis ($\theta = 0$) and apoapsis ($\theta = 180°$) as expected. Transverse velocity can be computed directly from the ratio of h and r

$$v_\theta = r\dot{\theta} = \frac{h}{r} = \frac{h(1+e\cos\theta)}{p} \tag{2.70}$$

Again, the trajectory equation (2.45) has been substituted for r. Using $p = h^2/\mu$ in Eq. (2.70) yields

$$v_\theta = \frac{\mu}{h}(1+e\cos\theta) \tag{2.71}$$

Dividing Eq. (2.69) by Eq. (2.71) yields the tangent of the flight-path angle

$$\tan\gamma = \frac{e\sin\theta}{1+e\cos\theta} \tag{2.72}$$

Applying a calculator's or computer's inverse-tangent function to Eq. (2.72) will place the flight-path angle in the correct range (i.e., $-90° < \gamma < 90°$).

Finally, let us determine the locations in the orbit where velocity is minimum and maximum. Equation (2.22) shows that angular momentum h is the product of radius r and

Table 2.2 True anomaly and flight-path angle values on an elliptical orbit.

Orbital position	True anomaly	Flight-path angle	Dot product
Periapsis	$\theta = 0$	$\gamma = 0$	$\mathbf{r} \cdot \mathbf{v} = 0$
Approaching apoapsis	$0° < \theta < 180°$	$\gamma > 0$	$\mathbf{r} \cdot \mathbf{v} > 0$
Apoapsis	$\theta = 180°$	$\gamma = 0$	$\mathbf{r} \cdot \mathbf{v} = 0$
Approaching periapsis	$180° < \theta < 360°$	$\gamma < 0$	$\mathbf{r} \cdot \mathbf{v} < 0$

the transverse velocity v_θ (i.e., the velocity component perpendicular to r). Hence, the angular momentum at periapsis and apoapsis is

$$h = r_p v_p = r_a v_a \tag{2.73}$$

where v_p and v_a are the velocities at periapsis and apoapsis, respectively. Because h is constant and periapsis r_p is the *minimum* radius, the satellite's maximum velocity is at periapsis. Conversely, the satellite's slowest velocity occurs when it is at apoapsis or the farthest position in its orbit.

Table 2.2 summarizes the values (or range of values) for true anomaly and flight-path angle for positions within an elliptical orbit. True anomaly θ is a key element because it is needed to determine whether flight-path angle is positive or negative [remember that using Eq. (2.22) to compute flight-path angle does not determine its sign]. The fourth column in Table 2.2 is the dot product $\mathbf{r} \cdot \mathbf{v} = r\dot{r}$ which can be used to determine the range for true anomaly. When $\mathbf{r} \cdot \mathbf{v} > 0$, radial velocity must be positive and the satellite is approaching apoapsis. If $\mathbf{r} \cdot \mathbf{v} < 0$, the satellite is approaching periapsis ($\dot{r} < 0$).

It is useful to summarize the orbital relationships that we have developed at this stage of the chapter (most of these relationships are valid for all conic sections).

1) Semimajor axis (a) determines total energy (this is true for all conic sections).
2) Parameter (p) determines the angular momentum magnitude h (this is true for all conic sections).
3) The three geometric characteristics (a, e, and p) are not independent. Knowledge of two characteristics can be used to determine the missing element. From a, e, and p we can determine total energy, angular momentum, and periapsis radius (this is true for all conic sections). For elliptical orbits we can determine the apoapsis radius.
4) Given any two of the three geometric characteristics (a, e, and p) and true anomaly θ (i.e., angular position in the orbit), we can determine radius r using the trajectory equation. Velocity magnitude v can be determined from the total energy. Flight-path angle γ can be determined from angular momentum or by using Eq. (2.72) (this is true for all conic sections).
5) Position and velocity vectors ($\mathbf{r},\mathbf{v}$) determine every orbital constant and the satellite's position in the orbit. Note that we calculate true anomaly using the trajectory equation where the proper quadrant for θ is determined by checking the sign of the dot product $\mathbf{r} \cdot \mathbf{v}$ (this is true for all conic sections).

The following examples illustrate many of these relationships for a satellite in an elliptical orbit.

Example 2.2 An Earth-orbiting satellite has semimajor axis $a = 7{,}758$ km and parameter $p = 7{,}634$ km. Determine (a) orbital energy, (b) angular momentum, and (c) whether or not the satellite will pass through the Earth's appreciable atmosphere (i.e., altitude less than 122 km).

a) We compute total energy using Eq. (2.63) with $\mu = 3.986(10^5)$ km^3/s^2

$$\xi = -\frac{\mu}{2a} = \boxed{-25.6896 \text{ km}^2/\text{s}^2}$$

Energy is negative because the satellite is following an elliptical orbit ($a > 0$).

b) We determine angular momentum using parameter p and Eq. (2.46)

$$h = \sqrt{\mu p} = \boxed{55{,}162.6 \text{ km}^2/\text{s}}$$

c) Computing the perigee radius r_p will determine the satellite's closest approach. We may use either Eq. (2.50) or Eq. (2.51). In either case, we need to determine the orbital eccentricity from a and p

$$e = \sqrt{1 - \frac{p}{a}} = 0.1264$$

This calculation also verifies that the orbit is elliptical. Perigee radius is

$$r_p = \frac{p}{1+e} = 6{,}777.2 \text{ km}$$

Perigee altitude is $r_p - R_E = 6{,}777.2 - 6{,}378 = 399.2$ km which is greater than 122 km. Therefore, this satellite will *not* pass through the Earth's appreciable atmosphere.

Figure 2.10 shows the elliptical orbit about the Earth with the proper scale. The Earth's atmosphere is the very thin shaded region surrounding the Earth in Figure 2.10. The

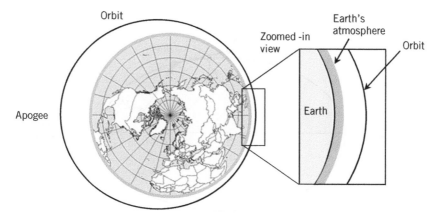

Figure 2.10 Elliptical orbit and Earth's atmosphere with zoomed-in view near perigee (Example 2.2).

zoomed-in view near perigee (shown on the right-hand side of Figure 2.10) illustrates the proximity between perigee and the upper atmosphere: the perigee pass is less than 280 km above the significant atmosphere. Computing the apogee radius r_a yields an apogee altitude of 2,361 km, which is a substantial distance above the Earth's atmosphere as shown on the left-hand side of Figure 2.10.

Example 2.3   The Chandra X-ray Observatory (CXO) is an Earth-orbiting satellite with perigee and apogee altitudes of 14,308 km and 134,528 km, respectively (Figure 2.11). Calculate (a) angular momentum of the orbit, (b) total energy of the orbit, and (c) radius, velocity, and flight-path angle at true anomaly $\theta = 120°$.

a)   First, let us compute the perigee and apogee *radii* from the given altitude information:

$$\text{Perigee radius: } r_p = 14{,}308\text{ km} + R_E = 20{,}686\text{ km}$$

$$\text{Apogee radius: } r_a = 134{,}528\text{ km} + R_E = 140{,}906\text{ km}$$

where $R_E = 6{,}378$ km is the Earth's radius. It is extremely important for the reader to remember that radius is the value used in the energy, momentum, and trajectory equations and *not* the altitude above a planet's surface!

   We can determine semimajor axis (a) and eccentricity (e) using perigee and apogee radii and Eqs. (2.64) and (2.65):

$$a = \frac{r_p + r_a}{2} = 80{,}796\text{ km}$$

$$e = \frac{r_a - r_p}{r_p + r_a} = 0.7440$$

   We can compute angular momentum directly from parameter p. Using Eq. (2.49), the parameter is determined to be

$$p = a\left(1 - e^2\right) = 36{,}075.81\text{ km}$$

Angular momentum is

$$h = \sqrt{\mu p} = \boxed{119{,}915.89\text{ km}^2/\text{s}}$$

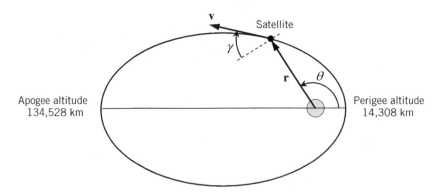

Figure 2.11 Chandra X-ray Observatory orbit (Example 2.3).

b) We can calculate total energy using semimajor axis $a = 80,796$ km

$$\xi = -\frac{\mu}{2a} = \boxed{-2.4667 \, \text{km}^2/\text{s}^2}$$

c) Using the trajectory equation (2.45) and the computed values of p and e, the radius at true anomaly $\theta = 120°$ is

$$r = \frac{p}{1 + e\cos\theta} = \boxed{57,444.31 \, \text{km}}$$

We can manipulate the energy equation (2.29) and obtain the velocity:

$$\xi = \frac{v^2}{2} - \frac{\mu}{r} \quad \rightarrow \quad v = \sqrt{2\left(\xi + \frac{\mu}{r}\right)} = \boxed{2.9907 \, \text{km/s}}$$

We can compute flight-path angle using either the angular momentum equation (2.22) or Eq. (2.72). First, let us determine γ using Eq. (2.22) and $h = 119,915.89 \, \text{km}^2/\text{s}$, $r = 57,444.31$ km, and $v = 2.9907$ km/s:

$$h = rv\cos\gamma \quad \rightarrow \quad \gamma = \cos^{-1}\left(\frac{h}{rv}\right) = \boxed{45.73°}$$

Flight-path angle is positive because $\theta = 120°$ (the CXO is approaching apogee; see Figure 2.11). Let us re-compute flight-path angle using Eq. (2.72) with $e = 0.7440$ and $\theta = 120°$:

$$\tan\gamma = \frac{e\sin\theta}{1 + e\cos\theta} = 1.025931 \quad \rightarrow \quad \gamma = 45.73° \text{ (same result)}$$

Example 2.4 The Apollo command module (CM) is being tracked by an Earth-based radar station during its return from the moon. The station determines that the CM orbit has a semimajor axis $a = 424,587$ km with eccentricity $e = 0.9849$.

a) Determine the flight-path angle and radial velocity of the CM when true anomaly is 330°.
b) Determine the vehicle's velocity and flight-path angle at the so-called "entry interface" (EI) or start of the atmospheric entry phase (EI is defined as 122 km above the Earth's surface).

a) We can compute flight-path angle using Eq. (2.72) with $e = 0.9849$ and $\theta = 330°$:

$$\tan\gamma = \frac{e\sin\theta}{1 + e\cos\theta} = -0.265766$$

Taking the inverse tangent, we find flight-path angle $\boxed{\gamma = -14.8832°.}$
One way to compute radial velocity is to use Eq. (2.69):

$$\dot{r} = \frac{\mu}{h}e\sin\theta$$

where angular momentum h is computed from parameter p

$$p = a(1 - e^2) = 12,725.72 \text{ km} \quad \rightarrow \quad h = \sqrt{\mu p} = 71,221.28 \text{ km}^2/\text{s}$$

Using these values, the radial velocity at $\theta = 330°$ is

$$\dot{r} = \frac{\mu}{h} e \sin \theta = \boxed{-2.7561 \text{ km/s}}$$

Radial velocity is negative because the CM is approaching Earth (or, approaching perigee).

Let us show an alternate solution method using the total energy. Because radial velocity is $v_r = v \sin \gamma$, we must determine the vehicle's velocity at true anomaly $\theta = 330°$. First, compute the energy from semimajor axis:

$$\xi = -\frac{\mu}{2a} = -0.469397 \text{ km}^2/\text{s}^2$$

Total energy is the sum of kinetic and potential energy. Potential energy depends on radius r which in turn can be determined using the trajectory equation (2.45) with $p = 12,725.72$ km, $e = 0.9849$, and $\theta = 330°$:

$$r = \frac{p}{1 + e \cos \theta} = 6,867.82 \text{ km}$$

Next, solve the energy equation (2.29) for velocity:

$$v = \sqrt{2 \left(\xi + \frac{\mu}{r} \right)} = 10.7303 \text{ km/s}$$

Therefore, the radial velocity is $v_r = v \sin \gamma = -2.7561$ km/s (same result).

b) We can write an energy equation by combining Eqs. (2.29) and (2.63)

$$\xi = \frac{v_{\text{EI}}^2}{2} - \frac{\mu}{r_{\text{EI}}} = -\frac{\mu}{2a} = -0.469397 \text{ km}^2/\text{s}^2$$

where r_{EI} and v_{EI} are the radius and velocity at entry interface, respectively. It should be clear that energy is constant along the return trajectory. Using $r_{\text{EI}} = 122 \text{ km} + R_E = 6,500$ km, the velocity at EI is

$$v_{\text{EI}} = \sqrt{2 \left(\xi + \frac{\mu}{r_{\text{EI}}} \right)} = \boxed{11.0321 \text{ km/s}}$$

The flight-path angle at EI can be determined from angular momentum, Eq. (2.22):

$$h = r_{\text{EI}} v_{\text{EI}} \cos \gamma_{\text{EI}} \quad \rightarrow \quad \gamma_{\text{EI}} = \cos^{-1} \left(\frac{h}{r_{\text{EI}} v_{\text{EI}}} \right)$$

Using the previously computed values for h, r_{EI}, and v_{EI}, we obtain

$$\boxed{\gamma_{\text{EI}} = -6.6841°}$$

Note that flight-path angle is negative at entry interface because the CM is approaching Earth (i.e., $\dot{r} < 0$).

Let us show an alternate solution for EI flight-path angle using Eq. (2.72):

$$\tan\gamma_{EI} = \frac{e\sin\theta_{EI}}{1 + e\cos\theta_{EI}}$$

We can obtain the true anomaly at EI (θ_{EI}) using the trajectory equation:

$$r_{EI} = \frac{p}{1 + e\cos\theta_{EI}} \quad \rightarrow \quad \cos\theta_{EI} = \frac{1}{e}\left(\frac{p}{r_{EI}} - 1\right) = 0.972487$$

The true anomaly at EI is $\theta_{EI} = -13.47°$ (or, 346.53°). True anomaly at EI is in the fourth quadrant because the CM is approaching perigee. Finally, the tangent of the flight-path angle is

$$\tan\gamma_{EI} = \frac{e\sin\theta_{EI}}{1 + e\cos\theta_{EI}} = -0.117192$$

The inverse-tangent operation yields $\gamma_{EI} = -6.6841°$ as previously computed.

One advantage of using Eq. (2.72) to determine the flight-path angle is that a quadrant check is unnecessary. However, we must correctly determine the satellite's true anomaly θ by using the trajectory equation (2.45) and additional information regarding the satellite's position in the orbit. In this example, we are told that the Apollo command module is returning to Earth and therefore it is approaching perigee ($180° < \theta < 360°$ and $\gamma < 0$). The reader should make a special note regarding this example!

2.5.3 Period of an Elliptical Orbit

We have not yet considered the period of an ellipse, or the time required for one orbital revolution. Figure 2.12 shows the differential area dA that is "swept" by the radius vector as the satellite moves counterclockwise in its orbit. The differential area swept by $\mathbf{r}$ through differential angle $d\theta$ is

$$dA = \frac{1}{2}r(rd\theta) \tag{2.74}$$

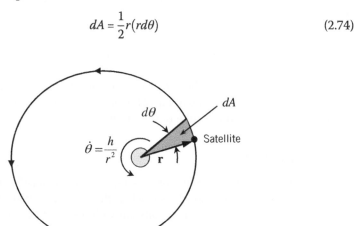

Figure 2.12 Differential area dA swept out by the radius vector.

Dividing both sides by differential time dt, we obtain the time rate of area, dA/dt

$$\frac{dA}{dt} = \frac{1}{2}r^2\frac{d\theta}{dt} \tag{2.75}$$

We know from the angular momentum equation (2.22) that the angular rate is $\dot{\theta} = h/r^2$ and therefore Eq. (2.75) becomes

$$\frac{dA}{dt} = \frac{1}{2}h \tag{2.76}$$

Because h is a constant, Eq. (2.76) shows that the time rate of area dA/dt swept by the radius vector is constant. Hence, Eq. (2.76) proves Kepler's second law: "equal areas are swept out by the radius vector in equal times."

We can determine the period of an elliptical orbit from Eq. (2.76) by separating variables

$$dA = \frac{1}{2}hdt \quad \text{or} \quad dt = \frac{2}{h}dA \tag{2.77}$$

Integrating both sides of Eq. (2.77) over one orbital revolution yields

$$T_{\text{period}} = \frac{2}{h}(\pi ab) \tag{2.78}$$

where T_{period} is the period and πab is the area of an ellipse. Equation (2.48) shows that the product ab is equal to $a^2\sqrt{1-e^2}$. Substituting this result and $h = \sqrt{\mu p}$ into Eq. (2.78), the period becomes

$$T_{\text{period}} = \frac{2\pi}{\sqrt{\mu p}}a^2\sqrt{1-e^2} \tag{2.79}$$

Equation (2.49) shows that $\sqrt{p} = \sqrt{a}\sqrt{1-e^2}$; substituting this result into Eq. (2.79) yields

$$T_{\text{period}} = \frac{2\pi}{\sqrt{\mu}}a^{3/2} \tag{2.80}$$

Equation (2.80) shows that the period of an elliptical orbit only depends on the semimajor axis a. Squaring both sides of Eq. (2.80) proves Kepler's third law: "the square of the period is proportional to the cube of the mean distance."

2.5.4 Circular Orbit

As previously mentioned, a circular orbit is a special case of an ellipse where eccentricity is zero. All of the orbital relationships (such as the energy, angular momentum, and trajectory equations) hold for circular orbits but with the condition that radius is constant and equal to the semimajor axis and parameter (i.e., $r = a = p$). Because periapsis is undefined for a circular orbit, we must measure the satellite's angular position relative to a fixed axis in the orbital plane.

We can determine a simple expression for circular velocity from the energy equations (2.29) and (2.63) with $a = r$

$$\text{Circle:} \quad \xi = \frac{v_c^2}{2} - \frac{\mu}{r} = -\frac{\mu}{2a} = -\frac{\mu}{2r} \tag{2.81}$$

where the subscript c indicates circular. Solving Eq. (2.81) for circular velocity, we obtain

$$v_c = \sqrt{\frac{\mu}{r}} \tag{2.82}$$

Equation (2.82) is the constant velocity of a satellite in a circular orbit with radius r. Because space missions frequently utilize circular orbits, we will often use this important relationship in this textbook. We can also derive the circular velocity from the balance between the centrifugal force and gravitational force:

$$m\frac{v_c^2}{r} = mg = m\frac{\mu}{r^2}$$

Canceling the satellite mass m in the above equation and solving for circular velocity, we obtain $v_c = \sqrt{\mu/r}$ which verifies Eq. (2.82).

We can determine the period of a circular orbit by equating the constant circular velocity to the distance traveled over one orbit (the circumference, $2\pi r$) divided by the period:

$$v_c = \sqrt{\frac{\mu}{r}} = \frac{2\pi r}{T_{\text{period}}} \tag{2.83}$$

Solving Eq. (2.83) for the period yields

$$\text{Circle:} \quad T_{\text{period}} = \frac{2\pi}{\sqrt{\mu}} r^{3/2} \tag{2.84}$$

This expression is identical to Eq. (2.80) with $a = r$.

2.5.5 Geocentric Orbits

We will briefly discuss a few common Earth-centered or *geocentric* orbits in this subsection. A *low-Earth orbit* (LEO) is a circular orbit with an altitude up to roughly 1,000 km. The lower bound on LEO altitude is determined by interaction with the upper atmosphere (and the subsequent aerodynamic drag) and the intended lifetime of the satellite. For example, the Apollo lunar missions began by injecting the upper stage of the Saturn V rocket into a 190-km altitude circular *parking orbit*. After the ground controllers verified that the spacecraft systems were operating as intended the upper stage was reignited to send the astronauts on a translunar orbit to the moon. Even at an altitude of 190 km, a satellite experiences enough aerodynamic drag such that it would lose energy over several days and eventually enter the dense atmosphere and be destroyed by aerodynamic heating (we will discuss aerodynamic drag on LEO satellites in Chapter 5). Therefore, a parking orbit is a temporary, intermediate orbit and not a final destination. Interplanetary spacecraft (such as the Mars Science Laboratory) are inserted into parking orbits before an upper rocket stage is fired to send the spacecraft on a trajectory to its planetary target.

Orbiting science platforms, such as the International Space Station (ISS) and the Hubble Space Telescope (HST), occupy LEOs. For example, the ISS and HST are in nearly circular LEOs with altitudes of about 410 and 540 km, respectively. The US Space Shuttle achieved circular LEO with altitudes ranging from roughly 300 to 500 km and was used to construct and service the ISS. For a typical Shuttle orbit with an altitude of 320 km (i.e., $r = 6,698$ km), Eq. (2.82) shows that the circular velocity is $v_c = 7.714$ km/s and Eq. (2.84) shows that its orbital period is about 91 min.

A *medium-Earth orbit* (MEO) has an altitude ranging from roughly 1,000 to 35,000 km. Navigation and communication satellites are often placed in MEO. One example is the Global Positioning System (GPS), which consists of a constellation of satellites in circular orbits at an altitude of 20,180 km ($r = 26,558$ km). Hence, a GPS satellite has a circular orbital velocity of 3.874 km/s and a period of 12 h.

A *geostationary-equatorial orbit* (GEO) is a circular orbit with angular velocity that matches the Earth's rotation rate about its axis. Because the Earth completes one revolution in one *sidereal day* (23 h, 56 min, and 4 s) the circular radius required for GEO can be computed from Eq. (2.84)

$$r_{GEO} = \left(\frac{T_{GEO}\sqrt{\mu}}{2\pi} \right)^{2/3}$$

where the GEO period is $T_{GEO} = 86,164$ s (or 23.934 h). GEO radius must be $r_{GEO} = 42,164$ km (or an altitude of 35,786 km above the Earth). The orbital plane of a GEO satellite is coincident with the Earth's equatorial plane (hence the "E" in the GEO acronym) so that it remains motionless (or stationary) to a ground-based observer and appears to "hover" overhead a particular geographic point on the equator. Communication and weather satellites are placed in GEO because they are always visible from a ground-based station and can always monitor the same geographic region. Using Eq. (2.82), we find that the circular orbital velocity for GEO is 3.075 km/s.

The reader may wonder why the period of GEO is 23 h, 56 min, and 4 s and not simply 24 h. A sidereal day (23.934 h) is the time required for the Earth to complete one revolution relative to an inertial frame. Hence, the *inertial* spin rate of the Earth is one revolution per sidereal day or $\omega_E = 7.292(10^{-5})$ rad/s. Therefore, a GEO satellite's rotation rate must be $\dot{\theta} = v_{GEO}/r_{GEO} = \omega_E$ in order to match the Earth's rotation. A *solar day* is the time required for the sun to reappear directly over the same meridian (line of longitude). In other words, a solar day (24 h) is the period between 12 o'clock noon on one day and 12 o'clock noon on the next day. Because we measure the 24 h solar day relative to the sun (and the Earth is moving in its orbit about the sun), the Earth actually completes more than one revolution in 24 h when the sun reappears directly overhead at noon.

The Geostationary Operational Environmental Satellite (GOES) system is a collection of GEO satellites used by the National Weather Service for weather monitoring and forecasting. The GOES system was established with the launch of GOES-1 in 1975. At the time of writing, GOES-13 (or GOES-East) is positioned at a longitude of 75° W (New York City is at 74° W) and GOES-15 (GOES-West) is located at 135° W (Honolulu, Hawaii is at 157.5° W). Of course, the GOES satellites "hover" over the equator at their respective longitudes. At GEO altitude each GOES satellite can view about 42% of the Earth's surface where the total viewing area is centered on its fixed longitude.

Russia uses the so-called *Molniya orbit* for its communication satellites. A Molniya orbit is highly elliptical with apogee and perigee altitudes of about 39,874 and 500 km, respectively. Eccentricity of a Molniya orbit is 0.741 and the period is 12 h. Molniya orbits are oriented so that apogee is located at a very high geographic latitude (we will discuss the three-dimensional orientation of orbits in Chapter 3). Because satellites in Molniya orbits spend most of their time near apogee, they are well suited to view northern latitudes such as Russia.

Example 2.5   Consider a US communication satellite that is destined for a geostationary-equatorial orbit (GEO). Initially, the satellite is placed in a 200-km altitude circular low-Earth orbit (LEO) by a launch vehicle. In order to reach GEO, the satellite follows a *geostationary transfer orbit* (GTO) as shown in Figure 2.13. The GTO is an ellipse that is tangent to the inner and outer circular orbits and therefore has a perigee radius equal to the LEO radius and an apogee radius equal to the GEO radius. Determine:

a) Perigee velocity on the GTO and LEO circular velocity.
b) Apogee velocity on the GTO and GEO circular velocity.
c) Period of the GTO.

a) We begin by defining the perigee and apogee radii for the GTO:

$$\text{GTO perigee:} \quad r_p = R_E + 200\,\text{km} = 6{,}578\,\text{km}$$

$$\text{GTO apogee:} \quad r_a = R_E + 35{,}786\,\text{km} = 42{,}164\,\text{km}$$

LEO altitude (200 km) is given in the problem while GEO altitude is known to be 35,786 km (see the previous section discussing GEO radius and the sidereal day). We can compute semimajor axis of the GTO from the perigee and apogee radii:

$$a = \frac{r_p + r_a}{2} = 24{,}371\,\text{km}$$

Total energy of the GTO is solely a function of a

$$\xi = -\frac{\mu}{2a} = -8.1778\,\text{km}^2/\text{s}^2$$

Knowledge of orbital energy allows us to determine velocity at a known radial distance using Eq. (2.29). Using perigee radius $r_p = 6{,}578$ km, the perigee velocity on the GTO is

$$v_p = \sqrt{2\left(\xi + \frac{\mu}{r_p}\right)} = \boxed{10.2390\,\text{km/s}}$$

The GTO perigee velocity v_p is shown in Figure 2.13.
    We compute circular velocity for LEO using Eq. (2.82) and radius $r_\text{LEO} = 6{,}578$ km

$$v_\text{LEO} = \sqrt{\frac{\mu}{r_\text{LEO}}} = \boxed{7.7843\,\text{km/s}}$$

A satellite in circular LEO and at GTO perigee has the *same* potential energy due to the common radius $r_\text{LEO} = r_p = 6{,}578$ km. However, the *total* energy of GTO is greater than LEO energy (to see this compare the GTO and LEO semimajor axes; see

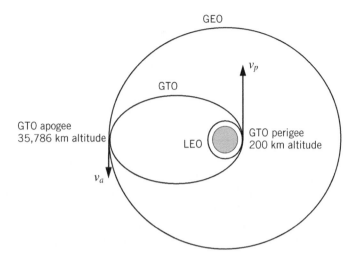

Figure 2.13 Geostationary transfer orbit (Example 2.5).

Figure 2.13). Therefore, the kinetic energy at GTO perigee must be greater than the kinetic energy of circular LEO; the calculations of v_p and v_{LEO} demonstrate this fact.

b) We can compute GTO apogee velocity using the energy equation with r_a = 42,164 km:

$$v_a = \sqrt{2\left(\xi + \frac{\mu}{r_a}\right)} = \boxed{1.5974 \text{ km/s}}$$

The GTO apogee velocity v_a is also shown in Figure 2.13. GTO apogee velocity is less than 16% of its perigee velocity.

We determine the circular velocity for GEO using radius r_{GEO} = 42,164 km

$$v_{GEO} = \sqrt{\frac{\mu}{r_{GEO}}} = \boxed{3.0747 \text{ km/s}}$$

This result makes sense because GEO energy is greater than GTO energy (again, see Figure 2.13 and compare the semimajor axes). Because GTO apogee and GEO have the same potential energy, the GEO kinetic energy *must* be greater than the kinetic energy at GTO apogee.

c) Orbital period is solely a function of semimajor axis. Using Eq. (2.80) the orbital period of the GTO is

$$T_{GTO} = \frac{2\pi}{\sqrt{\mu}} a^{3/2} = 37,863.5 \text{ s} = \boxed{10.52 \text{ h}}$$

In practice, an onboard rocket is fired in LEO to increase the satellite's velocity from v_{LEO} = 7.7843 km/s to the required GTO perigee velocity v_p = 10.2390 km/s. The satellite then coasts for one-half of a revolution (5.26 h) to reach GTO apogee where a second rocket burn is needed to increase the velocity from $v_a = 1.5974$ km/s to GEO circular velocity $v_{GEO} = 3.0747$ km/s. Orbital maneuvers and their associated propellant mass requirements are treated in Chapters 6 and 7.

Example 2.6 The Lunar Atmosphere and Dust Environment Explorer (LADEE) spacecraft was launched into a highly elliptical orbit by a Minotaur V booster in September 2013. After completing the fifth-stage burn of the Minotaur V at an altitude of 200 km, the LADEE spacecraft entered an elliptical orbit with perigee and apogee altitudes of 200 km and 278,000 km, respectively (see Figure 2.14). The LADEE spacecraft completed one revolution in this elliptical orbit before firing an onboard rocket at perigee to increase its orbital energy. Determine the eccentricity of the elliptical orbit and the coasting time between the Minotaur booster burnout and the thrusting maneuver at perigee.

First, we determine the perigee and apogee *radii* from the altitude information:

$$\text{Perigee: } r_p = R_E + 200\,\text{km} = 6{,}578\,\text{km}$$

$$\text{Apogee: } r_a = R_E + 278{,}000\,\text{km} = 284{,}378\,\text{km}$$

Using Eq. (2.65), the eccentricity is determined to be

$$e = \frac{r_a - r_p}{r_p + r_a} = \boxed{0.9548}$$

Hence, the "coasting orbit" for the LADEE spacecraft is highly elliptical.

We can compute the orbital period using Eq. (2.80) and knowledge of the semimajor axis. The semimajor axis is half of the sum of perigee and apogee radii:

$$a = \frac{r_p + r_a}{2} = 145{,}478\,\text{km}$$

The orbital period of the ellipse is

$$T_{\text{period}} = \frac{2\pi}{\sqrt{\mu}} a^{3/2} = 552{,}214\,\text{s} = 6.39\,\text{days}$$

Therefore, the total coasting time is one period, or

$$t_{\text{coast}} = T_{\text{period}} = \boxed{6.39\,\text{days}}$$

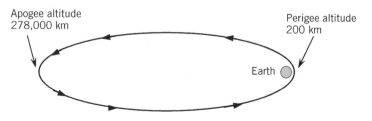

Apogee altitude
278,000 km

Perigee altitude
200 km

Earth

Figure 2.14 Elliptical orbit for the LADEE spacecraft (Example 2.6).

The LADEE spacecraft was placed in this intermediate elliptical orbit because the Minotaur V booster could not provide enough energy to achieve a direct orbit to the moon. LADEE's onboard propulsion system provided the final energy increase to send the spacecraft to the moon. We will analyze the LADEE orbital maneuvers in Chapter 7.

2.6 Parabolic Trajectory

Figure 2.15 shows a parabolic trajectory about the Earth. Recall that a parabola is an "open-ended" trajectory where the semimajor axis is infinite and eccentricity e is exactly unity. Equation (2.63) shows that the total energy ξ of a parabolic trajectory is zero because $a = \infty$. A parabola is the transitional conic section between an ellipse ($a > 0$, $e < 1$, and $\xi < 0$) and hyperbola ($a < 0$, $e > 1$, and $\xi > 0$). A parabolic trajectory is a one-way path to infinity. Figure 2.15 shows the scenario where the satellite is moving away from the central body and hence the flight-path angle γ is positive.

Let us substitute $e = 1$ into the trajectory equation (2.45) for a parabola:

$$\text{Parabola:} \qquad r = \frac{p}{1 + \cos\theta} \tag{2.85}$$

Equation (2.85) shows that radius r becomes infinite as true anomaly θ approaches $\pm 180°$ (note also that the periapsis radius of a parabola is always one-half the parameter; i.e., $r_p = p/2$).

Next, let us use Eq. (2.29) to determine the velocity on a parabola "at infinity" where $r \to \infty$ (or, $\theta \to 180°$):

$$\text{Parabola:} \qquad \xi = \frac{v_\infty^2}{2} - \overset{0}{\cancel{\frac{\mu}{r_\infty}}} = 0 \tag{2.86}$$

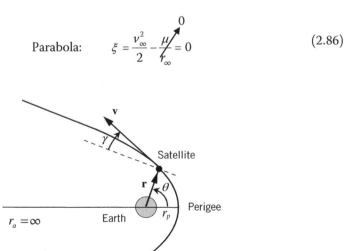

Figure 2.15 Parabolic trajectory about the Earth.

Note that we have used subscript ∞ in Eq. (2.86) to indicate conditions "at infinity." Because total energy ξ is zero on a parabolic trajectory, the satellite's velocity "at infinity" v_∞ is also zero! While this result seems very odd at first we must remember that in a "two-body universe" the central body is the *only* gravitational body that influences the satellite's motion. If a satellite is on a parabolic trajectory (with $\xi = 0$) it has just enough energy to "escape" the gravitational pull of the planet and reach a great distance (i.e., radius $r \to \infty$) where the potential energy is zero. Because total energy is zero at every point on the parabola, kinetic energy (and therefore velocity) is zero "at infinity." Of course, for real space missions, our solar system has N gravitational bodies (the sun, moon, and other planets) which continuously influence a satellite's motion. The reader must remember that the two-body problem is an approximation that allows us to develop analytical solutions for the satellite's motion through the use of conic sections.

Another way to think about this result is to remember that for our two-body problem we have fixed an inertial frame at the center of the sole celestial body and therefore a satellite's radius and velocity are relative to the gravitational body. For an Earth-orbiting satellite, the non-rotating "inertial" geocentric frame moves with the Earth along its orbit about the sun. When a geocentric satellite on a parabolic trajectory leaves Earth and reaches a very large radial distance ($r_\infty \approx \infty$), its velocity *relative to Earth* becomes zero. In this case, the satellite is now orbiting the sun in the same manner as the Earth. Hence, a parabolic trajectory is not useful in practice because it does not possess enough energy to move beyond Earth's orbit about the sun. The parabola represents the transitional conic section between closed orbits (ellipses, $\xi < 0$) and open-ended trajectories (hyperbolas, $\xi > 0$) that do have the excess energy to reach interplanetary targets.

In the previous discussion we called a parabolic trajectory an "escape" trajectory. The parabola has the minimum energy ($\xi = 0$) required for a satellite to theoretically reach an infinite distance from the gravitational body (i.e., to escape the central body). It is easy to determine the *escape velocity* v_{esc} required to achieve a parabolic trajectory. Using the energy equation, we may write

$$\text{Parabola:} \quad \xi = \frac{v_{esc}^2}{2} - \frac{\mu}{r} = 0 \tag{2.87}$$

where r is *any* radial position on the parabola. Solving Eq. (2.87) for the escape velocity, we obtain

$$v_{esc} = \sqrt{\frac{2\mu}{r}} \tag{2.88}$$

Equation (2.88) can be thought of as the "local escape velocity" because v_{esc} depends on the satellite's current radius r (i.e., the satellite's current potential energy). For example, the escape velocity for a satellite in LEO is much larger than the required escape velocity for a satellite in GEO because $r_{GEO} > r_{LEO}$. Equation (2.88) also shows that as radius becomes very large the local escape velocity goes to zero – the satellite has already achieved escape conditions!

Example 2.7 A satellite is in parabolic trajectory about the Earth with a perigee altitude of 500 km.

a) Compute the satellite's velocity at true anomalies $\theta = -60°$, $0°$, $90°$, $150°$, and $179°$.
b) Plot the satellite's velocity vs. true anomaly for $-179° \leq \theta \leq 179°$.

a) It is easy to compute the velocity from the energy equation for a parabola ($\xi = 0$) given the radius r:

$$\text{Parabola: } \xi = \frac{v^2}{2} - \frac{\mu}{r} = 0 \quad \rightarrow \quad v = \sqrt{\frac{2\mu}{r}}$$

We can obtain the radial position from the trajectory equation with $e = 1$:

$$\text{Parabola}: r = \frac{p}{1 + \cos\theta}$$

Because perigee radius is $r_p = p/2$ for a parabola, we can determine the parameter p

$$p = 2r_p = 2(R_E + 500\,\text{km}) = 13{,}756\,\text{km}$$

where $R_E = 6{,}378$ km. Using the constant p and the given true anomalies ($\theta = -60°, 0°, 90°, 150°$, and $179°$), we can compute the radius using the trajectory equation and the parabolic velocity using r and the energy equation. For example, using $\theta = -60°$ we have

$$r = \frac{13{,}756\,\text{km}}{1 + \cos(-60°)} = 9{,}170.7\,\text{km and } v = \sqrt{\frac{2\mu}{r}} = 9.3236\,\text{km/s}$$

Parabolic velocity values at the five true anomalies are shown in Table 2.3. Note that when $\theta = 90°$ the radius is equal to the parameter p, and that when $\theta = 179°$ the radial distance from the Earth is over 90 million km and the parabolic velocity is small as expected. For comparison, the average sun–Earth distance is almost 150 million km.

b) We can follow the same process as part (a) and compute the radial distances of the parabola corresponding to $-179° \leq \theta \leq 179°$ and then the parabolic velocities using the energy equation. Figure 2.16 shows parabolic velocity vs. true anomaly and the five values from Table 2.3 are shown with circular markers. Clearly, maximum velocity occurs at perigee (this is no surprise) and velocity is symmetric about the major axis.

Table 2.3 Radius and velocity on a parabolic trajectory (Example 2.7).

True anomaly, θ	Radius, r (km)	Velocity, v (km/s)
$-60°$	9,171	9.3236
$0°$	6,878	10.7660
$90°$	13,756	7.6127
$150°$	102,676	2.7864
$179°$	90,318,861	0.0939

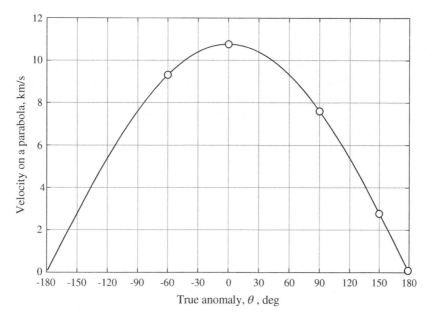

Figure 2.16 Parabolic velocity vs. true anomaly (Example 2.7).

Example 2.8 Determine the escape velocities for a satellite in a 300-km altitude circular LEO and for a satellite in GEO.

The LEO radius is $r_{LEO} = R_E + 300$ km $= 6{,}678$ km. Using Eq. (2.88), the local escape velocity is

$$\text{From LEO:} \quad v_{esc} = \sqrt{\frac{2\mu}{r_{LEO}}} = \boxed{10.9260 \text{ km/s}}$$

The GEO radius is $r_{GEO} = 42{,}164$ km. Hence, the local escape velocity at GEO is

$$\text{From GEO:} \quad v_{esc} = \sqrt{\frac{2\mu}{r_{GEO}}} = \boxed{4.3482 \text{ km/s}}$$

The local escape velocity at GEO is less than 40% of the escape speed at LEO.

As a side note, the circular orbital velocity in LEO is $v_{LEO} = 7.7258$ km/s and the circular orbital velocity in GEO is $v_{GEO} = 3.0747$ km/s. Therefore, we must increase LEO velocity by about 3.20 km/s to achieve escape speed while escape conditions are achieved at GEO by increasing velocity by 1.27 km/s. Firing an onboard rocket engine accelerates the satellite and creates the change in velocity. Although a GEO satellite requires less velocity change (or, added energy) to achieve an escape trajectory compared with a LEO satellite, the reader must note that much more energy is required to insert a satellite into GEO compared with inserting a satellite into a low-Earth orbit (after all, $a_{GEO} > a_{LEO}$). We will discuss orbital maneuvers using rocket propulsion in Chapters 6 and 7.

2.7 Hyperbolic Trajectory

Recall that a hyperbola is an open curve with negative semimajor axis, positive total energy, and eccentricity greater than unity. Examples of hyperbolic trajectories include spacecraft escaping the Earth's gravity at the onset of an interplanetary mission and a flyby encounter with a target planet. The hyperbola as a conic section has two branches (see Figure 2.7c) but only one branch represents the physical trajectory. Figure 2.17 shows a satellite on a hyperbolic trajectory. The arrival and departure paths of the hyperbola are along two straight-line asymptotes and the asymptotic velocity of the satellite at either end is computed from the energy equation (2.29)

$$\text{Hyperbola:} \quad \xi = \frac{v_\infty^2}{2} - \frac{\overset{0}{\cancel{\mu}}}{r_\infty} > 0 \tag{2.89}$$

or

$$v_\infty = \sqrt{2\xi} \tag{2.90}$$

This residual speed "at infinity" is called the *hyperbolic excess speed*. Referring again to Figure 2.17, the asymptotic velocity vectors are labeled $\mathbf{v}_\infty^-$ for the *arrival asymptote* and $\mathbf{v}_\infty^+$ for the *departure asymptote*. Because energy is constant on the hyperbola, the magnitudes of these two asymptotic velocities are equal (i.e., $v_\infty^- = v_\infty^+ = \sqrt{2\xi}$).

The angle between the asymptotes as shown in Figure 2.17 is called the *turning angle δ*. When a spacecraft encounters a gravitational body on a hyperbolic trajectory, the arrival hyperbolic velocity vector $\mathbf{v}_\infty^-$ is turned by angle δ to produce the departure hyperbolic velocity $\mathbf{v}_\infty^+$. The turning angle is solely a function of eccentricity

$$\delta = 2\sin^{-1}\left(\frac{1}{e}\right) \tag{2.91}$$

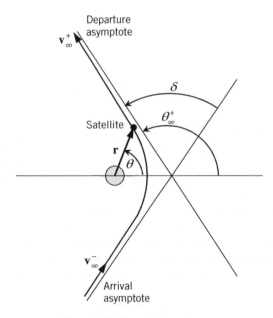

Figure 2.17 Hyperbolic trajectory.

Equation (2.91) can be derived from the geometry of a hyperbola presented in Figure 2.7c [i.e., $\sin(\delta/2) = a/c$ where $c/a = e$].

The true anomaly of the departure asymptote shown in Figure 2.17 is labeled θ_∞^+ and it is determined by evaluating the trajectory equation (2.45) at an infinite radial distance:

$$r_\infty = \frac{p}{1 + e\cos\theta_\infty^+} = \infty \tag{2.92}$$

Therefore, the denominator of Eq. (2.92) must be zero and we obtain

$$\theta_\infty^+ = \cos^{-1}\left(\frac{-1}{e}\right) \tag{2.93}$$

Because $e > 1$ the departure asymptotic true anomaly θ_∞^+ is always in the second quadrant (the *arrival* asymptotic true anomaly θ_∞^- is always in the third quadrant; note that $\cos\theta_\infty^- = \cos\theta_\infty^+$). Equations (2.91) and (2.93) show that when $e \approx 1$ both the turning angle δ and asymptotic true anomaly θ_∞ approach 180° (i.e., a "skinny hyperbola" similar to a parabola). At the other extreme when $e \gg 1$, the turning angle becomes very small and θ_∞^+ approaches 90° (i.e., a "flat hyperbola" with very little curvature).

While parabolic trajectories are of little interest to space mission designers, hyperbolic trajectories are essential for missions to interplanetary targets beyond Earth's gravitational influence. Recall that a fundamental assumption of the two-body problem (and the subsequent conic-section solution) is that the gravitational force of one celestial body is influencing the satellite's motion. Sending a robotic probe to Mars (for example) requires a hyperbolic departure trajectory from Earth so that the probe has excess speed "at infinity." Of course, after the satellite has "escaped" Earth's gravitational pull in a two-body sense and reached "infinity" (relative to Earth) it is primarily influenced by the sun's gravity. At this stage, we may analyze the satellite's motion as a two-body problem with the sun as the sole gravitational body. Eventually (if the trajectory is correctly planned), the satellite reaches the vicinity of the target planet (e.g., Mars) and we can analyze the arrival trajectory as another two-body problem. Because the probe is arriving "from infinity" with finite velocity (v_∞^-), the satellite will approach the target planet on a hyperbolic trajectory. This process of analyzing the entire interplanetary mission as a sequence of two-body problems is called the *patched-conic method* and will be discussed in detail in Chapter 10.

Example 2.9 The Mars Exploration Rover-A (MER-A) spacecraft was launched on June 10, 2003 by a Delta II booster rocket. At burnout conditions for the final rocket stage, the MER-A spacecraft was at an altitude of 225 km above the Earth's surface with a velocity of 11.4 km/s and a flight-path angle of 5°. Determine:

a) The departure hyperbolic excess speed v_∞^+.
b) The true anomaly of the departure asymptote θ_∞^+.

a) We can compute hyperbolic excess speed from the energy of the hyperbolic trajectory. Using the rocket burnout conditions (denoted by subscript "bo")

$$\text{Radius at burnout:} \quad r_{bo} = R_E + 225\,\text{km} = 6{,}603\,\text{km}$$

$$\text{Velocity at burnout:} \quad v_{bo} = 11.4\,\text{km/s}$$

Total energy on the hyperbola is

$$\xi = \frac{v_{bo}^2}{2} - \frac{\mu}{r_{bo}} = 4.6135 \, \text{km}^2/\text{s}^2$$

Using Eq. (2.90), the hyperbolic excess speed is

$$v_{\infty}^+ = \sqrt{2\xi} = \boxed{3.0376 \, \text{km/s}}$$

Note that we have inserted a superscript "+" to indicate the departure asymptotic speed.

b) Computing asymptotic true anomaly requires the eccentricity of the hyperbolic trajectory. We can compute the angular momentum of the hyperbola from the burn-out conditions:

$$h = r_{bo} v_{bo} \cos\gamma_{bo} = 74{,}987.76 \, \text{km}^2/\text{s}$$

where $\gamma_{bo} = 5°$ is the flight-path angle at burnout. Parameter p can be determined from angular momentum

$$p = \frac{h^2}{\mu} = 14{,}107.29 \, \text{km}$$

Recall that parameter is positive for all conic sections. We can calculate the semi-major axis of the hyperbolic trajectory from its energy

$$\xi = -\frac{\mu}{2a} \quad \rightarrow \quad a = -\frac{\mu}{2\xi} = -43{,}199.31 \, \text{km}$$

We see that $a < 0$ for a hyperbola because energy is positive. Finally, we employ Eq. (2.49) to determine eccentricity:

$$p = a(1 - e^2) \quad \rightarrow \quad e = \sqrt{1 - \frac{p}{a}} = 1.1518$$

Using Eq. (2.93), we find that the true anomaly of the departure asymptote is

$$\theta_{\infty}^+ = \cos^{-1}\left(\frac{-1}{e}\right) = \boxed{150.25°}$$

Example 2.10 In early 2007, the New Horizons spacecraft approached Jupiter on a hyperbolic trajectory with an asymptotic arrival velocity of $v_{\infty}^- = 18.427$ km/s. The space probe followed a "hyperbolic flyby trajectory" with a closest-approach distance of 32.25 R_J from the center of Jupiter where $R_J = 71{,}492$ km is Jupiter's equatorial radius. Calculate the spacecraft's velocity at periapsis and the turning angle δ from the hyperbolic flyby of Jupiter.

We can use the energy equation to determine the velocity at periapsis (or "perijove" for a trajectory about Jupiter):

$$\xi = \frac{v_{\infty}^2}{2} = \frac{v_p^2}{2} - \frac{\mu_J}{r_p} = 169.7772 \, \text{km}^2/\text{s}^2$$

Note that the hyperbolic excess speed (v_∞) solely determines the total energy of a hyperbolic trajectory. The perijove radius is $r_p = 32.25\ R_J = 2.3056(10^6)$ km. Using the energy equation, we determine the flyby velocity at perijove

$$v_p = \sqrt{2\left(\xi + \frac{\mu_J}{r_p}\right)} = \boxed{21.2001\ \text{km/s}}$$

Note that because Jupiter is the gravitational body we must use its gravitational parameter $\mu_J = 1.266865(10^8)$ km^3/s^2.

We need to determine the eccentricity of the hyperbolic trajectory in order to compute the turning angle. First, we compute the semimajor axis from the energy equation

$$\xi = -\frac{\mu_J}{2a} \quad \rightarrow \quad a = -\frac{\mu_J}{2\xi} = -373,096.4\ \text{km}$$

We can manipulate an expression for periapsis radius, Eq. (2.51), and solve it for eccentricity:

$$r_p = a(1-e) \quad \rightarrow \quad e = 1 - \frac{r_p}{a} = 7.1797$$

The turning angle is calculated using Eq. (2.91)

$$\delta = 2\sin^{-1}\left(\frac{1}{e}\right) = \boxed{16.0126°}$$

Figure 2.18 shows the Jupiter flyby (not to scale) as a "flat" hyperbolic trajectory with very little curvature and a small turning angle. It is rather intuitive that a very large periapsis

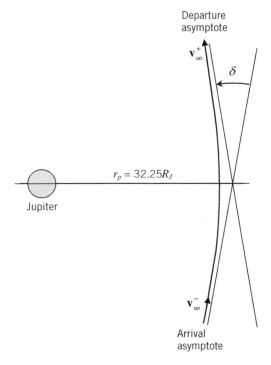

Figure 2.18 The New Horizons hyperbolic flyby of Jupiter (Example 2.10).

radius (as in this case) will result in a "flat" hyperbola with a small turning angle. Conversely, a very small periapsis radius will result in a "skinny" hyperbola with a large turning angle (in other words, flying deeper in the "gravity well" or closer to a planet will produce larger turning or angular motion). It is the "turning motion" of the hyperbolic flyby that produces the so-called *gravity assist maneuver* that alters the satellite's interplanetary trajectory without expending propellant. The New Horizons spacecraft used the Jupiter gravity assist outlined in this example to increase its speed (relative to the sun) and therefore reduce the flight time to its target (Pluto). We will discuss gravity assists in more detail in Chapter 10 in the context of interplanetary trajectories.

2.8 Summary

In this chapter, we developed the fundamental relationships for two-body motion. We did so by applying Newton's laws to a two-body system comprising a celestial gravitational body (e.g., the Earth) and a much smaller body (e.g., a satellite). Our result is a nonlinear vector differential equation that governs two-body motion. It is extremely important to reiterate the fundamental assumptions that lead to the two-body equation of motion: (1) the two bodies are point masses; (2) the mutual gravitational forces are the only forces acting on the two-body system; and (3) the mass of the smaller body (i.e., the satellite) is negligible relative to the central gravitational body. We demonstrated that total mechanical energy and angular momentum are constant for the two-body problem. These constants of motion lead to the so-called trajectory equation, which shows that the satellite's orbital path is a conic section: an ellipse, parabola, or hyperbola. In the end, we can relate the motion constants (energy and angular momentum) to geometric conic-section constants (i.e., semimajor axis, eccentricity, and parameter) for the orbit. Consequently, we can determine the satellite's radial position and velocity at any point in its orbit. As we will soon see in the subsequent chapters, the two-body orbital relationships developed in this chapter provide the foundation for analyzing and designing space missions.

Further Reading

Bate, R.R., Mueller, D.D., and White, J.E., *Fundamentals of Astrodynamics*, Dover, New York, 1971, Chapter 1.

Kaplan, M.H., *Modern Spacecraft Dynamics and Control*, John Wiley & Sons, Inc., Hoboken, NJ, 1976, Chapters 1 and 2.

Thomson, W.T., *Introduction to Space Dynamics*, Dover, New York, 1986, Chapter 4.

Vallado, D.A., *Fundamentals of Astrodynamics and Applications*, 4th edn, Microcosm Press, Hawthorne, CA, 2012, Chapter 1.

Wiesel, W.E., *Spaceflight Dynamics*, 3rd edn, Aphelion Press, Beavercreek, OH, 2010, Chapter 2.

Problems

For Earth-orbiting satellites, use R_E = 6,378 km for the radius of the Earth and μ = 3.986(10^5) km^3/s^2 for the gravitational parameter. For problems involving other celestial bodies (the moon, Mars, etc.) see Appendix A for their respective radii and gravitational parameters.

Conceptual Problems

2.1 An Earth-orbiting satellite has the following position and velocity vectors expressed in polar coordinates:

$$\mathbf{r} = 8{,}250\mathbf{u}_r \text{ km} \qquad \mathbf{v} = 1.2054\mathbf{u}_r + 7.0263\mathbf{u}_\theta \text{ km/s}$$

Determine the following:
a) Angular momentum (magnitude), h
b) Specific energy, ξ
c) Semimajor axis, a
d) Parameter, p
e) Eccentricity, e
f) Perigee and apogee radii, r_p and r_a
g) Flight-path angle, γ, at this instant
h) True anomaly, θ, at this instant.

2.2 Repeat Problem 2.1 for an Earth-orbiting satellite with the following position and velocity vectors expressed in polar coordinates:

$$\mathbf{r} = 9{,}104\mathbf{u}_r \text{ km} \qquad \mathbf{v} = -0.7004\mathbf{u}_r + 6.1422\mathbf{u}_\theta \text{ km/s}$$

2.3 Compute the eccentricity vector **e** using the Earth-orbiting satellite data in Problem 2.2 and show that its norm (magnitude) matches the eccentricity e as computed using the geometric parameters p and a.

2.4 An Earth-orbiting satellite has the following position and velocity vectors expressed in polar coordinates:

$$\mathbf{r} = 1{,}2426\mathbf{u}_r \text{ km} \qquad \mathbf{v} = 4.78\mathbf{u}_\theta \text{ km/s}$$

Determine the following parameters when the satellite is 2,000 km above the Earth's surface and approaching the Earth:
a) Orbital velocity
b) Flight-path angle, γ
c) True anomaly, θ.

2.5 An Earth-orbiting satellite has semimajor axis a = 9,180 km and eccentricity e = 0.12. Determine the radial position r, velocity v, and flight-path angle γ when the satellite is approaching Earth and 80° from perigee passage.

2.6 Develop an expression for eccentricity e in terms of specific energy ξ and angular momentum h.

2.7 At a particular instant in time, a tracking station determines that a space vehicle is at an altitude of 390.4 km with an inertial velocity of 9.7023 km/s and flight-path angle of 1.905°. Is this space vehicle in a closed orbit about the Earth or is it following an open-ended trajectory that will eventually "escape" Earth? Justify your answer.

2.8 An Earth-orbiting satellite has semimajor axis $a = 8{,}230$ km and eccentricity $e = 0.12$. Determine the satellite's maximum altitude above the Earth's surface.

2.9 Two satellites are being tracked by ground-based radar stations. Their altitusdes, inertial velocities, and flight-path angles at a particular instant in time are summarized in the following table:

Object	Altitude (km)	Velocity (km/s)	Flight-path angle
Satellite A	1,769.7560	6.8164	4.9665°
Satellite B	676.3674	7.8504	−6.8903°

Are these two satellites in the same orbit? Explain your answer.

2.10 An Earth-orbiting satellite has the following position and velocity vectors expressed in polar coordinates:

$$\mathbf{r} = 7{,}235\mathbf{u}_r \text{ km} \qquad \mathbf{v} = -0.204\mathbf{u}_r + 8.832\mathbf{u}_\theta \text{ km/s}$$

Determine the orbital period in minutes.

2.11 A tracking station determines that an Earth-observation satellite has perigee and apogee altitudes of 350 and 1,206 km, respectively. Determine the orbital period (in minutes) and the parameter p.

2.12 An Earth-orbiting satellite is at an altitude of 700 km with inertial velocity $v = 7.3944$ km/s and flight-path angle $\gamma = 0$. Is the satellite at perigee or apogee? Justify your answer.

2.13 The ratio of apoapsis and periapsis radii for a particular satellite orbit is $r_a/r_p = 1.6$. Determine the eccentricity of the orbit.

2.14 Derive an expression for the orbital period of a circular orbit in terms of its circular velocity speed v_c.

2.15 An Earth-observation satellite's closest approach is at an altitude of 300 km. If the satellite returns to its perigee position every 2 h determine the apogee altitude and the orbital eccentricity.

2.16 A launch vehicle experiences a malfunction in its guidance system. At burnout of its upper rocket stage, the vehicle is at an altitude of 250 km with an inertial velocity of 7.791 km/s and flight-path angle of 4.5°. Has the vehicle achieved a stable orbit? Explain your answer.

2.17 Figure P2.17 shows two satellites in Earth orbits: Satellite A is in a circular orbit with an altitude of 800 km, while Satellite B is in an elliptical orbit with a perigee altitude of 800 km. At the instant shown in Figure P2.17, Satellite B is passing through perigee while Satellite A lags behind Satellite B with an angular separation of 60°. Determine the *apogee* altitude of the elliptical orbit so that Satellites A and B occupy the same radial position after one revolution of Satellite B (in other words, Satellites A and B perform a rendezvous maneuver when Satellite B returns to perigee after one orbital revolution).

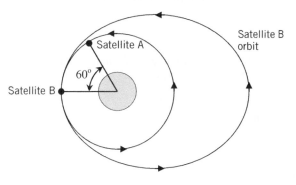

Figure P2.17

2.18 Derive an expression for the time rate of true anomaly, $\dot{\theta}$, as a function of parameter p, eccentricity e, and true anomaly θ.

2.19 A satellite is on a parabolic trajectory about the Earth. At 4,000 km above the Earth's surface, it has a flight-path angle of 25°. Determine the velocity and true anomaly of the satellite at this instant.

2.20 A satellite is approaching Earth on a parabolic trajectory with a velocity of 5.423 km/s. If the projected perigee altitude of the parabolic trajectory is 800 km, determine the radius, flight-path angle, and true anomaly at this instant.

2.21 An interplanetary spacecraft fires an onboard rocket in order to depart a low-Earth orbit. At engine cutoff (at an altitude of 200 km), the spacecraft has an inertial velocity of 11.814 km/s and zero flight-path angle. Determine:
a) Eccentricity of the hyperbolic departure trajectory
b) Velocity when the spacecraft is at a radial distance of 400,000 km
c) Flight-path angle when the spacecraft is at a radial distance of 400,000 km
d) True anomaly when the spacecraft is at a radial distance of 400,000 km
e) Hyperbolic excess speed, v_∞^+.

2.22 An interplanetary spacecraft is departing Earth on a hyperbolic trajectory with eccentricity $e = 1.4$ and semimajor axis $a = -16,900$ km. Determine:
a) Perigee altitude
b) Radial and transverse velocity components at true anomaly $\theta = 100°$
c) True anomaly of the departure asymptote, θ_∞^+
d) Hyperbolic turning angle, δ.

2.23 In March 2016, a spacecraft launched in early 2014 is approaching Earth on a hyperbolic trajectory for a gravity assist maneuver. Its hyperbolic excess speed on the arrival asymptote is $v_\infty^- = 2.78$ km/s and its projected perigee velocity is estimated by mission operators to be 10.9 km. Determine the perigee altitude and turning angle δ of the hyperbolic flyby.

MATLAB Problems

2.24 Write an M-file that will that will calculate the following characteristics of an Earth orbit (with the desired units):

Angular momentum, h (km^2/s)
Energy, ξ (km^2/s^2)
Semimajor axis, a (km)
Parameter, p (km)
Eccentricity, e
Period, T_{period} (h)
Perigee and apogee radii, r_p and r_a (km)
Flight-path angle, γ (deg)
True anomaly, θ (deg)

The inputs to the M-file are orbital radius r (in km), radial velocity v_r (in km/s), and transverse velocity v_θ (in km/s). The M-file should return an empty set (use open brackets [ ]) for characteristics that do not exist, such as period for a parabolic or hyperbolic trajectory. Test your M-file by solving Problem 2.1.

2.25 Write an M-file that will calculate a satellite's orbital "state" for a particular location in an Earth orbit. The desired outputs are radial position r (in km), velocity magnitude v (in km/s), and flight-path angle γ (in deg). The M-file inputs are semimajor axis a (in km), eccentricity e, and true anomaly θ (in deg). Test your M-file by solving Problem 2.4.

2.26 The second stage of a launch vehicle is approaching its main-engine cutoff (MECO). Suppose the vehicle's guidance system has the following simplified equations for orbital radius and velocity at MECO as a function of flight-path angle

$$r = 6,878 + 12\gamma \text{ km}, \qquad v = 7.613 - 1.5\gamma \text{ km/s}$$

where flight-path angle γ is in radians. Plot perigee altitude, semimajor axis, and eccentricity as a function of MECO flight-path angle for the range $-10° \leq \gamma \leq 10°$. Using these plots, determine the MECO flight-path angle that results in the maximum-energy elliptical orbit with a 200 km altitude perigee.

Mission Applications

2.27 GeoEye-1 is an Earth-observation satellite that provides high-resolution images for Google. The orbital period and eccentricity of GeoEye-1 are 98.33 min and 0.001027, respectively. Determine the perigee and apogee altitudes of GeoEye-1.

2.28 The Chandra X-ray Observatory (CXO) used a sequence of two transfer orbits to increase orbital energy (additional subsequent transfer orbits were used to eventually achieve the highly elliptical operational CXO orbit presented in Example 2.3). Figure P2.28 shows that the two transfer orbits are tangent at the perigee altitude of 300 km above the surface of the Earth. Determine the eccentricity and orbital period of each transfer orbit.

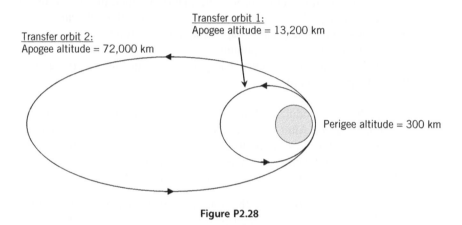

Figure P2.28

2.29 A US reconnaissance satellite is in an elliptical orbit with a period of 717.8 min. Ground-tracking stations determine that its perigee altitude is 2,052 km. What is the apogee altitude of this satellite?

2.30 The Apollo 17 command and service module (CSM) orbited the moon in a 116 km altitude circular orbit while two astronauts landed on the lunar surface. Determine the orbital velocity and period of the CSM.

2.31 The Meridian 4 is a Russian communication satellite that was launched in May 2011 on a Soyuz-2 rocket. The operational (target) orbit of the Meridian 4 satellite is an elliptical orbit with perigee and apogee altitudes of 998 and 39,724 km, respectively. The Meridian 4 satellite reached its target by following an elliptical transfer orbit that is tangent to the target orbit at apogee (Figure P2.31; not to scale). The perigee altitude of the transfer orbit is 203 km. Determine:
a) The perigee velocity on the transfer orbit
b) The transit time from perigee to apogee on the transfer orbit
c) The apogee velocity on the transfer orbit
d) The apogee velocity on the Meridian 4 target orbit.

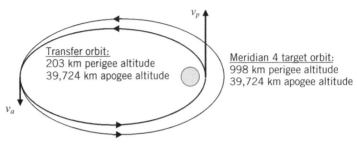

Figure P2.31

2.32 Lunar Orbiter 1 (1966) was the first US spacecraft to orbit the moon. It was initially inserted into a lunar orbit with angular momentum $h = 3,509.8 \text{ km}^2/\text{s}$ and specific energy $\xi = -0.886641 \text{ km}^2/\text{s}^2$. Determine the following:
   a) Periapsis ("perilune") and apoapsis ("apolune") altitudes
   b) Velocity at perilune and apolune
   c) Radial and transverse velocity components at true anomaly $\theta = 220°$.
   d) Orbital period (in min).

2.33 The Apollo lunar module (LM) used its ascent propulsion system (APS) to depart the moon's surface. After over 7 min of powered flight, the APS engine was shut down and the LM was at an altitude of 18 km above the moon with velocity $v = 1.687 \text{ km/s}$ and flight-path angle $\gamma = 0.4°$. Determine the periapsis ("perilune") and apoapsis ("apolune") altitudes of the LM's orbit after engine cutoff.

   Problems 2.34–2.36 involve the Mars Reconnaissance Orbiter (MRO) space-craft which approached the target planet in March 2006 and subsequently per-formed a propulsive maneuver to slow down and enter a closed orbit about Mars.

2.34 The MRO spacecraft approached Mars on a hyperbolic trajectory with eccentric-ity $e = 1.7804$ and asymptotic approach speed $v_\infty^- = 2.9572 \text{ km/s}$. Determine the altitude, velocity, and flight-path angle of the MRO spacecraft at its closest approach to Mars.

2.35 The MRO spacecraft fired its rocket engines at periapsis of the hyperbolic approach to slow the spacecraft's velocity to 4.5573 km/s for insertion into a closed orbit about Mars. Using the MRO hyperbolic trajectory information in Problem 2.34, determine the orbital period and eccentricity of the MRO space-craft after the orbit-insertion burn (the rocket burn did not change the periapsis radius – it is the same as the periapsis radius of the hyperbolic approach trajectory as determined in Problem 2.34).

2.36 The MRO used atmospheric drag at each periapsis pass ("aerobraking") to slow down and reduce the orbital energy. After the aerobraking phase (and a small pro-pulsive maneuver), the operational orbit for the MRO spacecraft has periapsis and apoapsis altitudes of 250 and 316 km above the surface of Mars, respectively. Determine the following:

    a) Semimajor axis, a

    b) Eccentricity, e

    c) Orbital period (in min)

    d) True anomaly, θ, when the altitude is 300 km

    e) Flight-path angle, γ, when the altitude is 300 km.

     Problems 2.37 and 2.38 involve the Lunar Atmosphere and Dust Environment Explorer (LADEE) spacecraft, which was launched in September 2013 (see Example 2.6).

2.37   The LADEE spacecraft was launched into a highly elliptical orbit about the Earth by a Minotaur V booster with perigee and apogee altitudes of 200 and 278,000 km, respectively (see Figure 2.14 and Example 2.6). Determine the altitude, velocity, and flight-path angle of the LADEE spacecraft at true anomaly $\theta = 300°$.

2.38   After a coasting translunar trajectory, the LADEE spacecraft was inserted into an elliptical orbit about the moon by performing a series of retrorocket propulsive burns. The orbital period of the LADEE spacecraft was 4 h and its orbital eccentricity was $e = 0.2761$ (Figure P2.38; not to scale). Determine the periapsis and apoapsis altitudes (or "perilune" and "apolune" altitudes) of the LADEE spacecraft in its lunar orbit.

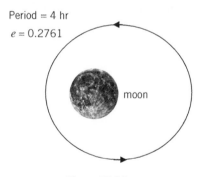

Period = 4 hr

$e = 0.2761$

moon

Figure P2.38

     Problems 2.39 and 2.40 involve the Stardust capsule which returned to Earth in January 2006 on a hyperbolic approach trajectory after sampling particles from the comet Wild-2.

2.39   When the Stardust capsule arrived at the "edge" of the Earth's atmosphere (the so-called "entry interface" altitude of 122 km), it had inertial velocity $v = 12.9$ km/s and flight-path angle $\gamma = -8.21°$. Determine the following:

    a) Specific energy, ξ

    b) Semimajor axis, a

    c) Eccentricity, e

    d) True anomaly, θ, at entry interface

    e) Arrival hyperbolic excess speed, v_∞^-

    f) Arrival asymptotic true anomaly, θ_∞^-.

2.40 Using the entry-interface state of the Stardust capsule from Problem 2.39, determine the velocity, flight-path angle, and true anomaly of the capsule when it was at radius $r = 384{,}400$ km (roughly the distance from the Earth to the moon).

2.41 The Pegasus launch vehicle reaches its second-stage burnout at an altitude of 192 km, inertial velocity $v = 5.49$ km/s, and flight-path angle $\gamma = 25.8°$. The launch vehicle then coasts in this orbit until it ignites its third stage when flight-path angle decreases to $2.2°$. Determine:
   a) Semimajor axis of the Pegasus orbit after second-stage burnout.
   b) Eccentricity of the Pegasus orbit after second-stage burnout.
   c) The altitude and velocity of the Pegasus launch vehicle when the third stage is ignited.

Problems 2.42 and 2.43 involve the Juno spacecraft which departed Earth in early August 2011 and arrived at Jupiter in early July 2016.

2.42 The Juno spacecraft approached Jupiter on a hyperbolic trajectory with eccentricity $e = 1.0172$ and semimajor axis $a = -4.384(10^6)$ km. Determine the asymptotic approach speed v_∞^- and radial distance from Jupiter at its closest approach.

2.43 The Juno spacecraft fired a retrorocket at its periapsis ("perijove") position to slow down and establish a highly elliptical orbit about Jupiter with semimajor axis $a = 4{,}092{,}211$ km and eccentricity $e = 0.981574$. Determine the orbital period and apoapsis ("apojove") radius of Juno's orbit.

3

Orbit Determination

3.1 Introduction

Orbit determination involves computing the defining characteristics of a satellite's orbit through observations. In Chapter 2, we introduced three of these defining orbital characteristics: semimajor axis a, eccentricity e, and true anomaly θ. Semimajor axis allows us to compute the total energy of the orbit whereas a and e allow us to calculate angular momentum magnitude. True anomaly θ defines the satellite's position in the orbit at a given instant. Hence, a, e, and θ provide information about the size and shape of the orbit and the satellite's location in the orbital plane. However, these three parameters provide no information regarding the three-dimensional orientation of the orbit. Clearly, the successful operation of an Earth-orbiting satellite requires careful planning of its orbit so that it passes over the desired geographic locations on Earth. One way to characterize an orbit in three dimensions is to determine the satellite's position and velocity vectors $\mathbf{r}$ and $\mathbf{v}$ in a Cartesian frame centered at the attracting gravitational body. Of course, the $\mathbf{r}$ and $\mathbf{v}$ vectors continuously change with time as the satellite moves in its orbit. Another way to fully characterize an orbit is to determine the so-called *classical orbital elements* (a, e, and θ are three of the orbital elements). We will see that knowledge of the orbital elements allows us to better visualize the orbit compared with knowledge of the vectors $(\mathbf{r},\mathbf{v})$.

This chapter begins by discussing how we can determine an orbit's defining characteristics (the orbital elements) from the satellite's Cartesian vectors $(\mathbf{r},\mathbf{v})$ at a given instant. The primary assumption for this orbit-determination problem, of course, is that the vectors $(\mathbf{r},\mathbf{v})$ are somehow available to us. The latter part of this chapter describes methods that use multiple satellite observations (or measurements) to determine the position and velocity vectors $(\mathbf{r},\mathbf{v})$ corresponding to a particular instant in time.

3.2 Coordinate Systems

In Chapter 2, we developed the two-body problem by using a Cartesian coordinate system fixed to the center of the gravitational body. Furthermore, we required that the body-centered XYZ coordinate system be an inertial frame (no rotation or acceleration) so that we could apply Newton's laws. We did not, however, specify a reference direction or

Space Flight Dynamics, First Edition. Craig A. Kluever.
© 2018 John Wiley & Sons Ltd. Published 2018 by John Wiley & Sons Ltd.
Companion website: www.wiley.com/go/Kluever/spaceflightmechanics

orientation of the Cartesian coordinates. Obviously, it is useful (and necessary) to define standard coordinate systems so that the space flight community has a common point of reference for defining a satellite's orbit.

In this textbook, we will primarily analyze satellite motion where the Earth and sun are the central gravitation bodies. We can make a long list of Earth-orbiting satellites used for communication, navigation, weather monitoring, and Earth science. Earth-orbiting satellites require an inertial Cartesian system fixed at the Earth's center. Interplanetary spacecraft, on the other hand, spend the majority of their transit time under the influence of the sun's gravity. Therefore, interplanetary missions require an inertial Cartesian system fixed at the center of the sun. We present the sun-centered coordinate system first because it establishes a standard reference direction that is also used for the Earth-centered system.

Figure 3.1 shows the *heliocentric-ecliptic* coordinate system. The origin of the Cartesian $(XYZ)_H$ axes is at the center of the sun. Earth's orbital plane about the sun is called the *ecliptic* and it defines the $X_H–Y_H$ plane. The $+Z_H$ axis is normal to the ecliptic and completes the right-handed Cartesian frame. The X_H axis is the intersection of the Earth's equatorial plane and the ecliptic, and its positive direction is from the Earth's center to the sun's center on the March (vernal) equinox as shown in Figure 3.1. The vernal equinox direction is shown in Figure 3.1 by the symbol for the constellation Aries (♈). The vernal equinox direction varies slightly due to third-body gravitational perturbations; therefore, it is not truly a fixed reference direction. For this reason, the $+X_H$ axis is fixed at the vernal equinox direction for a particular moment in time or *epoch*. A common heliocentric-ecliptic frame is the mean equator and equinox of January 1, 2000 or the *J2000 system*.

Figure 3.2 shows the *geocentric-equatorial* coordinate system for Earth-orbiting satellites. The Earth's equatorial plane is the fundamental $X–Y$ plane where the $+X$ axis points in the vernal equinox direction (i.e., the same direction as the $+X_H$ axis in the

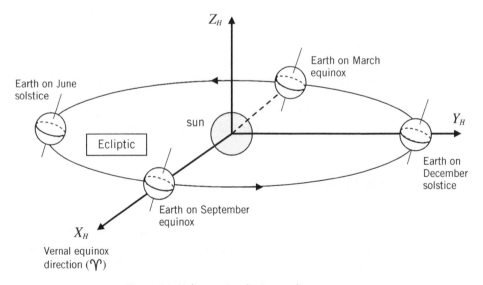

Figure 3.1 Heliocentric-ecliptic coordinate system.

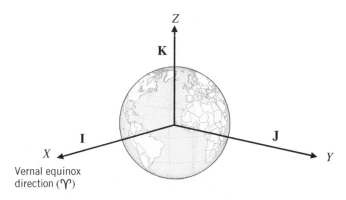

Z

K

I

X

J

Y

Vernal equinox
direction (♈)

Figure 3.2 Geocentric-equatorial or Earth-centered inertial coordinate system.

heliocentric-ecliptic system). The $+Z$ axis is along the direction of the North Pole. The geocentric-equatorial system is often called the *Earth-centered inertial* (ECI) frame. We will use unit vectors **IJK** to designate the directions of the XYZ axes as shown in Figure 3.2. It is important for the reader to remember that the ECI system, while it moves with the Earth's center, does *not* rotate.

The angle of obliquity ($\sim$23.4°) is the angle between the Earth's equatorial plane and the ecliptic plane. If we were to superimpose the ECI system (Figure 3.2) onto the Earth's center in Figure 3.1, the $+X_H$ and **I** directions would coincide. Rotating the ECI system about its **I** axis by the angle of obliquity would align the ECI and heliocentric-ecliptic frames.

3.3 Classical Orbital Elements

Recall that two-body orbital motion is governed by Eq. (2.12), which is a second-order vector differential equation. We can express the two-body equations of motion in state-variable form by using six scalar differential equations and therefore the complete solution requires six initial conditions: three initial position coordinates ($\mathbf{r}_0$) and three initial velocity coordinates ($\mathbf{v}_0$). Knowledge of a satellite's $\mathbf{r}$ and $\mathbf{v}$ vectors at any instant of time thus allows us to compute the orbital solution at any past or future time. We will call the collection of the position and velocity vectors ($\mathbf{r}, \mathbf{v}$) the satellite's *state vector*. For example, the state vector of a spacecraft on an interplanetary trajectory would consist of its position and velocity vectors expressed in the heliocentric-ecliptic coordinate system (Figure 3.1). Figure 3.3 shows the state vector ($\mathbf{r}, \mathbf{v}$) for an Earth-orbiting satellite. The satellite's $\mathbf{r}$ and $\mathbf{v}$ vectors are expressed as components along the **IJK** axes shown in Figure 3.3. It is very important for the reader to see that the state vector ($\mathbf{r}, \mathbf{v}$) contains six elements: three position coordinates and three velocity coordinates.

Although the state vector ($\mathbf{r}, \mathbf{v}$) completely characterizes the orbital motion, it does not provide an intuitive feel for the orbit. Obviously, all six elements of the state vector change as the satellite moves along its orbital path. It is easier to characterize and visualize an orbit by a set of six constants known as the *classical orbital elements*. Two orbital elements define the size and shape of the orbit; three orbital elements define the

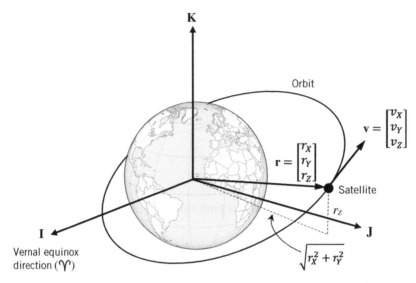

Figure 3.3 State vector (**r,v**) in the ECI coordinate system.

three-dimensional orientation of the orbit in space; the sixth element defines the position of the satellite in the orbit at a particular instant of time (or epoch). The six classical orbital elements are:

- semimajor axis, a
- eccentricity, e
- inclination, i
- longitude of the ascending node, Ω
- argument of periapsis, ω
- true anomaly at epoch, θ_0

As we discussed in Chapter 2, semimajor axis a and eccentricity e define the orbit's size (or energy) and shape. The next three orbital elements (i, Ω, and ω) define the orientation of the orbit in three-dimensional (**IJK**) space as seen in Figure 3.4. *Inclination i* is the angle between the **K** unit vector and the angular momentum vector **h**. Recall that angular momentum **h** is normal to the orbital plane. Inclination is the tilt of the orbital plane with respect to the equatorial (**IJ**) plane, and it varies from 0 to 180°. For $0° \leq i < 90°$, the orbit is said to be *direct* or *prograde* (easterly direction), for $90° < i \leq 180°$, the orbit is said to be *retrograde* (westerly orbit), and for $i = 90°$ the orbit is polar. The intersection between the orbital plane and the equatorial plane is called the *line of nodes* (see Figure 3.4), and the nodal line where the satellite is moving from the southern to northern hemisphere is called the *ascending node*, **n**, as shown in Figure 3.4. The *longitude of the ascending node* (Ω) is measured counterclockwise in the equatorial plane from the vernal equinox direction (**I**) to the ascending node **n**. The *argument of periapsis* (*perigee*, for Earth orbits) ω is measured in the orbital plane in the direction of satellite motion from the ascending node **n** to the periapsis (perigee) direction. The eccentricity vector, **e**, points from the focus to the periapsis direction as shown in Figure 3.4. True anomaly at epoch, θ_0, is the sixth orbital element, and it defines the angular location of the satellite in its orbit

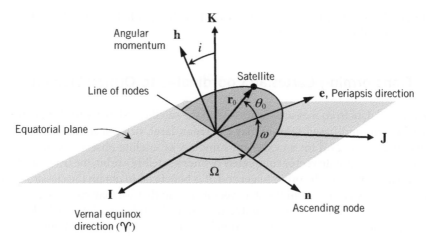

Figure 3.4 The four angular classical orbital elements: inclination i, longitude of the ascending node Ω, argument of periapsis ω, and true anomaly θ_0.

at epoch t_0. For two-body motion, the six classical orbital elements $(a, e, i, \Omega, \omega, \theta_0)$ are constants. The reader should keep in mind that true anomaly at epoch θ_0 is a constant because it defines the angular position of the satellite at the epoch time t_0. Clearly, true anomaly θ varies continuously from 0 to 2π for a closed orbit. As we shall see in Chapter 4, the location of the satellite (true anomaly θ) at any past or future instant of time can be obtained by using Kepler's time-of-flight equation.

The six elements $(a, e, i, \Omega, \omega, \theta_0)$ do not comprise the definitive set of orbital elements. Because semimajor axis a is infinite for a parabolic trajectory, the parameter p may be used as the first orbital element (note that $p > 0$ for all conic sections). The *longitude of periapsis*, ϖ, is sometimes used in place of argument of periapsis ω; it is defined as

$$\varpi \equiv \Omega + \omega \tag{3.1}$$

The longitude of periapsis is measured in two planes: first in the equatorial plane from the $\mathbf{I}$ axis to the ascending node $\mathbf{n}$, and then in the orbital plane to the periapsis direction $\mathbf{e}$. If the orbit is equatorial ($i = 0$), the ascending node $\mathbf{n}$ does not exist and longitude of periapsis ϖ is the angle from $\mathbf{I}$ to periapsis direction $\mathbf{e}$.

For an inclined circular orbit, the eccentricity vector $\mathbf{e}$ does not exist. Hence, ω is not defined and true anomaly does not have a line of reference. We may use the *argument of latitude* in place of ω and θ. Argument of latitude u is the angle measured in the orbital plane from the ascending node $\mathbf{n}$ to the radius vector $\mathbf{r}$. For an elliptical orbit, the argument of latitude at epoch is

$$u_0 \equiv \omega + \theta_0 \tag{3.2}$$

The *true longitude at epoch* is defined as

$$l_0 \equiv \Omega + \omega + \theta_0 \tag{3.3}$$

and may be used as the sixth element in place of true anomaly at epoch. True longitude is also measured in two planes from the $\mathbf{I}$ axis to the satellite's position vector $\mathbf{r}_0$. Note that for an equatorial orbit ($i = 0$), the angle Ω is undefined and hence l_0 is the angle in the equatorial plane from $\mathbf{I}$ to position vector $\mathbf{r}_0$. True longitude is always defined because the vectors $\mathbf{I}$ and $\mathbf{r}_0$ always exist.

3.4 Transforming Cartesian Coordinates to Orbital Elements

Observational data from measurements lead to the satellite's state vector $(\mathbf{r},\mathbf{v})$ in the ECI frame (these ground-based sensors may use radar, laser, or optical measurements; see Vallado [1; pp. 241–275] for an excellent summary of satellite observation techniques). While the Cartesian state vector completely defines a satellite's orbit, the $(\mathbf{r},\mathbf{v})$ coordinates do not aid in visualizing the orbit. In general, it is beneficial to obtain the classical orbital elements $(a, e, i, \Omega, \omega, \theta_0)$ from the state vector $(\mathbf{r},\mathbf{v})$ so that we may visualize a satellite's orbit. As an example, consider the highly elliptical, 12-h Molniya orbit used by Russia for communication satellites. Suppose the orbital elements for a particular Molniya orbit are

$$a = 26{,}565 \text{ km}$$

$$e = 0.7411$$

$$i = 63.4°$$

$$\Omega = 50°$$

$$\omega = -90°$$

$$\theta_0 = 180°$$

Figure 3.5 shows a Molniya orbit with these orbital elements. The orbital elements a and e tell us that the Molniya orbit is a "long and skinny" ellipse (we can compute the perigee and apogee radii using Eqs. (2.51) and (2.53) to find $r_p = 6{,}878$ km and $r_a = 46{,}252$ km). The

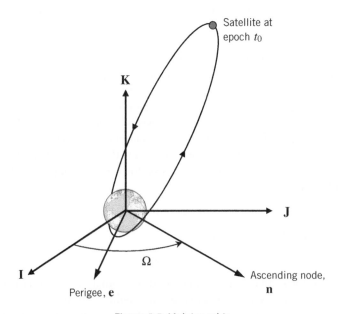

Figure 3.5 Molniya orbit.

large inclination ($i = 63.4°$) tells us that the orbital plane is tilted away from the equatorial (**IJ**) plane. Because the argument of perigee ω is $-90°$ (or $270°$), the perigee of the Molniya orbit is in the southern hemisphere. Finally, because the true anomaly at epoch is $\theta_0 = 180°$, we know that the satellite is at apogee; because $\omega = -90°$, the satellite's apogee is at its "highest point" (geographic latitude) on its passage through the northern hemisphere.

Now consider the state vector (in **IJK** coordinates) that corresponds to the satellite's position at epoch t_0 (apogee in this case):

$$
\mathbf{r}_0 = \begin{bmatrix} -15{,}865 \\ 13{,}312 \\ 41{,}357 \end{bmatrix} \text{ km}, \quad \mathbf{v}_0 = \begin{bmatrix} -0.9601 \\ -1.1443 \\ 0 \end{bmatrix} \text{ km/s}
$$

It is very difficult to visualize or ascertain the characteristics of the Molniya orbit directly from the state vector ($\mathbf{r}_0, \mathbf{v}_0$). Therefore, computing the six orbital elements from the six-dimensional state vector ($\mathbf{r}_0, \mathbf{v}_0$) is crucial to the orbit-determination process.

Before we proceed, it is important to stress that the state vector ($\mathbf{r}_0, \mathbf{v}_0$) expressed in a body-centered Cartesian frame and the six orbital elements (a, e, i, Ω, ω, θ_0) both completely define the two-body orbit. In other words, there is a unique mapping between the six-element state vector ($\mathbf{r}, \mathbf{v}$) and the six classical orbital elements. Furthermore, all six coordinates of the state vector ($\mathbf{r}, \mathbf{v}$) will change as the satellite moves from epoch t_0 to t_1 (unless time t_1 is k periods after epoch t_0). In contrast, the five orbital elements (a, e, i, Ω, ω) will remain constant for two-body motion and only the true anomaly θ will change as the satellite moves in its orbit.

Let us develop a systematic process for determining the classical orbital elements (a, e, i, Ω, ω, θ_0) from the state vector ($\mathbf{r}_0, \mathbf{v}_0$) at epoch t_0. We will follow the processes presented in Vallado [1; pp. 95–100] and Bate *et al.* [2; pp. 61–63]. The first step is to determine total energy from the magnitudes of the position and velocity vectors $\mathbf{r}_0$ and $\mathbf{v}_0$

$$
\xi = \frac{v_0^2}{2} - \frac{\mu}{r_0} \tag{3.4}
$$

where $r_0 = \|\mathbf{r}_0\|$ and $v_0 = \|\mathbf{v}_0\|$. Recall from Chapter 2 that semimajor axis is solely a function of energy:

$$
a = \frac{-\mu}{2\xi} \tag{3.5}
$$

Next, we compute the eccentricity vector $\mathbf{e}$ directly from state vector ($\mathbf{r}_0, \mathbf{v}_0$) by using Eq. (2.57):

$$
\mathbf{e} = \frac{1}{\mu}\left[\left(v_0^2 - \frac{\mu}{r_0}\right)\mathbf{r}_0 - (\mathbf{r}_0 \cdot \mathbf{v}_0)\mathbf{v}_0\right] \tag{3.6}
$$

We derived this expression for the eccentricity vector in Chapter 2 when we developed the trajectory equation (2.45). The vector $\mathbf{e}$ points from the center of the gravitational body to periapsis (see Figure 3.4) and its magnitude is the eccentricity of the orbit:

$$
e = \|\mathbf{e}\| \tag{3.7}
$$

Figure 3.4 shows that inclination is the angle between vectors $\mathbf{K}$ and $\mathbf{h}$ and therefore it can be computed from the dot (or scalar) product:

$$\cos i = \frac{\mathbf{K} \cdot \mathbf{h}}{h} \tag{3.8}$$

where the angular momentum vector $\mathbf{h}$ is computed from $\mathbf{r}_0$ and $\mathbf{v}_0$

$$\mathbf{h} = \mathbf{r}_0 \times \mathbf{v}_0 \tag{3.9}$$

Equation (3.8) is derived from the dot-product definition $\mathbf{A} \cdot \mathbf{B} = AB \cos \alpha$ where α is the angle between vectors $\mathbf{A}$ and $\mathbf{B}$ and A and B are their respective magnitudes. Note that the unity magnitude of vector $\mathbf{K}$ is not explicitly shown in Eq. (3.8). Because the range for inclination is $0° \leq i \leq 180°$, we may simply apply the inverse cosine operation to Eq. (3.8) without an additional quadrant check.

Longitude of the ascending node Ω is the angle between $\mathbf{I}$ and ascending node vector $\mathbf{n}$ (see Figure 3.4), and therefore its cosine is

$$\cos \Omega = \frac{\mathbf{I} \cdot \mathbf{n}}{\|\mathbf{n}\|} \tag{3.10}$$

(again, note that $\|\mathbf{I}\| = 1$). The node vector $\mathbf{n}$ is the cross product of vectors $\mathbf{K}$ and $\mathbf{h}$ (see Figure 3.4):

$$\mathbf{n} = \mathbf{K} \times \mathbf{h} \tag{3.11}$$

We cannot determine Ω by simply taking the inverse cosine of Eq. (3.10) because the resulting angle will always be in the first or second quadrant (i.e., between $0°$ and $180°$). Longitude of ascending node Ω ranges from 0 to $360°$; therefore, we need a quadrant check. From Figure 3.4, we see that a unit vector in the direction of node vector $\mathbf{n}$ may be expressed in terms of $\mathbf{I}$ and $\mathbf{J}$ components:

$$\frac{\mathbf{n}}{\|\mathbf{n}\|} = \cos \Omega \mathbf{I} + \sin \Omega \mathbf{J} \tag{3.12}$$

Note that dotting all terms in Eq. (3.12) with vector $\mathbf{I}$ yields Eq. (3.10) because $\mathbf{I} \cdot \mathbf{I} = 1$ and $\mathbf{I} \cdot \mathbf{J} = 0$. Similarly, we can dot all terms in Eq. (3.12) with vector $\mathbf{J}$ to yield

$$\sin \Omega = \frac{\mathbf{J} \cdot \mathbf{n}}{\|\mathbf{n}\|} \tag{3.13}$$

We must use Eqs. (3.10) and (3.13) together to determine the proper quadrant for Ω. For computer applications (such as MATLAB), the `atan2` function with input arguments $\sin \Omega$ and $\cos \Omega$ will place angle Ω in the proper quadrant. When using a hand-held calculator, the signs of $\sin \Omega$ and $\cos \Omega$ will determine the proper quadrant. For example, if $\sin \Omega = -0.5$ and $\cos \Omega = -0.866025$, we know that the longitude of the ascending node *must* be in the third quadrant, or $\Omega = 210°$. This simple example should serve as a strong warning to the reader: do not simply press the inverse-cosine (or inverse-sine) button on a calculator and assume that you have determined the correct angle!

Figure 3.4 shows that the argument of periapsis ω is the angle between $\mathbf{n}$ and $\mathbf{e}$ and therefore its cosine is

$$\cos \omega = \frac{\mathbf{n} \cdot \mathbf{e}}{\|\mathbf{n}\| e} \tag{3.14}$$

For reasons previously mentioned, we cannot determine ω by simply taking the inverse cosine of Eq. (3.14). The quadrant check for ω depends on the sign of e_Z, which is the $\mathbf{K}$ component of eccentricity vector $\mathbf{e}$. If $e_Z > 0$ (as shown in Figure 3.4), the vector $\mathbf{e}$ is in the northern hemisphere ("above" the equatorial plane) and argument of periapsis ranges from 0 to 180°. In this case, the "calculator" inverse-cosine operation of Eq. (3.14) produces the correct answer. If $e_Z < 0$ (periapsis is below the equatorial plane), the argument of periapsis is between 180° and 360°. In this case, the correct argument of periapsis is

$$\text{If } e_Z < 0: \qquad \omega = 360° - \cos^{-1}\left(\frac{\mathbf{n} \cdot \mathbf{e}}{\|\mathbf{n}\| e}\right) \tag{3.15}$$

The sixth orbital element is the true anomaly at epoch t_0 that corresponds to the state vector $(\mathbf{r}_0, \mathbf{v}_0)$. The cosine of θ_0 is determined by the dot product of the eccentricity vector $\mathbf{e}$ and the satellite's position vector $\mathbf{r}_0$ (see Figure 3.4)

$$\cos\theta_0 = \frac{\mathbf{e} \cdot \mathbf{r}_0}{e r_0} \tag{3.16}$$

True anomaly ranges from 0 to 360°; hence, we cannot rely solely on the inverse cosine function. The easiest quadrant check utilizes the sign of the dot product $\mathbf{r}_0 \cdot \mathbf{v}_0$. Recall from Chapter 2 that $\mathbf{r}_0 \cdot \mathbf{v}_0 = r_0 \dot{r}_0$ where $\dot{r}_0$ is the radial velocity component. Therefore, if $\mathbf{r}_0 \cdot \mathbf{v}_0 > 0$ the radius is increasing ($\dot{r}_0 > 0$) and the satellite is moving toward apoapsis and true anomaly is between 0° and 180°. On the other hand, if $\mathbf{r}_0 \cdot \mathbf{v}_0 < 0$ the radius is decreasing and the satellite is moving toward periapsis and θ_0 is between 180° and 360°. In this case, we must use

$$\text{If } \mathbf{r}_0 \cdot \mathbf{v}_0 < 0: \quad \theta_0 = 360° - \cos^{-1}\left(\frac{\mathbf{e} \cdot \mathbf{r}_0}{e r_0}\right) \tag{3.17}$$

It is worth repeating that orbital elements Ω, ω, and θ_0 can range from 0 to 360° and therefore an inverse cosine operation alone will not correctly determine the orbital element if the angle is between 180° and 360°. The correct quadrant must be determined by applying the quadrant checks denoted by Eqs. (3.13), (3.15), and (3.17). When determining inclination, one may simply take the inverse cosine of Eq. (3.8) because inclination is always between 0° and 180°.

It is easy to recognize that a computer program will readily perform these steps. MATLAB is well suited to carry out the vector manipulations (cross product, dot product, and vector norm) and the inverse-angle calculations with the atan2 function. The reader is encouraged to develop a general-purpose M-file that computes the classical orbital elements given the state vector $(\mathbf{r},\mathbf{v})$ as an input (see Problem 3.18). The following example illustrates this orbit-determination process.

Example 3.1 A tracking station determines the following state vector for an Earth-orbiting satellite in ECI coordinates:

$$\mathbf{r}_0 = \begin{bmatrix} 9,031.5 \\ -5,316.9 \\ -1,647.2 \end{bmatrix} \text{km}, \qquad \mathbf{v}_0 = \begin{bmatrix} -2.8640 \\ 5.1112 \\ -5.0805 \end{bmatrix} \text{km/s}$$

Determine the classical orbital elements at this epoch. Is this satellite in a Molniya orbit?

We will use $\mu = 3.986(10^5)$ km^3/s^2 for Earth-satellite example problems in this chapter. The reader can consult Appendix A for a more precise numerical value if desired. First, we determine the magnitude of the position and velocity vectors:

$$r_0 = \sqrt{9,031.5^2 + (-5,316.9)^2 + (-1,647.2)^2} = 10,609 \text{ km}$$

$$v_0 = \sqrt{(-2.8640)^2 + 5.1112^2 + (-5.0805)^2} = 7.7549 \text{ km/s}$$

Energy is

$$\xi = \frac{v_0^2}{2} - \frac{\mu}{r_0} = -7.5027 \text{ km}^2/\text{s}^2 = -\frac{\mu}{2a}$$

Therefore, semimajor axis is $\boxed{a = 26,563.6 \text{ km}}$

Using Eq. (3.6), the eccentricity vector is

$$\mathbf{e} = \frac{1}{\mu}\left[\left(v_0^2 - \frac{\mu}{r_0}\right)\mathbf{r}_0 - (\mathbf{r}_0 \cdot \mathbf{v}_0)\mathbf{v}_0\right]$$

The dot product $\mathbf{r}_0 \cdot \mathbf{v}_0$ is

$$\mathbf{r}_0 \cdot \mathbf{v}_0 = (9,031.5)(-2.8640) + (-5,316.9)(5.1112) + (-1,647.2)(-5.0805)$$
$$= -44,673.4 \text{ km}^2/\text{s}$$

The eccentricity vector is

$$\mathbf{e} = \begin{bmatrix} 0.1903 \\ 0.2718 \\ -0.6627 \end{bmatrix}$$

Eccentricity is the vector norm, $e = \|\mathbf{e}\| = \sqrt{0.1903^2 + 0.2718^2 + (-0.6627)^2} = \boxed{0.7411}$

Computing inclination requires the angular momentum vector:

$$\mathbf{h} = \mathbf{r}_0 \times \mathbf{v}_0 = \begin{vmatrix} \mathbf{I} & \mathbf{J} & \mathbf{K} \\ 9,031.5 & -5,316.9 & -1,647.2 \\ -2.8640 & 5.1112 & -5.0805 \end{vmatrix} = \begin{bmatrix} 35,432 \\ 50,602 \\ 30,934 \end{bmatrix} \text{ km}^2/\text{s}$$

Using Eq. (3.8), the cosine of inclination is

$$\cos i = \frac{\mathbf{K} \cdot \mathbf{h}}{h} = \frac{30,934}{69,086} = 0.4478$$

where the dot product $\mathbf{K} \cdot \mathbf{h} = h_Z$. Taking the inverse cosine, we obtain $\boxed{i = 63.4°}$

Computing the longitude of the ascending node (Ω) and argument of perigee (ω) requires the ascending node vector:

$$n = K \times h = \begin{vmatrix} I & J & K \\ 0 & 0 & 1 \\ 35{,}432 & 50{,}602 & 30{,}934 \end{vmatrix} = \begin{bmatrix} -50{,}602 \\ 35{,}432 \\ 0 \end{bmatrix}$$

Equations (3.10) and (3.13) allow us to compute the cosine and sine of Ω

$$\cos\Omega = \frac{I \cdot n}{\|n\|} = \frac{-50{,}602}{61{,}774} = -0.8192 \qquad \sin\Omega = \frac{J \cdot n}{\|n\|} = \frac{35{,}432}{61{,}774} = 0.5736$$

Because $\cos\Omega < 0$ and $\sin\Omega > 0$, the ascending node is in the second quadrant. The longitude of ascending node is $\boxed{\Omega = 145°}$

Using Eq. (3.14), the cosine of ω is

$$\cos\omega = \frac{n \cdot e}{\|n\| e} = \frac{0.8570}{(61{,}774)(0.7411)} = 10^{-5}$$

The inverse cosine of 10^{-5} is 90°. However, we must check the **K** component of the eccentricity vector **e** in order to determine whether ω is +90° or −90°. Because $e_Z = -0.6627 < 0$, the perigee is in the southern hemisphere and Eq. (3.15) shows that the argument of perigee is

$$\omega = 360° - 90° = \boxed{270° \ (\text{or} -90°)}$$

Finally, we compute the cosine of true anomaly at epoch using Eq. (3.16)

$$\cos\theta_0 = \frac{e \cdot r_0}{e r_0} = \frac{(0.1903)(9{,}031.5) + (0.2718)(-5{,}316.9) + (-0.6627)(-1{,}647.2)}{(0.7411)(10{,}609)}$$

$$= \frac{1{,}365}{7{,}863} = 0.1736$$

Because $r_0 \cdot v_0 < 0$ (see computations for **e**), the satellite is approaching perigee and we must use Eq. (3.17) to compute true anomaly:

$$\theta_0 = 360° - \cos^{-1}\left(\frac{e \cdot r_0}{e r_0}\right) = 360° - 80° = \boxed{280°}$$

If we compare these calculated orbital elements to the description of the Molniya orbit at the beginning of this section, we see that the orbit in this example is indeed a Molniya orbit. The critical orbital features for a Molniya orbit are $a = 26{,}564$ km (i.e., the period is about 12 h), $e = 0.7411$, $i = 63.4°$, and $\omega = -90°$ (perigee in southern hemisphere). The argument of perigee must be −90° so that the Molniya apogee is at the highest possible northern latitude for good observation of Russia. Because the apogee altitude is very large ($\sim$39,874 km) and the perigee altitude is low ($\sim$500 km), the apogee speed is relatively slow ($\sim$1.5 km/s) and the perigee speed is very fast ($\sim$10 km/s). Therefore, a satellite in a Molniya orbit will spend most of its time near apogee where it can send and receive signals with ground stations in Russia.

3.5   Transforming Orbital Elements to Cartesian Coordinates

The previous section presented a systematic process for determining the orbital elements from the Cartesian coordinates. This process represents the initial orbit-determination problem where ground-based observations yield the state vector $(\mathbf{r}_0, \mathbf{v}_0)$ and the desired end result is the set of orbital elements $(a, e, i, \Omega, \omega, \theta_0)$ at epoch t_0. As we shall see in later chapters, it is sometimes advantageous to perform orbital calculations using a satellite's Cartesian coordinates $(\mathbf{r}, \mathbf{v})$ instead of orbital elements. One important example is the orbit-transfer problem; that is, determining the orbit that passes from initial position vector $\mathbf{r}_1$ to a desired terminal position vector $\mathbf{r}_2$ with a specified flight time. For this reason, we must develop an inverse process that transforms the orbital elements $(a, e, i, \Omega, \omega, \theta)$ to the Cartesian state vector $(\mathbf{r}, \mathbf{v})$.

We will demonstrate the transformation process using an Earth-orbiting satellite and the ECI coordinate system (Figures 3.2 and 3.3). Hence, we will denote the satellite's ECI Cartesian coordinates as $\mathbf{r}_{ECI}$ and $\mathbf{v}_{ECI}$ for added clarity. The procedure we develop, however, may be applied to the heliocentric-ecliptic Cartesian frame shown in Figure 3.1. Converting orbital elements to $(\mathbf{r}_{ECI}, \mathbf{v}_{ECI})$ requires a two-step process:

1) Use the orbital elements to develop a state vector $(\mathbf{r}_{PQW}, \mathbf{v}_{PQW})$ expressed in the *perifocal coordinate system*, **PQW**.
2) Transform the perifocal state vector $(\mathbf{r}_{PQW}, \mathbf{v}_{PQW})$ to the ECI state vector $(\mathbf{r}_{ECI}, \mathbf{v}_{ECI})$.

Figure 3.6 shows the perifocal coordinate system. "Perifocal" indicates that the principle axis points from the focus to the periapsis direction and Figure 3.6 shows that the x_e axis (with unit vector **P**) does indeed point toward perigee. The orbital plane is the fundamental $(x_e y_e)$ plane and the y_e axis (with unit vector **Q**) is 90° from perigee in the direction of the satellite's motion. The z_e axis (with unit vector **W**) is perpendicular to the orbital plane (or, along the angular momentum vector **h**). The reader should note that the **PQW** frame is an orthogonal Cartesian system.

Figure 3.7 shows the **PQ** plane (i.e., orbital plane) and the satellite at an arbitrary position $\mathbf{r}_{PQW}$ corresponding to true anomaly θ. The **W** axis (not shown in Figure 3.7) is equal to the cross product $\mathbf{P} \times \mathbf{Q}$ and points out of the page. Because the satellite's motion

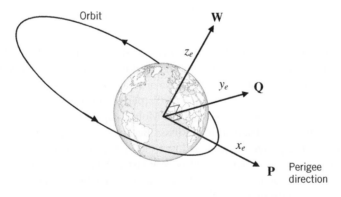

Figure 3.6 Perifocal coordinate system, **PQW**.

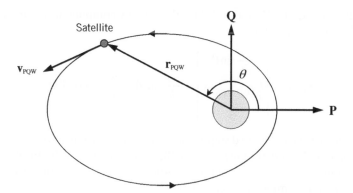

Figure 3.7 State vector ($\mathbf{r}_{PQW}$,$\mathbf{v}_{PQW}$) in the perifocal coordinate system.

is contained in the orbital plane, its position and velocity vectors $\mathbf{r}_{PQW}$ and $\mathbf{v}_{PQW}$ are composed of **P** and **Q** components only. The position vector in **PQW** coordinates is

$$\mathbf{r}_{PQW} = \begin{bmatrix} r\cos\theta \\ r\sin\theta \\ 0 \end{bmatrix} = r\cos\theta\mathbf{P} + r\sin\theta\mathbf{Q} \tag{3.18}$$

where we can determine the satellite's current radius from the trajectory equation (2.45):

$$r = \frac{p}{1 + e\cos\theta} \tag{3.19}$$

Note that the parameter p is determined from the first two classical orbital elements [i.e., $p = a(1 - e^2)$]. The velocity vector in **PQW** can be obtained by taking the first time derivative of Eq. (3.18)

$$\mathbf{v}_{PQW} = \dot{\mathbf{r}}_{PQW} = \begin{bmatrix} \dot{r}\cos\theta - r\dot{\theta}\sin\theta \\ \dot{r}\sin\theta + r\dot{\theta}\cos\theta \\ 0 \end{bmatrix} \tag{3.20}$$

$$= (\dot{r}\cos\theta - r\dot{\theta}\sin\theta)\mathbf{P} + (\dot{r}\sin\theta + r\dot{\theta}\cos\theta)\mathbf{Q}$$

The derivatives of unit vectors **P** and **Q** are zero because they have fixed magnitudes and directions. Recall that in Chapter 2 we developed expressions for radial and transverse velocity components $\dot{r}$ and $r\dot{\theta}$; see Eqs. (2.69) and (2.71). These velocity component expressions are repeated below:

$$\text{Radial velocity:} \qquad \dot{r} = \frac{\mu}{h}e\sin\theta \tag{3.21}$$

$$\text{Transverse velocity:} \qquad r\dot{\theta} = \frac{\mu}{h}(1 + e\cos\theta) \tag{3.22}$$

We can substitute Eqs. (3.21) and (3.22) for the velocity components in Eq. (3.20). After some simplification, we obtain

$$
\mathbf{v}_{\mathrm{PQW}} = \begin{bmatrix} \dfrac{-\mu}{h}\sin\theta \\[2mm] \dfrac{\mu}{h}(e+\cos\theta) \\[2mm] 0 \end{bmatrix} = \dfrac{-\mu}{h}\sin\theta\,\mathbf{P} + \dfrac{\mu}{h}(e+\cos\theta)\mathbf{Q} \tag{3.23}
$$

At this point we have completed step 1: using three orbital elements (a, e, and θ) in Eqs. (3.18), (3.19), and (3.23) determines the satellite's state vector in the **PQW** coordinate system (of course, angular momentum is solely a function of the parameter, i.e., $h = \sqrt{p\mu}$, and p is determined by elements a and e).

The second (and final) step involves a coordinate transformation from the **PQW** frame to the **IJK** (or ECI) frame. Here we use the three angular orbital elements (i, Ω, and ω) that define the orientation of the perifocal frame relative to the ECI frame. The resulting coordinate transformation (from **PQW** to **IJK**) will have the form

$$
\mathbf{r}_{\mathrm{ECI}} = \widetilde{\mathbf{R}}\mathbf{r}_{\mathrm{PQW}} \tag{3.24}
$$

$$
\mathbf{v}_{\mathrm{ECI}} = \widetilde{\mathbf{R}}\mathbf{v}_{\mathrm{PQW}} \tag{3.25}
$$

where $\widetilde{\mathbf{R}}$ is a 3×3 *rotation matrix* that transforms a vector in the perifocal **PQW** frame to the geocentric-equatorial **IJK** frame. It is useful to summarize a few key points associated with the coordinate transformation performed by matrix $\widetilde{\mathbf{R}}$ and Eqs. (3.24) and (3.25):

1) Rotation matrix $\widetilde{\mathbf{R}}$ transforms *any* vector expressed in **PQW** coordinates to **IJK** coordinates, whether that vector is position $\mathbf{r}_{\mathrm{PQW}}$ or velocity $\mathbf{v}_{\mathrm{PQW}}$.
2) The rotation matrix $\widetilde{\mathbf{R}}$ does not change the magnitude of the vector after the transformation; that is, $\|\mathbf{r}_{\mathrm{ECI}}\| = \|\mathbf{r}_{\mathrm{PQW}}\|$ and $\|\mathbf{v}_{\mathrm{ECI}}\| = \|\mathbf{v}_{\mathrm{PQW}}\|$.
3) The rotation matrix $\widetilde{\mathbf{R}}$ depends on a sequence of three angular rotations required to align the two coordinate systems. The order of the sequence of rotations is important.

We will present the basic steps for developing the rotation matrix $\widetilde{\mathbf{R}}$. For additional details, the reader may consult the excellent discussions in Vallado [1; pp. 159–174] and Bate *et al.* [2; pp. 74–83].

3.5.1 Coordinate Transformations

Let us begin with an example of a simple transformation where the two coordinate frames differ by a single rotation about a primary axis. Figure 3.8 shows the rotation of orthogonal axes XYZ about its $+Z$ axis through angle χ to produce the orthogonal "primed" axes $X'Y'Z'$. The angular rotation χ is positive as defined by the "right-hand rule"; that is, curling the fingers of the right hand about the Z axis so that the thumb points "up" in Figure 3.8. Therefore, the new Z' is aligned with the original Z axis and the new X' and Y' axes are in the same plane as the original X–Y plane. Now, suppose we have the position vector $\mathbf{r}_{XYZ} = 5\mathbf{I} + 5\mathbf{K}$ as shown in Figure 3.8 where **IJK** are unit vectors along the original XYZ coordinate system. The same position vector expressed in the $X'Y'Z'$ coordinate system is

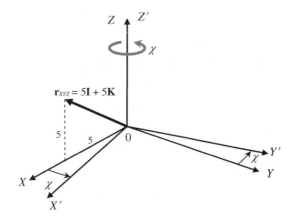

Figure 3.8 Coordinate transformation: positive rotation about the *Z* axis.

$$\mathbf{r}_{X'Y'Z'} = \begin{bmatrix} \cos\chi & \sin\chi & 0 \\ -\sin\chi & \cos\chi & 0 \\ 0 & 0 & 1 \end{bmatrix} \mathbf{r}_{XYZ} \tag{3.26}$$

Using the position vector $\mathbf{r}_{XYZ} = 5\mathbf{I} + 5\mathbf{K}$ shown in Figure 3.8, we have

$$\mathbf{r}_{X'Y'Z'} = \begin{bmatrix} \cos\chi & \sin\chi & 0 \\ -\sin\chi & \cos\chi & 0 \\ 0 & 0 & 1 \end{bmatrix} \begin{bmatrix} 5 \\ 0 \\ 5 \end{bmatrix}$$

Carrying out the matrix-vector multiplication, we obtain $\mathbf{r}_{X'Y'Z'} = 5\cos\chi\mathbf{I}' - 5\sin\chi\mathbf{J}' + 5\mathbf{K}'$ where $\mathbf{I}'\mathbf{J}'\mathbf{K}'$ are unit vectors along the new "primed" coordinate system $X'Y'Z'$. The Z-components of $\mathbf{r}_{X'Y'Z'}$ and $\mathbf{r}_{XYZ}$ are identical because the rotation to establish $X'Y'Z'$ was about the $+Z$ axis [note that the elements of the third row and third column of the rotation matrix in Eq. (3.26) are zero except for the "1" in the lower right corner that preserves the Z component]. The reader should note that the vector norms of $\mathbf{r}_{XYZ}$ and $\mathbf{r}_{X'Y'Z'}$ are both $\sqrt{50}$ which shows that the magnitude of the position vector did not change after the coordinate transformation.

Let us rewrite the coordinate transformation denoted by Eq. (3.26) as

$$\mathbf{r}_{X'Y'Z'} = \mathbf{C}(\chi)\mathbf{r}_{XYZ} \tag{3.27}$$

where $\mathbf{C}(\chi)$ is the rotation matrix associated with a positive rotation about the $+Z$ axis through angle χ

$$\mathbf{C}(\chi) = \begin{bmatrix} \cos\chi & \sin\chi & 0 \\ -\sin\chi & \cos\chi & 0 \\ 0 & 0 & 1 \end{bmatrix} \tag{3.28}$$

Rotation matrix $\mathbf{C}(\chi)$ is an *orthogonal matrix*; that is, its rows and columns are orthogonal unit vectors [the rows (or columns) are orthogonal because the dot product of any two rows (or columns) is zero]. Because matrix $\mathbf{C}(\chi)$ is orthogonal, it has the properties

$$\mathbf{C}(\chi)\mathbf{C}^T(\chi) = \mathbf{C}^T(\chi)\mathbf{C}(\chi) = \mathbf{I} \tag{3.29}$$

$$\mathbf{C}^{-1}(\chi) = \mathbf{C}^T(\chi) \tag{3.30}$$

where superscript T indicates the transpose of matrix $\mathbf{C}(\chi)$ and $\mathbf{I}$ is the 3×3 identity matrix. Multiplying both sides of Eq. (3.27) by $\mathbf{C}^{-1}(\chi) = \mathbf{C}^T(\chi)$ yields

$$\mathbf{r}_{XYZ} = \mathbf{C}^T(\chi)\mathbf{r}_{X'Y'Z'} \tag{3.31}$$

which is the transformation from the $X'Y'Z'$ frame to the XYZ frame.

A coordinate transformation may involve a positive rotation about the $+X$ axis. Figure 3.9 shows the rotation of Cartesian frame XYZ about its $+X$ axis through angle α to produce the frame $X'Y'Z'$. A position vector $\mathbf{r}_{XYZ}$ expressed in the XYZ coordinate system may be transformed using the rotation matrix

$$\mathbf{r}_{X'Y'Z'} = \begin{bmatrix} 1 & 0 & 0 \\ 0 & \cos\alpha & \sin\alpha \\ 0 & -\sin\alpha & \cos\alpha \end{bmatrix} \mathbf{r}_{XYZ} \tag{3.32}$$

$$= \mathbf{A}(\alpha)\mathbf{r}_{XYZ}$$

The rotation matrix $\mathbf{A}(\alpha)$ is orthogonal and the X component of $\mathbf{r}_{XYZ}$ is unaltered by the transformation.

Figure 3.10 shows the rotation of Cartesian frame XYZ about its $+Y$ axis through angle β to produce the frame $X'Y'Z'$. Transformation of vector $\mathbf{r}_{XYZ}$ from the XYZ coordinate system to $X'Y'Z'$ is

$$\mathbf{r}_{X'Y'Z'} = \begin{bmatrix} \cos\beta & 0 & -\sin\beta \\ 0 & 1 & 0 \\ \sin\beta & 0 & \cos\beta \end{bmatrix} \mathbf{r}_{XYZ} \tag{3.33}$$

$$= \mathbf{B}(\beta)\mathbf{r}_{XYZ}$$

Equations (3.32), (3.33), and (3.27) are the coordinate transformations from the XYZ frame to a new frame $X'Y'Z'$ created after a single positive rotation about a principle axis.

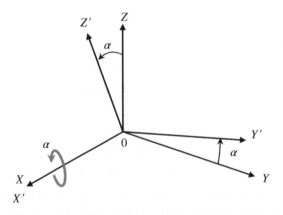

Figure 3.9 Coordinate transformation: positive rotation about the X axis.

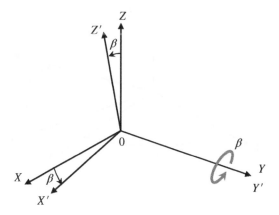

Figure 3.10 Coordinate transformation: positive rotation about the *Y* axis.

Remember that our goal is to develop a transformation between the ECI (or **IJK**) coordinate system and the perifocal **PQW** system. In general, the **IJK** system can be aligned with the **PQW** system after three successive rotations, as Figure 3.11 illustrates. The first rotation is about the **K** (or $+Z$) axis through the ascending node angle Ω (Figure 3.11a). After this first rotation, the **I** (or $+X$) axis becomes the X' axis (i.e., the ascending node vector **n**) and the $X'Y'$ axes remain in the equatorial plane. The second rotation is about the intermediate X' axis through the inclination i (Figure 3.11b). This rotation establishes the angular momentum vector **h** (note that the Z'' axis in Figure 3.11b is aligned with the **W** axis of the perifocal system). The third and final rotation is about the Z'' (or **W**) axis through the argument of perigee ω (Figure 3.11c) in order to establish the **P** axis (perigee direction) and the perifocal **PQW** coordinate system. Next, let us apply the appropriate rotation matrices to the position vector $\mathbf{r}_{ECI}$ that is expressed in **IJK** coordinates:

$$\text{First rotation about} +Z \text{ axis:} \quad \mathbf{r}_{X'Y'Z'} = \mathbf{C}(\Omega)\mathbf{r}_{ECI} \tag{3.34}$$

$$\text{Second rotation about} +X' \text{axis:} \quad \mathbf{r}_{X''Y''Z''} = \mathbf{A}(i)\mathbf{r}_{X'Y'Z'} \tag{3.35}$$

$$\text{Third rotation about} +Z'' \text{ axis:} \quad \mathbf{r}_{PQW} = \mathbf{C}(\omega)\mathbf{r}_{X''Y''Z''} \tag{3.36}$$

Note that there is no Y-axis rotation in our sequence of rotations from **IJK** to **PQW**. Finally, we can combine Eqs. (3.34)–(3.36) to yield the direct computation of $\mathbf{r}_{PQW}$ from **IJK** coordinates $\mathbf{r}_{ECI}$:

$$\mathbf{r}_{PQW} = \mathbf{C}(\omega)\mathbf{A}(i)\mathbf{C}(\Omega)\mathbf{r}_{ECI} \tag{3.37}$$

Equation (3.37) transforms the position vector $\mathbf{r}_{ECI}$ (expressed in **IJK** coordinates) to position vector $\mathbf{r}_{PQW}$ (expressed in **PQW** coordinates). The order of multiplying the rotation matrices is important and must be preserved: the first rotation $\mathbf{C}(\Omega)$ is the *last* matrix in the left-to-right matrix multiplication presented in Eq. (3.37).

Remember that our overall goal is to derive the state vector $(\mathbf{r},\mathbf{v})$ in the ECI (or **IJK**) coordinate system from the orbital elements. Our first step involved computing the position and velocity vectors $\mathbf{r}_{PQW}$ and $\mathbf{v}_{PQW}$ in perifocal coordinates [see Eqs. (3.18) and (3.23)]. Therefore, we know the left-hand side of Eq. (3.37). We can solve Eq. (3.37) for $\mathbf{r}_{ECI}$ by multiplying both sides by the matrix inverse of $\mathbf{C}(\omega)\mathbf{A}(i)\mathbf{C}(\Omega)$

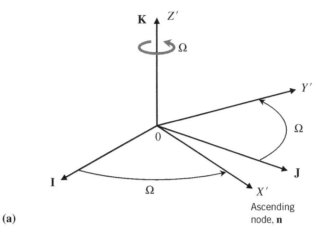

(a)

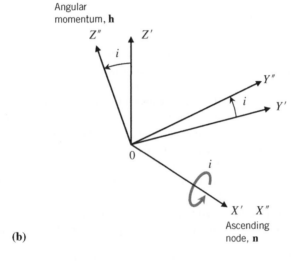

(b)

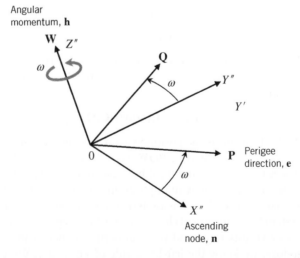

(c)

Figure 3.11 Three rotations of **IJK** to establish **PQW**: (a) positive rotation about **K** through angle Ω to establish the ascending node vector; (b) positive rotation about X' through angle i to establish the angular momentum vector; and (c) positive rotation about Z'' (**W**) through angle ω to establish the **PQW** frame.

$$\mathbf{r}_{ECI} = [\mathbf{C}(\omega)\mathbf{A}(i)\mathbf{C}(\Omega)]^{-1}\mathbf{r}_{PQW} \tag{3.38}$$

Comparing Eq. (3.38) with Eq. (3.24) we see that the inverse of matrix $\mathbf{C}(\omega)\mathbf{A}(i)\mathbf{C}(\Omega)$ is the overall rotation matrix $\widetilde{\mathbf{R}}$ that transforms *any* vector expressed in the **PQW** frame to the ECI (**IJK**) frame. Because the product of three orthogonal matrices is also orthogonal, the inverse of matrix $\mathbf{C}(\omega)\mathbf{A}(i)\mathbf{C}(\Omega)$ is its transpose. Hence, the overall rotation matrix is

$$\widetilde{\mathbf{R}} = \left[\begin{bmatrix} \cos\omega & \sin\omega & 0 \\ -\sin\omega & \cos\omega & 0 \\ 0 & 0 & 1 \end{bmatrix} \begin{bmatrix} 1 & 0 & 0 \\ 0 & \cos i & \sin i \\ 0 & -\sin i & \cos i \end{bmatrix} \begin{bmatrix} \cos\Omega & \sin\Omega & 0 \\ -\sin\Omega & \cos\Omega & 0 \\ 0 & 0 & 1 \end{bmatrix} \right]^{T}$$

Carrying out the matrix multiplications and transposing the result, we obtain the overall rotation matrix from **PQW** to **IJK**:

$$\widetilde{\mathbf{R}} = \begin{bmatrix} c_\Omega c_\omega - s_\Omega s_\omega c_i & -c_\Omega s_\omega - s_\Omega c_\omega c_i & s_\Omega s_i \\ s_\Omega c_\omega + c_\Omega s_\omega c_i & -s_\Omega s_\omega + c_\Omega c_\omega c_i & -c_\Omega s_i \\ s_\omega s_i & c_\omega s_i & c_i \end{bmatrix} \tag{3.39}$$

We have used the short-hand notation $c_\alpha = \cos\alpha$ and $s_\alpha = \sin\alpha$ for the cosine and sine of the three rotation angles. Multiplying rotation matrix $\widetilde{\mathbf{R}}$ and *any* vector expressed in the **PQW** frame will transform it to **IJK** coordinates.

Let us summarize the steps for transforming the classical orbital elements to the state vector $(\mathbf{r},\mathbf{v})$ in the ECI (or **IJK**) frame:

1) Using Eqs. (3.18) and (3.23), determine the satellite's position and velocity vectors in the **PQW** frame, $\mathbf{r}_{PQW}$ and $\mathbf{v}_{PQW}$.
2) Use Eq. (3.39) to determine the overall rotation matrix from **PQW** to **IJK** and then use $\widetilde{\mathbf{R}}$ in Eqs. (3.24) and (3.25) to compute the state vector $(\mathbf{r}_{ECI},\mathbf{v}_{ECI})$ in the ECI coordinate system.

This process is a bit cumbersome for calculations with a hand-held calculator. However, it is easy to see that MATLAB is well-suited to carry out the matrix-vector multiplications. We encourage the reader to develop a general-purpose M-file that will compute $(\mathbf{r}_{ECI},\mathbf{v}_{ECI})$ given an arbitrary set of six classical orbital elements.

Example 3.2 Let us consider again the Molniya orbit presented in Example 3.1. A satellite in a Molniya orbit has the following orbital elements at epoch time t_0

$$a = 26,564 \text{ km}$$
$$e = 0.7411$$
$$i = 63.4°$$
$$\Omega = 200°$$
$$\omega = -90°$$
$$\theta_0 = 30°$$

Determine the satellite's state vector at this epoch in the Earth-centered inertial (ECI) coordinate system.

First, we use semimajor axis a and eccentricity e to compute parameter p and angular momentum h:

$$p = a(1 - e^2) = 11{,}974.3 \text{ km}, \quad h = \sqrt{p\mu} = 69{,}086.5 \text{ km}^2/\text{s}$$

Orbital radius at $\theta_0 = 30°$ is determined by the trajectory equation

$$r = \frac{p}{1 + e\cos\theta} = 7{,}293.3 \text{ km}$$

The satellite's position vector in the **PQW** frame is computed using Eq. (3.18)

$$\mathbf{r}_{\text{PQW}} = \begin{bmatrix} r\cos\theta_0 \\ r\sin\theta_0 \\ 0 \end{bmatrix} = \begin{bmatrix} 6{,}316.21 \\ 3{,}646.67 \\ 0 \end{bmatrix} \text{ km} = 6{,}316.21\mathbf{P} + 3{,}646.67\mathbf{Q} \text{ km}$$

Equation (3.23) allows us to calculate the satellite's velocity vector in **PQW**

$$\mathbf{v}_{\text{PQW}} = \begin{bmatrix} \dfrac{-\mu}{h}\sin\theta_0 \\ \dfrac{\mu}{h}(e + \cos\theta_0) \\ 0 \end{bmatrix} = \begin{bmatrix} -2.8848 \\ 9.2724 \\ 0 \end{bmatrix} \text{ km/s} = -2.8848\mathbf{P} + 9.2724\mathbf{Q} \text{ km/s}$$

The first step is complete. Next, use Eq. (3.39) to determine the overall rotation matrix from **PQW** to **IJK**

$$\widetilde{\mathbf{R}} = \begin{bmatrix} c_\Omega c_\omega - s_\Omega s_\omega c_i & -c_\Omega s_\omega - s_\Omega c_\omega c_i & s_\Omega s_i \\ s_\Omega c_\omega + c_\Omega s_\omega c_i & -s_\Omega s_\omega + c_\Omega c_\omega c_i & -c_\Omega s_i \\ s_\omega s_i & c_\omega s_i & c_i \end{bmatrix} = \begin{bmatrix} -0.1531 & -0.9397 & -0.3058 \\ 0.4208 & -0.3420 & 0.8402 \\ -0.8942 & 0 & 0.4478 \end{bmatrix}$$

where we have used $c_\Omega = \cos\Omega = -0.9397$, $s_\omega = \sin\omega = -1$, $c_i = \cos i = 0.4478$, and so on. Finally, the position and velocity vectors in ECI coordinates are

$$\mathbf{r}_{\text{ECI}} = \widetilde{\mathbf{R}}\mathbf{r}_{\text{PQW}} = \begin{bmatrix} -0.1531 & -0.9397 & -0.3058 \\ 0.4208 & -0.3420 & 0.8402 \\ -0.8942 & 0 & 0.4478 \end{bmatrix} \begin{bmatrix} 6{,}316.21 \\ 3{,}646.67 \\ 0 \end{bmatrix} = \begin{bmatrix} -4{,}394.0 \\ 1{,}410.3 \\ -5{,}647.7 \end{bmatrix} \text{ km}$$

Or, $\mathbf{r}_{\text{ECI}} = -4{,}394.0\mathbf{I} + 1{,}410.3\mathbf{J} - 5{,}647.7\mathbf{K}$ km

$$\mathbf{v}_{\text{ECI}} = \widetilde{\mathbf{R}}\mathbf{v}_{\text{PQW}} = \begin{bmatrix} -0.1531 & -0.9397 & -0.3058 \\ 0.4208 & -0.3420 & 0.8402 \\ -0.8942 & 0 & 0.4478 \end{bmatrix} \begin{bmatrix} -2.8848 \\ 9.2724 \\ 0 \end{bmatrix}$$

$$= \begin{bmatrix} -8.2715 \\ -4.3852 \\ 2.5794 \end{bmatrix} \text{ km/s}$$

Or, $\mathbf{v}_{ECI} = -8.2715\mathbf{I} - 4.3852\mathbf{J} + 2.5794\mathbf{K}$ km/s

As a check, the reader can compare the vector norms of $\mathbf{r}_{ECI}$ and $\mathbf{r}_{PQW}$ (both should equal $r = 7{,}293.3$ km) and the vector norms of $\mathbf{v}_{ECI}$ and $\mathbf{v}_{PQW}$ (both should equal 9.7108 km/s).

3.6 Ground Tracks

The *ground track* of a satellite is the locus of points where the position vector $\mathbf{r}$ intersects the surface of the planet. For an Earth-orbiting satellite, the ground track is the projection of the orbit onto the surface of the rotating Earth. Ground tracks are obviously useful in that they clearly show the geographic regions where the satellite passes overhead. Figure 3.12 shows a ground track for an inclined geocentric satellite. If the Earth did not rotate, the ground track would be a great circle formed by the intersection of the orbital plane and the stationary Earth. Hence, for a non-rotating Earth, the ground track would retrace its path and pass over the same geographic regions every orbital revolution. For two-body motion, the orbital plane remains inertially fixed and therefore the ground track passes over a different geographic region each orbit because the Earth rotates beneath it. Figure 3.12 shows the orbital inclination i (i.e., the angle of the ground track at the equatorial crossing) and the *azimuth angle β* of the satellite's velocity vector $\mathbf{v}$ at its current location in the orbit. Azimuth angle (or heading angle) β is measured clockwise from north to the projection of the velocity vector $\mathbf{v}$. Angle ϕ in Figure 3.12 is the latitude of the satellite at the given instant. Latitude is zero when the satellite crosses the equator and is at its maximum magnitude when the satellite is moving due east (i.e., $\beta = 90°$). The ascending node vector $\mathbf{n}$ points from the Earth's center to the point where the ground track crosses the equatorial plane as the satellite is moving in a northerly direction.

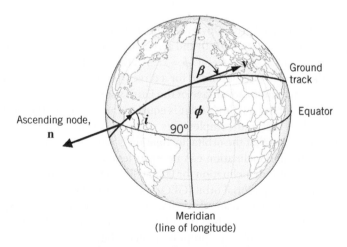

Figure 3.12 Satellite ground track.

We may apply Napier's rules to the right spherical triangle shown in Figure 3.12 to obtain the relationship

$$\cos i = \sin\beta\cos\phi \qquad (3.40)$$

The reader should note that the left-hand side of Eq. (3.40) is constant (inclination is a constant) while both azimuth β and latitude ϕ change as the satellite traces its ground track. We may summarize a few key points surrounding Eq. (3.40) and the relationship between inclination, latitude, and azimuth:

1) When $\phi = 0$ (equatorial crossing), azimuth angle $\beta = 90° - i$ for an *ascending* pass; $\beta = 90° + i$ for a *descending* pass.
2) Azimuth angle is in the first quadrant ($0° < \beta < 90°$) for an ascending equatorial crossing and in the second quadrant ($90° < \beta < 180°$) for a descending equatorial crossing.
3) When the satellite is moving easterly ($\beta = 90°$), the magnitude of the latitude is equal to the inclination, or $|\phi| = i$. At this point the satellite is either at its maximum northern latitude ($\phi = i$) or its maximum southern latitude ($\phi = -i$).
4) For a polar orbit ($i = 90°$), $\beta = 0$ when the satellite is moving northerly and $\beta = 180°$ when the satellite is moving southerly. Latitude ranges from $-90°$ (South Pole) to $90°$ (North Pole) for each pass.

Item 3 is of particular interest for launching rockets to a prograde (easterly) orbit. Launch sites have an easterly velocity component due to the Earth's rotation rate:

$$\mathbf{v}_{\text{site}} = \boldsymbol{\omega}_E \times \mathbf{r}_{\text{site}} = \begin{vmatrix} \mathbf{I} & \mathbf{J} & \mathbf{K} \\ 0 & 0 & \omega_E \\ r_{\text{site}_X} & r_{\text{site}_Y} & r_{\text{site}_Z} \end{vmatrix} \qquad (3.41)$$

where $\boldsymbol{\omega}_E$ is the Earth's angular velocity vector, $\mathbf{r}_{\text{site}}$ is the Earth-surface position vector of the launch site (in **IJK**), and $\mathbf{v}_{\text{site}}$ is the inertial velocity of the launch site in **IJK** coordinates. Using a spherical Earth, the easterly velocity of the Earth's surface is

$$v_{\text{site}} = \omega_E R_E \cos\phi \qquad (3.42)$$

where R_E is the radius of the Earth. Obviously, the surface velocity of the Earth due to its rotation is greatest at the equator ($\phi = 0$) and zero at the poles ($\phi = \pm90°$). For US launch vehicles departing from Cape Canaveral ($\phi = 28.5°$), the easterly direction of the launch site is about 0.41 km/s. Satellites intended for a direct (prograde) orbit are launched due east in order to take maximum advantage of the Earth's surface velocity. Equation (3.40) shows that inclination of the satellite's orbit is equal to the latitude of the launch site ϕ if the rocket is launched east and completes its powered ascent with azimuth $\beta = 90°$. Equation (3.40) also shows that the orbital inclination is *minimized* when the launch azimuth is 90° (east) and that inclination can never be less than the latitude of the launch site. Therefore, the minimum inclination for US satellites launched from Cape Canaveral is 28.5°. A geostationary-equatorial orbit (GEO) was identified in Chapter 2 as a circular orbit with zero inclination and an angular velocity that matches the Earth's spin rate ω_E. Consequently, GEO satellites launched from Cape Canaveral ultimately require a 28.5° "plane-change" maneuver involving a propulsive rocket burn. We shall see in Chapter 7 that changing orbital inclination is expensive in terms of propellant mass. It is for these reasons that the European Space Agency has established its launch facility near Kourou in French Guiana where the latitude is 5.2°.

As a quick example of azimuth and inclination effects, launch from Cape Canaveral with azimuth $\beta = 120°$ (southeast) leads to an orbital inclination of 40.4°. Launching with azimuth $\beta = 60°$ (northeast) also produces $i = 40.4°$. However, a satellite launched with $\beta = 120°$ will begin its ground track on a *descending* arc (moving north to south), while a satellite launched with $\beta = 60°$ will be on an *ascending* arc (south to north) toward its peak latitude $\phi_{max} = i = 40.4°$.

Example 3.3   A Falcon 9 rocket is launched from Cape Canaveral (latitude $\phi = 28.5°$) in order to send supplies to the International Space Station (ISS). The ISS has an orbital inclination $i = 51.65°$. Determine the launch azimuth of the Falcon 9 so that its upper stage is placed in an orbit with the correct inclination and initially moving toward the equator after orbital insertion.

To begin, solve Eq. (3.40) for the sine of azimuth angle β

$$\sin\beta = \frac{\cos i}{\cos\phi} = \frac{\cos(51.65°)}{\cos(28.5°)} = 0.7060$$

Therefore, azimuth is $\beta = \sin^{-1}(0.7060) = 44.912°$ or 135.088°. Remember that we measure azimuth β clockwise from north to the velocity vector (see Figure 3.12). Therefore, launching with $\beta = 44.912°$ (northeast) places the satellite on an *ascending* arc that is away from the equator. For launch with $\beta = 135.088°$, the satellite is initially moving on a descending arc toward the equator as required in the problem statement.

Hence, the proper launch azimuth is $\boxed{\beta = 135.088°}$

As previously mentioned at the beginning of this section, plotting a satellite's ground track is a useful representation of its geographic coverage. Figure 3.13 shows three

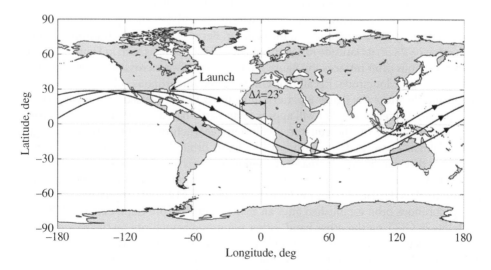

Figure 3.13 Satellite ground tracks on a Mercator map.

ground tracks of a circular Earth orbit projected on to a Mercator map. Note that the first ground track originates at Cape Canaveral (eastern coast of the Florida peninsula at latitude 28.5° N, 80.5° W) with a 90° azimuth angle. Each ground track represents a single orbital revolution of the satellite about the Earth. Remember that a two-body orbit is fixed relative to an inertial (non-rotating) frame and therefore the ground track would retrace its path if the Earth did not rotate. The effect of Earth's rotation is apparent in Figure 3.13: the Earth rotates easterly beneath the satellite's orbit and therefore the ground track is displaced westward on the Mercator map. The westward longitude displacement between two successive ground tracks can be used to estimate the orbital period:

$$T_{period} = \frac{\Delta\lambda}{\omega_E} \tag{3.43}$$

where $\Delta\lambda$ is the ground track's longitude displacement after one orbital revolution and ω_E is Earth's rotation rate (approximately 360°/24 h or 15°/h). It is probably easiest to measure the longitude shift $\Delta\lambda$ by observing successive ground tracks at equatorial crossings (ascending or descending nodes). Figure 3.13 shows that the longitude displacement at equatorial crossings is approximately $\Delta\lambda = 23°$ and therefore the orbital period is approximately $T_{period} = 23°/(15°/h) = 1.5333$ h = 92 min. Using Eq. (2.80), we can compute the semimajor axis from the approximate period to find $a \approx 6{,}750$ km (recall that the precise value for Earth's inertial rotation rate is one revolution every sidereal day, or $\omega_E = 15.041$ deg/h).

Increasing the semimajor axis (and orbital period) increases the westward longitude displacement $\Delta\lambda$. An Earth orbit with semimajor axis $a \approx 26{,}610$ km has a 12-h period and longitude shift $\Delta\lambda = 180°$. If semimajor axis is greater than 26,610 km, the period exceeds 12 h and consequently $\Delta\lambda > 180°$ and successive ground tracks actually appear to be displaced to the east! We can continue to increase the semimajor axis until the period is approximately 24 h ($a > 42{,}000$ km) and $\Delta\lambda = 360°$. An Earth orbit with its period equal to the Earth's rotation rate is called a *geosynchronous orbit*. Figure 3.14 shows the ground track of a circular geosynchronous orbit with inclination $i = 28.5°$. The ground track is a "figure 8" where the ascending and descending nodes are located at the same geographic longitude on the equator. Because the geosynchronous orbit has inclination $i = 28.5°$, its latitude ranges from 28.5° N to 28.5° S as seen in Figure 3.14. The geosynchronous orbit retraces the figure-8 pattern and always crosses the equator at the same geographic longitude. Recall that GEO is a geosynchronous orbit with zero inclination and therefore the figure-8 pattern in Figure 3.14 is collapsed to a single point on the equator. A satellite in GEO will appear to an Earth-based observer to "hover" over a particular longitude on the equator. Achieving geosynchronous orbit typically begins with a launch phase that establishes a low-Earth orbit (LEO) followed by an orbital maneuver that significantly increases the semimajor axis (orbital maneuvers will be presented in Chapter 7). Because the orbital maneuver requires time, the geosynchronous ground track is displaced from the launch site. Figure 3.14 could represent an inclined geosynchronous orbit established after an easterly launch from Cape Canaveral. Placing a satellite in GEO, on the other hand, either requires a launch site on the equator or an orbital maneuver that changes the inclination to zero. As we have previously noted, inclination-change maneuvers involve rocket burns that require substantial propellant mass that ultimately diminishes the payload mass delivered to the target orbit.

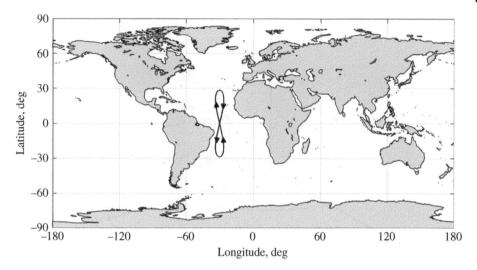

Figure 3.14 Ground track of a geosynchronous orbit ($i = 28.5°$) on a Mercator map.

3.7 Orbit Determination from One Ground-Based Observation

The overall objective of this chapter is to introduce the reader to preliminary orbit-determination methods. The two-body orbit is completely characterized by the classical orbital elements (a, e, i, Ω, ω, θ_0) at epoch t_0. Section 3.4 has shown that we can systematically determine the six classical orbital elements from the six-element state vector ($\mathbf{r}_0, \mathbf{v}_0$) corresponding to epoch t_0. Therefore, the orbit-determination problem for an Earth satellite relies on obtaining the satellite's position and velocity vectors expressed in the ECI coordinate system. Observational data from ground-based sensors (such as radar or optical measurements) are used to compute the state vector ($\mathbf{r}_0, \mathbf{v}_0$) in the ECI frame.

3.7.1 Topocentric-Horizon Coordinate System

In this section, we will present the most straightforward (and perhaps least realistic) orbit-determination method that relies on six independent Earth-based measurements taken together at one time instant. The basic idea is that an Earth-based station measures the position and velocity vectors of a satellite *relative* to a moving coordinate system fixed with the station. This relative state vector is then transformed to the ECI coordinate system to produce the desired ($\mathbf{r}_{ECI}, \mathbf{v}_{ECI}$). Figure 3.15 shows the station-based *topocentric-horizon* coordinate system which consists of three orthogonal axes: the **S** axis points south along a local meridian, **E** points east along a line of latitude, and **Z** points straight up (in the opposite direction of Earth's center). The horizon plane formed by the **S** and **E** axes is the fundamental plane. We will refer to the topocentric-horizon coordinate system as the **SEZ** frame (just as we refer to the ECI coordinate system as the **IJK** frame). The fictitious tracking station shown in Figure 3.15 is located in Eastern Europe and the **SEZ** frame is fixed to the geographic location of the station. Hence, the **SEZ**

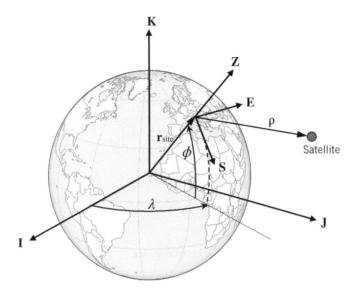

Figure 3.15 Topocentric-horizon (**SEZ**) coordinate system.

frame rotates and moves relative to the inertial **IJK** coordinate system. The latitude of the tracking station in Figure 3.15 is the fixed angle ϕ whereas the station's longitude angle λ is continuously changing due to the Earth's rotation:

$$\lambda(t) = \lambda_0 + \omega_E(t - t_0) \tag{3.44}$$

We measure angle λ_0 from the inertial **I** axis in the equatorial plane to the local meridian of the tracking station at epoch t_0. It is important for the reader to note that the longitude angle λ defined by Eq. (3.44) is *not* the geographic longitude of the ground station measured east or west of the Greenwich meridian (the *geographic* latitude and longitude of our fictitious station shown in Figure 3.15 is approximately 47° N, 20° E).

Figure 3.15 shows the satellite's position vector $\boldsymbol{\rho}$ relative to the **SEZ** frame whose origin is fixed at the tracking station. Let us assume that the ground-based radar station can measure the magnitude of the line-of-sight (LOS) range to the satellite and the elevation and azimuth angles that define the LOS direction to the satellite. Figure 3.16 shows the ground-fixed station at the origin of the **SEZ** frame. *Elevation angle σ is measured from the horizontal plane to the LOS vector $\boldsymbol{\rho}$.* The elevation angle is always between zero (i.e., the satellite is on the horizon) and +90° (i.e., the satellite is directly overhead). *Azimuth angle β is measured clockwise from north to the projection of vector $\boldsymbol{\rho}$ onto the horizontal plane* (note that azimuth angle for a satellite ground track in Section 3.6 is also measured clockwise from north). When the satellite is directly overhead, we have $\sigma = 90°$ and azimuth is undefined. The position of the satellite relative to the tracking station is

$$\boldsymbol{\rho} = \rho(-\cos\sigma\cos\beta\mathbf{S} + \cos\sigma\sin\beta\mathbf{E} + \sin\sigma\mathbf{Z}) \tag{3.45}$$

where ρ is the magnitude of the LOS range vector $\boldsymbol{\rho}$ as measured by the radar site. In summary, the radar station makes three independent measurements (ρ, σ, and β) and Eq. (3.45) allows the computation of $\boldsymbol{\rho}$, the position vector of the satellite relative to the ground-based station.

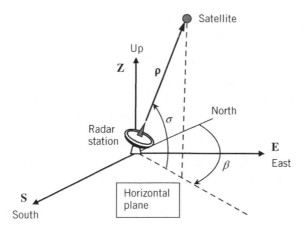

Figure 3.16 Line-of-sight range measured in the **SEZ** coordinate system.

3.7.2 Inertial Position Vector

Referring back to Figure 3.15, we can see that the position vector from the center of the Earth to the satellite is the vector sum

$$\mathbf{r}_{ECI} = \mathbf{r}_{site} + \boldsymbol{\rho} \tag{3.46}$$

where $\mathbf{r}_{site}$ is the position vector from the Earth's center to the tracking station and $\boldsymbol{\rho}$ is the position vector of the satellite relative to the station (note that $\mathbf{r}_{ECI}$ is not shown in Figure 3.15 so that the figure can remain uncluttered). Ultimately, we want the satellite position vector $\mathbf{r}_{ECI}$ in **IJK** (inertial) coordinates. From Figure 3.15, it is easy to see that the inertial position of the radar site in ECI coordinates is

$$\mathbf{r}_{site} = R_E(\cos\phi\cos\lambda\mathbf{I} + \cos\phi\sin\lambda\mathbf{J} + \sin\phi\mathbf{K}) \tag{3.47}$$

where R_E is the radius of a spherical Earth. Of course, the Earth is not a perfect sphere and tracking stations are located at various altitudes relative to sea level. For now, we will proceed with this spherical-Earth model for computing the ground-station vector $\mathbf{r}_{site}$. Later in this section, we will introduce an ellipsoid-Earth model and redefine vector $\mathbf{r}_{site}$.

We cannot yet use Eq. (3.46) to determine $\mathbf{r}_{ECI}$ because $\mathbf{r}_{site}$ is expressed in **IJK** coordinates [see Eq. (3.47)] and $\boldsymbol{\rho}$ is expressed in **SEZ** coordinates [see Eq. (3.45)]. One solution is to transform $\boldsymbol{\rho}$ from the **SEZ** frame to the **IJK** frame. The **IJK** frame in Figure 3.15 can be brought in alignment with the **SEZ** frame using two successive rotations: (1) first, rotate the **IJK** frame about its **K** axis by $+\lambda$ to align the **I** axis with the ground station's local meridian; and (2) secondly, rotate the intermediate coordinate system about its intermediate y-axis through angle $(90° - \phi)$ to establish the **SEZ** frame. Therefore, the transformation of vector $\boldsymbol{\rho}_{ECI}$ from the **IJK** frame to the **SEZ** frame is

$$\boldsymbol{\rho}_{SEZ} = \mathbf{B}(90° - \phi)\mathbf{C}(\lambda)\boldsymbol{\rho}_{ECI} \tag{3.48}$$

where $\mathbf{C}(\lambda)$ is the z-axis rotation matrix and $\mathbf{B}(90° - \phi)$ is the y-axis rotation matrix. Equation (3.48) transforms a station-relative position vector expressed in the **IJK** frame $(\boldsymbol{\rho}_{ECI})$ to the **SEZ** frame. However, we already have the relative position vector

$\boldsymbol{\rho}$ expressed in the **SEZ** frame; see Eq. (3.45). Therefore, we need to apply the matrix inverse to Eq. (3.48) to obtain $\boldsymbol{\rho}_{ECI}$

$$\boldsymbol{\rho}_{ECI} = [\mathbf{B}(90° - \phi)\mathbf{C}(\lambda)]^{-1}\boldsymbol{\rho}_{SEZ} \tag{3.49}$$

Because the rotation matrices are orthogonal, the matrix transpose is equal to the matrix inverse. Using Eqs. (3.28) and (3.33) for the two rotation matrices in Eq. (3.49), the overall **SEZ**-to-**IJK** rotation matrix becomes

$$\mathbf{D} = \left[\begin{bmatrix} \cos(90° - \phi) & 0 & -\sin(90° - \phi) \\ 0 & 1 & 0 \\ \sin(90° - \phi) & 0 & \cos(90° - \phi) \end{bmatrix} \begin{bmatrix} \cos\lambda & \sin\lambda & 0 \\ -\sin\lambda & \cos\lambda & 0 \\ 0 & 0 & 1 \end{bmatrix} \right]^{T} \tag{3.50}$$

Carrying out the matrix multiplication and transposing the result, we obtain

$$\mathbf{D} = \begin{bmatrix} \sin\phi\cos\lambda & -\sin\lambda & \cos\phi\cos\lambda \\ \sin\phi\sin\lambda & \cos\lambda & \cos\phi\sin\lambda \\ -\cos\phi & 0 & \sin\phi \end{bmatrix} \tag{3.51}$$

Equation (3.51) presents the rotation matrix $\mathbf{D}$ that transforms *any* vector expressed in the **SEZ** frame to the **IJK** coordinate system. Applying the $\mathbf{D}$ matrix to the station-relative position vector, we obtain

$$\boldsymbol{\rho}_{ECI} = \mathbf{D}\boldsymbol{\rho}_{SEZ} \tag{3.52}$$

where $\boldsymbol{\rho}_{SEZ}$ is determined by Eq. (3.45).

Let us summarize the entire process for determining a satellite's ECI position vector $\mathbf{r}_{ECI}$ from three independent ground-station measurements taken at time t:

1) Use Eq. (3.44) to determine the current longitude angle $\lambda(t)$ of the ground station relative to the inertial **I** axis. This step requires knowledge of the station's longitude angle λ_0 at reference time t_0.
2) Use Eq. (3.47) along with the tracking station's longitude and latitude (λ, ϕ) to determine the position vector of the station $\mathbf{r}_{site}$ in the **IJK** frame [remember that Eq. (3.47) assumes that the station is located on a spherical Earth with radius R_E].
3) With knowledge of the station's measurements for range magnitude ρ, elevation angle σ, and azimuth angle β, use Eq. (3.45) to determine vector $\boldsymbol{\rho}$ in the **SEZ** frame.
4) Compute the **SEZ**-to-**IJK** rotation matrix $\mathbf{D}$ using Eq. (3.51).
5) Use Eq. (3.52) to transform position vector $\boldsymbol{\rho}$ to the **IJK** frame.
6) Finally, use Eq. (3.46) to compute the satellite's Earth-centered position vector $\mathbf{r}_{ECI}$ from the vector addition of $\mathbf{r}_{site}$ and $\boldsymbol{\rho}_{ECI}$.

Processing these steps solves half of the orbit-determination problem in that we now have a single ECI position vector $\mathbf{r}_{ECI}$ corresponding to epoch t. We need the satellite's velocity vector $\mathbf{v}_{ECI}$ in order to complete the orbit-determination process.

3.7.3 Inertial Velocity Vector

The satellite's inertial velocity vector can be determined from its relative velocity vector and its inertial position. Let us assume that the ground-fixed station can provide three additional independent *rate* measurements: LOS range-rate $\dot{\rho}$, and the angular rates $\dot{\sigma}$ and $\dot{\beta}$. Stations can determine the magnitude of range rate using the frequency shift (or Doppler effect) between the transmitted and received signals. The elevation and azimuth angular rates can be determined from the mechanical servo-drives mounted on the tracking antenna. The satellite's velocity vector *relative to the station* is

$$\dot{\boldsymbol{\rho}} = \dot{\rho}_S \mathbf{S} + \dot{\rho}_E \mathbf{E} + \dot{\rho}_Z \mathbf{Z} \tag{3.53}$$

where the **SEZ** velocity components are obtained by differentiating all terms in Eq. (3.45):

$$\dot{\rho}_S = -\dot{\rho}\cos\sigma\cos\beta + \rho\dot{\sigma}\sin\sigma\cos\beta + \rho\dot{\beta}\cos\sigma\sin\beta \tag{3.54}$$

$$\dot{\rho}_E = \dot{\rho}\cos\sigma\sin\beta - \rho\dot{\sigma}\sin\sigma\sin\beta + \rho\dot{\beta}\cos\sigma\cos\beta \tag{3.55}$$

$$\dot{\rho}_Z = \dot{\rho}\sin\sigma + \rho\dot{\sigma}\cos\sigma \tag{3.56}$$

Clearly, $\dot{\boldsymbol{\rho}}$ is the satellite's velocity relative to the rotating **SEZ** frame. Because we want the *inertial* velocity of the satellite relative to the fixed **IJK** frame, we must add the inertial velocity of the satellite's location in the moving frame:

$$\mathbf{v}_{\text{ECI}} = \dot{\boldsymbol{\rho}} + \boldsymbol{\omega}_E \times \mathbf{r}_{\text{ECI}} \tag{3.57}$$

Equation (3.57) is an application of the Coriolis theorem [see Eq. (C.17) in Appendix C] and may be rewritten as

$$\mathbf{v}_{\text{ECI}} = \dot{\mathbf{r}}_{\text{ECI}} = \left.\frac{d\mathbf{r}_{\text{ECI}}}{dt}\right|_{\text{fix}} = \left.\frac{d\mathbf{r}_{\text{ECI}}}{dt}\right|_{\text{rot}} + \boldsymbol{\omega}_E \times \mathbf{r}_{\text{ECI}} \tag{3.58}$$

where $d\mathbf{r}_{\text{ECI}}/dt|_{\text{fix}}$ is the time derivative of position vector $\mathbf{r}_{\text{ECI}}$ with respect to the inertial (non-rotating) **IJK** frame and $d\mathbf{r}_{\text{ECI}}/dt|_{\text{rot}}$ is the time derivative of $\mathbf{r}_{\text{ECI}}$ with respect to the rotating **SEZ** frame. Taking the time derivative of Eq. (3.46), we obtain $d\mathbf{r}_{\text{ECI}}/dt|_{\text{rot}} = \dot{\boldsymbol{\rho}}$ because the radar station's position vector expressed in the rotating **SEZ** frame, $\mathbf{r}_{\text{site}} = R_E \mathbf{Z}$, is constant.

We can summarize the process for determining a satellite's ECI velocity vector $\mathbf{v}_{\text{ECI}}$ from three independent ground-station rate measurements taken at time t:

1) With knowledge of the LOS range and range rate (ρ and $\dot{\rho}$) and the LOS angles and their rates (σ, β, $\dot{\sigma}$, and $\dot{\beta}$), use Eqs. (3.54)–(3.56) to compute the relative velocity vector $\dot{\boldsymbol{\rho}}_{\text{SEZ}}$ in **SEZ** coordinates.

2) Using the **SEZ**-to-**IJK** rotation matrix **D**, transform $\dot{\boldsymbol{\rho}}_{\text{SEZ}}$ to the **IJK** frame:

$$\dot{\boldsymbol{\rho}}_{\text{ECI}} = \mathbf{D}\dot{\boldsymbol{\rho}}_{\text{SEZ}}$$

3) Using Eq. (3.57) and position vector $\mathbf{r}_{\text{ECI}}$ (in **IJK** coordinates), compute the satellite's inertial velocity vector in the **IJK** frame

$$\mathbf{v}_{\text{ECI}} = \dot{\boldsymbol{\rho}}_{\text{ECI}} + \boldsymbol{\omega}_E \times \mathbf{r}_{\text{ECI}}$$

where the Earth's angular velocity vector is $\boldsymbol{\omega}_E = \omega_E \mathbf{K}$.

In summary, we use the six independent ground-station measurements (ρ, σ, β, $\dot{\rho}$, $\dot{\sigma}$, and $\dot{\beta}$) to compute the station-relative vectors $\boldsymbol{\rho}$ and $\dot{\boldsymbol{\rho}}$ expressed in the **SEZ** frame. The satellite's geocentric inertial vectors $\mathbf{r}_{\text{ECI}}$ and $\mathbf{v}_{\text{ECI}}$ are determined by utilizing the **SEZ**-to-**IJK** rotation matrix **D** and the term $\boldsymbol{\omega}_E \times \mathbf{r}_{\text{ECI}}$ that is associated with the rotating **SEZ** frame. Finally, we should reiterate that the rotation matrix **D** depends on the station's current longitude λ relative to the fixed **I** axis. Furthermore, the inertial position of the ground station, $\mathbf{r}_{site}$, is modeled using a spherical Earth.

Example 3.4 The New Boston Air Force Station (NBAFS) in New Hampshire is located at latitude $\phi = 42.9°$ N. At epoch t_0 the NBAFS measures the LOS range, elevation and azimuth angles, and their rates:

$\rho = 668.3$ km, $\sigma = 62.5°$, $\beta = 135.4°$, $\dot{\rho} = 2.39$ km/s, $\dot{\sigma} = -0.65$ deg/s, $\dot{\beta} = -0.38$ deg/s

The longitude of the NBAFS as measured from the inertial **I** axis is 240.7° at epoch t_0. Determine the satellite's state vector at this epoch in the ECI coordinate system.

First, we determine the ECI position vector using the steps following Eq. (3.52) with longitude $\lambda = 240.7°$. Using the spherical Earth model (3.47) with $R_E = 6{,}378.137$ km, the geocentric position of the NBAFS at this instant is

$$\mathbf{r}_{site} = R_E(\cos\phi\cos\lambda\mathbf{I} + \cos\phi\sin\lambda\mathbf{J} + \sin\phi\mathbf{K}) = \begin{bmatrix} -2{,}286.522 \\ -4{,}074.533 \\ 4{,}341.731 \end{bmatrix} \text{ km}$$

The satellite's position vector relative to the station is computed using Eq. (3.45)

$$\boldsymbol{\rho}_{SEZ} = \rho(-\cos\sigma\cos\beta\mathbf{S} + \cos\sigma\sin\beta\mathbf{E} + \sin\sigma\mathbf{Z})$$
$$= 219.723\mathbf{S} + 216.675\mathbf{E} + 592.789\mathbf{Z} \text{ km}$$

Next, we use the **SEZ**-to-**IJK** rotation matrix **D** to transform $\boldsymbol{\rho}_{SEZ}$ to the ECI frame. Using Eq. (3.51), the **D** matrix is

$$\mathbf{D} = \begin{bmatrix} \sin\phi\cos\lambda & -\sin\lambda & \cos\phi\cos\lambda \\ \sin\phi\sin\lambda & \cos\lambda & \cos\phi\sin\lambda \\ -\cos\phi & 0 & \sin\phi \end{bmatrix} = \begin{bmatrix} -0.33313 & 0.87207 & -0.35849 \\ -0.59364 & -0.48938 & -0.63883 \\ -0.73254 & 0 & 0.68072 \end{bmatrix}$$

The station-relative position vector in **IJK** coordinates is

$$\boldsymbol{\rho}_{ECI} = \mathbf{D}\boldsymbol{\rho}_{SEZ} = \begin{bmatrix} -96.752 \\ -615.162 \\ 242.569 \end{bmatrix} \text{ km}$$

Finally, add $\mathbf{r}_{site}$ to $\boldsymbol{\rho}_{ECI}$ to compute the satellite's inertial geocentric position vector:

$$\mathbf{r}_{ECI} = \mathbf{r}_{site} + \boldsymbol{\rho}_{ECI} = \begin{bmatrix} -2{,}383.274 \\ -4{,}689.696 \\ 4{,}584.299 \end{bmatrix} \text{ km}$$

Next, we use Eqs. (3.54)–(3.56) to determine the **SEZ** components of the station-relative velocity:

$$\dot{\rho}_S = -\dot{\rho}\cos\sigma\cos\beta + \rho\dot{\sigma}\sin\sigma\cos\beta + \rho\dot{\beta}\cos\sigma\sin\beta = 4.1371 \text{ km/s}$$
$$\dot{\rho}_E = \dot{\rho}\cos\sigma\sin\beta - \rho\dot{\sigma}\sin\sigma\sin\beta + \rho\dot{\beta}\cos\sigma\cos\beta = 6.9541 \text{ km/s}$$
$$\dot{\rho}_Z = \dot{\rho}\sin\sigma + \rho\dot{\sigma}\cos\sigma = -1.3808 \text{ km/s}$$

Using the rotation matrix $\mathbf{D}$, transform $\dot{\boldsymbol{\rho}}_{SEZ}$ to the **IJK** frame:

$$\dot{\boldsymbol{\rho}}_{ECI} = \mathbf{D}\dot{\boldsymbol{\rho}}_{SEZ} = \begin{bmatrix} 5.1813 \\ -4.9770 \\ -3.9706 \end{bmatrix} \text{ km/s}$$

Finally, use Eq. (3.57) with $\omega_E = 7.29212(10^{-5})$ **K** rad/s to compute the satellite's inertial velocity in the **IJK** frame:

$$\mathbf{v}_{ECI} = \dot{\boldsymbol{\rho}}_{ECI} + \boldsymbol{\omega}_E \times \mathbf{r}_{ECI}$$

$$= \begin{bmatrix} 5.1813 \\ -4.9770 \\ -3.9706 \end{bmatrix} + \begin{vmatrix} \mathbf{I} & \mathbf{J} & \mathbf{K} \\ 0 & 0 & 7.29212(10^{-5}) \\ -2{,}383.274 & -4{,}689.696 & 4{,}584.299 \end{vmatrix}$$

Carrying out the cross product and adding vectors we obtain

$$\mathbf{v}_{ECI} = \begin{bmatrix} 5.5233 \\ -5.1508 \\ -3.9706 \end{bmatrix} \text{ km/s}$$

The satellite's state vector at epoch t_0 is $(\mathbf{r}_{ECI}, \mathbf{v}_{ECI})$.

3.7.4 Ellipsoidal Earth Model

Accurately determining the geocentric position vector of the ground station ($\mathbf{r}_{site}$) is crucial to computing the satellite's geocentric position vector $\mathbf{r}_{ECI}$. Recall that Eq. (3.47) uses a spherical-Earth model with constant radius R_E. While a spherical model is simple to use, it will result in position errors on the order of 10 km. Our Earth is a "flattened" sphere where its equatorial radius is about 21 km greater than its polar radius. An ellipsoid of revolution about the polar axis is a simple but more accurate representation of the shape of the Earth. Figure 3.17 shows the geometry of the ellipsoidal Earth model where a_E is the semimajor (equatorial) axis and b_E is the semiminor (polar) axis. We may think of the ellipse in Figure 3.17 as the plane produced by a vertical slice of the solid ellipsoid along the meridian that contains the ground station. Therefore, the **K** axis points toward the North Pole and the ζ axis lies in the equatorial plane and points from the Earth's center to ground station's meridian intersection with the equator. Figure 3.17 also shows two distinct definitions of latitude. The *geocentric latitude* ϕ' is measured from the equatorial plane to the line from the Earth's center to the point on the surface of the ellipsoid. The *geodetic latitude* ϕ is measured from the equatorial plane to the line normal to the ellipsoid's surface as shown in Figure 3.17. When we use "latitude" in common language and as a coordinate on a map, it is the geodetic latitude ϕ. The reference ellipsoid shown in Figure 3.17 represents the Earth's "mean sea level," and therefore the topographic

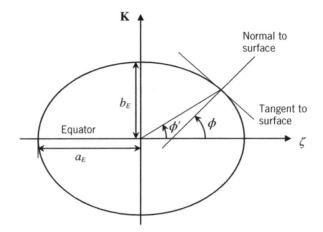

Figure 3.17 Ellipsoidal Earth model.

location of a ground station will include its altitude above sea level (or, altitude above the reference ellipsoid).

We will use the following values for the ellipsoidal Earth model: semimajor axis a_E = 6,378.137 km and semiminor axis b_E = 6,356.752 km. We compute the eccentricity of the Earth ellipse e_E using the ratio of the minor and major axes:

$$\frac{b_E}{a_E} = \sqrt{1 - e_E^2} \qquad (3.59)$$

Therefore, the eccentricity of the Earth ellipse is e_E = 0.08182.

Recall that our overall objective in this subsection is to determine the geocentric position vector of the ground station $\mathbf{r}_{\text{site}}$ (in **IJK** coordinates) using an oblate ellipsoid to model the Earth's mean sea level. The solution to this problem requires that we determine the **K**-axis and ζ-axis projections of a point on the ellipsoid that is defined by (geodetic) latitude ϕ. We will only present the result here; the interested reader may consult Vallado [1; pp. 134–138] or Bate *et al.* [2; pp. 93–98] for the complete derivation. The geocentric vector to a ground station located relative to an ellipsoidal Earth model is

$$\mathbf{r}_{\text{site}} = \left(\frac{a_E}{\sqrt{1 - e_E^2 \sin^2 \phi}} + h_{\text{site}} \right) \cos\phi (\cos\lambda \mathbf{I} + \sin\lambda \mathbf{J}) + \left(\frac{a_E (1 - e_E^2)}{\sqrt{1 - e_E^2 \sin^2 \phi}} + h_{\text{site}} \right) \sin\phi \mathbf{K}$$

$$(3.60)$$

where h_{site} is the altitude of the ground station above the reference ellipsoid (i.e., mean sea level). Remember that longitude λ is the angle in the equatorial plane from the inertial **I** axis to the ground station's meridian at the epoch when the measurements are taken (see Figure 3.15). Note that if e_E = 0, the ellipsoid becomes a sphere and Eq. (3.60) is reduced to the spherical Earth model, Eq. (3.47), with radius $R_E = a_E + h_{\text{site}}$.

Example 3.5   Repeat Example 3.4 using an ellipsoidal Earth. The New Boston Air Force Station is 390 m above sea level.

The ground station's ECI position vector is computed using Eq. (3.60) with $h_{site} = 0.39$ km, $a_E = 6{,}378.137$ km, $e_E = 0.08182$, $\lambda = 240.7°$, and $\phi = 42.9°$:

$$\mathbf{r}_{site} = \left(\frac{a_E}{\sqrt{1 - e_E^2 \sin^2 \phi}} + h_{site} \right) \cos\phi \left(\cos\lambda \mathbf{I} + \sin\lambda \mathbf{J} \right) + \left(\frac{a_E \left(1 - e_E^2 \right)}{\sqrt{1 - e_E^2 \sin^2 \phi}} + h_{site} \right) \sin\phi \mathbf{K}$$

Or,

$$\mathbf{r}_{site} = \begin{bmatrix} -2{,}290.216 \\ -4{,}081.117 \\ 4{,}319.635 \end{bmatrix} \text{ km}$$

Comparing this result with the ground station vector from Example 3.4, we observe component differences on the order of 4–22 km. Recall that in Example 3.4 the ground station is assumed to be located at sea level on a spherical Earth.

The satellite's position vector relative to the station (expressed in the **IJK** frame) depends on the measurements ρ, σ, and β and the latitude and longitude of the station. Therefore, $\boldsymbol{\rho}_{ECI}$ is unchanged from the value computed in Example 3.4 and the ECI position vector of the satellite is the vector sum of $\mathbf{r}_{site}$ (ellipsoid model) and $\boldsymbol{\rho}_{ECI}$:

$$\mathbf{r}_{ECI} = \mathbf{r}_{site} + \boldsymbol{\rho}_{ECI} = \boxed{\begin{bmatrix} -2{,}386.968 \\ -4{,}696.279 \\ 4{,}562.204 \end{bmatrix} \text{ km}}$$

The component differences between the inertial position vectors computed using the ellipsoidal and spherical Earth models differ by 4–23 km and these dispersions are solely due to the computation of the ground station vector $\mathbf{r}_{site}$.

The relative velocity vector $\dot{\boldsymbol{\rho}}$ (in either frame) is unchanged from the value computed in Example 3.4. However, the cross product $\boldsymbol{\omega}_E \times \mathbf{r}_{ECI}$ differs from its computation in Example 3.4 because the inertial position vector $\mathbf{r}_{ECI}$ (computed above) is slightly different. Carrying out the vector operations in Eq. (3.57) yields

$$\mathbf{v}_{ECI} = \dot{\boldsymbol{\rho}}_{ECI} + \boldsymbol{\omega}_E \times \mathbf{r}_{ECI}$$

$$= \begin{bmatrix} 5.1813 \\ -4.9770 \\ -3.9706 \end{bmatrix} + \begin{vmatrix} \mathbf{I} & \mathbf{J} & \mathbf{K} \\ 0 & 0 & 7.29212(10^{-5}) \\ -2{,}386.968 & -4{,}696.279 & 4{,}562.204 \end{vmatrix}$$

Or,

$$\mathbf{v}_{ECI} = \boxed{\begin{bmatrix} 5.5237 \\ -5.1511 \\ -3.9706 \end{bmatrix} \text{ km/s}}$$

Comparing the inertial velocities computed in Examples 3.4 and 3.5, we see that the largest component difference is less than 0.5 m/s.

3.8 Orbit Determination from Three Position Vectors

The previous section presented an orbit-determination method that uses six independent measurements at a single observation time to compute the six elements of the state vector $(\mathbf{r},\mathbf{v})$. Three time-rate measurements (range rate $\dot{\rho}$, and angular rates $\dot{\sigma}$, $\dot{\beta}$) are required to determine the inertial velocity vector $\mathbf{v}$ in the ECI frame. Oftentimes the angular rate measurements are unreliable which degrades the orbit-determination process.

In this section, we present a geometric method for orbit determination that is based on three successive inertial position vectors. This method utilizes concepts from vector calculus pioneered by Josiah Willard Gibbs at the end of the nineteenth century and is therefore known as the Gibbs method. Figure 3.18 shows the basic premise of the Gibbs method: using three successive geocentric position vectors $(\mathbf{r}_1, \mathbf{r}_2, \mathbf{r}_3)$ determine the satellite's inertial geocentric velocity vector $\mathbf{v}_2$. Each position vector $\mathbf{r}_i$ shown in Figure 3.18 is in the ECI frame and can be computed from three topocentric-horizon measurements $(\rho_i, \sigma_i, \beta_i)$ at observation time t_i (of course, we also need to know the ground station's latitude ϕ and longitude λ_i at each observation time t_i). After we determine the ECI velocity vector $\mathbf{v}_2$ corresponding to the middle observation, we can determine the orbital elements from the state vector $(\mathbf{r}_2,\mathbf{v}_2)$ using the methods we previously described in this chapter.

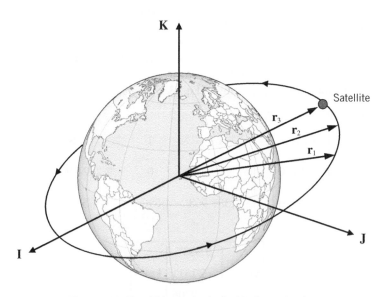

Figure 3.18 The Gibbs method of orbit determination.

The derivation of the Gibbs method presented here follows Bate *et al.* [2; pp. 109–115]. The basis of the Gibbs method is that the three position vectors ($\mathbf{r}_1$, $\mathbf{r}_2$, $\mathbf{r}_3$) must lie in the same plane. Therefore, the third position vector $\mathbf{r}_3$ must be a linear combination of the two previous vectors $\mathbf{r}_1$ and $\mathbf{r}_2$. This linear dependence may be written as

$$c_1\mathbf{r}_1 + c_2\mathbf{r}_2 + c_3\mathbf{r}_3 = 0 \tag{3.61}$$

where c_1, c_2, and c_3 are constants. Next, take the cross product of Eq. (3.61) with *each* of the three position vectors to produce three vector equations:

$$c_1\mathbf{r}_1 \times \mathbf{r}_1 + c_2\mathbf{r}_2 \times \mathbf{r}_1 + c_3\mathbf{r}_3 \times \mathbf{r}_1 = 0 \tag{3.62a}$$

$$c_1\mathbf{r}_1 \times \mathbf{r}_2 + c_2\mathbf{r}_2 \times \mathbf{r}_2 + c_3\mathbf{r}_3 \times \mathbf{r}_2 = 0 \tag{3.62b}$$

$$c_1\mathbf{r}_1 \times \mathbf{r}_3 + c_2\mathbf{r}_2 \times \mathbf{r}_3 + c_3\mathbf{r}_3 \times \mathbf{r}_3 = 0 \tag{3.62c}$$

Of course, the cross product $\mathbf{r}_i \times \mathbf{r}_i = 0$. Rearranging Eq. (3.62a)–(3.62c) and noting that $\mathbf{r}_i \times \mathbf{r}_j = -\mathbf{r}_j \times \mathbf{r}_i$, we obtain

$$c_2\mathbf{r}_1 \times \mathbf{r}_2 = c_3\mathbf{r}_3 \times \mathbf{r}_1 \tag{3.63a}$$

$$c_1\mathbf{r}_1 \times \mathbf{r}_2 = c_3\mathbf{r}_2 \times \mathbf{r}_3 \tag{3.63b}$$

$$c_1\mathbf{r}_3 \times \mathbf{r}_1 = c_2\mathbf{r}_2 \times \mathbf{r}_3 \tag{3.63c}$$

The next step involves the projection of each position vector $\mathbf{r}_i$ onto the periapsis direction. Figure 3.4 shows that the dot product of position vector $\mathbf{r}$ and eccentricity vector $\mathbf{e}$ is

$$\mathbf{r} \cdot \mathbf{e} = re\cos\theta \tag{3.64}$$

We can also obtain the result $re\cos\theta$ from the trajectory equation:

$$r = \frac{p}{1 + e\cos\theta} \quad \text{or} \quad r + re\cos\theta = p \tag{3.65}$$

Therefore, $\mathbf{r} \cdot \mathbf{e} = re\cos\theta = p - r$. Taking the dot product of each term in Eq. (3.61) with the eccentricity vector $\mathbf{e}$ and substituting $\mathbf{r}_i \cdot \mathbf{e} = p - r_i$, we obtain

$$c_1(p - r_1) + c_2(p - r_2) + c_3(p - r_3) = 0 \tag{3.66}$$

Multiplying all terms in Eq. (3.66) by the vector product $\mathbf{r}_3 \times \mathbf{r}_1$ yields

$$(\mathbf{r}_3 \times \mathbf{r}_1)c_1(p - r_1) + (\mathbf{r}_3 \times \mathbf{r}_1)c_2(p - r_2) + (\mathbf{r}_3 \times \mathbf{r}_1)c_3(p - r_3) = 0 \tag{3.67}$$

Finally, we make use of equation set (3.63) by substituting Eq. (3.63c) for the term $c_1\mathbf{r}_3 \times \mathbf{r}_1$ and Eq. (3.63a) for the term $c_3\mathbf{r}_3 \times \mathbf{r}_1$. After these substitutions, Eq. (3.67) only involves one unknown constant c_2

$$(\mathbf{r}_2 \times \mathbf{r}_3)c_2(p - r_1) + (\mathbf{r}_3 \times \mathbf{r}_1)c_2(p - r_2) + (\mathbf{r}_1 \times \mathbf{r}_2)c_2(p - r_3) = 0 \tag{3.68}$$

Or, after factoring out the single constant c_2 from Eq. (3.68), we obtain

$$c_2[(\mathbf{r}_2 \times \mathbf{r}_3)(p - r_1) + (\mathbf{r}_3 \times \mathbf{r}_1)(p - r_2) + (\mathbf{r}_1 \times \mathbf{r}_2)(p - r_3)] = 0 \tag{3.69}$$

Constant c_2 can be ignored in Eq. (3.69) because the bracketed term must equal the zero vector. Finally, we can group all terms involving parameter p on the left-hand side and all terms involving positions r_i on the right-hand side to yield

$$p[(\mathbf{r}_2 \times \mathbf{r}_3) + (\mathbf{r}_3 \times \mathbf{r}_1) + (\mathbf{r}_1 \times \mathbf{r}_2)] = r_1(\mathbf{r}_2 \times \mathbf{r}_3) + r_2(\mathbf{r}_3 \times \mathbf{r}_1) + r_3(\mathbf{r}_1 \times \mathbf{r}_2) \tag{3.70}$$

Equation (3.70) allows us to compute the parameter p from the three position vectors. Let us define the auxiliary vectors $\mathbf{D}$ and $\mathbf{N}$ as the left- and right-hand sides of Eq. (3.70):

$$\mathbf{D} = (\mathbf{r}_2 \times \mathbf{r}_3) + (\mathbf{r}_3 \times \mathbf{r}_1) + (\mathbf{r}_1 \times \mathbf{r}_2) \tag{3.71}$$

$$\mathbf{N} = r_1(\mathbf{r}_2 \times \mathbf{r}_3) + r_2(\mathbf{r}_3 \times \mathbf{r}_1) + r_3(\mathbf{r}_1 \times \mathbf{r}_2) \tag{3.72}$$

Therefore, Eq. (3.70) becomes $p\mathbf{D} = \mathbf{N}$ and the parameter is $p = N/D$ where $N = \|\mathbf{N}\|$ and $D = \|\mathbf{D}\|$.

The next steps involve determining expressions for the perifocal-frame unit vectors $\mathbf{P}$, $\mathbf{Q}$, and $\mathbf{W}$. Note that vectors $\mathbf{N}$ and $\mathbf{D}$ both involve the two cross products ($\mathbf{r}_1 \times \mathbf{r}_2$ and $\mathbf{r}_2 \times \mathbf{r}_3$) that are in the direction of the angular momentum vector $\mathbf{h}$, and one cross product ($\mathbf{r}_3 \times \mathbf{r}_1$) that is opposite of $\mathbf{h}$. Therefore, both $\mathbf{N}$ and $\mathbf{D}$ are in the direction of vector $\mathbf{h}$ and the perifocal $\mathbf{W}$ axis can be defined as

$$\mathbf{W} = \frac{\mathbf{N}}{N} \tag{3.73}$$

The periapsis direction is the unit vector along the eccentricity vector, $\mathbf{P} = \mathbf{e}/e$, and the $\mathbf{Q}$ axis is defined by the cross product of unit vectors $\mathbf{W}$ and $\mathbf{P}$

$$\mathbf{Q} = \mathbf{W} \times \mathbf{P} = \frac{\mathbf{N} \times \mathbf{e}}{Ne} \tag{3.74}$$

Or,

$$Ne\mathbf{Q} = \mathbf{N} \times \mathbf{e} \tag{3.75}$$

Substituting Eq. (3.72) for vector $\mathbf{N}$ in Eq. (3.75), we obtain

$$Ne\mathbf{Q} = r_1(\mathbf{r}_2 \times \mathbf{r}_3) \times \mathbf{e} + r_2(\mathbf{r}_3 \times \mathbf{r}_1) \times \mathbf{e} + r_3(\mathbf{r}_1 \times \mathbf{r}_2) \times \mathbf{e} \tag{3.76}$$

Each right-hand side term in Eq. (3.76) is a vector triple product:

$$(\mathbf{a} \times \mathbf{b}) \times \mathbf{c} = (\mathbf{a} \cdot \mathbf{c})\mathbf{b} - (\mathbf{b} \cdot \mathbf{c})\mathbf{a}$$

Using the vector triple product, Eq. (3.76) becomes

$$Ne\mathbf{Q} = r_1(\mathbf{r}_2 \cdot \mathbf{e})\mathbf{r}_3 - r_1(\mathbf{r}_3 \cdot \mathbf{e})\mathbf{r}_2 + r_2(\mathbf{r}_3 \cdot \mathbf{e})\mathbf{r}_1 - r_2(\mathbf{r}_1 \cdot \mathbf{e})\mathbf{r}_3 + r_3(\mathbf{r}_1 \cdot \mathbf{e})\mathbf{r}_2 - r_3(\mathbf{r}_2 \cdot \mathbf{e})\mathbf{r}_1 \tag{3.77}$$

Recalling Eqs. (3.64) and (3.65), the six dot products in Eq. (3.77) are $\mathbf{r}_i \cdot \mathbf{e} = p - r_i$; Eq. (3.77) becomes

$$Ne\mathbf{Q} = r_1(p - r_2)\mathbf{r}_3 - r_1(p - r_3)\mathbf{r}_2 + r_2(p - r_3)\mathbf{r}_1 - r_2(p - r_1)\mathbf{r}_3 + r_3(p - r_1)\mathbf{r}_2 - r_3(p - r_2)\mathbf{r}_1 \tag{3.78}$$

After some algebra, Eq. (3.78) becomes

$$Ne\mathbf{Q} = p[(r_2 - r_3)\mathbf{r}_1 + (r_3 - r_1)\mathbf{r}_2 + (r_1 - r_2)\mathbf{r}_3] \tag{3.79}$$

Let us define the right-hand side bracketed vector in Eq. (3.79) as auxiliary vector $\mathbf{S}$

$$\mathbf{S} = (r_2 - r_3)\mathbf{r}_1 + (r_3 - r_1)\mathbf{r}_2 + (r_1 - r_2)\mathbf{r}_3 \tag{3.80}$$

Hence, unit vector $\mathbf{Q} = \mathbf{S}/S$. The magnitude of each side of Eq. (3.79) is

$$Ne = pS \tag{3.81}$$

Substituting $p = N/D$ into Eq. (3.81), we obtain

$$e = \frac{S}{D} \tag{3.82}$$

Now we have obtained the orbit's size and shape (parameter p and eccentricity e) in terms of the magnitudes of the three auxiliary vectors $\mathbf{N}$, $\mathbf{D}$, and $\mathbf{S}$. However, we can use these three vectors to determine the ECI velocity vector corresponding to the middle observation. To show this, we recall an intermediate step in our derivation of the trajectory equation in Chapter 2. Repeat Eq. (2.40) with the middle observation position and velocity vectors $\mathbf{r}_2$ and $\mathbf{v}_2$

$$\mathbf{v}_2 \times \mathbf{h} = \mu\left(\frac{\mathbf{r}_2}{r_2} + \mathbf{e}\right) \tag{3.83}$$

Next, cross Eq. (3.83) with angular momentum $\mathbf{h}$

$$\mathbf{h} \times (\mathbf{v}_2 \times \mathbf{h}) = \mu\left(\frac{\mathbf{h} \times \mathbf{r}_2}{r_2} + \mathbf{h} \times \mathbf{e}\right) \tag{3.84}$$

The left-hand side is a vector triple product and the result is $h^2\mathbf{v}_2$ because $\mathbf{h}$ and $\mathbf{v}_2$ are orthogonal. Next, we substitute $\mathbf{h} = h\mathbf{W}$ and $\mathbf{e} = e\mathbf{P}$ into Eq. (3.84) to yield

$$h^2\mathbf{v}_2 = \mu\left(\frac{h\mathbf{W} \times \mathbf{r}_2}{r_2} + h\mathbf{W} \times e\mathbf{P}\right) \tag{3.85}$$

Substituting $\mathbf{Q} = \mathbf{W} \times \mathbf{P}$ and solving Eq. (3.85) for velocity yields

$$\mathbf{v}_2 = \frac{\mu}{h}\left(\frac{\mathbf{W} \times \mathbf{r}_2}{r_2} + e\mathbf{Q}\right) \tag{3.86}$$

The final substitutions involve $h = \sqrt{p\mu} = \sqrt{N\mu/D}$, $e = S/D$, and unit vectors $\mathbf{W} = \mathbf{D}/D$ and $\mathbf{Q} = \mathbf{S}/S$. Applying these substitutions Eq. (3.86) becomes

$$\mathbf{v}_2 = \frac{1}{r_2}\sqrt{\frac{\mu}{ND}}(\mathbf{D} \times \mathbf{r}_2) + \sqrt{\frac{\mu}{ND}}\mathbf{S} \tag{3.87}$$

Equation (3.87) is our final result: the ECI velocity vector of the middle observation as a function of vectors $\mathbf{r}_2$, $\mathbf{N}$, $\mathbf{D}$, and $\mathbf{S}$. Of course, $\mathbf{N}$, $\mathbf{D}$, and $\mathbf{S}$ are determined by the three ECI position vectors $\mathbf{r}_1$, $\mathbf{r}_2$, and $\mathbf{r}_3$.

The Gibbs method fails if the three position vectors are *not* coplanar. We may use the following test:

$$\varepsilon = \frac{\mathbf{r}_1 \cdot (\mathbf{r}_2 \times \mathbf{r}_3)}{r_1\|\mathbf{r}_2 \times \mathbf{r}_3\|} \tag{3.88}$$

Scalar quantity ε is the cosine of the angle between position vector $\mathbf{r}_1$ and the cross-product result $\mathbf{r}_2 \times \mathbf{r}_3$. Because $\mathbf{r}_2 \times \mathbf{r}_3$ is along the angular momentum vector $\mathbf{h}$ (which is orthogonal to $\mathbf{r}_1$), the scalar ε should be *exactly* zero if all three vectors are coplanar. However, measurement errors and/or sensor noise will lead to three position vectors that are not exactly coplanar. A reasonable test is to ensure that $|\varepsilon| < \cos 88°$ (or $|\varepsilon| < 0.0349$) for coplanar position vectors.

The Gibbs method has numerical instability if the position vectors are too closely spaced. Vallado [1; p. 459] recommends a 1° angular separation as a lower threshold. We can easily compute the angular separation using the dot product and inverse cosine function:

$$\theta_{12} = \cos^{-1} \left(\frac{\mathbf{r}_1 \cdot \mathbf{r}_2}{r_1 r_2} \right) \quad \text{and} \quad \theta_{23} = \cos^{-1} \left(\frac{\mathbf{r}_2 \cdot \mathbf{r}_3}{r_2 r_3} \right) \tag{3.89}$$

We may think of the separation angles θ_{12} and θ_{23} as the changes in true anomaly between two successive position vectors.

We may now summarize the Gibbs method as follows:

1) Compute ε using Eq. (3.88) and the three position vectors $\mathbf{r}_1$, $\mathbf{r}_2$, and $\mathbf{r}_3$. If $|\varepsilon| < 0.0349$, then the three position vectors are essentially coplanar and the Gibbs method can be used.
2) Using Eq. (3.89), compute the angular separations between position vectors. If either θ_{12} or θ_{23} is less than 1°, the Gibbs method may produce inaccurate results.
3) Using the three position vectors $\mathbf{r}_1$, $\mathbf{r}_2$, and $\mathbf{r}_3$, compute the auxiliary vectors $\mathbf{D}$, $\mathbf{N}$, and $\mathbf{S}$ using Eqs. (3.71), (3.72), and (3.80).
4) Compute the ECI velocity vector of the middle observation, $\mathbf{v}_2$, using Eq. (3.87).
5) Use the techniques of Section 3.4 to compute the orbital elements from the state vector $(\mathbf{r}_2, \mathbf{v}_2)$.

The following example illustrates the Gibbs method for orbit determination.

Example 3.6   A ground station makes LOS measurements of a satellite at three observation times and determines the following three ECI position vectors:

$$\mathbf{r}_1 = \begin{bmatrix} -11,052.902 \\ -12,938.738 \\ 8,505.244 \end{bmatrix} \text{km} \quad \mathbf{r}_2 = \begin{bmatrix} -10,378.257 \\ -15,955.205 \\ 14,212.351 \end{bmatrix} \text{km} \quad \mathbf{r}_3 = \begin{bmatrix} -9,336.222 \\ -17,747.079 \\ 18,337.068 \end{bmatrix} \text{km}$$

The ground-station operators suspect that this satellite is in a Molniya orbit. Compute the satellite's state vector for the middle observation and determine if the satellite is indeed in a Molniya orbit.

The Gibbs method relies on the magnitudes of each position vector and the three cross products. The three magnitudes are

$$r_1 = 19,024.110 \text{ km}, \quad r_2 = 23,754.320 \text{ km}, \quad \text{and} \quad r_3 = 27,173.000 \text{ km}$$

The three cross products are

$$\mathbf{r}_1 \times \mathbf{r}_2 = \begin{bmatrix} -48,186,974.358 \\ 68,818,114.713 \\ 42,069,769.035 \end{bmatrix} \text{km}^2$$

$$\mathbf{r}_2 \times \mathbf{r}_3 = \begin{bmatrix} -40,343,963.066 \\ 57,617,140.253 \\ 35,222,410.926 \end{bmatrix} km^2$$

$$\mathbf{r}_3 \times \mathbf{r}_1 = \begin{bmatrix} 86,315,281.358 \\ -123,270,969.423 \\ -75,357,794.605 \end{bmatrix} km^2$$

First, we must use Eq. (3.88) to check the coplanar condition:

$$\varepsilon = \frac{\mathbf{r}_1 \cdot (\mathbf{r}_2 \times \mathbf{r}_3)}{r_1 \| \mathbf{r}_2 \times \mathbf{r}_3 \|} = -8.54(10^{-9})$$

The magnitude of the ε parameter is extremely small and hence the three position vectors are coplanar. The next step is to determine the angular spacing between the position vectors. Using Eq. (3.89), we obtain

$$\theta_{12} = \cos^{-1}\left(\frac{\mathbf{r}_1 \cdot \mathbf{r}_2}{r_1 r_2}\right) = \cos^{-1}\left(\frac{442,029,587.85}{451,904,790.21}\right) = 12.00°$$

$$\theta_{23} = \cos^{-1}\left(\frac{\mathbf{r}_2 \cdot \mathbf{r}_3}{r_2 r_3}\right) = \cos^{-1}\left(\frac{640,664,841.65}{645,476,124.86}\right) = 7.00°$$

The angular separations are greater than 1° and therefore the Gibbs method should provide sufficient accuracy. Next, we compute the auxiliary vectors $\mathbf{D}$, $\mathbf{N}$, and $\mathbf{S}$ using Eqs. (3.71), (3.72), and (3.80):

$$\mathbf{D} = (\mathbf{r}_2 \times \mathbf{r}_3) + (\mathbf{r}_3 \times \mathbf{r}_1) + (\mathbf{r}_1 \times \mathbf{r}_2) = \begin{bmatrix} -2,215,656.066 \\ 3,164,285.542 \\ 1,934,385.356 \end{bmatrix} km^2$$

$$\mathbf{N} = r_1(\mathbf{r}_2 \times \mathbf{r}_3) + r_2(\mathbf{r}_3 \times \mathbf{r}_1) + r_3(\mathbf{r}_1 \times \mathbf{r}_2) = \begin{bmatrix} -26,531,841,547.5 \\ 37,891,405,572.8 \\ 23,163,695,418.9 \end{bmatrix} km^3$$

$$\mathbf{S} = (r_2 - r_3)\mathbf{r}_1 + (r_3 - r_1)\mathbf{r}_2 + (r_1 - r_2)\mathbf{r}_3 = \begin{bmatrix} -2,622,648.754 \\ -1,836,395.621 \\ -3.881 \end{bmatrix} km^2$$

Finally, use Eq. (3.87) to compute the ECI velocity vector corresponding to the middle observation:

$$\mathbf{v}_2 = \frac{1}{r_2}\sqrt{\frac{\mu}{ND}}(\mathbf{D} \times \mathbf{r}_2) + \sqrt{\frac{\mu}{ND}}\mathbf{S} = \begin{bmatrix} 0.7610 \\ -1.8108 \\ 3.8337 \end{bmatrix} km/s$$

Hence, the state vector for the middle observation is

$$\mathbf{r}_2 = \begin{bmatrix} -10,378.257 \\ -15,955.205 \\ 14,212.351 \end{bmatrix} \text{km}, \quad \mathbf{v}_2 = \begin{bmatrix} 0.7610 \\ -1.8108 \\ 3.8337 \end{bmatrix} \text{km/s}$$

We can now compute the six orbital elements from this state vector using the methods of Section 3.4. However, the Gibbs method allows us to take a few short cuts. For example, parameter p can be computed from magnitudes N and D

$$p = \frac{N}{D} = 11,974.710 \text{ km}$$

Eccentricity is the ratio of S and D; see Eq. (3.82)

$$e = \frac{S}{D} = 0.7411$$

Semimajor axis can be computed from p and e

$$a = \frac{p}{1-e^2} = 26,565 \text{ km}$$

Referring back to Section 3.4, we see that these values of semimajor axis a and eccentricity e match the values for a Molniya orbit. To complete the orbit-determination process, we can use Eq. (3.8) to compute inclination:

$$\cos i = \frac{\mathbf{K} \cdot \mathbf{h}}{h}$$

where the angular momentum is $\mathbf{h} = \mathbf{r}_2 \times \mathbf{v}_2$. Carrying out the calculations, we find that $\cos i = 0.447759$, or $i = 63.4°$ which is the inclination of a Molniya orbit.

The last critical element is the argument of perigee as computed by Eq. (3.14)

$$\cos \omega = \frac{\mathbf{n} \cdot \mathbf{e}}{\|\mathbf{n}\| e}$$

We must compute the ascending node vector using Eq. (3.11)

$$\mathbf{n} = \mathbf{K} \times \mathbf{h}$$

and the eccentricity vector using Eq. (3.6) and state vector $(\mathbf{r}_2, \mathbf{v}_2)$

$$\mathbf{e} = \frac{1}{\mu} \left[\left(v_2^2 - \frac{\mu}{r_2} \right) \mathbf{r}_2 - (\mathbf{r}_2 \cdot \mathbf{v}_2) \mathbf{v}_2 \right]$$

The reader can carry out these vector manipulations to find that

$$\mathbf{n} = \begin{bmatrix} -50,603.214 \\ -35,432.759 \\ 0 \end{bmatrix} \quad \text{and} \quad \mathbf{e} = \begin{bmatrix} -0.1903 \\ 0.2718 \\ -0.6627 \end{bmatrix}$$

Hence, $\cos \omega = -1.45(10^{-6}) \approx 0$. The argument of perigee is either $+90°$ or $-90°$. Because the $\mathbf{K}$ component of the eccentricity vector $\mathbf{e}$ is negative, the perigee is south of the equatorial plane and the argument of perigee is $\omega = -90°$. We have now determined that the satellite is indeed in a Molniya orbit (the reader should note that the longitude of the ascending node Ω is not needed to define a specific Molniya orbit unless we are given the epoch time t_2 and requirements for the orbit's apogee to pass over a particular geographic longitude).

3.9 Survey of Orbit-Determination Methods

The two previous sections have presented two orbit-determination methods. The first method relied on six independent measurements at one time instant (three position and three time-rate measurements). The result was a one-to-one mapping from the six measurements to the six-element state vector. The second method (the Gibbs method) used three ECI position vectors taken at three observation times to obtain the ECI velocity vector corresponding to the middle observation.

Our discussion of orbit-determination methods has only scratched the surface because a multitude of other methods exist. Vallado [1; pp. 433–498], Bate *et al.* [2; pp. 117–131], and Tapley *et al.* [3] provide excellent, detailed descriptions of alternate orbit-determination methods. They also provide algorithms for various orbit-determination techniques. We conclude this chapter with a survey of these methods for solving the orbit-determination problem.

3.9.1 Orbit Determination Using Angles-Only Measurements

Early orbit-determination techniques relied on optical tracking with telescopes. A single observation consists of two angular measurements that define the LOS vector from the observation site to the satellite. Figure 3.19 shows the two LOS angles: the *right ascension* α and *declination* δ measured in the *topocentric-equatorial* coordinate system. The origin of the topocentric-equatorial system is fixed to the ground-based observation site

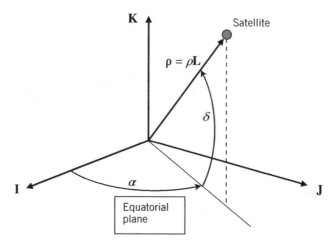

Figure 3.19 Line-of-sight angles measured in topocentric-equatorial system.

(like the topocentric-horizon or **SEZ** system) but its fundamental plane is the equatorial plane. Hence, we may think of the topocentric-equatorial system as the geocentric ECI system translated to the observation site's location on the Earth's surface. Right ascension α is measured in the equatorial plane from the inertial **I** axis (vernal equinox direction) to the projection of the LOS vector. Declination δ is measured from the equatorial plane to the LOS vector. Figure 3.19 shows that the observation site can compute the LOS unit vector **L** (in **IJK** coordinates) from angles α and δ

$$\mathbf{L} = \begin{bmatrix} \cos\delta \cos\alpha \\ \cos\delta \sin\alpha \\ \sin\delta \end{bmatrix} \tag{3.90}$$

The reader should note that **L** is the LOS unit vector to the satellite and is not the site-relative position vector $\boldsymbol{\rho}$. As the name implies, "angles-only" methods do not rely on range measurements (however, if range were available the relative position vector would be $\boldsymbol{\rho} = \rho\mathbf{L}$ as shown in Figure 3.19).

Because a single observation consists of two angular measurements (α_i, δ_i), a minimum of three "angles-only" measurements at three different epochs are required to provide six independent quantities needed for the orbit solution. Figure 3.20 shows three LOS unit vectors $\mathbf{L}_1$, $\mathbf{L}_2$, and $\mathbf{L}_3$ to a satellite at three observation times (again it is important for the reader to note that vectors $\mathbf{L}_i$ in Figure 3.20 are LOS unit vectors and not position vectors). Of course, the three LOS vectors can be computed using Eq. (3.90) with three sets of angular measurements (α_1, δ_1), (α_2, δ_2), and (α_3, δ_3).

Laplace developed the first angles-only orbit-determination method in 1780 and his method uses a series of successive differentiations of the LOS data to determine the state vector $(\mathbf{r}_2, \mathbf{v}_2)$ corresponding to the middle observation. To show the basis of Laplace's

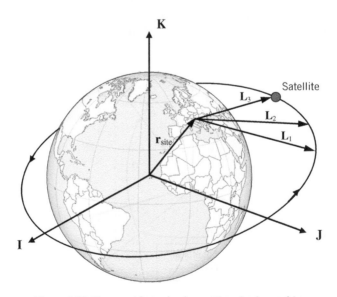

Figure 3.20 Three angles-only observations $\mathbf{L}_1$, $\mathbf{L}_2$, and $\mathbf{L}_3$.

method, we can express the satellite's geocentric position vector $\mathbf{r}$ in terms of LOS unit vector $\mathbf{L}$

$$\mathbf{r} = \mathbf{r}_{site} + \rho\mathbf{L} \tag{3.91}$$

where $\mathbf{r}_{site}$ is the ECI position vector of the observation site (see Figure 3.20). Of course, range ρ is not measured and is not known. Two time derivatives of Eq. (3.91) yield the two-body equation of motion $\ddot{\mathbf{r}} = -\mu\mathbf{r}/r^3$ as a function of the first and second time derivatives of $\mathbf{r}_{site}$, ρ, and $\mathbf{L}$. The first and second time derivatives of LOS vector $\mathbf{L}$ are obtained by differentiating interpolating formulas for $\mathbf{L}(t)$. Hence, we also require the "time tags" t_1, t_2, and t_3 associated with the three observations to compute $\dot{\mathbf{L}}$ and $\ddot{\mathbf{L}}$. By using the three LOS vectors $\mathbf{L}_i$ and their time tags to compute $\dot{\mathbf{L}}$ and $\ddot{\mathbf{L}}$, we are able to solve for the unknowns ρ, $\dot{\rho}$, and $\ddot{\rho}$ and eventually obtain the ECI state vector $(\mathbf{r}_2, \mathbf{v}_2)$ for the middle observation. Vallado [1; pp. 433–439] and Bate *et al.* [2; pp. 117–122] present the details of Laplace's angles-only method.

3.9.2 Orbit Determination Using Three Position Vectors

Radar and laser tracking stations can provide range information and this additional measurement (along with the LOS angles) allows the calculation of the site-relative position vector $\boldsymbol{\rho}$. Adding the station's geocentric position vector $\mathbf{r}_{site}$ to the relative position $\boldsymbol{\rho}$ (transformed to the ECI frame) produces the satellite's geocentric position vector $\mathbf{r}$. Section 3.8 presented the Gibbs method for orbit determination. The Gibbs method requires three sequential observations (three ECI position vectors $\mathbf{r}_1$, $\mathbf{r}_2$, and $\mathbf{r}_3$) along an orbital pass as inputs and obtains the orbit by enforcing a geometric constraint requiring all three vectors to be coplanar. As noted in Section 3.8, the Gibbs method becomes inaccurate and unreliable for closely spaced observations. The Herrick–Gibbs method alleviates this problem by computing the approximate velocity vector corresponding to the middle observation using a Taylor series expansion. Because an expansion is used for $\mathbf{r}(t)$, the Herrick–Gibbs method requires the three "time tags" (t_1, t_2, and t_3) associated with each observation. Furthermore, this method is most accurate for closely spaced observations (less than $3°$ apart), but produces erroneous results for widely spaced position vectors. Vallado [1; pp. 461–467] presents the details of the Herrick–Gibbs method.

3.9.3 Orbit Determination from Two Position Vectors and Time

The orbit can be determined from the knowledge of two position vectors ($\mathbf{r}_1$ and $\mathbf{r}_2$) and the time interval (or "time-of-flight") between the two observations. Figure 3.21 shows the orbit-determination scenario. Johann Heinrich Lambert first solved this problem in 1761; today we call it *Lambert's problem*. Because Lambert's problem relies on time-of-flight calculations between two positions in an orbit, we shall postpone a detailed discussion until Chapter 4.

Several solution techniques for Lambert's problem have been discovered and presented in the literature since Lambert's original formulation in 1761. Many solutions to Lambert's problem rely on computation of the so-called *Lagrangian coefficients* that relate the position and velocity vectors between two time instants:

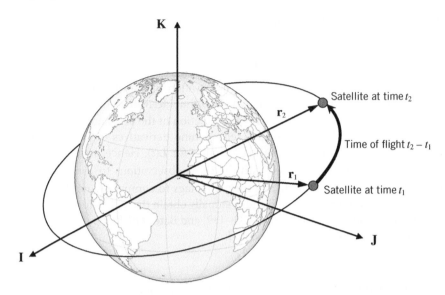

Figure 3.21 Orbit determination using two position vectors and flight time.

$$\mathbf{r}_2 = f\mathbf{r}_1 + g\mathbf{v}_1 \tag{3.92a}$$

$$\mathbf{v}_2 = \dot{f}\mathbf{r}_1 + \dot{g}\mathbf{v}_1 \tag{3.92b}$$

The four scalar coefficients $f, g, \dot{f}$, and $\dot{g}$ in Eq. (3.92) are the Lagrangian coefficients, and they are used to propagate the position and velocity vectors ahead in time. These four coefficients are expressed as functions of known values (position vectors $\mathbf{r}_1$ and $\mathbf{r}_2$, and flight time $t_2 - t_1$) and an unknown orbital parameter (such as semimajor axis a or parameter p). Iterating on the unknown orbital parameter determines the appropriate numerical values of $f, g, \dot{f}$, and $\dot{g}$. Detailed equations for the Lagrangian coefficients are derived and presented in Chapter 4 (Section 4.5). In Chapter 4 (Section 4.6), we present one iterative method for solving Lambert's problem.

Solving Lambert's problem (by some iterative technique) is a fundamental problem in orbital mechanics. The orbital solution to Lambert's problem has many important applications such as orbit determination, orbital maneuvering (Chapter 7), and interplanetary mission design (Chapter 10).

3.9.4 Statistical Orbit Determination

The orbit-determination methods discussed thus far compute the state vector $(\mathbf{r}, \mathbf{v})$ from a minimal set of observations that include LOS angles, position vectors, and time. However, a large number of observations are recorded during the operation of a satellite. The accuracy of the orbit solution is improved by considering all observations instead of a small subset (such as three sets of LOS angles or three position vectors). Carl Friedrich Gauss developed a method to determine the theoretical orbit passing through all observation data such that the sum of squares of the residuals (the difference between the actual observations and the theoretical orbit) is minimized. This technique, known as the method of least squares, is the basis of estimation theory. In 1959, Peter Swerling developed a sequential algorithm for orbit determination. This technique is based on the least-squares method and uses multiple satellite observations to estimate its orbit. In 1960, Rudolf E. Kalman developed a recursive algorithm that produces optimal estimates of a system's state vector over time. Kalman's method also utilizes a least-squares approach. These estimation algorithms use a series of measurements that are corrupted by sensor noise and other stochastic (random) inaccuracies that can be quantified by statistical parameters. This algorithm has come to be known as the *Kalman filter* and is used to provide the "best" estimate of a satellite's state vector (**r**,**v**) given a series of noisy observations. The Kalman filter also provides a statistical measure of confidence in the estimated state via the covariance matrix, which contains statistical information for the closeness of fit between the observations and the estimated orbit. Both the state estimate and covariance matrix are propagated ahead in time by the Kalman filter through a recursive scheme that does not require reprocessing all past measurement information. Therefore, the Kalman filter has been used to perform onboard navigation duties for spacecraft including the US Space Shuttle. Tapley *et al.* [3] provide an in-depth discussion of Kalman filtering and the statistical orbit-determination problem.

3.10 Summary

This chapter has presented the six constant *classical orbital elements* that completely define or determine a two-body orbit: (1) semimajor axis a; (2) eccentricity e; (3) inclination i; (4) longitude of the ascending node Ω; (5) argument of periapsis ω; and (6) true anomaly at epoch θ_0. Semimajor axis, eccentricity, and true anomaly were discussed in Chapter 2; the first two elements determine the size and shape of the conic section whereas true anomaly θ_0 pinpoints the satellite's position in the orbit at a particular time instant (or epoch). Orbital elements i, Ω, and ω determine the orbit's three-dimensional orientation in a Cartesian coordinate frame. The six classical orbital elements have a unique mapping with the satellite's position and velocity vectors expressed in an inertial Cartesian frame, that is, the so-called *state vector* $(\mathbf{r}_0, \mathbf{v}_0)$. The satellite's state vector $(\mathbf{r}_0, \mathbf{v}_0)$ at epoch t_0 can be transformed to the corresponding orbital elements $(a, e, i, \Omega, \omega, \theta_0)$ by computing the total energy, angular momentum vector, eccentricity vector, and ascending node vector. Similarly, we can transform the six orbital elements to the six-dimensional state vector in the Cartesian frame. It is very important for the reader to note that a two-body orbit is *uniquely defined* by the six classical orbital elements *or* the six-dimensional state vector $(\mathbf{r}_0, \mathbf{v}_0)$.

The classical orbital elements allow us to visualize the size and shape of an orbit and the orientation of its orbital plane and periapsis direction. These characteristics offer a distinct advantage over the Cartesian state vector $(\mathbf{r}_0, \mathbf{v}_0)$. However, the orbit-determination process begins by obtaining the satellite's position and velocity vectors from observational data derived from ground-based sensor measurements. In this chapter, we presented two orbit-determination methods. The first method derives the state vector $(\mathbf{r}_0, \mathbf{v}_0)$ from six simultaneous ground-based measurements (three position coordinates and three time-rate coordinates). The second method determines the state vector by using three position vectors taken at three observation times. Finally, we ended this chapter with a brief survey of orbit-determination methods that use a combination of observations that include position vectors, LOS angles, and observation times.

References

1 Vallado, D.A., *Fundamentals of Astrodynamics and Applications*, 4th edn, Microcosm Press, Hawthorne, CA, 2013.
2 Bate, R.R., Mueller, D.D., and White, J.E., *Fundamentals of Astrodynamics*, Dover, New York, 1971.
3 Tapley, B.D., Schutz, B.E., and Born, G.H., *Statistical Orbit Determination*, Elsevier, Oxford, UK, 2004.

Problems

Conceptual Problems

3.1 An Earth-orbiting satellite has the following position and velocity vectors in the ECI frame

$$\mathbf{r} = \begin{bmatrix} -6,796 \\ 4,025 \\ 3,490 \end{bmatrix} \text{km}, \qquad \mathbf{v} = \begin{bmatrix} -3.7817 \\ -6.0146 \\ 1.1418 \end{bmatrix} \text{km/s}$$

Determine the six classical orbital elements.

3.2 A geocentric satellite has the following ECI position and velocity vectors at epoch t_0

$$\mathbf{r} = 8,207\mathbf{I} + 7,114\mathbf{J} + 5,253\mathbf{K} \text{ km}, \qquad \mathbf{v} = -3.0347\mathbf{I} + 2.7144\mathbf{J} + 3.2901\mathbf{K} \text{ km/s}$$

Determine the classical orbital elements.

3.3 A geocentric satellite has the following ECI position and velocity vectors at epoch t_0

$$\mathbf{r} = 6,93\mathbf{I} + 5,696\mathbf{J} - 4,586\mathbf{K} \text{ km}, \qquad \mathbf{v} = -7.8456\mathbf{I} - 2.0905\mathbf{J} - 2.1124\mathbf{K} \text{ km/s}$$

Determine the classical orbital elements.

3.4 An Earth-orbiting satellite has the following ECI position and velocity vectors at epoch t_0

$$\mathbf{r} = 6,678\mathbf{J} \text{ km}, \qquad \mathbf{v} = -7.725835\mathbf{I} \text{ km/s}$$

Determine the semimajor axis a, eccentricity e, and inclination i. In addition, show that longitude of the ascending node Ω and argument of perigee ω are undefined and compute an appropriate angle that defines the location of the satellite in its orbit.

3.5 A geocentric satellite has the following ascending node and eccentricity vectors expressed in the ECI frame:

$$\mathbf{n} = -19,858\mathbf{I} - 9,260\mathbf{J}, \qquad \mathbf{e} = 0.0854\mathbf{I} - 0.0441\mathbf{J} + 0.0277\mathbf{K}$$

When the satellite is at perigee, its position vector in the ECI frame is

$$\mathbf{r} = 7,993\mathbf{I} - 4,124\mathbf{J} + 2,590\mathbf{K} \text{ km}$$

Determine:
a) Semimajor axis, a
b) Velocity at perigee, v_p
c) Longitude of the ascending node, Ω
d) Argument of perigee, ω.

3.6 An Earth-orbiting satellite has semimajor axis $a = 12,400$ km and eccentricity $e = 0.14$. Determine its position and velocity vectors in the perifocal frame for true anomaly $\theta = 200°$.

3.7 A geocentric satellite has orbital elements $i = 28.5°$, $\Omega = 152°$, and $\omega = 180°$. Determine its position and velocity vectors in the perifocal frame if it is currently at apogee with an altitude of 2,337 km and inertial velocity of 6.592 km/s.

3.8 A reconnaissance satellite is in an elliptical orbit about the Earth with a period of 717.8 min, perigee altitude of 2,052 km, and inclination of 63.0°. Its longitude of the ascending node and argument of perigee are 116.2° and 270°, respectively. Determine the ECI state vector $(\mathbf{r}, \mathbf{v})$ of the satellite when its true anomaly is 200°. Use the magnitudes of the $\mathbf{r}$ and $\mathbf{v}$ vectors to compute the orbital energy, semimajor axis, and orbital period of the satellite in order to verify your calculations.

3.9 A geocentric satellite has the following orbital elements:

$$a = 9,056 \text{ km}$$
$$e = 0.142$$
$$i = 7.2°$$
$$\Omega = 200°$$
$$\omega = 60°$$
$$\theta_0 = 320°$$

Determine:
a) The satellite's position and velocity vectors in the perifocal frame at this instant.
b) The satellite's state vector $(\mathbf{r}, \mathbf{v})$ at this epoch in ECI coordinates.

3.10 Prove that any geocentric satellite with an argument of perigee $\omega = \pm 90°$ that is currently at perigee or apogee must have a zero $\mathbf{K}$-axis velocity component.

3.11 At time t_0, an Earth-orbiting satellite has position vector $\mathbf{r}_0 = 7,643\mathbf{K}$ km in an ECI coordinate frame. Prove that the satellite *must* be in polar orbit.

3.12 At a particular instant, an Earth-orbiting satellite has position vector $\mathbf{r} = 7{,}705\mathbf{I} + 4448\mathbf{J}$ km in ECI coordinates and its inertial velocity vector has azimuth angle $\beta = 120°$. Determine the satellite's inclination and longitude of the ascending node.

3.13 A geocentric satellite has the following orbital elements: $a = 6{,}778$ km, $e = 0$, $i = 54°$, $\Omega = 78°$, and argument of latitude at epoch $u_0 = 180°$. Compute the azimuth angle β corresponding to the satellite's velocity vector at this instant.

3.14 A satellite is in a circular orbit about the Earth. At a particular instant, the azimuth (or heading) angle of the satellite's inertial velocity vector is $\beta = 102°$ and the satellite is directly over Columbia, Missouri (longitude = $92.33°$ W, latitude = $38.95°$ N). Compute the maximum and minimum latitudes of the satellite's ground track.

3.15 A satellite is in an inclined circular orbit about the Earth. The geographic longitude of the satellite's equatorial crossing on an ascending arc for its 126th orbit is $\lambda_{126} = 56.31°$ E. The geographic longitude of its 128th ascending-arc equatorial crossing is $\lambda_{128} = 31.76°$ W. Determine the altitude of the circular orbit.

3.16 An engineer receives four position vectors in the ECI frame:

$$\mathbf{r}_1 = \begin{bmatrix} 1{,}044.1 \\ 7{,}539.6 \\ 1{,}731.8 \end{bmatrix} \text{ km, } \mathbf{r}_2 = \begin{bmatrix} -229.6 \\ 7{,}389.3 \\ 2{,}491.0 \end{bmatrix} \text{ km, } \mathbf{r}_3 = \begin{bmatrix} -577.6 \\ 7{,}540.5 \\ 1{,}908.7 \end{bmatrix} \text{ km, } \mathbf{r}_4 = \begin{bmatrix} -3{,}658.7 \\ 6{,}553.0 \\ 2{,}002.8 \end{bmatrix} \text{ km}$$

Three of these position vectors represent Satellite A at three instants of time while one position vector represents Satellite B. Which is the position vector for Satellite B?

3.17 A ground-based tracking station determines three ECI position vectors for a geocentric satellite:

$$\mathbf{r}_1 = \begin{bmatrix} -2{,}858.9 \\ -6{,}310.2 \\ -204.3 \end{bmatrix} \text{ km, } \mathbf{r}_2 = \begin{bmatrix} -1{,}401.4 \\ -6{,}832.1 \\ -918.7 \end{bmatrix} \text{ km, } \mathbf{r}_3 = \begin{bmatrix} -46.1 \\ -6{,}978.8 \\ -1{,}508.8 \end{bmatrix} \text{ km}$$

Use Gibbs' method to determine the satellite's inertial velocity $\mathbf{v}_2$ associated with the middle observation.

MATLAB Problems

3.18 Write an M-file that will calculate the six orbital elements $(a, e, i, \Omega, \omega, \theta_0)$ given the satellite's current geocentric state vector $(\mathbf{r}_0, \mathbf{v}_0)$ as the input. The input position and velocity vectors are in the ECI frame with units of kilometers and kilometers per second, respectively. The semimajor axis a output should have units of kilometers, while all output angles should have units of degrees. If the orbit is equatorial, then the M-file should return an empty set for longitude of the ascending node; that is, $\Omega = [\,]$. For elliptical equatorial orbits, the M-file should return the longitude of perigee (ϖ) in place of the argument of perigee. For inclined circular orbits, the M-file should return the argument of latitude (u) in place of true anomaly θ_0 (in this case the argument of perigee is undefined, i.e., ω = [ ]). Finally, for circular equatorial orbits set Ω = [ ] and ω = [ ] (undefined) and replace true anomaly with true longitude at epoch l_0. Test your M-file by solving Problem 3.1.

3.19 Write an M-file that will calculate the geocentric state vector (**r**,**v**) given the six orbital elements ($a, e, i, \Omega, \omega, \theta$) as inputs. The input semimajor axis a is in kilometers and all four input angles are in degrees. The output position and velocity vectors are in the ECI frame with units of kilometers and kilometers per second, respectively. Test your M-file by solving Problem 3.9.

3.20 Write an M-file that computes the satellite's state vector (**r**,**v**) given the six measurements in the topocentric-horizon (SEZ) frame: line-of-sight (LOS) range ρ, elevation angle σ, azimuth angle β, and the three time derivatives of these parameters. LOS range should be in units of kilometers, the two input angles are in degrees, and the three input rates should be in kilometers per second and degrees per second, respectively. Other inputs to M-file include geodetic latitude and longitude of the ground-based tracking station (in degrees), altitude of the tracking station above sea level (in m), and a flag that allows the user to choose between the spherical or ellipsoidal Earth model. The six outputs should be the satellite's **r** and **v** vectors expressed in the ECI frame in kilometers and kilometers per second, respectively. Test your M-file by solving Examples 3.4 and 3.5.

3.21 Write an M-file that performs Gibbs' method of orbit determination. The three inputs are three successive satellite position vectors $\mathbf{r}_1$, $\mathbf{r}_2$, and $\mathbf{r}_3$ expressed in the ECI frame with units of kilometers. The single output is the satellite's inertial velocity vector $\mathbf{v}_2$ associated with the middle observation (expressed in the ECI frame with units of kilometers per second). Test your M-file by solving Example 3.6.

Mission Applications

Problems 3.22 and 3.23 involve the Stardust capsule, which returned to Earth in January 2006 on a hyperbolic approach trajectory after sampling particles from the comet Wild-2.

3.22 A ground-based tracking station determined the Stardust capsule's position and velocity vectors (in the ECI frame) as it approached Earth:

$$\mathbf{r} = \begin{bmatrix} 219{,}469 \\ 99{,}139 \\ -87{,}444 \end{bmatrix} \text{km,} \quad \mathbf{v} = \begin{bmatrix} -5.7899 \\ -2.2827 \\ 2.3722 \end{bmatrix} \text{km/s}$$

Determine the six orbital elements of the Stardust capsule's orbit at this epoch.

3.23 A ground-based tracking station determines the three ECI position vectors of the Stardust capsule during its return to Earth:

$$\mathbf{r}_1 = \begin{bmatrix} 208,668.2 \\ 94,880.1 \\ -83,019.1 \end{bmatrix} km, \quad \mathbf{r}_2 = \begin{bmatrix} 63,942.2 \\ 37,491.1 \\ -23,787.6 \end{bmatrix} km, \quad \mathbf{r}_3 = \begin{bmatrix} 35,211.9 \\ 25,745.8 \\ -12,098.5 \end{bmatrix} km$$

Use Gibbs' method to determine the semimajor axis, eccentricity, and (theoretical) perigee altitude of the Stardust capsule (note that the Stardust capsule entered the Earth's atmosphere before it reached its perigee altitude).

3.24 The orbital elements of the Hubble Space Telescope (HST) at a particular epoch are

$$a = 6,922.3 \text{ km}, \quad e = 0.001143, \quad i = 50.75°$$
$$\Omega = 193.89°, \quad \omega = 85.41°, \quad \theta = 10.23°$$

Determine the state vector ($\mathbf{r}, \mathbf{v}$) of the HST (in ECI coordinates) at this instant.

Problems 3.25–3.27 involve the Lunar Atmosphere and Dust Environment Explorer (LADEE) spacecraft, which was launched in September 2013 and was eventually placed in an orbit about the moon.

3.25 A ground-tracking station determines the position and velocity vectors of the LADEE spacecraft on the first leg of its orbit transfer to the moon (see Figure 2.14 and Example 2.6). These vectors (in the ECI frame) are

$$\mathbf{r} = \begin{bmatrix} -201,283 \\ -107,796 \\ -37,306 \end{bmatrix} km, \quad \mathbf{v} = \begin{bmatrix} -0.6325 \\ -0.5385 \\ 0.1252 \end{bmatrix} km/s$$

a) Prove that the LADEE spacecraft is between perigee and apogee at this instant of time.
b) Determine the six classical orbital elements of LADEE's orbit at this epoch.

3.26 When the LADEE spacecraft reached the vicinity of the moon, it was placed in a highly elliptical lunar orbit by firing its onboard rocket. Some time after the orbit-insertion burn, LADEE's orbital elements (measured relative to a Cartesian *moon-centered inertial frame*) were determined to be

$$a = 9,732 \text{ km}, \quad e = 0.7803, \quad i = 15.7°$$
$$\Omega = 137.7°, \quad \omega = 324.3°, \quad \theta = 183.3°$$

Determine the state vector ($\mathbf{r}, \mathbf{v}$) of the LADEE spacecraft in a moon-centered inertial coordinate system (let $\mathbf{I}_m \mathbf{J}_m \mathbf{K}_m$ be unit vectors along the moon-centered inertial frame).

3.27 After coasting in a highly elliptical lunar orbit for three revolutions, the LADEE spacecraft again fired its onboard rocket to decrease its orbital energy and eccentricity. The resulting orbital elements (measured relative to a Cartesian moon-centered inertial frame) after the rocket burn are

$$a = 2{,}953 \text{ km}, \quad e = 0.2761, \quad i = 15.7°$$
$$\Omega = 137.7°, \qquad \omega = 324.3°, \quad \theta = 0°$$

Determine the state vector $(\mathbf{r}, \mathbf{v})$ of the LADEE spacecraft in a moon-centered inertial coordinate system (let $\mathbf{I}_m \mathbf{J}_m \mathbf{K}_m$ be unit vectors along the moon-centered inertial frame).

3.28 A French military micro-satellite is part of its early warning missile-detection system. The satellite has perigee and apogee altitudes of 434 and 17,430 km, respectively, an inclination of 2.0°, a longitude of the ascending node of 190.6°, and an argument of perigee of −89.4°. Determine the satellite's state vector $(\mathbf{r}_0, \mathbf{v}_0)$ in ECI coordinates at the epoch when the satellite is at apogee.

3.29 The Baikonur Cosmodrome launch facility in Kazakhstan has a latitude of 45.9° N and longitude of 63.3° E. It has been used to launch Proton rockets that have delivered payloads to the International Space Station (ISS). The ISS has an orbital inclination of 51.65°.
a) Determine the Proton rocket's azimuth angle if the engine cutoff conditions after launch result in a descending arc.
b) Compute the azimuth angle of the velocity vector of the ISS as it crosses the equator on a descending arc (moving from north to south).

Problems 3.30–3.32 involve the Very Long Baseline Array (VLBA) tracking station that is part of the Mauna Kea Observatory in Hawaii. The VLBA is located at latitude $\phi = 19.8°$ N.

3.30 At instant t_0, the longitude of the VLBA as measured from the inertial **I** axis is 283.5°. The VLBA tracks a satellite and determines its line-of-sight range and elevation and azimuth angles at epoch t_0:

$$\rho = 1{,}298.4 \text{ km}, \quad \sigma = 62.7°, \quad \beta = 158.2°$$

Determine the inertial position vector of the satellite at instant t_0 in ECI coordinates. Use a spherical Earth model.

3.31 The VLBA station is tracking a high-altitude satellite. At instant t_0, it measures the line-of-sight range and the elevation and azimuth angles:

$$\rho = 36{,}669.78 \text{ km}, \quad \sigma = 57.037°, \quad \beta = 227.915°$$

The values of ρ, σ, and β are constant (i.e., relative velocity vector $\dot{\boldsymbol{\rho}}_{SEZ}$ is zero). The VLBA is at longitude $\lambda = 32.1°$ at t_0. Show that the VLBA station is currently tracking a satellite in a geostationary-equatorial orbit (GEO). Assume that the Earth is a perfect sphere.

3.32 The VLBA station is tracking the Hubble Space Telescope (HST). At epoch t_0, the VLBA station measures the line-of-sight range, elevation and azimuth angles, and their rates:

$$\rho = 807.7 \text{ km}, \qquad \sigma = 44.4°, \qquad \qquad \beta = 148.5°$$

$$\dot{\rho} = 0.4027 \text{ km/s}, \quad \dot{\sigma} = -0.0331 \text{ deg/s}, \quad \dot{\beta} = -0.703 \text{ deg/s}$$

The longitude of the VLBA as measured from the inertial **I** axis is 120.4° at epoch t_0. Determine the state vector of the HST at this epoch in the ECI coordinate system using an ellipsoidal Earth model (the VLBA is at an altitude of 3,719 m above sea level). In addition, compute the orbital period, eccentricity, and inclination of the HST from this single observation.

4

Time of Flight

4.1 Introduction

The previous chapters dealt with the fundamental characteristics of a two-body orbit (energy and angular momentum). In addition, we showed how to determine a satellite's radius and velocity given its angular position (true anomaly) in the orbit. Chapter 3 focused on orbit determination; that is, how to compute the orbital elements given a set of observations or measurements. However, we chose to present the orbit-determination methods that *did not* rely on the elapsed time or "flight time" between observations.

This chapter introduces and discusses the "time of flight" (TOF); that is, computation of the transit time between two positions in an orbit. Kepler determined the fundamental relationship between orbital position and TOF. This relationship is known as *Kepler's equation* and it serves as the starting point for this chapter. Kepler's equation allows us to *propagate* a two-body orbit, or predict a satellite's orbital position at an arbitrary time in the future. Finally, the ability to relate TOF to orbital position leads to additional orbit-determination methods as we shall see in the latter part of this chapter when we present Lambert's problem.

4.2 Kepler's Equation

It is possible to derive an expression that relates TOF and orbital position by using either analytical techniques (e.g., calculus) or geometrical methods. Kepler developed a TOF expression by using a geometric approach (of course, calculus did not exist in Kepler's time). We will present the basis of Kepler's geometric TOF derivation here but we will eventually develop a flight-time equation using integral calculus. Of course, both methods lead to the same result: Kepler's equation for TOF.

4.2.1 Time of Flight Using Geometric Methods

Recall that in Chapter 2 we showed that the rate of area swept out by the radius vector is a constant:

$$\frac{dA}{dt} = \frac{h}{2} \tag{4.1}$$

Space Flight Dynamics, First Edition. Craig A. Kluever.
© 2018 John Wiley & Sons Ltd. Published 2018 by John Wiley & Sons Ltd.
Companion website: www.wiley.com/go/Kluever/spaceflightmechanics

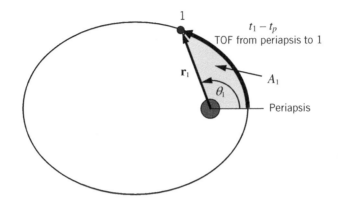

Figure 4.1 Swept area A_1 after time-of-flight from periapsis to position 1.

Hence, the ratio of "swept area" and the corresponding TOF is always equal to the constant $h/2$. Referring to Figure 4.1, we can form the ratio of swept area A_1 and the TOF $t_1 - t_p$ and equate it to the ratio of *total* area of an ellipse A_{ellipse} and orbital period T_{period}

$$\frac{h}{2} = \frac{A_1}{t_1 - t_p} = \frac{A_{\text{ellipse}}}{T_{\text{period}}} \tag{4.2}$$

Solving Eq. (4.2) for the TOF during the sweep of area A_1, we obtain

$$t_1 - t_p = \frac{A_1}{A_{\text{ellipse}}} T_{\text{period}} \tag{4.3}$$

In Eq. (4.3), $t_1 - t_p$ is the satellite's transit time from periapsis to position 1 on the elliptical orbit shown in Figure 4.1. Clearly, the area ratio A_1/A_{ellipse} is less than one and therefore the TOF $t_1 - t_p$ is less than one period. It should also be clear to the reader that determining the swept area A_1 in Figure 4.1 establishes the TOF from periapsis to point 1.

Kepler defined an auxiliary angle that allows the calculation of the swept area A_1 in Figure 4.1. Figure 4.2 shows an elliptical orbit circumscribed with an "auxiliary circle" with radius a where the center of the circle is point O (intersection of the major and minor axes). The attracting body is at point F (the focus) and the satellite's position in the orbit is point Y (its angular position is denoted by true anomaly θ). If we extend a vertical line (perpendicular to the major axis) through the satellite's position Y, this vertical line intersects the auxiliary circle at point X as seen in Figure 4.2. The angle E in Figure 4.2 is the *eccentric anomaly* and is measured from the major axis in the direction of motion in the same manner as true anomaly θ. Kepler used eccentric anomaly E to compute the swept area A_1 (the reader can likely see how area A_1 in Figure 4.2 can be calculated from the circular sector and triangular areas). Bate *et al.* [1; pp. 182–185] and Vallado [2; pp. 43–45] present the geometrical TOF derivation in terms of swept area A_1.

4.2.2 Time of Flight Using Analytical Methods

Our analytical TOF derivation begins with the cosine of eccentric anomaly E. Referring again to Figure 4.2, we see that $\cos E$ is

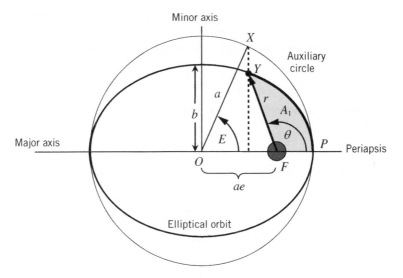

Figure 4.2 Eccentric anomaly, *E*, for an elliptical orbit.

$$\cos E = \frac{ae + r\cos\theta}{a} \tag{4.4}$$

Multiplying Eq. (4.4) by *e/e*, we obtain

$$\cos E = \frac{ae^2 + re\cos\theta}{ae} \tag{4.5}$$

The trajectory equation (2.45) provides the position of the satellite

$$r = \frac{p}{1 + e\cos\theta} = \frac{a(1 - e^2)}{1 + e\cos\theta} \tag{4.6}$$

Equation (4.6) shows that $r + re\cos\theta = a - ae^2$, or $ae^2 + re\cos\theta = a - r$. Using this last result to replace the numerator of Eq. (4.5), we obtain

$$\cos E = \frac{a - r}{ae} \tag{4.7}$$

Next, take the time derivative of Eq. (4.7) to yield

$$-ae\dot{E}\sin E = -\dot{r} \tag{4.8}$$

Of course, we remember that orbital elements *a* and *e* are constants. Recall that in Chapter 2 we developed an expression for the satellite's radial velocity component

$$\dot{r} = \frac{\mu}{h}e\sin\theta \tag{4.9}$$

where angular momentum is $h = \sqrt{\mu p} = \sqrt{\mu a(1 - e^2)}$. Using Eq. (4.9) in Eq. (4.8) (and canceling *e* from both sides), we obtain

$$a\dot{E}\sin E = \frac{\mu}{\sqrt{\mu a(1 - e^2)}}\sin\theta \tag{4.10}$$

We are almost at the point where we can separate variables and integrate Eq. (4.10). However, true anomaly θ is a function of eccentric anomaly E. Referring again to Figure 4.2, we can write an expression for the length of the dashed vertical line:

$$a \sin E = \left(\frac{a}{b}\right) r \sin \theta \tag{4.11}$$

where the ellipse-to-circle vertical ratio is $b/a = \sqrt{1 - e^2}$. Next, solve Eq. (4.7) for the satellite's radial position

$$r = a(1 - e \cos E) \tag{4.12}$$

Using these substitutions, we can solve Eq. (4.11) for the sine of true anomaly

$$\sin \theta = \frac{\sqrt{1 - e^2} \sin E}{1 - e \cos E} \tag{4.13}$$

Substituting Eq. (4.13) into Eq. (4.10) (and some algebra) yields

$$\dot{E} = \frac{dE}{dt} = \sqrt{\frac{\mu}{a^3}} \frac{1}{1 - e \cos E} \tag{4.14}$$

Separating variables (and placing dt on the left-hand side), Eq. (4.14) becomes

$$dt = \sqrt{\frac{a^3}{\mu}} (1 - e \cos E) dE \tag{4.15}$$

Equation (4.15) is easily integrated from periapsis to the satellite's orbital position. Therefore, the lower integration bounds at periapsis are time t_p and eccentric anomaly $E_p = 0$. Integrating Eq. (4.15), we obtain

$$t_1 - t_p = \sqrt{\frac{a^3}{\mu}} (E - e \sin E) \tag{4.16}$$

Equation (4.16) is *Kepler's equation* and it determines the TOF from periapsis to the satellite's current position (position 1 in Figure 4.1) as a function of semimajor axis a, eccentricity e, and eccentric anomaly E. Kepler introduced another angle called the *mean anomaly*, M, defined as the parenthetical term in the TOF equation (4.16):

$$M = E - e \sin E \tag{4.17}$$

The reader should note that mean anomaly M is indeed an angle (with units of radians) but it is *defined* by the nonlinear function (4.17). Hence, mean anomaly is not a physical angle that can be visualized in any way. Using mean anomaly, we express Eq. (4.16) as

$$M = n(t_1 - t_p) = E - e \sin E \tag{4.18}$$

Equation (4.18) is another form of Kepler's equation. The dimensions of Eq. (4.18) are easy to identify: angle M (in rad) is the product of the constant angular rate n (in rad/s) and TOF $t_1 - t_p$ (in s). Although we cannot visualize mean anomaly M, Figure 4.2 and Eq. (4.17) show that at periapsis $\theta_p = E_p = M_p = 0$ and at apoapsis $\theta_a = E_a = M_a = 180°$. The angular velocity term n in Eq. (4.18) is called the *mean motion*. Comparing Eqs. (4.16) and (4.18), we see that mean motion is

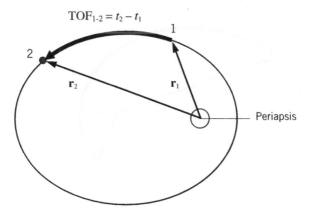

Figure 4.3 Time of flight from position 1 to position 2.

$$n = \sqrt{\frac{\mu}{a^3}} \tag{4.19}$$

The constant mean motion n is the time rate of mean anomaly, or $n = dM/dt$. One way to think of the mean motion n is to divide one orbital revolution ($dM = 2\pi$) by the orbital period ($dt = T_{\text{period}}$):

$$n = \frac{dM}{dt} = \frac{2\pi}{T_{\text{period}}} = \frac{2\pi}{2\pi a^{3/2}/\sqrt{\mu}} = \sqrt{\frac{\mu}{a^3}} \tag{4.20}$$

This simple calculation verifies Eq. (4.19).

Equation (4.16) or (4.18) determines the TOF from periapsis to a prescribed angular position defined by eccentric anomaly E. Figure 4.3 shows a more general case where we desire the elapsed TOF between positions 1 and 2. Using Kepler's equation (4.18), the TOF from position 1 to position 2 is

$$\text{TOF}_{1\text{-}2} = t_2 - t_1 = \frac{1}{n}(M_2 - M_1) \tag{4.21}$$

where the two mean anomalies are $M_1 = E_1 - e\sin E_1$ and $M_2 = E_2 - e\sin E_2$, respectively. Here we use $\text{TOF}_{1\text{-}2}$ to denote TOF from position 1 to position 2. One way to interpret Eq. (4.21) is the difference in the elapsed times from periapsis to each respective position, that is,

$$\text{TOF}_{1\text{-}2} = (t_2 - t_p) - (t_1 - t_p) = t_2 - t_1 \tag{4.22}$$

Note that Eq. (4.21) results in the orbital period when $M_1 = 0$ (periapsis) and $M_2 = 2\pi$ (one revolution):

$$\text{One revolution: } \text{TOF}_{1\text{-}2} = \frac{1}{\sqrt{\mu/a^3}}(2\pi - 0) = \frac{2\pi}{\sqrt{\mu}}a^{3/2} = T_{\text{period}} \tag{4.23}$$

In order to use Kepler's equation in the form of Eq. (4.21) with $t_2 > t_1$, mean anomaly M_2 must be greater than mean anomaly M_1. Figure 4.4 shows a counter scenario where

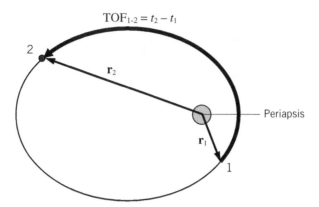

Figure 4.4 Time of flight with passage through periapsis.

$M_2 < M_1$: the initial eccentric anomaly E_1 (and hence M_1) is greater than 180° (when expressed as a positive angle) and the terminal eccentric anomaly E_2 (and M_2) is less than 180°. For scenarios where the satellite passes through periapsis en route from position 1 to 2, the simple solution is to add 2π to mean anomaly M_2 so that it is greater than M_1.

4.2.3 Relating Eccentric and True Anomalies

Kepler's TOF equation requires the eccentric anomaly E. True anomaly θ is a known quantity if we have knowledge of the six orbital elements or the position and velocity vectors $\mathbf{r}$ and $\mathbf{v}$. Therefore, we must establish the relationship between true anomaly θ and eccentric anomaly E. Let us repeat Eq. (4.4), the cosine of E

$$\cos E = \frac{ae + r\cos\theta}{a}$$

Substituting Eq. (4.12) for radial position r, we obtain

$$\cos E = \frac{ae + a(1 - e\cos E)\cos\theta}{a} = e + \cos\theta - e\cos E\cos\theta \tag{4.24}$$

Solving Eq. (4.24) for cosine of E yields

$$\cos E = \frac{e + \cos\theta}{1 + e\cos\theta} \tag{4.25}$$

We cannot rely on Eq. (4.25) alone to compute E because the inverse cosine operation always places the angle in the first or second quadrant, that is, between periapsis and apoapsis. The sine of eccentric anomaly can be calculated using Eq. (4.11); substituting $b/a = \sqrt{1 - e^2}$ and the trajectory equation $r = a(1 - e^2)/(1 + e\cos\theta)$ into Eq. (4.11), we obtain

$$a\sin E = \frac{1}{\sqrt{1 - e^2}}\frac{a(1 - e^2)}{1 + e\cos\theta}\sin\theta \tag{4.26}$$

Simplifying Eq. (4.26) yields

$$\sin E = \frac{\sqrt{1-e^2}\,\sin\theta}{1+e\cos\theta} \tag{4.27}$$

The reader should carefully and correctly use Eqs. (4.25) and (4.27) to determine the proper quadrant for E. Recall that the inverse cosine function on a calculator always places the angular solution in the range $[0, \pi]$ (first and second quadrants) while the inverse sine function places the solution in $[-\pi/2, \pi/2]$ (first and fourth quadrants). As a quick example, consider the case where $\cos E = -0.8660$ and $\sin E = -0.5$. Because $\cos E$ is negative, the eccentric anomaly is in the second or third quadrant, or $90° < E < 270°$. Because $\sin E$ is negative, E is in the third or fourth quadrant, or $180° < E < 360°$. Hence, the third quadrant satisfies both $\cos E < 0$ and $\sin E < 0$ and therefore the correct eccentric anomaly is $E = 210°$. When using computer programs such as MATLAB, the $\texttt{atan2}$ function will determine the correct quadrant for E given $\sin E$ and $\cos E$ as the two inputs.

As we shall soon see, there is a need to determine true anomaly θ from eccentric anomaly E. We can multiply both sides of Eq. (4.25) by the denominator term $1 + e\cos\theta$ to obtain

$$\cos E(1 + e\cos\theta) = e + \cos\theta \tag{4.28}$$

Solving for cosine of true anomaly yields

$$\cos\theta = \frac{\cos E - e}{1 - e\cos E} \tag{4.29}$$

We have already derived an expression for the sine of true anomaly; Eq. (4.13) is repeated below:

$$\sin\theta = \frac{\sqrt{1-e^2}\,\sin E}{1 - e\cos E} \tag{4.13}$$

Equations (4.29) and (4.13) establish the proper quadrant for true anomaly θ given eccentric anomaly E.

It is possible to develop a single equation that converts θ to E (and vice versa) without any quadrant ambiguity. To begin, write the tangent half-angle formula and use Eqs. (4.29) and (4.13) to substitute for $\cos\theta$ and $\sin\theta$

$$\tan\frac{\theta}{2} = \frac{\sin\theta}{1+\cos\theta} = \frac{\frac{\sqrt{1-e^2}\sin E}{1-e\cos E}}{1+\frac{\cos E - e}{1-e\cos E}} \tag{4.30}$$

$$= \frac{\sqrt{1-e^2}\,\sin E}{1 - e\cos E + \cos E - e} \tag{4.31}$$

Because the denominator of Eq. (4.31) is $(1-e)(1+\cos E)$, the right-hand side of Eq. (4.31) includes the tangent of $E/2$

$$\tan\frac{\theta}{2} = \sqrt{\frac{1+e}{1-e}}\,\tan\frac{E}{2} \tag{4.32}$$

Of course, we can solve Eq. (4.32) for the tangent half-angle of eccentric anomaly:

$$\tan\frac{E}{2} = \sqrt{\frac{1-e}{1+e}}\tan\frac{\theta}{2} \qquad (4.33)$$

Equations (4.32) and (4.33) can be used to convert between true and eccentric anomalies without a quadrant check. Because Eqs. (4.32) and (4.33) both require a single inverse-trigonometric calculator operation, these equations are likely the preferred option for converting eccentric anomaly to (or from) true anomaly. The reader should note, however, that the inverse tangent operation places the angular solution in $[-\pi/2, \pi/2]$. Therefore, when either true or eccentric anomaly is in the third or fourth quadrant, the inverse-tangent "calculator" operation using either Eq. (4.32) or (4.33) will result in a negative angle between $-\pi$ and 0 rad.

Example 4.1   Figure 4.5 shows a Molniya orbit; a 12-h, highly eccentric orbit used by Russia for communication satellites. Perigee and apogee altitudes of the Molniya orbit are 500 km and 39,873 km, respectively. Determine the satellite's TOF from perigee to an altitude of 30,000 km for two cases: (a) Case 1: flight-path angle $\gamma > 0$ at 30,000 km altitude; and (b) Case 2: flight-path angle $\gamma < 0$ at 30,000 km altitude.

Kepler's equation requires knowledge of mean motion n (which is determined from semimajor axis) and eccentricity. The orbit's perigee and apogee radii are

$$r_p = 500\,\text{km} + R_E = 6{,}878\,\text{km} \qquad \text{and} \qquad r_a = 39{,}873\,\text{km} + R_E = 46{,}251\,\text{km}$$

where $R_E = 6{,}378$ km. The semimajor axis of the Molniya orbit is

$$a = \frac{1}{2}(r_p + r_a) = \frac{1}{2}(6{,}878\,\text{km} + 46{,}251\,\text{km}) = 26{,}564.5\,\text{km}$$

The mean motion is $n = \sqrt{\mu/a^3} = 1.4582(10^{-4})$ rad/s
Eccentricity is

$$e = \frac{r_a - r_p}{r_a + r_p} = 0.7411$$

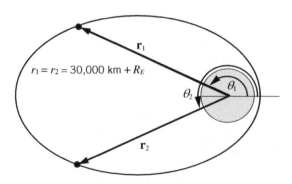

Figure 4.5 Molniya orbit (Example 4.1).

a) Case 1: positive flight-path angle at 30,000 km altitude

Because flight-path angle is positive, the satellite has positive radial velocity and is between perigee and apogee ($\mathbf{r_1}$ in Figure 4.5). We can use the trajectory equation to solve for the cosine of true anomaly, θ_1

$$r = \frac{p}{1 + e \cos \theta_1} = 30{,}000 \text{ km} + R_E = 36{,}378 \text{ km}$$

Using $p = a(1 - e^2) = 11{,}975.2$ km, we obtain

$$\cos \theta_1 = \frac{1}{e}\left(\frac{p}{r} - 1\right) = -0.9052$$

Because the satellite has not yet reached apogee, true anomaly is in the second quadrant and therefore $\theta_1 = 154.85°$.

Next, determine the eccentric anomaly using Eq. (4.33):

$$\tan\frac{E_1}{2} = \sqrt{\frac{1-e}{1+e}} \tan\frac{\theta_1}{2} = 1.7286, \quad \text{so } E_1 = 119.90° \ (= 2.0927 \text{ rad})$$

Finally, use Kepler's equation (4.18) to obtain the TOF to position 1:

$$t_1 - t_p = \frac{1}{n}(E_1 - e \sin E_1) = \boxed{9{,}945.2 \text{ s} = 2.763 \text{ h}}$$

Remember that E_1 must be in radians in Kepler's equation (note that n has units of rad/s).

b) Case 2: negative flight-path angle at 30,000 km altitude

Because flight-path angle is negative, the satellite has passed apogee and therefore its true anomaly is between 180° and 360° (i.e., $\mathbf{r_2}$ in Figure 4.5). The altitudes of positions 1 and 2 are the same (30,000 km), and therefore the trajectory equation yields $\cos \theta_2 = -0.9052$. However, true anomaly is in the third quadrant: $\theta_2 = 205.15°$ (see Figure 4.5). Eccentric anomaly of position 2 is

$$\tan\frac{E_2}{2} = \sqrt{\frac{1-e}{1+e}} \tan\frac{\theta_2}{2} = -1.7286, \quad \text{so } E_2 = -119.90° = 240.10° \ (= 4.1905 \text{ rad})$$

Note that we added 360° to our inverse-tangent (calculator) result so that E_2 is positive (of course $E_2 = -119.90°$ is the same angle as $E_2 = 240.10°$).

Using Kepler's equation with E_2, we obtain the TOF from perigee to position 2:

$$t_2 - t_p = \frac{1}{n}(E_2 - e \sin E_2) = \boxed{33{,}143.5 \text{ s} = 9.207 \text{ h}}$$

Example 4.2   Figure 4.6 shows the Molniya orbit discussed in Example 4.1. Determine the TOF from position $\mathbf{r_1}$ ($\theta_1 = 230°$) to position $\mathbf{r_2}$ ($\theta_2 = 120°$).

Kepler's equation (4.21) provides the flight time between two arbitrary positions in the orbit:

$$\text{TOF}_{1-2} = t_2 - t_1 = \frac{1}{n}(M_2 - M_1)$$

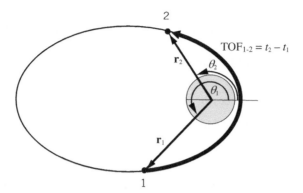

Figure 4.6 Molniya orbit (Example 4.2).

We know from Example 4.1 that mean motion for the Molniya orbit is $n = 1.4582(10^{-4})$ rad/s. The mean anomalies are

$$M_1 = E_1 - e\sin E_1 \quad \text{and} \quad M_2 = E_2 - e\sin E_2$$

We also know from Example 4.1 that eccentricity is $e = 0.7411$. Using Eq. (4.33), we find that the eccentric anomaly of position 1 is

$$\tan\frac{E_1}{2} = \sqrt{\frac{1-e}{1+e}}\,\tan\frac{\theta_1}{2} = -0.8270, \quad \text{so } E_1 = -79.18° = 280.82°\,(= 4.9012\,\text{rad})$$

Hence the mean anomaly of position 1 is $M_1 = E_1 - e\sin E_1 = 5.6291$ rad.

The eccentric anomaly of position 2 is

$$\tan\frac{E_2}{2} = \sqrt{\frac{1-e}{1+e}}\,\tan\frac{\theta_2}{2} = 0.6679, \quad \text{so } E_2 = 67.48°\,(= 1.1778\,\text{rad})$$

However, because the satellite passes through perigee on the way to position 2 (see Figure 4.6), we must add 2π to eccentric anomaly E_2 so that $E_2 > E_1$ (and $M_2 > M_1$). Using $E_2 = 1.1778$ rad $+ 2\pi = 7.4609$ rad, the associated mean anomaly is $M_2 = E_2 - e\sin E_2 = 6.7764$ rad.

Finally, the flight time from 1 to 2 is

$$\text{TOF}_{1-2} = t_2 - t_1 = \frac{1}{n}(M_2 - M_1) = \boxed{7{,}867.5\,\text{s} = 2.185\,\text{h}}$$

As a side note, the TOF calculations are correct if we express eccentric anomaly E_1 as a negative angle $(E_1 = -79.18° = -1.3820\,\text{rad})$ with the second eccentric anomaly $E_2 = 67.48° = 1.1778$ rad. With this convention, we maintain $E_2 > E_1$ (and $M_2 > M_1$) without adding 2π to E_2. However, it is this author's opinion that maintaining positive eccentric and mean anomalies in all calculations (and hence adding 2π when the satellite passes through periapsis) is a systematic approach that will not yield negative flight times.

4.3 Parabolic and Hyperbolic Time of Flight

4.3.1 Parabolic Trajectory Flight Time

We can determine transit time on a parabolic trajectory by using integral calculus. Our analytical derivation begins with the angular momentum expressed as the product of radial position r and transverse velocity $v_\theta = r\dot{\theta}$

$$h = r^2\dot{\theta} = r^2\frac{d\theta}{dt} \tag{4.34}$$

Separating variables and substituting the trajectory equation $r = (h^2/\mu)/(1+e\cos\theta)$ yields

$$hdt = \frac{h^4/\mu^2}{(1+e\cos\theta)^2}d\theta \tag{4.35}$$

or,

$$\frac{\mu^2}{h^3}dt = \frac{1}{(1+e\cos\theta)^2}d\theta \tag{4.36}$$

Integrating both sides of Eq. (4.36) from periapsis ($\theta = 0$) to position 1 on the orbit yields

$$\int_{t_p}^{t}\frac{\mu^2}{h^3}dt = \int_{0}^{\theta}\frac{1}{(1+e\cos\theta)^2}d\theta \tag{4.37}$$

Note that the lower bound on the left-hand side integral is t_p because we are computing TOF from periapsis to a position on the parabola. For a parabola, we have $e = 1$ and Eq. (4.37) becomes

$$\int_{t_p}^{t}\frac{\mu^2}{h^3}dt = \int_{0}^{\theta}\frac{1}{(1+\cos\theta)^2}d\theta \tag{4.38}$$

The analytical integral of Eq. (4.38) is a relatively simple expression:

$$\frac{\mu^2}{h^3}(t-t_p) = \frac{1}{2}\tan\frac{\theta}{2} + \frac{1}{6}\tan^3\frac{\theta}{2} \tag{4.39}$$

Let's define the *parabolic anomaly* $B \equiv \tan(\theta/2)$ and express angular momentum in terms of parameter by using $h = \sqrt{p\mu}$. With these substitutions, Eq. (4.39) becomes

$$2\sqrt{\frac{\mu}{p^3}}(t-t_p) = B + \frac{1}{3}B^3 \tag{4.40}$$

Equation (4.40) is called *Barker's equation*. It determines the TOF from periapsis to a position on the parabolic trajectory denoted by true anomaly θ. Note that parabolic anomaly B is not an angle (unlike eccentric anomaly E or mean anomaly M) but that it is dimensionless. The reader should also note that the term $\sqrt{\mu/p^3}$ in Barker's equation (4.40) has the appropriate units of per second and has a similar form as mean motion $n = \sqrt{\mu/a^3}$ in Kepler's equation.

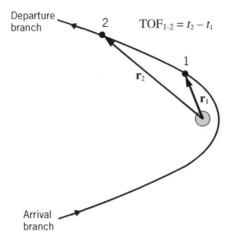

Figure 4.7 Time of flight on a parabolic trajectory.

Figure 4.7 shows the TOF between two arbitrary positions on a parabolic trajectory. For this scenario, we may express Eq. (4.40) as

$$t_2 - t_1 = \frac{1}{2} \sqrt{\frac{p^3}{\mu}} \left[\left(B_2 + \frac{1}{3} B_2^3 \right) - \left(B_1 + \frac{1}{3} B_1^3 \right) \right] \tag{4.41}$$

where parabolic anomalies B_1 and B_2 are computed from their respective true anomalies. When the satellite is *approaching* periapsis ($\pi < \theta < 2\pi$, or the "arrival branch"), parabolic anomaly B is always negative. When the satellite is moving *away* from periapsis ($0 < \theta < \pi$, or the "departure branch"), parabolic anomaly B is always positive. Of course, $B = 0$ when the satellite is at periapsis. The following example illustrates TOF calculations for a parabola.

Example 4.3 Figure 4.8 shows a geocentric satellite on a parabolic trajectory with a perigee altitude of 25,500 km. Compute the TOF between position vectors $\mathbf{r}_1$ and $\mathbf{r}_2$ with true anomalies $\theta_1 = 315°$ and $\theta_2 = 90°$, respectively.

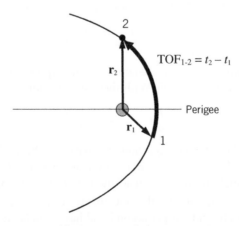

Figure 4.8 Time of flight on a parabolic trajectory (Example 4.3).

Barker's equation (4.41) provides the transit time between two positions on a parabolic orbit. First, we determine the parameter p from the perigee radius:

$$r_p = \frac{p}{1+e} = \frac{p}{2} = 25{,}500\,\text{km} + R_E = 31{,}878\,\text{km}$$

Therefore, $p = 63{,}756$ km.

Next, compute the two parabolic anomalies from the respective true anomalies:

$$\text{Position 1: } \theta_1 = 315°\colon \quad B_1 = \tan\frac{\theta_1}{2} = -0.4142$$

$$\text{Position 2: } \theta_2 = 90°\colon \quad B_2 = \tan\frac{\theta_2}{2} = 1$$

Note that because parabolic anomaly B is computed using a tangent half-angle we are assured that $B_2 > B_1$. This condition is guaranteed whether or not the satellite passes through periapsis.

Finally, use Eq. (4.41) to compute the parabolic flight time between $\mathbf{r}_1$ and $\mathbf{r}_2$:

$$t_2 - t_1 = \frac{1}{2}\sqrt{\frac{p^3}{\mu}}\left[\left(B_2 + \frac{1}{3}B_2^3\right) - \left(B_1 + \frac{1}{3}B_1^3\right)\right] = \boxed{22{,}581.8\,\text{s} = 6.273\,\text{h}}$$

4.3.2 Hyperbolic Trajectory Flight Time

It is possible to derive a TOF equation for a hyperbolic trajectory using either analytical or geometrical methods. We will not present either derivation here; the interested reader can consult Vallado [2; pp. 52–57] and Curtis [3; pp. 165–169] for details.

Flight time from periapsis to an arbitrary position on a hyperbola (defined by true anomaly θ) is

$$t - t_p = \sqrt{\frac{-a^3}{\mu}}(e\sinh F - F) \tag{4.42}$$

where F is the *hyperbolic anomaly*. Recall that semimajor axis a is negative for a hyperbolic trajectory (energy is positive) and therefore the term $\sqrt{-a^3/\mu}$ in Eq. (4.42) is a real number.

Hyperbolic sine of F can be expressed as a function of true anomaly and eccentricity:

$$\sinh F = \frac{\sqrt{e^2 - 1}\sin\theta}{1 + e\cos\theta} \tag{4.43}$$

We can compute hyperbolic anomaly F by simply applying the inverse hyperbolic sine operation to Eq. (4.43) because there is no quadrant ambiguity. However, if a calculator does not possess the inverse hyperbolic sine function, the following expression may be used

$$F = \ln\left(z + \sqrt{1 + z^2}\right) \tag{4.44}$$

where $z = \sinh F$. Whenever a satellite is on the "departure branch" of a hyperbola $(0 < \theta < \pi)$, hyperbolic anomaly F is positive; whenever a satellite is on the "arrival branch" $(\pi < \theta < 2\pi)$, F is negative. The reader should recall that true anomaly on a hyperbolic trajectory has asymptotic limits:

$$\theta_\infty = \cos^{-1}\left(\frac{-1}{e}\right) \tag{4.45}$$

Hence, on the departure branch we have $0 < \theta < \theta_\infty$ (and $F > 0$), and on the arrival branch we have $-\theta_\infty < \theta < 0$ (and $F < 0$).

Time of flight between two arbitrary positions on a hyperbola is

$$t_2 - t_1 = \sqrt{\frac{-a^3}{\mu}}\left[(e\sinh F_2 - F_2) - (e\sinh F_1 - F_1)\right] \tag{4.46}$$

Equation (4.46) is reduced to the hyperbolic flight time from periapsis equation (4.42) because $\sinh F_1 = F_1 = 0$ when $\theta_1 = 0$.

Example 4.4 The Mars Exploration Rover-A (MER-A) spacecraft Spirit departed Earth on June 10, 2003 on a hyperbolic trajectory. An upper-stage rocket was fired to increase the geocentric velocity to 11.4 km/s at an altitude of 225 km above the Earth. Figure 4.9 shows that the flight-path angle is zero after upper-stage burnout (Figure 4.9 is not to scale). Determine the flight time from upper-stage burnout to the lunar-orbit crossing at a geocentric radius of 384,400 km.

Because flight-path angle is zero, the upper-stage burnout point in Figure 4.9 is perigee (note that in Example 2.9 the burnout flight-path angle for the MER-A spacecraft was $\gamma_{bo} = 5°$; in this example we assume that the upper-stage burnout is at perigee). We can

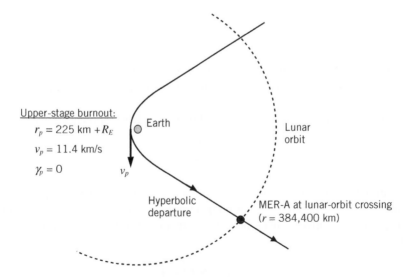

Figure 4.9 MER-A hyperbolic departure trajectory (Example 4.4).

denote velocity and radius at perigee as $v_p = 11.4\,\text{km/s}$ and $r_p = 225\,\text{km} + R_E = 6{,}603\,\text{km}$. Next, we compute the characteristics of the MER-A hyperbolic departure trajectory:

$$\text{Energy: } \xi = \frac{v_p^2}{2} - \frac{\mu}{r_p} = 4.6135\,\text{km}^2/\text{s}^2 = \frac{-\mu}{2a}$$

$$\text{Semimajor axis: } a = \frac{-\mu}{2\xi} = -43{,}199.3\,\text{km}$$

$$\text{Angular momentum: } h = r_p v_p = 75{,}274.2\,\text{km}^2/\text{s} = \sqrt{p\mu}$$

$$\text{Parameter: } p = \frac{h^2}{\mu} = 14{,}215.3\,\text{km} = a\left(1 - e^2\right)$$

$$\text{Eccentricity: } e = \sqrt{1 - \frac{p}{a}} = 1.1528$$

These calculations confirm that the orbit is a hyperbola.

We need the true anomaly at lunar-orbit crossing in order to determine the hyperbolic anomaly F in the TOF equation (4.42). Using the trajectory equation evaluated at the lunar-orbit radial distance,

$$r = \frac{p}{1 + e\cos\theta} = 384{,}400\,\text{km}$$

we find that true anomaly is $\theta = 146.65°$ (see Figure 4.9). Using Eq. (4.43) to compute $\sinh F$ and F yields

$$\sinh F = \frac{\sqrt{e^2 - 1}\,\sin\theta}{1 + e\cos\theta} = 8.5275 \text{ and } F = 2.8399$$

Using a, e, $\sinh F$, and F in the hyperbolic TOF equation, we obtain

$$t - t_p = \sqrt{\frac{-a^3}{\mu}}(e\sinh F - F) = \boxed{99{,}423.6\,\text{s} = 27.62\,\text{h} = 1.15\,\text{days}}$$

These calculations show that the MER-A spacecraft required slightly more than one day to cross the moon's orbit on its Earth-departure hyperbolic trajectory.

Example 4.5 The Mars Exploration Rover-A (MER-A) spacecraft Spirit approached Mars on a hyperbolic trajectory. Figure 4.10 shows the hyperbolic approach where position 1 has a radial distance $r_1 = 60{,}000\,\text{km}$ from Mars and position 2 has a radial distance $r_2 = 3{,}521\,\text{km}$. Position 2 is the so-called "entry interface" (EI) where the spacecraft begins to enter the upper Martian atmosphere at an altitude of approximately 125 km (of course, two-body motion does *not* hold for flight beyond EI; comparing Figure 4.10 with Figure 2.17 shows that we are only considering the *arrival* branch of the hyperbola). The velocity and flight-path angle of MER-A at position 2 are $v_2 = 5.4\,\text{km/s}$ and $\gamma_2 = -11.5°$, respectively. Determine the flight time from r_1 to r_2 (EI).

As with Example 4.4, we begin by computing the characteristics of the hyperbolic trajectory from the given data (note that the gravitational parameter for Mars is $\mu_M = 42{,}828\,\text{km}^3/\text{s}^2$).

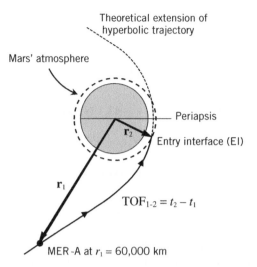

Figure 4.10 MER-A hyperbolic arrival trajectory (Example 4.5).

Energy: $\xi = \dfrac{v_2^2}{2} - \dfrac{\mu_M}{r_2} = 2.4164\,\text{km}^2/\text{s}^2 = \dfrac{-\mu_M}{2a}$

Semimajor axis: $a = \dfrac{-\mu_M}{2\xi} = -8{,}861.9\,\text{km}$

Angular momentum: $h = r_2 v_2 \cos\gamma_2 = 18{,}631.7\,\text{km}^2/\text{s} = \sqrt{p\mu_M}$

Parameter: $p = \dfrac{h^2}{\mu_M} = 8{,}105.5\,\text{km} = a\left(1-e^2\right)$

Eccentricity: $e = \sqrt{1 - \dfrac{p}{a}} = 1.3837$

In order to determine the hyperbolic anomalies at positions 1 and 2, we need the respective true anomalies at these locations in the trajectory. Using the trajectory equation at position 1, we find

Position 1: $r_1 = \dfrac{p}{1+e\cos\theta_1} = 60{,}000\,\text{km} \;\rightarrow\; \theta_1 = \cos^{-1}\left[\dfrac{1}{e}\left(\dfrac{p}{r_1}-1\right)\right] = -128.69°$

Note that θ_1 is negative because the spacecraft is approaching Mars. The trajectory equation is applied at position 2 (entry interface) to yield

Position 2: $r_2 = \dfrac{p}{1+e\cos\theta_2} = 3{,}522\,\text{km} \;\rightarrow\; \theta_2 = \cos^{-1}\left[\dfrac{1}{e}\left(\dfrac{p}{r_2}-1\right)\right] = -19.78°$

As before, θ_2 is negative because the MER-A spacecraft is approaching Mars (recall that flight-path angle γ_2 is negative).

Using Eq. (4.43) to compute $\sinh F$ and F at both locations yields

Position 1: $\sinh F_1 = \dfrac{\sqrt{e^2-1}\sin\theta_1}{1+e\cos\theta_1} = -5.5260$ and $F_1 = -2.4107$

Position 2: $\sinh F_2 = \dfrac{\sqrt{e^2-1}\ \sin\theta_2}{1+e\cos\theta_2} = -0.1406$ and $F_2 = -0.1402$

Finally, using the hyperbolic TOF equation (4.46), we obtain

$$t_2-t_1 = \sqrt{\dfrac{-a^3}{\mu_M}}[(e\sinh F_2 - F_2) - (e\sinh F_1 - F_1)] = \boxed{20{,}886\,\text{s} = 5.8\,\text{h}}$$

The MER-A spacecraft required nearly 6 h to reach Mars' atmospheric entry point from a radial distance of 60,000 km.

As a side note, we can use the energy equation to compute the hyperbolic arrival velocity "at infinity"

$$\text{Energy: } \xi = \dfrac{v_\infty^2}{2} = 2.4164\,\text{km}^2/\text{s}^2 \ \rightarrow\ v_\infty = 2.20\,\text{km/s}$$

This is the Mars-relative velocity of the MER-A spacecraft as it enters Mars' gravity field. In Chapter 10, we will address this issue further when we discuss interplanetary trajectories.

4.4 Kepler's Problem

The previous sections and example problems have demonstrated that TOF between two points on a conic section (ellipse, parabola, or hyperbola) is relatively easy to compute if the orbital characteristics (a and e) and orbital positions (θ_1 and θ_2) are known. For example, the right-hand side of Kepler's equation (4.21)

$$t_2-t_1 = \sqrt{\dfrac{a^3}{\mu}}[(E_2 - e\sin E_2) - (E_1 - e\sin E_1)] \tag{4.47}$$

is completely determined if we know a, e, E_1, and E_2. Of course, eccentric anomalies E_1 and E_2 can be computed from true anomalies θ_1 and θ_2 and eccentricity e.

The inverse problem is not so straightforward. Suppose we know the orbital elements a and e and the initial orbital position indicated by true anomaly θ_1. We wish to determine the future (or *propagated*) angular position of the satellite θ_2 after transit time t_2-t_1. Equation (4.21) presents the flight time in terms of the change in mean anomaly

$$t_2-t_1 = \dfrac{1}{n}(M_2 - M_1) \tag{4.48}$$

Solving Eq. (4.48) for mean anomaly M_2 yields

$$M_2 = M_1 + n(t_2 - t_1) \tag{4.49}$$

Because we know the initial mean anomaly M_1 (determined using true anomaly θ_1) and flight time t_2-t_1, we can use Eq. (4.49) to calculate the propagated mean anomaly M_2. Recall that mean anomaly is a function of eccentric anomaly

$$M_2 = E_2 - e\sin E_2 \tag{4.50}$$

Equation (4.50) is, of course, a form of Kepler's equation. However, we cannot solve Eq. (4.50) for eccentric anomaly E_2 because it is a transcendental equation [i.e., we cannot manipulate Eq. (4.50) so that E_2 appears by itself on one side of the equals sign]. Solving Eq. (4.50) is known as *Kepler's problem*: determining eccentric anomaly E (position in the orbit) from a known value of mean anomaly M.

Because no closed-form solution to Eq. (4.50) exists, we must resort to an iterative numerical search. To begin, let us pose Eq. (4.50) as a root-solving problem (in addition, let us drop the subscript 2 for convenience):

$$f(E) = E - e\sin E - M = 0 \qquad (4.51)$$

Perhaps the most straightforward approach is to use Newton's method to iteratively search for the root of Eq. (4.51):

$$E_{k+1} = E_k - \frac{f(E_k)}{f'(E_k)} \qquad (4.52)$$

where $f'(E_k)$ is the derivative of Eq. (4.51) with respect to E

$$f'(E) = \frac{df}{dE} = 1 - e\cos E \qquad (4.53)$$

We can now summarize an iterative algorithm for solving Kepler's problem:

1) Determine a, e, and n from the orbit characteristics. Compute the *propagated* mean anomaly M using Eq. (4.49) and the known initial mean anomaly and TOF.
2) Guess a trial (starting) value of eccentric anomaly E_k. Set the iteration index at $k = 1$.
3) Use Eq. (4.51) to evaluate $f(E_k)$ and check for convergence: if $|f(E_k)| < \varepsilon$, then the iteration has converged and skip to step 5.
4) Use Eq. (4.52) to select a new trial value for eccentric anomaly [making use of Eq. (4.53) to evaluate the derivative $f'(E_k)$]. Update the iteration index $k = k + 1$ and return to step 3.
5) Compute the true anomaly θ of the propagated position from the converged value of eccentric anomaly E.

The tolerance for convergence may be set at $\varepsilon = 10^{-8}$ and convergence typically occurs after three to five iterations for orbits with $e < 0.8$.

A good initial guess for E (step 2) will reduce the total number of iterations. Figure 4.11 shows mean anomaly M as a function of eccentric anomaly E for a range of eccentricity. Figure 4.11 and Eq. (4.50) show that for nearly circular orbits ($e \approx 0$), $E = M$ is a good starting guess for Newton's method. In fact, Newton's method will always converge from the starting guess $E = M$. Battin [4; pp. 199–200] presents a power-series solution for eccentric anomaly that was originally developed by Lagrange. The power series to second order is

$$E = M + e\sin M + \frac{e^2}{2}\sin 2M \qquad (4.54)$$

Equation (4.54) provides a better starting guess (compared with $E = M$) for elliptical orbits and typically reduces the numerical search by one to two iterations. As a quick example, consider the case where eccentric anomaly $E = 5$ rad, eccentricity $e = 0.7$, and

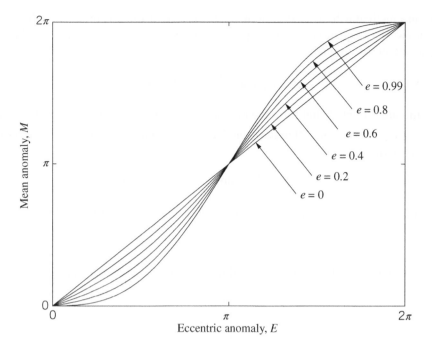

Figure 4.11 Mean anomaly as a function of eccentric anomaly.

the corresponding mean anomaly is $M = E - e\sin E = 5.6712$ rad. Hence, the initial guess $E = M$ (=5.6712 rad) exhibits a 13% error. The initial guess using Eq. (4.54) is $E = 5.0387$ rad, which corresponds to an error of less than 1%.

Example 4.6   Consider again the Molniya orbit discussed in Examples 4.1 and 4.2. If a satellite is located at true anomaly $\theta_1 = 260°$ at time t_1, determine the position r_2 and true anomaly θ_2 of the satellite 50 min later.

From Example 4.1, we know the semimajor axis, mean motion, and eccentricity of the Molniya orbit:

$$a = 26{,}564.5\,\text{km}, \quad n = \sqrt{\mu/a^3} = 1.4582\left(10^{-4}\right)\text{rad/s}, \quad e = 0.7411$$

Using Eq. (4.33), we obtain the initial eccentric anomaly E_1 from θ_1 and e

$$\tan\frac{E_1}{2} = \sqrt{\frac{1-e}{1+e}}\tan\frac{\theta_1}{2} = -0.4596, \quad \text{so } E_1 = -49.36°, \text{ or } E_1 = -0.8615\,\text{rad}$$

Hence, the corresponding initial mean anomaly is $M_1 = E_1 - e\sin E_1 = -0.2992$ rad. The propagated mean anomaly is

$$M_2 = M_1 + n(t_2 - t_1) = -0.2992\,\text{rad} + (1.4582(10^{-4})\,\text{rad/s})(50\,\text{min})(60\,\text{s/min})$$

$$= 0.1383\,\text{rad}$$

Therefore, we can now express Kepler's equation in terms of the propagated position 2:

$$M_2 = E_2 - e\sin E_2$$

Newton's method will be used to numerically search for the eccentric anomaly E_2 that satisfies Kepler's equation. We use Eq. (4.54) to determine an initial guess for the propagated eccentric anomaly

$$E_2^{(1)} = M_2 + e \sin M_2 + \frac{e^2}{2} \sin 2M_2 = 0.3155 \, \text{rad}$$

Here the superscript (1) indicates the iteration index while the subscript 2 indicates the propagated position (point 2) in the orbit. Next, we use Eq. (4.51) to test if $E_2^{(1)}$ is a solution (or root):

$$f\left(E_2^{(1)}\right) = E_2^{(1)} - e \sin E_2^{(1)} - M_2 = -0.0528$$

Clearly, $E_2^{(1)}$ is not the solution. Equation (4.53) is used to compute the derivative

$$f'\left(E_2^{(1)}\right) = 1 - e \cos E_2^{(1)} = 0.2955$$

and Newton's method, Eq. (4.52), is used to compute the next trial value of E_2

$$E_2^{(2)} = E_2^{(1)} - \frac{f\left(E_2^{(1)}\right)}{f'\left(E_2^{(1)}\right)} = 0.4940 \, \text{rad}$$

The error function using the updated E_2 is

$$f\left(E_2^{(2)}\right) = E_2^{(2)} - e \sin E_2^{(2)} - M_2 = 0.0043$$

The error function $f(E_2)$ has been significantly reduced after the first Newton-iteration step. Newton's method is repeated until convergence is achieved in four iterations. Table 4.1 summarizes the numerical search.

The eccentric anomaly of the satellite 50 min after the given initial position is $E_2 = 0.4815$ rad. Because we want true anomaly of the propagated position, we employ Eq. (4.32) to convert E_2 to θ_2

$$\tan\frac{\theta_2}{2} = \sqrt{\frac{1+e}{1-e}} \tan\frac{E_2}{2} = 0.6367 \quad \rightarrow \quad \theta_2 = 2\tan^{-1}(0.6367) = 1.1339 \, \text{rad}$$

Table 4.1 Numerical solution of Kepler's problem (Example 4.6).

Iteration, k	Trial eccentric anomaly, $E_2^{(k)}$ (rad)	Error function, $f\left(E_2^{(k)}\right)$
1	0.315452	−0.052767
2	0.494041	0.004325
3	0.481597	$2.7001(10^{-5})$
4	0.481518	$1.0623(10^{-9})$

Therefore, $\boxed{\theta_2 = 64.97°}$

Radius at the propagate position is determined using the trajectory equation:

$$r_2 = \frac{p}{1+e\cos\theta_2} = \frac{a(1-e^2)}{1+e\cos\theta_2} = \boxed{9{,}116.1\,\text{km}}$$

4.5   Orbit Propagation Using Lagrangian Coefficients

The previous sections have presented TOF expressions for elliptical orbits and parabolic and hyperbolic trajectories. These equations have assumed that the orbit's energy and angular momentum (i.e., a and e) are known. A more fundamental problem is determining the orbit, given two known position vectors $\mathbf{r}_1$ and $\mathbf{r}_2$ and the TOF between them. As mentioned in Chapter 3, this important problem was first solved by Johann Heinrich Lambert in 1761 and is known as *Lambert's problem*. Solving Lambert's problem leads to methods for orbit determination (Chapter 3), orbital maneuvering (Chapter 7), and interplanetary mission design (Chapter 10). We will discuss Lambert's problem in more detail in the following section of this chapter.

In this section, we present a method for propagating an orbit ahead in time by using the so-called *Lagrangian coefficients*. This technique is important because it is frequently used in algorithms that solve Lambert's problem. We will present one solution method that uses Lagrangian coefficients in Section 4.6.

It may be useful to begin with the intended result: we wish to compute the position and velocity vectors of a satellite at a future time t as a *linear* combination of the position and velocity vectors at some *initial* time $t_0 = 0$, that is,

$$\mathbf{r}(t) = f\mathbf{r}_0 + g\mathbf{v}_0 \tag{4.55a}$$

$$\mathbf{v}(t) = \dot{f}\mathbf{r}_0 + \dot{g}\mathbf{v}_0 \tag{4.55b}$$

In Eq. (4.55), the initial state of the satellite is $(\mathbf{r}_0,\mathbf{v}_0)$ and the propagated (predicted) state at time t is $(\mathbf{r}(t),\mathbf{v}(t))$. Recall from Chapter 3 that the six coordinates of the combined $(\mathbf{r},\mathbf{v})$ vectors can be transformed into the six classical orbital elements, including the true anomaly that pertains to the particular instant in time. The four scalar coefficients f, g, $\dot{f}$, and $\dot{g}$ in Eq. (4.55) are the Lagrangian coefficients. Clearly, the coefficients $\dot{f}$ and $\dot{g}$ are time derivatives of f and g because the time derivative of Eq. (4.55a) yields Eq. (4.55b), or $\dot{\mathbf{r}}(t) = \mathbf{v}(t)$. We will now derive equations for these four coefficients.

It turns out that the four Lagrangian coefficients are not independent. To show this, use Eq. (4.55) to compute the cross product of $\mathbf{r}(t)$ and $\mathbf{v}(t)$

$$\mathbf{r}(t) \times \mathbf{v}(t) = (f\mathbf{r}_0 + g\mathbf{v}_0) \times (\dot{f}\mathbf{r}_0 + \dot{g}\mathbf{v}_0) \tag{4.56}$$

The left-hand side of Eq. (4.56) is $\mathbf{h}$; the right-hand side is expanded below

$$\mathbf{h} = f\dot{f}\mathbf{r}_0 \times \mathbf{r}_0 + f\dot{g}\mathbf{r}_0 \times \mathbf{v}_0 - \dot{f}g\mathbf{r}_0 \times \mathbf{v}_0 + g\dot{g}\mathbf{v}_0 \times \mathbf{v}_0 \tag{4.57}$$

Because the right-hand side cross products are $\mathbf{r}_0 \times \mathbf{r}_0 = \mathbf{0}$, $\mathbf{r}_0 \times \mathbf{v}_0 = \mathbf{h}$, and $\mathbf{v}_0 \times \mathbf{v}_0 = \mathbf{0}$, Eq. (4.57) becomes

$$\mathbf{h} = f\dot{g}\mathbf{h} - \dot{f}g\mathbf{h} \tag{4.58}$$

Hence, the scalar coefficients of $\mathbf{h}$ in Eq. (4.57) must satisfy

$$1 = f\dot{g} - \dot{f}g \tag{4.59}$$

Therefore, we can use Eq. (4.59) and three known Lagrangian coefficients to determine the fourth coefficient.

To begin our analysis, let us express the initial position and velocity vectors in the perifocal (or **PQW**) frame [see Eqs. (3.18) and (3.23)]:

$$\mathbf{r}_0 = r_0 \cos\theta_0 \mathbf{P} + r_0 \sin\theta_0 \mathbf{Q} \tag{4.60a}$$

$$\mathbf{v}_0 = \frac{-\mu}{h}\sin\theta_0 \mathbf{P} + \frac{\mu}{h}(e + \cos\theta_0)\mathbf{Q} \tag{4.60b}$$

Recall that the unit vector $\mathbf{P}$ is along the periapsis direction, $\mathbf{Q}$ is in the orbital plane and 90° from periapsis in the direction of motion, and $\mathbf{W}$ is normal to the orbit (along the angular momentum vector $\mathbf{h}$). We wish to invert Eq. (4.60) and solve for unit vectors $\mathbf{P}$ and $\mathbf{Q}$ in terms of the initial state $(\mathbf{r}_0, \mathbf{v}_0)$. One way to do this is to group the four terms on the right-hand side of Eq. (4.60) in a two-dimensional matrix:

$$\mathbf{C} = \begin{bmatrix} r_0 \cos\theta_0 & r_0 \sin\theta_0 \\ \dfrac{-\mu}{h}\sin\theta_0 & \dfrac{\mu}{h}(e + \cos\theta_0) \end{bmatrix} \tag{4.61}$$

The inverse of matrix $\mathbf{C}$ is

$$\mathbf{C}^{-1} = \frac{1}{\det(\mathbf{C})}\begin{bmatrix} \dfrac{\mu}{h}(e + \cos\theta_0) & -r_0 \sin\theta_0 \\ \dfrac{\mu}{h}\sin\theta_0 & r_0 \cos\theta_0 \end{bmatrix} \tag{4.62}$$

where the determinant of $\mathbf{C}$ is angular momentum h [the reader should note that the cross product $\mathbf{r}_0 \times \mathbf{v}_0$ with $\mathbf{r}_0$ and $\mathbf{v}_0$ expressed in terms of their $\mathbf{P}$ and $\mathbf{Q}$ components leads to $\mathbf{r}_0 \times \mathbf{v}_0 = \mathbf{h} = \det(\mathbf{C})\mathbf{W}$]. Using Eq. (4.62), we can write the $\mathbf{P}$ and $\mathbf{Q}$ unit vectors in terms of $\mathbf{r}_0$ and $\mathbf{v}_0$:

$$\mathbf{P} = \frac{\mu}{h^2}(e + \cos\theta_0)\mathbf{r}_0 - \frac{r_0}{h}\sin\theta_0 \mathbf{v}_0 \tag{4.63a}$$

$$\mathbf{Q} = \frac{\mu}{h^2}\sin\theta_0 \mathbf{r}_0 + \frac{r_0}{h}\cos\theta_0 \mathbf{v}_0 \tag{4.63b}$$

Now, write an expression for $\mathbf{r}(t)$ and $\mathbf{v}(t)$ vectors at an arbitrary future time using perifocal coordinates

$$\mathbf{r}(t) = r\cos\theta \mathbf{P} + r\sin\theta \mathbf{Q} \tag{4.64a}$$

$$\mathbf{v}(t) = \frac{-\mu}{h}\sin\theta \mathbf{P} + \frac{\mu}{h}(e + \cos\theta)\mathbf{Q} \tag{4.64b}$$

Equation (4.64) has the same format as Eq. (4.60); however, Eq. (4.64) uses the *propagated* radius and true anomaly r and θ, respectively. Substituting Eq. (4.63) for $\mathbf{P}$ and $\mathbf{Q}$ in Eq. (4.64) and collecting terms yields

$$\mathbf{r}(t) = \frac{\mu r \cos\theta(e + \cos\theta_0) + \mu r \sin\theta(\sin\theta_0)}{h^2}\mathbf{r}_0 + \frac{-r\cos\theta(r_0\sin\theta_0) + r\sin\theta(r_0\cos\theta_0)}{h}\mathbf{v}_0$$

$$(4.65a)$$

$$v(t) = \frac{-\mu^2\sin\theta(e + \cos\theta_0) + \mu^2(e + \cos\theta)(\sin\theta_0)}{h^3}\mathbf{r}_0$$

$$+ \frac{\mu r_0 \sin\theta(\sin\theta_0) + \mu r_0(e + \cos\theta)(\cos\theta_0)}{h^2}\mathbf{v}_0$$

$$(4.65b)$$

Comparing Eqs. (4.65) and (4.55), we can identify the four Lagrangian coefficients:

$$f = \frac{\mu r \cos\theta(e + \cos\theta_0) + \mu r \sin\theta(\sin\theta_0)}{h^2} \tag{4.66}$$

$$g = \frac{-r\cos\theta(r_0\sin\theta_0) + r\sin\theta(r_0\cos\theta_0)}{h} \tag{4.67}$$

$$\dot{f} = \frac{-\mu^2\sin\theta(e + \cos\theta_0) + \mu^2(e + \cos\theta)(\sin\theta_0)}{h^3} \tag{4.68}$$

$$\dot{g} = \frac{\mu r_0 \sin\theta(\sin\theta_0) + \mu r_0(e + \cos\theta)(\cos\theta_0)}{h^2} \tag{4.69}$$

Note that all four Lagrangian coefficients depend on products of sine and cosine of θ_0 and θ. Therefore, these combinations may be written as trigonometric functions of differences in true anomaly (i.e., $\Delta\theta = \theta - \theta_0$). Furthermore, we may substitute $h = \sqrt{p\mu}$, $e = (p/r - 1)/\cos\theta$, and $e = (p/r_0 - 1)/\cos\theta_0$ (from the trajectory equation) in order to express the Lagrangian coefficients in terms of p. Using these substitutions, Eqs. (4.66)–(4.69) are reduced to (somewhat) simpler expressions:

$$f = 1 - \frac{r}{p}(1 - \cos\Delta\theta) \tag{4.70}$$

$$g = \frac{rr_0\sin\Delta\theta}{\sqrt{p\mu}} \tag{4.71}$$

$$\dot{f} = \sqrt{\frac{\mu}{p}}\left(\frac{1 - \cos\Delta\theta}{p} - \frac{1}{r} - \frac{1}{r_0}\right)\tan\frac{\Delta\theta}{2} \tag{4.72}$$

$$\dot{g} = 1 - \frac{r_0}{p}(1 - \cos\Delta\theta) \tag{4.73}$$

where $\Delta\theta = \theta - \theta_0$ is the change in true anomaly between position vectors $\mathbf{r}$ and $\mathbf{r}_0$. Note that when $\Delta\theta = 0$ (i.e., no propagation ahead in time), Eqs. (4.70)–(4.73) show that $f = 1$, $g = 0$, $\dot{f} = 0$, and $\dot{g} = 1$. Hence, when $\Delta\theta = 0$ we obtain $\mathbf{r}(t) = f\mathbf{r}_0 + g\mathbf{v}_0 = \mathbf{r}_0$ and $v(t) = \dot{f}\mathbf{r}_0 + \dot{g}\mathbf{v}_0 = \mathbf{v}_0$ as expected.

We can now summarize our orbit-propagation method:

1) Given the initial state $(\mathbf{r}_0, \mathbf{v}_0)$, compute the parameter p from angular momentum $(\mathbf{h} = \mathbf{r}_0 \times \mathbf{v}_0)$ and semimajor axis a from energy $(\xi = v_0^2/2 - \mu/r_0 = -\mu/2a)$. Determine eccentricity e from p and a $(e = \sqrt{1 - p/a})$.
2) Determine the initial true anomaly θ_0 from the trajectory equation $r_0 = p/(1 + e\cos\theta_0)$.

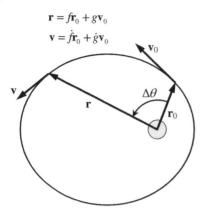

$$\mathbf{r} = f\mathbf{r}_0 + g\mathbf{v}_0$$
$$\mathbf{v} = \dot{f}\mathbf{r}_0 + \dot{g}\mathbf{v}_0$$

Figure 4.12 Orbit propagation from state $(\mathbf{r}_0, \mathbf{v}_0)$ to $(\mathbf{r}, \mathbf{v})$.

3) Select the propagated (future) true anomaly θ and compute the propagated radius using the trajectory equation $r = p/(1 + e\cos\theta)$.
4) Compute the four Lagrangian coefficients using Eqs. (4.70)–(4.73) with knowledge of the four values $\Delta\theta = \theta - \theta_0$, r_0, r, and p.
5) Calculate the propagated state $(\mathbf{r}, \mathbf{v})$ using the Lagrangian coefficients in Eq. (4.55).

Figure 4.12 shows orbit propagation from $(\mathbf{r}_0, \mathbf{v}_0)$ to $(\mathbf{r}, \mathbf{v})$ using the Lagrangian coefficients.

It is useful to step back and reexamine the propagation method previously summarized. Because we know the initial state $(\mathbf{r}_0, \mathbf{v}_0)$, we can compute the corresponding six orbital elements $(a, e, i, \Omega, \omega, \theta_0)$. When we select the propagated (future) true anomaly θ (step 3), we can simply compute the propagated $(\mathbf{r}, \mathbf{v})$ vectors by using the six orbital elements $(a, e, i, \Omega, \omega, \theta)$ and the transformation methods outlined in Chapter 3. So, orbit propagation using the Lagrangian coefficients provides an alternative technique (and perhaps a more compact method) for determining the future state $(\mathbf{r}, \mathbf{v})$ from the initial state $(\mathbf{r}_0, \mathbf{v}_0)$ for a selected orbit position θ. The key here is that the orbit is known and we have selected the future angular position θ. The more fundamental (and more difficult) problem is Lambert's problem: determine the orbit given two position vectors $\mathbf{r}_0$ and $\mathbf{r}$ and the corresponding TOF.

As previously mentioned, Lambert's problem will be addressed in the next section. One way to formulate Lambert's problem is to express the Lagrangian coefficients in terms of change in eccentric anomaly ΔE and TOF. To do so, we start by rewriting the initial position vector in perifocal coordinates, Eq. (4.60), in terms of initial eccentric anomaly E_0

$$\mathbf{r}_0 = (a\cos E_0 - ae)\mathbf{P} + a\sqrt{1 - e^2}\sin E_0\mathbf{Q} \tag{4.74}$$

where the $\mathbf{P}$ coordinate $(r_0\cos\theta_0)$ is obtained from Eq. (4.4) and the $\mathbf{Q}$ coordinate $(r_0\sin\theta_0)$ is the product of Eqs. (4.12) and (4.13). Next, take the time derivative of Eq. (4.74)

$$\mathbf{v}_0 = \dot{\mathbf{r}}_0 = -a\dot{E}_0\sin E_0\mathbf{P} + a\dot{E}_0\sqrt{1 - e^2}\cos E_0\mathbf{Q} \tag{4.75}$$

Recall that we determined the time rate of eccentric anomaly in the derivation of Kepler's equation in Section 4.2. Equation (4.14) is repeated below (with the appropriate subscript)

$$\dot{E}_0 = \sqrt{\frac{\mu}{a^3}}\frac{1}{1 - e\cos E_0} \tag{4.76}$$

Equation (4.12) shows that $1 - e\cos E_0 = r_0/a$. Substituting this result into Eq. (4.76), we obtain a simple expression for $\dot{E}_0$

$$\dot{E}_0 = \frac{1}{r_0}\sqrt{\frac{\mu}{a}} \tag{4.77}$$

Substituting Eq. (4.77) for $\dot{E}_0$ in Eq. (4.75) yields

$$\mathbf{v}_0 = \frac{-\sqrt{\mu a}}{r_0}\sin E_0 \mathbf{P} + \frac{\sqrt{\mu a}}{r_0}\sqrt{1-e^2}\cos E_0 \mathbf{Q} \tag{4.78}$$

The propagated state $(\mathbf{r}, \mathbf{v})$ at a future eccentric anomaly E is computed using Eqs. (4.74) and (4.78)

$$\mathbf{r} = (a\cos E - ae)\mathbf{P} + a\sqrt{1-e^2}\sin E \mathbf{Q} \tag{4.79a}$$

$$\mathbf{v} = \frac{-\sqrt{\mu a}}{r}\sin E \mathbf{P} + \frac{\sqrt{\mu a}}{r}\sqrt{1-e^2}\cos E \mathbf{Q} \tag{4.79b}$$

The subsequent steps follow the same procedure that was outlined at the beginning of this section: (1) invert Eqs. (4.74) and (4.78) and solve for the $\mathbf{P}$ and $\mathbf{Q}$ unit vectors; (2) substitute the $\mathbf{P}$ and $\mathbf{Q}$ vectors into Eq. (4.79); and (3) express the trigonometric products as sine and cosine of the difference in eccentric anomaly $\Delta E = E - E_0$. We will not present the details of these steps. The resulting Lagrangian coefficients in terms of ΔE are

$$f = 1 - \frac{a}{r_0}(1 - \cos\Delta E) \tag{4.80}$$

$$g = (t - t_0) - \sqrt{\frac{a^3}{\mu}}(\Delta E - \sin\Delta E) \tag{4.81}$$

$$\dot{f} = \frac{-\sqrt{\mu a}\sin\Delta E}{r r_0} \tag{4.82}$$

$$\dot{g} = 1 - \frac{a}{r}(1 - \cos\Delta E) \tag{4.83}$$

where $\Delta E = E - E_0$ is the change in eccentric anomaly between position vectors $\mathbf{r}$ and $\mathbf{r}_0$. Equations (4.80)–(4.83) determine the Lagrangian coefficients in terms of the difference in eccentric anomaly ΔE. Note that the g coefficient, Eq. (4.81), depends on the flight time $t - t_0$ from position $\mathbf{r}_0$ to $\mathbf{r}$. In addition, note that when $\Delta E = 0$ and $t - t_0 = 0$ (i.e., no propagation), Eqs. (4.80)–(4.83) show that $f = 1$, $g = 0$, $\dot{f} = 0$, and $\dot{g} = 1$ as expected.

In summary, we may compute the four Lagrangian coefficients using the eccentric anomaly difference ΔE [i.e., Eqs. (4.80)–(4.83)], or the true anomaly difference $\Delta\theta$ [i.e., Eqs. (4.70)–(4.73)]. Both sets of equations yield identical results as long as ΔE

corresponds to $\Delta\theta$ and eccentricity e. The following two examples illustrate orbit propagation with the Lagrangian coefficients computed using $\Delta\theta$ and ΔE.

Example 4.7 A satellite has the following initial position and velocity vectors in the Earth-centered inertial (ECI) frame:

$$\mathbf{r}_0 = \begin{bmatrix} -15,634 \\ 4,689 \\ 7,407 \end{bmatrix} \text{km}, \quad \mathbf{v}_0 = \begin{bmatrix} -4.6954 \\ -2.3777 \\ 0.6497 \end{bmatrix} \text{km/s}$$

Propagate the orbit and determine the position and velocity vectors $(\mathbf{r}, \mathbf{v})$ corresponding to a change in true anomaly of $\Delta\theta = 80°$.

We can compute the Lagrangian coefficients using Eqs. (4.70)–(4.73) and true anomaly difference $\Delta\theta$. To do so, we need to calculate the parameter p and radii r_0 and r. Because the radius is computed from the trajectory equation, we require eccentricity e and the initial and propagated true anomalies, θ_0 and θ. Therefore, we begin by computing the basic orbital characteristics (energy and angular momentum) from the initial state $(\mathbf{r}_0, \mathbf{v}_0)$:

$$\text{Initial radius: } r_0 = \|\mathbf{r}_0\| = 17,924.1 \text{ km}$$

$$\text{Initial velocity: } v_0 = \|\mathbf{v}_0\| = 5.3031 \text{ km/s}$$

$$\text{Energy: } \xi = \frac{v_0^2}{2} - \frac{\mu}{r_0} = -8.1771 \text{ km}^2/\text{s}^2$$

$$\text{Semimajor axis: } a = \frac{-\mu}{2\xi} = 24,373.0 \text{ km}$$

$$\text{Angular momentum: } \mathbf{h} = \mathbf{r}_0 \times \mathbf{v}_0 = [20,658 \quad -24,621 \quad 59,190]^T \text{ km}^2/\text{s}$$

$$\text{Parameter: } p = \frac{h^2}{\mu} = 11,380.8 \text{ km}$$

$$\text{Eccentricity: } e = \sqrt{1 - \frac{p}{a}} = 0.7301$$

We need to compute the propagated radius (magnitude r) using the trajectory equation and propagated true anomaly $\theta = \theta_0 + \Delta\theta$. Because we know r_0, p, and e, we can compute the initial true anomaly from the trajectory equation:

$$r_0 = \frac{p}{1 + e\cos\theta_0} \quad \rightarrow \quad \cos\theta_0 = \frac{1}{e}\left(\frac{p}{r_0} - 1\right) = -0.5$$

Because $\cos\theta_0 < 0$, the initial true anomaly θ_0 could be in the second or third quadrant. The position and velocity dot product yields $\mathbf{r}_0 \cdot \mathbf{v}_0 = r_0\dot{r}_0 = 67,071 \text{ km}^2/\text{s} > 0$, and therefore θ_0 is in the second quadrant because radial velocity is positive. Therefore, initial true anomaly is $\theta_0 = \cos^{-1}(-0.5) = 120°$ and propagated true anomaly is $\theta = \theta_0 + \Delta\theta = 200°$. Propagated radius is $r = p/(1 + e\cos\theta) = 36,253.4$ km.

Now we have all four values required for the Lagrangian coefficients. Using Eqs. (4.70)–(4.73), we can calculate

$$f = 1 - \frac{r}{p}(1 - \cos\Delta\theta) = -1.632334$$

$$g = \frac{rr_0 \sin\Delta\theta}{\sqrt{p\mu}} = 9{,}501.2766\,\text{s}$$

$$\dot{f} = \sqrt{\frac{\mu}{p}}\left(\frac{1-\cos\Delta\theta}{p}\cdot\frac{1}{r} - \frac{1}{r_0}\right)\tan\frac{\Delta\theta}{2} = -5.345888\left(10^{-5}\right)\text{s}^{-1}$$

$$\dot{g} = 1 - \frac{r_0}{p}(1 - \cos\Delta\theta) = -0.301453$$

Finally, the propagated state is computed by using the Lagrangian coefficients in Eq. (4.55):

$$\mathbf{r} = f\mathbf{r}_0 + g\mathbf{v}_0 = \begin{bmatrix} -19{,}092 \\ -30{,}245 \\ -5{,}918 \end{bmatrix} \text{km}$$

$$\mathbf{v} = \dot{f}\mathbf{r}_0 + \dot{g}\mathbf{v}_0 = \begin{bmatrix} 2.2512 \\ 0.4661 \\ -0.5918 \end{bmatrix} \text{km/s}$$

The reader should note that the Lagrangian coefficients f and $\dot{g}$ are dimensionless by observing their respective defining equations and the propagation equations shown above. Furthermore, the units of the Lagrangian coefficient $\dot{f}$ are per second, which is the time derivative of coefficient f. The units of g are seconds, which is the time integral of coefficient $\dot{g}$.

As a final note, the interested reader can determine the classical orbital elements by transforming either state $(\mathbf{r}_0, \mathbf{v}_0)$ or $(\mathbf{r}, \mathbf{v})$ using the methods presented in Chapter 3. This particular orbit is (approximately) a geostationary transfer orbit (GTO) with a perigee altitude of 200 km and apogee altitude reaching geostationary orbit (note that $e = 0.7301$). The remaining classical orbital elements are $i = 28.5°$, $\Omega = 40°$, and $\omega = 0$ (i.e., the perigee direction is along the ascending node). Figure 4.13 shows the GTO for this example. Note the initial position at $\theta_0 = 120°$ and propagated position at $\theta = 200°$.

Example 4.8 Repeat the orbit-propagation problem outlined in Example 4.7 using the Lagrangian coefficients defined by the change in eccentric anomaly, ΔE.

Here the Lagrangian coefficients are determined by Eqs. (4.80)–(4.83) and hence we need a, r_0, r, ΔE, and TOF $t - t_0$. We calculated the first three parameters in Example 4.7. Equation (4.33) will yield the eccentric anomalies corresponding to the initial and propagated true anomalies, $\theta_0 = 120°$ and $\theta = 200°$:

Initial position: $\quad \tan\dfrac{E_0}{2} = \sqrt{\dfrac{1-e}{1+e}}\tan\dfrac{\theta_0}{2} = 0.684102 \quad \rightarrow \quad E_0 = 1.1200\,\text{rad}$

Propagated position: $\quad \tan\dfrac{E}{2} = \sqrt{\dfrac{1-e}{1+e}}\tan\dfrac{\theta}{2} = -2.239944 \quad \rightarrow \quad E = 3.9814\,\text{rad}$

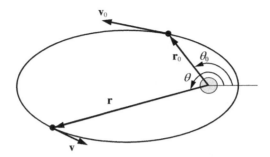

Figure 4.13 Geostationary transfer orbit propagation from state $(\mathbf{r_0}, \mathbf{v_0})$ to $(\mathbf{r}, \mathbf{v})$ (Example 4.7).

Note that the "calculator" inverse-tangent operation for the propagated position would yield $E = -2.3018$ rad (third quadrant) which is less than E_0. Because $\Delta E = E - E_0 > 0$, we added 2π to obtain $E = 3.9814$ rad (third quadrant). Using these values, the difference in eccentric anomaly is $\Delta E = E - E_0 = 2.7814$ rad (or 159.4°).

Next, we use Kepler's equation (4.21) to compute the TOF from $\mathbf{r_0}$ to $\mathbf{r}$

$$t - t_0 = \frac{1}{n}(M - M_0)$$

The mean motion is $n = \sqrt{\mu/a^3} = 1.6592(10^{-4})$ rad/s, and the two mean anomalies are

Initial mean anomaly: $\quad M_0 = E_0 - e\sin E_0 = 0.5195$ rad

Propagated mean anomaly: $\quad M = E - e\sin E = 4.5249$ rad

and, therefore, the flight time is $t - t_0 = 24{,}140.5$ s (or 6.7 h). Using Eqs. (4.80)–(4.83), we obtain the four Lagrangian coefficients:

$$f = 1 - \frac{a}{r_0}(1 - \cos\Delta E) = -1.632334$$

$$g = (t - t_0) - \sqrt{\frac{a^3}{\mu}}(\Delta E - \sin\Delta E) = 9{,}501.2766 \text{ s}$$

$$\dot{f} = \frac{-\sqrt{\mu a}\sin\Delta E}{r r_0} = -5.345888(10^{-5}) \text{ s}^{-1}$$

$$\dot{g} = 1 - \frac{a}{r}(1 - \cos\Delta E) = -0.301453$$

The above Lagrangian coefficients are identical to the $f, g, \dot{f}$, and $\dot{g}$ coefficients computed in Example 4.7 using the difference in true anomaly. Of course, this result should have been anticipated because $\Delta E = 159.4°$ was computed using the true anomalies corresponding to $\Delta\theta = 80°$ in Example 4.7. In other words, both $\Delta E = 159.4°$ and $\Delta\theta = 80°$ represent the same change in orbital position from $\mathbf{r_0}$ to $\mathbf{r}$. The propagated position and velocity vectors $(\mathbf{r}, \mathbf{v})$ are computed using the Lagrangian coefficients in Eq. (4.55) and the results are identical to the solution of Example 4.7.

Examples 4.7 and 4.8 show that orbit propagation is probably easier to perform by using the Lagrangian coefficients defined in terms of difference in true anomaly $\Delta\theta$. Calculating the coefficients in terms of ΔE requires the extra steps of computing eccentric anomalies from true anomalies and the TOF. The real advantage of the ΔE formulation is the case where flight time between $\mathbf{r}_0$ and $\mathbf{r}$ is known. This scenario is Lambert's problem, which we describe in the next section.

4.6 Lambert's Problem

Figure 4.14a presents the scenario for Lambert's problem: given two position vectors $\mathbf{r}_1$ and $\mathbf{r}_2$ and the flight time between them, determine the orbit. Of course, we can determine all orbital characteristics if we can compute the appropriate velocity vector $\mathbf{v}_1$ or $\mathbf{v}_2$. Most solutions to Lambert's problem use the Lagrangian coefficients and Eq. (4.55a) and (4.55b), which are repeated here with the slight change in subscripts to match Figure 4.14a

$$\mathbf{r}_2 = f\mathbf{r}_1 + g\mathbf{v}_1 \tag{4.84a}$$

$$\mathbf{v}_2 = \dot{f}\mathbf{r}_1 + \dot{g}\mathbf{v}_1 \tag{4.84b}$$

The reader should note the difference between the orbit-propagation method presented in Section 4.5 and Lambert's problem. In Section 4.5, the initial state $(\mathbf{r}_1, \mathbf{v}_1)$ is known and hence all orbital elements can be determined from energy, the angular momentum vector, and the eccentricity vector. For Lambert's problem, the two position vectors $\mathbf{r}_1$ and $\mathbf{r}_2$ and corresponding TOF are known and the orbit must be determined. As we shall see, solving Lambert's problem requires an iterative search because the expressions for the Lagrangian coefficients are transcendental functions of the unknown parameters.

Two possible paths (with the *same* flight time) exist between position vectors $\mathbf{r}_1$ and $\mathbf{r}_2$. Figure 4.14a shows the "short-way" transfer from $\mathbf{r}_1$ to $\mathbf{r}_2$ where the transfer angle $\Delta\theta$ is less than 180°. It is possible to determine a different orbit from $\mathbf{r}_1$ to $\mathbf{r}_2$ (in the opposite direction of the short-way transfer) where the transfer angle is greater than 180°. Figure 4.14b shows the "long-way" transfer from $\mathbf{r}_1$ to $\mathbf{r}_2$ with the same TOF as the short-way transfer in Figure 4.14a. Figure 4.15 combines Figures 4.14a and 4.14b and presents the short-way and long-way orbits on the same diagram. The short-way path is in the counter-clockwise direction while the long-way path is clockwise. Figure 4.15

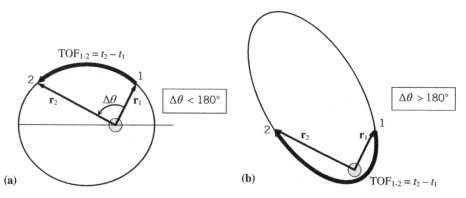

Figure 4.14 Lambert's problem: (a) "short-way" transfer $\Delta\theta < 180°$ and (b) "long-way" transfer $\Delta\theta > 180°$.

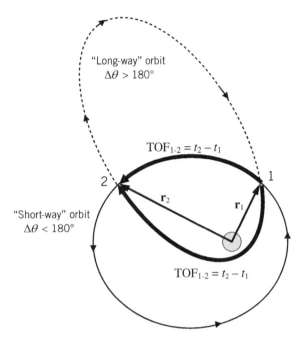

Figure 4.15 Short-way and long-way transfers between $\mathbf{r_1}$ and $\mathbf{r_2}$ with the same time of flight.

shows that two very different orbits provide a transfer from $\mathbf{r_1}$ to $\mathbf{r_2}$ with the same TOF $t_2 - t_1$. The short-way orbit has less energy (smaller semimajor axis) than the long-way orbit (dashed line) because a satellite on the short-way path travels a shorter distance in the same flight time. In some scenarios, the long-way path may become a hyperbolic trajectory in order to complete the transfer in the same flight time as the short-way path. The reader should also note that the short-way path is not always in the counter-clockwise direction (as shown in Figures 4.14a and 4.15); the directions of the short-way and long-way transfers depend on the position vectors $\mathbf{r_1}$ and $\mathbf{r_2}$.

We can compute the transfer angle or difference in true anomaly from the dot product of the two position vectors:

$$\cos\Delta\theta = \frac{\mathbf{r_1}\cdot\mathbf{r_2}}{r_1 r_2} \tag{4.85}$$

The "calculator" inverse cosine operation always places the angle in the first or second quadrant (i.e., a short-way transfer). If the long-way path is desired, then the inverse-cosine operation of Eq. (4.85) must be subtracted from 2π. The reader should also note that the angular momentum vector $\mathbf{h}$ for the short-way path in Figure 4.15 is directed out of the page while the angular momentum vector of the long-way path is into the page.

Many different techniques for solving Lambert's problem have been developed, including Lambert's original formulation, Gauss' method, and Battin's method. Vallado [2; pp. 467–498] presents a very good overview of these various methods including algorithms for solving Lambert's problem. Here we will formulate Lambert's problem and show one iterative technique for obtaining the solution. Using the orbit-propagation equations (4.84a) and (4.84b), we can write expressions for the unknown initial and terminal velocity vectors in terms of $\mathbf{r}_1$ and $\mathbf{r}_2$

$$\mathbf{v}_1 = \frac{1}{g}(\mathbf{r}_2 - f\mathbf{r}_1) \tag{4.86}$$

$$\mathbf{v}_2 = \dot{f}\mathbf{r}_1 + \frac{\dot{g}}{g}(\mathbf{r}_2 - f\mathbf{r}_1) \tag{4.87}$$

Clearly, determining the Lagrangian coefficients is the key to obtaining $\mathbf{v}_1$ and $\mathbf{v}_2$. Next, equate the two equation sets that define the Lagrangian coefficients in terms of $\Delta\theta$ and ΔE; that is, Eqs. (4.70)–(4.73) and Eqs. (4.80)–(4.83) with a change in subscript notation to indicate positions 1 and 2:

$$f = 1 - \frac{r_2}{p}(1 - \cos\Delta\theta) = 1 - \frac{a}{r_1}(1 - \cos\Delta E) \tag{4.88}$$

$$g = \frac{r_1 r_2 \sin\Delta\theta}{\sqrt{p\mu}} = (t_2 - t_1) - \sqrt{\frac{a^3}{\mu}}(\Delta E - \sin\Delta E) \tag{4.89}$$

$$\dot{f} = \sqrt{\frac{\mu}{p}}\left(\frac{1 - \cos\Delta\theta}{p} - \frac{1}{r_1} - \frac{1}{r_2}\right)\tan\frac{\Delta\theta}{2} = \frac{-\sqrt{\mu a}\sin\Delta E}{r_1 r_2} \tag{4.90}$$

$$\dot{g} = 1 - \frac{r_1}{p}(1 - \cos\Delta\theta) = 1 - \frac{a}{r_2}(1 - \cos\Delta E) \tag{4.91}$$

Equations (4.88)–(4.91) are composed of seven variables: r_1, r_2, a, p, $\Delta\theta$, ΔE, and $t_2 - t_1$. Of these seven variables, four are known: r_1, r_2, $\Delta\theta$, and TOF $t_2 - t_1$. Therefore, we have three unknown values: semimajor axis a, parameter p, and difference in eccentric anomaly ΔE. At first glance, it appears that we have an overdetermined system of four equations and three unknowns. However, recall that the four Lagrangian coefficients are not independent because of the condition $f\dot{g} - \dot{f}g = 1$ that was obtained by computing the angular momentum vector of the propagated state $(\mathbf{r}, \mathbf{v})$ [see Eqs. (4.56)–(4.59) for details]. In truth, we have three independent Lagrangian-coefficient equations and three unknowns. The difficulty is that these equations are transcendental functions of the unknown quantities. Therefore, a closed-form solution cannot be determined. We require an iterative search method for solution.

A basic iterative algorithm for solving Lambert's problem follows:

1) Given position vectors $\mathbf{r}_1$ and $\mathbf{r}_2$ (and direction of travel, i.e. short-way or long-way path), determine the radius magnitudes r_1 and r_2 and transfer angle $\Delta\theta$ using Eq. (4.85).
2) Guess a trial value of one of the three unknown quantities (a, p, or ΔE).
3) Use the f and $\dot{f}$ equations (4.88) and (4.90) to determine the other two remaining unknown values.
4) Use the g equation (4.89) to determine the TOF for the trial value.
5) Adjust the iteration parameter (go back to step 3) until the computed flight time in step 4 matches the actual TOF.

As previously stated, we will present one method for solving Lambert's problem. Before demonstrating one solution technique, it is very important to note that the

Lagrangian-coefficient expressions (4.80)–(4.83) hold for elliptical orbits only (note the existence of eccentric anomaly ΔE which only pertains to ellipses). In order to accommodate hyperbolic orbits, we require equations for the Lagrangian coefficients in terms of change in hyperbolic anomaly ΔF [the expressions in terms of $\Delta\theta$, Eqs. (4.70)–(4.73), hold for all orbits]. It is for this reason that a universal-variable formulation is often employed in order to develop Lagrangian-coefficient equations that are valid for elliptical, parabolic, and hyperbolic orbits. We will not pursue universal variables here; the interested reader may consult References [1–3] for methods that solve Lambert's problem using universal variables.

One technique for solving Lambert's problem is the p-iteration method. As the name implies, we guess a trial value of parameter and iterate on p until the computed TOF matches the actual flight time. Although a complete p-iteration algorithm must accommodate hyperbolic orbits (and hence express the Lagrangian coefficients in terms of ΔF), our discussion here will only consider elliptical orbits. The p-iteration method is selected because it is somewhat more intuitive than other methods and therefore relatively easy to comprehend. Again, the objective here is to present an introduction to Lambert's solution. The formulation of the p-iteration method is from Bate *et al.* [1; pp. 241–251].

Because p is our iteration variable, we must determine the two other unknown variables, a and ΔE, from the Lagrangian-coefficient equations. To do so requires manipulation of the f and $\dot{f}$ equations (4.88) and (4.90). These steps are not shown here (see Bate *et al.* [1] for details). Semimajor axis a can be determined from a trial value of p using

$$a = \frac{mkp}{(2m - l^2)p^2 + 2klp - k^2} \tag{4.92}$$

where the auxiliary variables k, l, and m are constants that are functions of the known values

$$k = r_1 r_2 (1 - \cos\Delta\theta) \tag{4.93}$$

$$l = r_1 + r_2 \tag{4.94}$$

$$m = r_1 r_2 (1 + \cos\Delta\theta) \tag{4.95}$$

Clearly, the constants k, l, and m can be computed directly from $\mathbf{r}_1$ and $\mathbf{r}_2$, that is, the given information for Lambert's problem. Equation (4.92) shows that semimajor axis becomes infinite (i.e., a parabolic orbit) when the denominator term is zero. The two roots of the quadratic denominator in Eq. (4.92) are

$$p_{\min} = \frac{k}{l + \sqrt{2m}} \tag{4.96}$$

$$p_{\max} = \frac{k}{l - \sqrt{2m}} \tag{4.97}$$

Here we use the "min" and "max" subscripts to denote the parameter limits for *elliptical transfers*. Therefore, if $p_{\min} < p < p_{\max}$, semimajor axis (4.92) is positive and the orbit between $\mathbf{r}_1$ and $\mathbf{r}_2$ is an ellipse. As p approaches the limits $p_{\min}$ or $p_{\max}$, the transfer approaches a parabolic trajectory. For $p > p_{\max}$, the transfer is hyperbolic (however, we will not consider hyperbolic transfers here). Because we are only considering elliptical transfers, we can restrict our p-iteration search between $p_{\min}$ and $p_{\max}$ (of course, a general p-iteration method must consider elliptical and hyperbolic transfers).

Next, we compute the f, g, and $\dot{f}$ coefficients using Eqs. (4.88)–(4.90) for the trial value of p and the known difference in true anomaly $\Delta\theta$

$$f = 1 - \frac{r_2}{p}(1 - \cos\Delta\theta)$$

$$g = \frac{r_1 r_2 \sin\Delta\theta}{\sqrt{p\mu}}$$

$$\dot{f} = \sqrt{\frac{\mu}{p}}\left(\frac{1 - \cos\Delta\theta}{p} - \frac{1}{r_1} - \frac{1}{r_2}\right)\tan\frac{\Delta\theta}{2}$$

We can determine $\cos\Delta E$ and $\sin\Delta E$ using the right-hand sides of Eqs. (4.88) and (4.90) with the numerical values of the f and $\dot{f}$ coefficients (computed above) and the trial semi-major axis a that has been computed using Eq. (4.92)

$$\cos\Delta E = 1 - \frac{r_1}{a}(1 - f) \tag{4.98}$$

$$\sin\Delta E = \frac{-r_1 r_2 \dot{f}}{\sqrt{\mu a}} \tag{4.99}$$

Both sine and cosine of ΔE are required to resolve the correct quadrant (e.g., using MATLAB's `atan2` function). The reader should ensure that ΔE is always positive, $0 < \Delta E < 2\pi$, so that the TOF calculation is correct. The trial TOF is computed from the g coefficient and Eq. (4.89)

$$t_2 - t_1 = g + \sqrt{\frac{a^3}{\mu}}(\Delta E - \sin\Delta E) \tag{4.100}$$

If the trial flight time computed using Eq. (4.100) matches the actual TOF, then the trial value of p is correct and we have determined the correct orbit from $\mathbf{r}_1$ to $\mathbf{r}_2$. If there is any error in flight time, then parameter p must be adjusted until the flight-time error is driven to a negligible value. One way to adjust p between iterations is to use the secant search method:

$$p_{i+1} = p_i - \tau_i \frac{p_i - p_{i-1}}{\tau_i - \tau_{i-1}} \tag{4.101}$$

where p_i is the parameter for the ith iteration, and τ_i is the difference between the flight time computed using Eq. (4.100) and the actual (or desired) flight time. Note that the fraction term on the right-hand side of Eq. (4.101) is the inverse of the finite-difference approximation of the derivative $d\tau/dp$. Therefore, the secant method is essentially New-ton's root-solving algorithm where the derivative term is replaced by a finite difference. After p is updated using Eq. (4.101), the p-iteration algorithm must update a [Eq. (4.92)], ΔE [Eqs. (4.98) and (4.99)], and TOF [Eq. (4.100)]. Convergence to the correct orbit occurs when $|\tau_i| < \varepsilon$ where ε is an acceptable (small) TOF error. When the converged solution is obtained, we can compute the two velocity vectors using the Lagrangian coef-ficients and Eqs, (4.86) and (4.87). The following example illustrates the solution to Lam-bert's problem using the p-iteration method.

Example 4.9 A ground station determines two position vectors for a weather-forecasting satellite operated by the National Oceanic and Atmospheric Administration (NOAA). The two position vectors in the ECI frame are

$$\mathbf{r}_1 = \begin{bmatrix} -5,655.144 \\ -3,697.284 \\ -2,426.687 \end{bmatrix} \text{km}, \quad \mathbf{r}_2 = \begin{bmatrix} 5,891.286 \\ 2,874.322 \\ -2,958.454 \end{bmatrix} \text{km}$$

The flight time between position vectors is 63 min. Use the p-iteration method to determine the NOAA satellite's orbit.

The difference in true anomaly is computed using Eq. (4.85)

$$\cos\Delta\theta = \frac{\mathbf{r}_1 \cdot \mathbf{r}_2}{r_1 r_2} = -0.712062$$

Therefore, the two possible values are $\Delta\theta = 135.40°$ (short way) or $\Delta\theta = 224.60°$ (long way). Because the flight time is relatively "large" (63 min) and both arcs are significant, it is not apparent which path is correct (as a counter example, it would be easy to select the short-way path if the flight time was 6 min and we had to choose between $\Delta\theta = 20°$ or $\Delta\theta = 340°$). Let us begin with the short-way path, $\Delta\theta = 135.40°$. The auxiliary constants are computed using Eqs. (4.93)–(4.95):

$$k = r_1 r_2 (1 - \cos\Delta\theta) = 8.839433(10^7) \text{ km}^2$$

$$l = r_1 + r_2 = 14,370.85 \text{ km}$$

$$m = r_1 r_2 (1 + \cos\Delta\theta) = 1.486631(10^7) \text{ km}^2$$

We see that k, l, and m are the same constants, whether we use the short-way or long-way path. Equations (4.96) and (4.97) provide the lower and upper bounds on p for an elliptical transfer:

$$p_{min} = \frac{k}{l + \sqrt{2m}} = 4,459.042 \text{ km}$$

$$p_{max} = \frac{k}{l - \sqrt{2m}} = 9,911.799 \text{ km}$$

We can select the first trial value of parameter closer to p_{min}

$$p_1 = 0.7 p_{min} + 0.3 p_{max} = 6,094.869 \text{ km}$$

Using Eq. (4.92), the corresponding trial semimajor axis is

$$a = \frac{mkp}{(2m - l^2)p^2 + 2klp - k^2} = 7,255.803 \text{ km}$$

The three independent Lagrangian coefficients are computed using r_1, r_2, $\Delta\theta$, and the trial value of p

$$f = 1 - \frac{r_2}{p}(1 - \cos\Delta\theta) = -1.020182$$

$$g = \frac{r_1 r_2 \sin\Delta\theta}{\sqrt{p\mu}} = 735.466534 \text{ s}$$

$$\dot{f} = \sqrt{\frac{\mu}{p}} \left(\frac{1 - \cos\Delta\theta}{p} - \frac{1}{r_1} - \frac{1}{r_2} \right) \tan\frac{\Delta\theta}{2} = 5.049967 \left(10^{-5}\right) s^{-1}$$

Coefficients $f, \dot{f}$, and the trial value of a are used to solve for change in eccentric anomaly:

$$\cos\Delta E = 1 - \frac{r_1}{a}(1-f) = -0.998824 \quad \text{and} \quad \sin\Delta E = \frac{-r_1 r_2 \dot{f}}{\sqrt{\mu a}} = -0.048482$$

Using atan2 and ensuring a positive value, we obtain $\Delta E = 3.190094$ rad. The trial TOF is computed using Eq. (4.100) and g, ΔE, and a

$$t_2 - t_1 = g + \sqrt{\frac{a^3}{\mu}} (\Delta E - \sin\Delta E) = 3,905.86 \, s = 65.0977 \, min$$

Hence the TOF error is $\tau_1 = 2.0977$ min for the current iterate. We cannot use the secant method for the second iteration (we do not yet have past-iteration data), so we select a second trial value of p that is closer to p_{max}:

$$p_2 = 0.3p_{min} + 0.7p_{max} = 8,275.972 \, km$$

The p-iteration algorithm recalculates a, f, g, $\dot{f}$, ΔE, and the trial flight time using the above sequence of equations. The trial flight time is 27.52 min and the corresponding error is $\tau_2 = -35.48$ min which is significantly worse than the first trial. However, the two trial values of p were chosen arbitrarily. The third trial value of p is computed using the secant search (4.101)

$$p_3 = p_2 - \tau_2 \frac{p_2 - p_1}{\tau_2 - \tau_1} = 6,216.625 \, km$$

This procedure repeats until the flight-time error is less than 10^{-4} min (0.006 s). Table 4.2 summarizes the p-iteration scheme for the short-way transfer with $\Delta\theta = 135.40°$. The converged value for parameter is $p = 6,144.013$ km and the corresponding semimajor axis is $a = 7,193.6$ km. We can compute the eccentricity of the short-way solution:

Table 4.2 p-iteration trials for the short-way transfer (Example 4.9).

Iteration	Trial p (km)	Trial $t_2 - t_1$ (min)	Flight-time error (min)
1	6,094.869	65.0977	2.0977
2	8,275.972	27.5198	−35.4802
3	6,216.626	60.1459	−2.8541
4	6,036.473	67.7868	4.7868
5	6,149.333	62.7814	−0.2186
6	6,144.404	62.9840	−0.0160
7	6,144.013	63.0001	$5.82(10^{-5})$

Table 4.3 *p*-iteration trials for the long-way transfer (Example 4.9).

Iteration	Trial p (km)	Trial $t_2 - t_1$ (min)	Flight-time error (min)
1	6,094.869	37.4175	−25.5825
2	8,275.972	134.6874	71.6874
3	6,668.512	48.3316	−14.6684
4	6,941.555	55.3307	−7.6693
5	7,240.742	65.0128	2.0128
6	7,178.545	62.7892	−0.2108
7	7,184.442	62.9948	−0.0052
8	7,184.592	63.0000	$1.40(10^{-5})$

$$e = \sqrt{1 - \frac{p}{a}} = 0.3820$$

Hence, the short-way orbit is not circular. Furthermore, the perigee radius of the short-way orbit is $r_p = p/(1 + e) = 4{,}445.8$ km which is less than the radius of the Earth. Clearly, the short-way orbit is not feasible.

We can repeat the *p*-iteration steps for the long-way path where $\Delta\theta = 224.60°$. As previously stated, the k, l, and m constants and limits p_{min} and p_{max} remain the same as those computed for the short-way path. The iterations corresponding to the long-way path are summarized in Table 4.3. The semimajor axis and eccentricity of the converged orbit solution are $a = 7{,}184.60$ km and $e = 0.001$, respectively, which indicate a near-circular orbit. Hence, the NOAA satellite follows the long-way path between positions $\mathbf{r}_1$ and $\mathbf{r}_2$. Figure 4.16 shows the long-way transfer on the nearly circular orbit.

The Lagrangian coefficients associated with the converged long-way orbit are

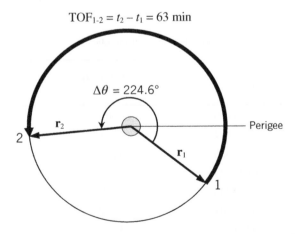

$$\text{TOF}_{1\text{-}2} = t_2 - t_1 = 63 \text{ min}$$

$$\Delta\theta = 224.6°$$

Figure 4.16 Long-way transfer for a NOAA weather satellite (Example 4.9).

$$f = -0.713771, \quad g = -677.398216\,\text{s}, \quad \dot{f} = 7.273210\left(10^{-4}\right)\text{s}^{-1}, \quad \text{and } \dot{g} = -0.710752$$

We can use Eqs. (4.86) and (4.87) to determine the initial and terminal velocity vectors in the ECI frame:

$$\mathbf{v}_1 = \frac{1}{g}(\mathbf{r}_2 - f\mathbf{r}_1) = \begin{bmatrix} -2.7381 \\ -0.3474 \\ 6.9244 \end{bmatrix} \text{km/s}$$

$$\mathbf{v}_2 = \dot{f}\mathbf{r}_1 + \frac{\dot{g}}{g}(\mathbf{r}_2 - f\mathbf{r}_1) = \begin{bmatrix} -2.1670 \\ -2.4422 \\ -6.6865 \end{bmatrix} \text{km/s}$$

Finally, we can use either state, $(\mathbf{r}_1, \mathbf{v}_1)$ or $(\mathbf{r}_2, \mathbf{v}_2)$, to determine the remaining orbital elements. We find that the initial and terminal true anomalies are $\theta_1 = -40°$ and $\theta_2 = 184.6°$ as shown in Figure 4.16. The inclination of the NOAA satellite is $i = 98.77°$, which is a nearly polar orbit.

As a final note for this example, let us observe the trends in the flight time and orbital path at the limiting values of parameter. Figure 4.17 shows TOF error (trial flight time minus actual flight time) for short-way and long-way paths with $p_{min} < p < p_{max}$. For short-way paths with $p \to p_{min}$, the transfer becomes a very thin, long ellipse with a very large apogee distance as shown in Figure 4.18a. Because the satellite passes through apogee on the short-way path with $p \to p_{min}$ (a "lofted" transfer; see Figure 4.18a), the flight

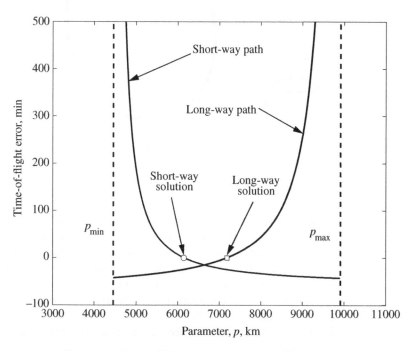

Figure 4.17 Time-of-flight errors vs. parameter (Example 4.9).

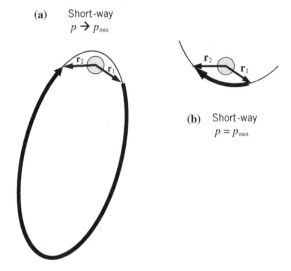

Figure 4.18 Short-way paths: (a) $p \rightarrow p_{\min}$ and (b) $p = p_{\max}$ (Example 4.9).

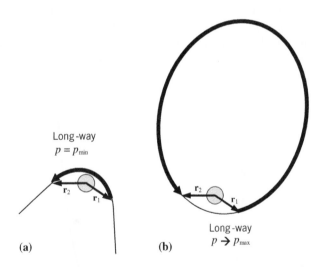

Figure 4.19 Long-way paths: (a) $p = p_{\min}$ and (b) $p \rightarrow p_{\max}$ (Example 4.9).

time is very long and hence the TOF error is very large as seen in Figure 4.17. As p becomes larger and approaches $p_{\max}$, the satellite passes through perigee on its short-way path and hence the flight time becomes smaller. When $p = p_{\max}$, the transfer is a parabola and the short-way path passes through perigee as shown in Figure 4.18b. For $p > p_{\max}$, the short-way transfer becomes hyperbolic and the flight time continues to diminish. Figure 4.19 shows the long-way paths for $p = p_{\min}$ (parabola) and $p \rightarrow p_{\max}$ (elliptical transit through apogee). When using the limiting values for p, the apse directions and flight times of the long-way paths are essentially reversed when compared with the short-way paths.

Example 4.9 used a secant search to iterate on p; other root-solving methods (such as the bisection method or Brent's method) may be used. As a practical matter, it may be useful to check the flight time at the limiting values of p (e.g., $p = 1.001p_{min}$ and $p = 0.999p_{max}$) to ensure that the actual flight time is bracketed. Another search option is to employ a preliminary "brute-force" search where the flight-time error is computed for a small number of equally spaced values of p between p_{min} and p_{max}. A good initial guess for p can then be determined by interpolating the TOF error data and the subsequent secant search should rapidly converge to the solution.

It is important to restate that this section has presented an algorithm for solving Lambert's problem when the orbit transfer is an ellipse. Extending these results to hyperbolic trajectories requires expressing the Lagrangian coefficients in terms of difference in hyperbolic anomaly ΔF or using universal variables to accommodate all possible orbits with one set of equations. We will solve Lambert's problem to obtain interplanetary transfers. As we shall see in Chapter 10, the transfer between two planetary orbits (such as an Earth–Mars transfer) is an ellipse with the sun as the primary gravitational body. Therefore, we may use our "ellipse-only" formulation of Lambert's problem (and the p-iteration method) to design trajectories for interplanetary missions.

4.7 Summary

In this chapter, we formulated the relationship between flight time and position in an orbit. This relationship, first developed by Johann Kepler, is known as *Kepler's equation*. For elliptical orbits, TOF is related to eccentric anomaly, which is an auxiliary angle developed by Kepler. Eccentric anomaly is a function of the satellite's angular position in the orbit (true anomaly) and the orbital eccentricity. Therefore, if we know the orbital characteristics (semimajor axis and eccentricity) and two angular positions, we can compute the corresponding flight time using Kepler's equation. However, if we want to determine the angular position of a satellite at a future time, the solution is not so straightforward because Kepler's equation is transcendental in eccentric anomaly and no closed-form solution exists. This orbit-propagation scenario is known as *Kepler's problem* and it must be solved using a numerical iteration scheme such as Newton's method. We also developed the so-called Lagrangian coefficients that can be used to propagate an orbit that is expressed in terms of position and velocity vectors. The real advantage of the Lagrangian coefficients is that they serve as a foundation for solving *Lambert's problem*, which involves determining the orbit that passes between two known position vectors with a specified flight time. As with Kepler's problem, the solution to Lambert's problem requires numerical iteration. The ability to solve Lambert's problem provides an invaluable tool for orbit determination and interplanetary mission design (Chapter 10).

References

1 Bate, R.R., Mueller, D.D., and White, J.E., *Fundamentals of Astrodynamics*, Dover, New York, 1971.

2 Vallado, D.A., *Fundamentals of Astrodynamics and Applications*, 4th edn, Microcosm Press, Hawthorne, CA, 2013.

3 Curtis, H.D., *Orbital Mechanics for Engineering Students*, 3rd edn, Butterworth-Heinemann, Oxford, 2014.

4 Battin, R.H., *An Introduction to the Mathematics and Methods of Astrodynamics, Revised Edition*, AIAA Education Series, Reston, VA, 1999.

Problems

Conceptual Problems

4.1 A geocentric satellite is at perigee where its altitude is 250 km and its velocity is 9.045 km/s. Determine the flight time to reach true anomaly $\theta_1 = 100°$.

4.2 An Earth-observation satellite has perigee and apogee altitudes of 350 and 1,206 km, respectively. Determine the satellite's flight time from true anomaly $\theta_1 = 270°$ to $\theta_2 = 90°$.

4.3 A launch vehicle experiences a partial failure during the ascent phase and consequently its upper-stage engine is shut down 94 s too soon. The premature engine shutdown occurs at an altitude of 185 km with an inertial velocity $v = 7.226$ km/s and flight-path angle $\gamma = 1.5°$. Determine the flight time from engine shutdown to the entry-interface altitude of 122 km when the vehicle begins to reenter the Earth's atmosphere.

4.4 A space probe is departing Earth orbit on a parabolic trajectory. Determine the flight time from perigee (where the altitude is 350 km) to the position on the parabolic path where it crosses geostationary orbit (i.e., the radial distance is 42,164 km).

4.5 Repeat Problem 4.4 for the case where the space probe is departing Earth orbit on a *hyperbolic* trajectory. The hyperbolic excess speed of the departure asymptote is $v_\infty^+ = 2.51$ km/s. Determine the flight time from perigee (where the altitude is 350 km) to the position on the hyperbolic path where it crosses geostationary orbit (i.e., the radial distance is 42,164 km).

4.6 A geocentric satellite is following a parabolic trajectory with a perigee altitude of 800 km. The true anomaly of the satellite at its current position is $\theta_1 = 300°$. Determine the satellite's flight time from its current position to a radial distance of 500,000 km.

4.7 An interplanetary spacecraft is returning to Earth for a "gravity-assist maneuver" (to be analyzed in Chapter 10). The spacecraft is approaching Earth and at time t_1 its radial position, velocity, and flight-path angle are $r_1 = 88,071$ km, $v_1 = 4.107$ km/s, and $\gamma_1 = -77.1°$, respectively. Determine the flight time from its current position to perigee.

4.8 An Earth-observation satellite is in an orbit with perigee and apogee altitudes of 500 and 1,600 km, respectively. If the satellite is currently at true anomaly $\theta_1 = 320°$, determine its true anomaly 30 min later.

4.9 An Earth-orbiting satellite is currently at perigee (altitude = 350 km) with a velocity of 8.12 km/s. Determine its radius, velocity, and flight-path angle 4.2 h after perigee passage.

4.10 A geocentric satellite orbit has semimajor axis $a = 9,875$ km and eccentricity $e = 0.305$. If the satellite is currently at true anomaly $\theta_1 = 200°$, determine its true anomaly 7.3 h later.

Problems 4.11–4.13 involve a geocentric satellite with the following position and velocity vectors (in the ECI frame) at epoch time t_0:

$$\mathbf{r}_0 = \begin{bmatrix} -6,796 \\ 4,025 \\ 3,490 \end{bmatrix} \text{km}, \quad \mathbf{v}_0 = \begin{bmatrix} -3.7817 \\ -6.0146 \\ 1.1418 \end{bmatrix} \text{km/s}$$

4.11 Determine the flight time from the state vector $(\mathbf{r}_0, \mathbf{v}_0)$ to state vector $(\mathbf{r}_1, \mathbf{v}_1)$ for a $200°$ increase in true anomaly (i.e., $\Delta\theta = 200°$).

4.12 Propagate the orbit using the Lagrangian coefficients and determine the position and velocity vectors $(\mathbf{r}, \mathbf{v})$ corresponding to a change in true anomaly of $\Delta\theta = 200°$.

4.13 Repeat Problem 4.12 by propagating the orbit using the Lagrangian coefficients that are defined by the change in eccentric anomaly, ΔE.

4.14 A ground station determines two position vectors of a geocentric satellite. The two position vectors in the ECI frame are

$$\mathbf{r}_1 = \begin{bmatrix} -6,241.629 \\ -1,434.658 \\ 2,745.447 \end{bmatrix} \text{km}, \quad \mathbf{r}_2 = \begin{bmatrix} 2,833.781 \\ -7,414.259 \\ 2,568.063 \end{bmatrix} \text{km}$$

The flight time between the position vectors is $t_2 - t_1 = 27$ min. Carry out the calculations of the p-iteration method to show that $p = 7,778.228$ km satisfies Lambert's problem for a *short-way* transfer. Determine the satellite's velocity vectors $\mathbf{v}_1$ and $\mathbf{v}_2$, eccentricity e, and inclination i.

4.15 Repeat Problem 4.14 for the *long-way* transfer, and show that $p = 3,305.535$ km satisfies Lambert's problem. Explain why the 27 min long-way transfer is *not* feasible.

MATLAB Problems

4.16 Write an M-file that will solve Kepler's problem and determine a geocentric satellite's propagated position in an elliptical orbit for a desired time of flight. The inputs to the M-file are semimajor axis a (in km), eccentricity e, current true anomaly θ_1 (in degrees), and flight time $t_2 - t_1$ (in min). The outputs are the states at the propagated orbital position: radius r_2 (in km), velocity v_2 (in km/s), flight-path angle γ_2 (in degrees), and true anomaly θ_2 (in degrees). Test your M-file by solving Example 4.6.

4.17 Write an M-file that will propagate an orbit using the Lagrangian coefficients. The inputs to the M-file are the satellite's initial state vector $(\mathbf{r}_0,\mathbf{v}_0)$ in the ECI frame (with units of kilometers and kilometers per second, respectively) and the change in true anomaly $\Delta\theta$ (in degrees). The outputs are the satellite's propagated state vector $(\mathbf{r},\mathbf{v})$ in the ECI frame (in km and km/s), the initial and propagated true anomalies (θ_0 and θ, in degrees), and the corresponding time of flight (in min). The M-file should be able to accommodate elliptical and hyperbolic orbits. Test your M-file by solving Example 4.7.

4.18 Write an M-file that will solve Lambert's problem using the p-iteration method. The inputs are the satellite's initial and target position vectors in ECI coordinates ($\mathbf{r}_1$ and $\mathbf{r}_2$, in km), the flight time between $\mathbf{r}_1$ and $\mathbf{r}_2$ (in min), and a flag that indicates either the short-way or long-way transfer. The outputs are the corresponding velocity vectors $\mathbf{v}_1$ and $\mathbf{v}_2$ (in km/s), the converged parameter p (in km), eccentricity e, and semimajor axis a (in km). Test your M-file by solving Example 4.9.

Mission Applications

4.19 The Pegasus launch vehicle reaches its second-stage burnout at an altitude of 192 km, inertial velocity $v = 5.49$ km/s, and flight-path angle $\gamma = 25.8°$. The launch vehicle then coasts in this orbit for 432 s (7.2 min) at which point it ignites its third stage. Determine the altitude, velocity, and flight-path angle of the Pegasus vehicle at third-stage ignition.

4.20 Figure P4.20 shows an intermediate transfer orbit used by the Chandra X-ray Observatory (CXO). Perigee and apogee altitudes of the CXO transfer orbit are 300 and 72,000 km, respectively. If the CXO is currently at true anomaly $\theta_1 = 45°$, determine the flight time for it to return to perigee.

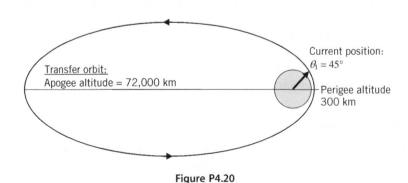

Figure P4.20

Problems 4.21 and 4.22 involve the Stardust capsule, which returned to Earth in January 2006 on a hyperbolic approach trajectory after sampling particles from the comet Wild-2.

4.21 A ground-based station is tracking the Stardust capsule during its return to Earth. The station determines the capsule's position and velocity vectors in the ECI frame at epoch time t_0:

$$\mathbf{r}_0 = \begin{bmatrix} 219{,}469 \\ 99{,}139 \\ -87{,}444 \end{bmatrix} \text{km}, \quad \mathbf{v}_0 = \begin{bmatrix} -5.7899 \\ -2.2827 \\ 2.3722 \end{bmatrix} \text{km/s}$$

Propagate the Stardust capsule's trajectory using the Lagrangian coefficients and determine its ECI position and velocity vectors $(\mathbf{r}, \mathbf{v})$ corresponding to a change in true anomaly of $\Delta\theta = 90°$. What is the altitude of the Stardust capsule at the propagated orbital position? In addition, compute the time of flight between these two positions.

4.22 When the Stardust capsule arrived at the "edge" of the Earth's atmosphere (the so-called "entry interface" altitude of 122 km), it had inertial velocity $v = 12.9$ km/s and flight-path angle $\gamma = -8.21°$. Determine the flight time of the Stardust capsule from a radial position of 300,000 km to entry interface (i.e., 122 km altitude).

4.23 A ground-based station is tracking a military satellite that is in a near-equatorial, high-eccentricity orbit. At epoch time t_1, the satellite's ECI position vector is

$$\mathbf{r}_1 = \begin{bmatrix} 6{,}379.7 \\ -3{,}504.9 \\ -207.7 \end{bmatrix} \text{km}$$

The station determines the satellite's ECI position vector 67 min later:

$$\mathbf{r}_2 = \begin{bmatrix} 3{,}854.4 \\ 18{,}422.6 \\ 519.2 \end{bmatrix} \text{km}$$

Carry out the *p*-iteration algorithm calculations to show that $p = 10{,}596.1$ km satisfies Lambert's problem for the flight time between these two position vectors. In addition, compute the satellite's eccentricity, semimajor axis, and period.

Problems 4.24 and 4.25 involve the translunar trajectory of the Lunar Atmosphere and Dust Environment Explorer (LADEE) spacecraft (see Figure 2.14 and Example 2.6). Shortly after departing Earth orbit, a ground-based tracking station determines LADEE's state vector in ECI coordinates at epoch time t_0

$$\mathbf{r}_0 = \begin{bmatrix} 2{,}449.5 \\ 5{,}822.9 \\ -5{,}020.5 \end{bmatrix} \text{km}, \quad \mathbf{v}_0 = \begin{bmatrix} -6.0947 \\ 1.9272 \\ -7.4294 \end{bmatrix} \text{km/s}$$

4.24 Determine LADEE's radial distance, velocity, and flight-path angle 24 h after epoch time t_0.

4.25 Determine LADEE's geocentric position and velocity vectors $(\mathbf{r},\mathbf{v})$ after a change in true anomaly of $\Delta\theta = 120°$ and compute the corresponding flight time.

Problems 4.26 and 4.27 involve the Mars Reconnaissance Orbiter (MRO) spacecraft that was placed in a near-circular orbit about Mars in late 2006. At epoch time t_0, the MRO state vector is

$$\mathbf{r}_0 = \begin{bmatrix} 544.0 \\ 1,855.4 \\ 3,138.2 \end{bmatrix} \text{km}, \quad \mathbf{v}_0 = \begin{bmatrix} 2.0700 \\ -2.4510 \\ 1.1081 \end{bmatrix} \text{km/s}$$

The state vector is measured relative to a Mars-centered inertial coordinate system.

4.26 Determine the flight time for the MRO spacecraft to transit from $(\mathbf{r}_0,\mathbf{v}_0)$ to true anomaly $\theta = 60°$.

4.27 Determine the MRO state vector $(\mathbf{r},\mathbf{v})$ 8 h after epoch t_0.

Problems 4.28–4.30 involve the Juno spacecraft, which departed Earth in early August 2011 and arrived at Jupiter in early July 2016.

4.28 The Juno spacecraft approached Jupiter on a hyperbolic trajectory with eccentricity $e = 1.0172$ and semimajor axis $a = -4.384(10^6)$ km. Determine the flight time from a radial distance of $1.07(10^6)$ km (roughly the orbit of Jupiter's moon Ganymede) to periapsis ("perijove") passage.

4.29 The Juno spacecraft fired a retrorocket at its periapsis ("perijove") position to slow down and establish a highly elliptical (and nearly polar) orbit about Jupiter with semimajor axis $a = 4,092,211$ km and eccentricity $e = 0.981574$ (the period of this orbit is 53.5 days). Juno used a highly elliptical orbit so that it was well above Jupiter's harsh radiation environment for the majority of its orbit. Determine the total flight time that the Juno spacecraft's altitude is less than 100,000 km in a single orbit (Jupiter's polar radius is 66,854 km).

4.30 Mission operators planned a second retrorocket burn at perijove for October 2016. This burn would slow down Juno and reduce its orbital period from 53.5 to 14 days (however, the rocket burn was canceled due to a valve malfunction). The (planned) smaller orbit has semimajor axis $a = 1,674,498$ km and eccentricity $e = 0.954970$. Determine the total flight time that the Juno spacecraft's altitude is less than 100,000 km in a single orbit (Jupiter's polar radius is 66,854 km). Compare this total radiation exposure time for the 14-day orbit to the exposure time in Problem 4.29 for a 53.5-day orbit.

5

Non-Keplerian Motion

5.1 Introduction

Chapters 2–4 presented the fundamentals of two-body orbital motion. Recall that two major assumptions lead to the two-body problem: (1) the gravitational body is spherically symmetric; and (2) no forces other than gravity act on the satellite. The principle conclusions gleaned from the two-body problem are:

1) A satellite's path is a conic section (circle, ellipse, parabola, or hyperbola) with the gravitational body at the focus.
2) A satellite's total energy and angular momentum relative to the central body remain constant along its orbital path.
3) The satellite's motion is contained in a plane that is fixed in inertial space.
4) The classical orbital elements (a, e, i, Ω, ω) are constants.

Unfortunately, the two-body problem only exists in theory. It is rather easy to "poke holes" in the assumptions for two-body motion. For example, the Earth is not a perfect sphere with homogenous mass distribution. Furthermore, it is clear that third-body gravitational forces (due to the sun, moon, and planets) *always* act on a satellite. Finally, other forces arising from atmospheric drag, solar radiation pressure, and onboard thrust may act on the satellite. These effects are *perturbations* that cause the satellite's motion to deviate from the theoretical two-body motion. In many cases, the perturbing forces are small relative to the central body's gravity and therefore the satellite's orbit slowly deviates from the theoretical two-body motion over a long time scale. However, in other scenarios (such as an Earth orbit with a perigee pass through the atmosphere), the perturbing forces may be large enough to produce rapid changes in the orbital elements.

It is useful to define a few terms for this chapter. We will call a satellite's orbital path that is governed by the two-body problem *Keplerian motion*. Therefore, Keplerian motion has the characteristics previously listed (i.e., the orbit is a conic section; the orbital plane is fixed in space, etc.). A satellite follows *non-Keplerian motion* if the two-body problem does *not* hold because of the presence of a non-spherical central body or additional forces due to other gravitational bodies, drag, or rocket thrust. It is important to note that theoretical Keplerian motion is an idealization that does not exist.

This preliminary discussion is not intended to disregard our "analytical tool-kit" that has been developed in Chapters 2–4. Quite the contrary, the two-body problem leads to extremely useful analytical expressions that allow rapid orbital calculations. We will

Space Flight Dynamics, First Edition. Craig A. Kluever.
© 2018 John Wiley & Sons Ltd. Published 2018 by John Wiley & Sons Ltd.
Companion website: www.wiley.com/go/Kluever/spaceflightmechanics

continue to use Keplerian motion for evaluating orbital maneuvers (Chapters 7, 8) and analyzing interplanetary missions (Chapter 10). For many space mission scenarios, two-body motion provides an excellent approximation of the satellite's true motion. The goal of this chapter is to present an introduction to orbital perturbations and their effects on the theoretical Keplerian motion. For some particular perturbing forces, we will develop analytical expressions for changes in the orbital elements over time.

5.2 Special Perturbation Methods

Let us begin our discussion of orbital perturbations by repeating the governing equation of motion for the two-body problem:

$$\ddot{\mathbf{r}} + \frac{\mu}{r^3}\mathbf{r} = 0 \tag{5.1}$$

where $\mathbf{r}$ is the position vector of the satellite relative to an inertial frame fixed to the center of the gravitational body. Writing the two-body equation in terms of absolute acceleration, we obtain

$$\ddot{\mathbf{r}} = -\frac{\mu}{r^3}\mathbf{r} = \frac{\mu}{r^2}\left(\frac{-\mathbf{r}}{r}\right) \tag{5.2}$$

Equation (5.2) reflects the two major assumptions of the two-body problem, that is, the only force acting on the satellite is the gravity of the central body with (acceleration) magnitude μ/r^2 and unit-vector direction $-\mathbf{r}/r$. The magnitude and direction of the gravitational acceleration are consistent with a spherically symmetric gravitational body.

Next, let us consider the two-body equation (5.2) with a *perturbing* vector

$$\ddot{\mathbf{r}} = -\frac{\mu}{r^3}\mathbf{r} + \mathbf{a}_P \tag{5.3}$$

where $\mathbf{a}_P$ is the perturbing acceleration acting on the satellite. For an Earth-orbiting satellite, $\mathbf{a}_P$ could be computed from the vector sum of the appropriate perturbing forces (third-body gravity, atmospheric drag, onboard thrust, etc.) divided by the satellite's mass. Of course, the perturbation acceleration $\mathbf{a}_P$ must be expressed in a non-rotating coordinate system such as the Earth-centered inertial (ECI) frame. Once we have determined the perturbing acceleration $\mathbf{a}_P$, we may obtain the "true" non-Keplerian motion of the satellite by two successive integrations: integrating Eq. (5.3) produces the velocity vector $\dot{\mathbf{r}} = \mathbf{v}$; integrating velocity $\mathbf{v}$ produces the satellite's position vector $\mathbf{r}$.

Special perturbation methods numerically integrate the perturbed two-body equations of motion in order to determine the satellite's position $\mathbf{r}$ and velocity $\mathbf{v}$ at later times. This method is conceptually simple and relatively easy to implement with today's computers. However, the perturbed orbit obtained through numerical integration is a *specific* (or *special*) case that is only valid for the given initial conditions and parameters that model the perturbing acceleration $\mathbf{a}_P$. *Cowell's formulation* (or *Cowell's method*) is a special perturbation method and obtains the perturbed orbit by direct numerical integration of Eq. (5.3). *General perturbation methods* reformulate the perturbed equations of motion using simplifying approximations so that they may be *analytically* integrated. We treat special perturbations in this section and discuss general perturbations in Section 5.3.

It is useful to briefly define *conservative* and *non-conservative* forces (you may know these terms from a university physics course). Total energy is constant for a satellite that is solely under the influence of a conservative force. Gravitational forces from the central body and perturbing third bodies (sun, moon, etc.) are conservative forces. We can determine the absolute acceleration of a conservative force from the gradient of a scalar potential function. Non-conservative forces increase or decrease the satellite's total energy. Atmospheric drag, solar radiation pressure, and thrust are examples of non-conservative forces.

5.2.1 Non-Spherical Central Body

The first perturbation we will consider is the non-spherical shape of the central gravitational body. Recall that in Chapter 3, we used an ellipsoidal model for the Earth: that is, a "flattened" sphere where the equatorial radius is about 21 km greater than the polar radius. In addition, the Earth has uneven mass distribution that varies with latitude and longitude. Spherical harmonics are used to model a planet's surface and subsequent gravitational field. We will provide a brief introduction to this topic by focusing on the spherical harmonics that represent the "oblate ellipsoid" shape of the Earth.

We may model the Earth's (or any planet's) gravitation field using a scalar *potential function* $U(r, \lambda, \phi')$ that depends on radius r, longitude λ, and geocentric latitude ϕ' (recall that *geocentric* latitude is measured from the equatorial plane to a line from the center of the gravitational body to the surface; see Figure 3.17). For the Earth, we will call $U(r, \lambda, \phi')$ the *geopotential function*. The gravitational acceleration of an Earth-orbiting satellite is the gradient of the geopotential function. To show this, let us consider the *two-body* geopotential function

$$U_{2b}(r) = \frac{\mu}{r} \tag{5.4}$$

Note that we have used the subscript "2b" to indicate a "two-body" potential function. The two-body geopotential function U_{2b} is the negative potential energy and only depends on radius r (there is no dependence on latitude or longitude because the two-body problem assumes that the Earth is a homogeneous sphere). The absolute acceleration due to gravity is the gradient of the geopotential function

$$\ddot{\mathbf{r}} = \nabla U_{2b}(r) \tag{5.5}$$

where the "del" or vector differential operator for the ECI Cartesian frame is

$$\nabla = \frac{\partial}{\partial x}\mathbf{I} + \frac{\partial}{\partial y}\mathbf{J} + \frac{\partial}{\partial z}\mathbf{K} \tag{5.6}$$

Recall that **IJK** are unit vectors associated with the ECI coordinate system. The satellite's position vector **r** has ECI components (x, y, z), or

$$\mathbf{r} = x\mathbf{I} + y\mathbf{J} + z\mathbf{K} \tag{5.7}$$

and the magnitude of the position vector is $r = \sqrt{x^2 + y^2 + z^2}$. Therefore, the two-body geopotential (5.4) is rewritten as

$$U_{2b} = \frac{\mu}{\sqrt{x^2 + y^2 + z^2}} \tag{5.8}$$

The gradient of the two-body geopotential is computed by applying the "del" operator defined by Eq. (5.6):

$$\nabla U_{2b} = \frac{-\mu x}{\left(x^2 + y^2 + z^2\right)^{3/2}}\mathbf{I} + \frac{-\mu y}{\left(x^2 + y^2 + z^2\right)^{3/2}}\mathbf{J} + \frac{-\mu z}{\left(x^2 + y^2 + z^2\right)^{3/2}}\mathbf{K} \tag{5.9}$$

Note that the common denominator term in Eq. (5.9) is r^3. Furthermore, we can substitute $\mathbf{r} = x\mathbf{I} + y\mathbf{J} + z\mathbf{K}$ so that Eq. (5.9) becomes

$$\nabla U_{2b} = -\frac{\mu}{r^3}\mathbf{r} \tag{5.10}$$

Using this result as the right-hand side of Eq. (5.5), we obtain

$$\ddot{\mathbf{r}} = -\frac{\mu}{r^3}\mathbf{r} \tag{5.11}$$

Equation (5.11) is the governing equation of motion for the two-body problem. This simple exercise shows that the governing equation for two-body (Keplerian) motion can be derived from the two-body geopotential function $U_{2b} = \mu/r$.

We desire a more accurate representation of the Earth's gravitational field (in particular, the gravitational field of an oblate, flattened sphere). Let us express the *total* geopotential function $U(r, \lambda, \phi')$ as the sum of a two-body potential function and a *disturbing* potential function

$$U(r,\lambda,\phi') = \frac{\mu}{r} + R(r,\lambda,\phi') \tag{5.12}$$

It should be clear that μ/r is the two-body potential in Eq. (5.12) and $R(r, \lambda, \phi')$ is the disturbing potential function. The disturbing potential function $R(r, \lambda, \phi')$ represents perturbations due to a non-spherical Earth with an uneven mass distribution. Hence, it depends on radius, longitude, and geocentric latitude. It is possible to express the disturbing potential function R in terms of *spherical harmonics* or periodic functions on the surface of a unit sphere. These spherical harmonics consist of *zonal harmonics* (bands of latitude), *sectoral harmonics* (sections of longitude), and *tesseral harmonics* ("checkerboard tiles" that depend on latitude and longitude). Characterizing the disturbing potential R using a complete set of spherical harmonic functions is beyond the scope of this textbook (the interested reader may consult Vallado [1; pp. 538–550] for details). Instead, let us focus on a total geopotential function that is axially symmetric about the $\mathbf{K}$ axis and therefore only depends on radius and latitude:

$$U(r,\phi') = \frac{\mu}{r}\left[1 - \sum_{k=2}^{\infty} J_k \left(\frac{R_E}{r}\right)^k P_k(\sin\phi')\right] \tag{5.13}$$

where J_k are the zonal harmonic coefficients, R_E is the equatorial radius of the Earth, and P_k is a Legendre polynomial of order k. The input to the Legendre polynomial is $\sin\phi' = z/r$. The Legendre polynomials represent the "harmonic fluctuations" of the Earth's surface relative to a spherical shape as latitude varies. The dimensionless zonal coefficients J_k represent the "dips" and "bulges" of the Earth's surface (relative to a sphere) and they are empirically determined from satellite observations. Zonal coefficient J_2 models the Earth's "bulge" at its equator and it is nearly 1000 times larger than all other J_k coefficients. Therefore, if we only consider the J_2 coefficient, Eq. (5.13) becomes

$$U(r,\phi') = \frac{\mu}{r}\left[1 - J_2\left(\frac{R_E}{r}\right)^2 P_2(\sin\phi')\right] \tag{5.14}$$

Equation (5.14) is the total geopotential function for an *oblate Earth* or a "flattened" sphere. This geopotential function has axial symmetry about the Earth's polar axis (i.e., the Earth's mass is equally distributed with longitude), and the second-order Legendre polynomial $P_2(\sin\phi')$ models the Earth's mass bulge at its equator. The reader should note that Eq. (5.14) represents the simplest possible non-spherical model of the Earth because we have neglected zonal harmonics greater than order 2 as well as all sectoral and tesseral harmonics. Finally, the reader should note that if we set $J_2 = 0$ (no oblateness) in Eq. (5.14), then we have $U(r) = \mu/r$ and we are back to a spherical Earth and two-body (Keplerian) motion.

The reader may have some difficulty comprehending or visualizing the geopotential function Eq. (5.14) that models Earth's equatorial bulge. In order to enhance our understanding of the geopotential function, let us plot $U(r, \phi')$ for various radial distances from the Earth's center. First, we must present the second-order Legendre polynomial

$$P_2(u) = \frac{1}{2}\left(3u^2 - 1\right) \tag{5.15}$$

where the input is $u = \sin\phi'$. In order to evaluate Eq. (5.14), we also need numerical values for Earth's equatorial radius and Earth's second zonal harmonic coefficient J_2. Let us use $R_E = 6{,}378.14$ km, $J_2 = 0.0010826267$, and $\mu = 3.986(10^5)$ km^3/s^2 (Earth's gravitational parameter). Using these values in Eq. (5.14), we can evaluate the geopotential function $U(r, \phi')$ for a fixed radius r and geocentric latitude ϕ' ranging from $-90°$ (South Pole) to $+90°$ (North Pole) [of course we must use Eq. (5.15) to evaluate the second-order Legendre polynomial as latitude varies]. Figure 5.1 shows how the oblate geopotential

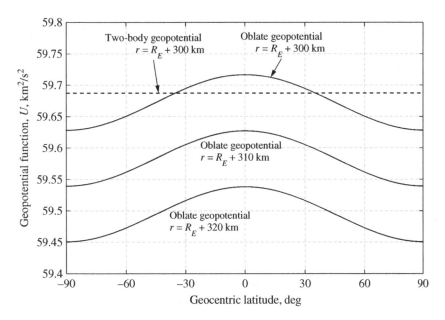

Figure 5.1 Oblate geopotential function [Eq. (5.14)] vs. latitude: polar low-Earth orbit.

function varies with latitude for a polar low-Earth orbit (LEO). It is clear that $U(r, \phi')$ exhibits a "bulge" at the equator ($\phi' = 0$) and a minimum value at the poles. Figure 5.1 also shows that the "strength" of the oblate geopotential function diminishes as radial distance increases. The two-body geopotential function evaluated at $r_{LEO} = R_E +$ 300 km [i.e., $U(r_{LEO}) = \mu/r_{LEO}$] is shown in Figure 5.1 as the dashed line. Because the two-body geopotential corresponds to a spherical, homogeneous Earth, it does not exhibit any variation with latitude. Figure 5.2 shows the oblate geopotential function for circular polar orbits with radii approximately equal to the radius of geostationary orbit, that is, 42,164 km. Note that the geopotential function's "equatorial bulge" is significantly reduced for near-geostationary orbits (i.e., the geopotential function appears to behave more like the two-body potential). Therefore, we can expect that an oblate-Earth gravity model will exhibit a more pronounced effect on satellites in low-altitude orbits as compared with high-altitude orbits.

The next step is to take the gradient of Eq. (5.14) in order to determine the satellite's absolute acceleration in an oblate-Earth gravity field. After substituting Eq. (5.15) for the Legendre polynomial, Eq. (5.14) becomes

$$U(r,\phi') = \frac{\mu}{r}\left[1 - \frac{J_2}{2}\left(\frac{R_E}{r}\right)^2\left(\frac{3z^2}{r^2} - 1\right)\right] \tag{5.16}$$

Note that we have used $u = \sin\phi' = z/r$ in the Legendre polynomial (5.15). We must also substitute $r = \sqrt{x^2 + y^2 + z^2}$ and $r^2 = x^2 + y^2 + z^2$ into Eq. (5.16) so that the geopotential function U is in terms of Cartesian coordinates (x,y,z). The satellite's absolute acceleration due to the oblate-Earth gravity field is the gradient of Eq. (5.16):

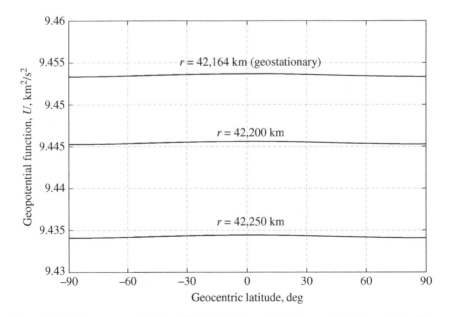

Figure 5.2 Oblate geopotential function [Eq. (5.14)] vs. latitude: geostationary orbital radius.

$$\ddot{\mathbf{r}} = \nabla U(r,\phi') = \frac{\partial U}{\partial x}\mathbf{I} + \frac{\partial U}{\partial y}\mathbf{J} + \frac{\partial U}{\partial z}\mathbf{K} \tag{5.17}$$

After taking the partial derivatives (and performing some algebra), we can write the three acceleration components in the ECI frame:

$$\ddot{x} = \frac{\partial U}{\partial x} = \frac{-\mu x}{r^3}\left[1 - J_2\frac{3}{2}\left(\frac{R_E}{r}\right)^2\left(\frac{5z^2}{r^2} - 1\right)\right] \tag{5.18}$$

$$\ddot{y} = \frac{\partial U}{\partial y} = \frac{-\mu y}{r^3}\left[1 - J_2\frac{3}{2}\left(\frac{R_E}{r}\right)^2\left(\frac{5z^2}{r^2} - 1\right)\right] \tag{5.19}$$

$$\ddot{z} = \frac{\partial U}{\partial z} = \frac{-\mu z}{r^3}\left[1 - J_2\frac{3}{2}\left(\frac{R_E}{r}\right)^2\left(\frac{5z^2}{r^2} - 3\right)\right] \tag{5.20}$$

Equations (5.18)–(5.20) are the absolute acceleration components of a satellite orbiting an oblate spheroid Earth. These equations fit the form of Eq. (5.3), the perturbed two-body equations of motion. Note that the terms outside the brackets in Eqs. (5.18)–(5.20) are the two-body gravitational components of $-\mu\mathbf{r}/r^3$. The terms involving J_2 are the components of the perturbing acceleration $\mathbf{a}_P$.

Now we can apply the special perturbation method to a satellite orbiting a non-spherical (oblate) Earth. Numerically integrating Eqs. (5.18)–(5.20) will yield the velocity components, $\mathbf{v} = [\dot{x}\ \dot{y}\ \dot{z}]^T$, and numerically integrating $\mathbf{v}$ will yield the position vector, $\mathbf{r} = [x\ y\ z]^T$. The following example demonstrates the special perturbation technique for determining the non-Keplerian motion of an Earth satellite.

Example 5.1 Use the special perturbation method to obtain the non-Keplerian motion of a LEO that is perturbed by Earth-oblateness (J_2) effects. The initial orbital elements at time $t = 0$ are

> Semimajor axis a_0 = 8,059 km
> Eccentricity e_0 = 0.15
> Inclination i_0 = 20°
> Longitude of the ascending node Ω_0 = 60°
> Argument of perigee ω_0 = 30°
> True anomaly θ_0 = 50°

The special perturbation method requires that we numerically integrate the satellite's perturbed equations of motion. Equations (5.18)–(5.20) are the satellite's absolute acceleration components due to central-body gravity perturbed by the Earth-oblateness (or J_2) effect. We can use a numerical integration algorithm (such as a Runge–Kutta scheme) to integrate Eqs. (5.18)–(5.20) from time $t = 0$ to an arbitrary final time $t = t_1$. Of course, we must integrate these three acceleration equations to obtain the ECI velocity vector $\mathbf{v}$, and integrate the three velocity components to obtain the ECI position vector $\mathbf{r}$. The numerical integration must begin at the initial ECI state vector $(\mathbf{r}_0,\mathbf{v}_0)$ associated with the initial LEO orbital elements. Using the coordinate transformation algorithm presented in Section 3.5, the initial Cartesian coordinates are

$$\mathbf{r}_0 = \begin{bmatrix} -5,134.41 \\ 4,405.01 \\ 2,420.05 \end{bmatrix} \text{km}, \quad \mathbf{v}_0 = \begin{bmatrix} -5.5265 \\ -5.5142 \\ 0.7385 \end{bmatrix} \text{km/s}$$

Here we use MATLAB's M-file ode45.m to numerically integrate Eqs. (5.18)–(5.20) starting from the initial state vector $(\mathbf{r}_0, \mathbf{v}_0)$. The final end time is set at 10 h (note that because semimajor axis is 8,059 km, the orbital period is 120 min = 2 h). The numerical values of the constants used here are $J_2 = 0.0010826267$, $R_E = 6,378.14$ km, and $\mu = 3.986004(10^5)$ km^3/s^2. Numerical integration produces ECI vectors $\mathbf{r}(t)$ and $\mathbf{v}(t)$.

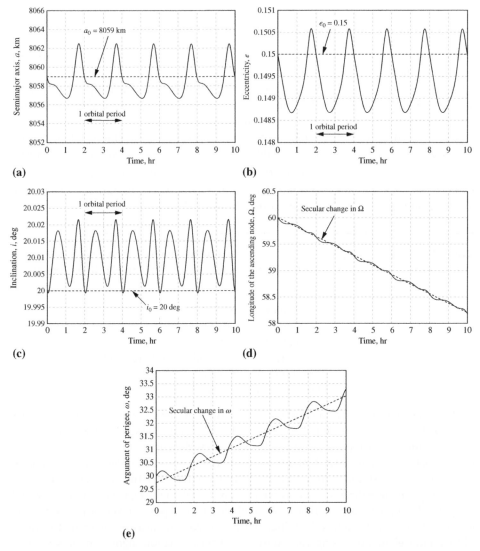

Figure 5.3 LEO with J_2 perturbation: (a) semimajor axis; (b) eccentricity; (c) inclination; (d) longitude of the ascending node; and (e) argument of perigee (Example 5.1).

Because the histories of these Cartesian coordinates provide very little insight into how the orbit is affected by the J_2 perturbation, the state vector $(\mathbf{r},\mathbf{v})$ simulation data are transformed to orbital elements using the methods presented in Section 3.4.

Figure 5.3 presents the time histories of the five classical orbital elements after numerical integration of the perturbed equations of motion. Figure 5.3a shows that semimajor axis a is perturbed by the Earth's oblateness and varies during an orbital revolution (recall that the orbital period is 2 h). However, semimajor axis returns to its initial value (a_0 = 8,059 km) at the end of every 2 h orbit. This behavior shows that the J_2 perturbation is conservative because its net effect on energy is zero over an orbital revolution. Figure 5.3b shows that eccentricity also exhibits oscillations during each 2-h orbit and Figure 5.3c shows that inclination has two periodic cycles each revolution. Figures 5.3a–c show that Earth's oblateness causes very small periodic variations in semimajor axis, eccentricity, and inclination and that the net change is zero for these three elements over each orbital revolution. The oblateness perturbation, however, causes the longitude of the ascending node Ω and argument of perigee ω to drift over time as illustrated in Figures 5.3d and 5.3e. Periodic fluctuations in Ω and ω are evident, but these oscillations are superimposed on to a linear function with time. The *secular* changes in Ω and ω are the linear variations with time and are illustrated by the dashed lines in Figures 5.3d and 5.3e. Figure 5.3d shows that Ω diminishes by about 1.8° over 10 h and hence the ascending node vector $\mathbf{n}$ is drifting *westward* at an average rate of 0.18 deg/h or 4.32 deg/day. Figure 5.3e shows that the argument of perigee ω increases by about 3.25° over 10 h. Therefore, the perigee direction $\mathbf{e}$ is drifting away from the ascending node $\mathbf{n}$ at an average rate of 0.325 deg/h or 7.8 deg/day.

Example 5.1 is a demonstration of the special perturbation method where the only perturbation is due to a non-spherical (oblate) Earth. Numerical integration of the perturbed equations of motion illustrates that Earth oblateness causes periodic variations in the orbital elements a, e, and i but with a net zero change after a full revolution. The longitude of ascending node Ω and argument of perigee ω show periodic and *secular* changes. The dashed lines in Figures 5.3d and 5.3e show the secular changes (or linear drift) in Ω and ω. Because Ω shows a steady drift rate, the orbital plane is rotating about the Earth's pole. The linear drift in ω indicates that the apse line is rotating in the orbital plane about the angular momentum vector $\mathbf{h}$.

We should reiterate that Example 5.1 has demonstrated non-Keplerian motion where the only perturbation is due to an oblate Earth. Furthermore, this example illustrates the "special" part of the special perturbation method; that is, we gleaned the oblateness effect only after performing numerical integration and plotting the results. The secular changes in Ω and ω pertain to the "specific" or "special" initial orbit presented in Example 5.1. At this point, we cannot make general statements regarding the secular drift rates for Ω and ω. *General perturbation methods* use analytical techniques to develop "general" expressions that convey the effects of perturbations. We will revisit and characterize the Earth-oblateness effect with general perturbation methods in the next section.

5.3   General Perturbation Methods

General perturbation methods seek analytical solutions for a perturbed orbit. Analytical solutions are tractable if a series expansion replaces the perturbing acceleration.

Therefore, the non-Keplerian motion produced by general perturbations is an approximate solution. However, the advantage of general perturbations is that the analytical expressions allow us to rapidly compute the perturbed orbit instead of relying on numerical integration. In addition, general perturbation methods give us insight to how a particular perturbation (such as an oblate Earth) alters the Keplerian (or two-body) motion.

5.3.1 Lagrange's Variation of Parameters

We will present a brief overview of a method known as the *variation of parameters*. Originally developed by Leonhard Euler and improved by Joseph-Louis Lagrange, the variation of parameters (or variation of orbital elements) consists of six first-order differential equations

$$\frac{d\boldsymbol{\alpha}}{dt} = \mathbf{f}(\boldsymbol{\alpha}, t) \tag{5.21}$$

where $\boldsymbol{\alpha}$ is the 6×1 vector of the orbital elements

$$\boldsymbol{\alpha} = \begin{bmatrix} a & e & i & \Omega & \omega & \sigma \end{bmatrix}^T \tag{5.22}$$

The sixth orbital element is angle $\sigma = -nt_p$, where t_p is the time instant (a constant) when the satellite is at periapsis. Recall from Chapter 4 that n is the angular velocity known as the mean motion

$$n = \sqrt{\frac{\mu}{a^3}} \tag{5.23}$$

We use σ as the sixth orbital element instead of θ_0 (true anomaly at a known instant of time). For the two-body problem, all orbital elements are constant. Hence, the right-hand side of Eq. (5.21) is zero (i.e., $d\boldsymbol{\alpha}/dt = \mathbf{0}$). When perturbations are present, the right-hand side vector $\mathbf{f}(\boldsymbol{\alpha}, t)$ in Eq. (5.21) is not a null vector, and we have non-Keplerian motion where the orbital elements vary with time. Lagrange developed the variation of parameters for conservative perturbations.

We will only show an outline of the variation of parameters approach based on its discussion in Battin [2; pp. 476–483]. Our goal is to develop the right-hand sides of the first-order differential equations for $d\boldsymbol{\alpha}/dt$. To begin, let us rewrite the satellite's absolute acceleration using the geopotential function U

$$\ddot{\mathbf{r}} = \nabla U \tag{5.24}$$

Recall for a non-spherical central body, the geopotential is the sum of the two-body potential function U_{2b} and the disturbing function R

$$U = \frac{\mu}{r} + R \tag{5.25}$$

Next, apply the del operator to U and substitute the result in Eq. (5.24) to yield

$$\ddot{\mathbf{r}} = -\frac{\mu}{r^3}\mathbf{r} + \nabla R \tag{5.26}$$

Equation (5.26) is the satellite's acceleration due to a two-body gravity field that is disturbed by the function R. It should be clear that Eq. (5.26) becomes the two-body problem when the disturbing function R vanishes (i.e., no perturbations).

The position and velocity vector solutions for either perturbed or unperturbed (two-body) motion can be expressed as

$$\mathbf{r} = \mathbf{r}(t, \boldsymbol{\alpha}) \tag{5.27}$$

$$\mathbf{v} = \mathbf{v}(t, \boldsymbol{\alpha}) \tag{5.28}$$

In other words, position and velocity vectors are functions of the six orbital elements (collected in vector $\boldsymbol{\alpha}$) and time t. We know that the velocity vector $\mathbf{v}$ is the time derivative of the position vector $\mathbf{r}$. The time derivative of Eq. (5.27) is

$$\text{Non-Keplerian motion:} \quad \mathbf{v} = \frac{d\mathbf{r}}{dt} = \frac{\partial \mathbf{r}}{\partial t} + \frac{\partial \mathbf{r}}{\partial \boldsymbol{\alpha}} \frac{d\boldsymbol{\alpha}}{dt} \tag{5.29}$$

Equation (5.29) allows for non-Keplerian motion because it contains the perturbation term $d\boldsymbol{\alpha}/dt$. For two-body motion, $d\boldsymbol{\alpha}/dt = \mathbf{0}$ and Eq. (5.29) becomes

$$\text{Two-body motion:} \quad \mathbf{v} = \frac{\partial \mathbf{r}}{\partial t} \tag{5.30}$$

Now let us introduce the concept of *osculating orbital elements*. Consider a satellite with non-Keplerian motion caused by perturbations (e.g., an oblate Earth). Suppose at time instant t_1 we can determine the satellite's ECI state vector $(\mathbf{r}_1, \mathbf{v}_1)$. Given the state vector $(\mathbf{r}_1, \mathbf{v}_1)$, we can compute the classical orbital elements. These elements would describe the satellite's two-body motion that would exist if suddenly all perturbations vanished at time t_1. The instantaneous elements are called the osculating orbital elements. Because the perturbations persist, the orbital elements change with time. However, at time t_1 the non-Keplerian orbit and the fictitious two-body orbit (the *osculating orbit*) both contain the state vector $(\mathbf{r}_1, \mathbf{v}_1)$. Using the osculating orbit concept, we can equate the perturbed and unperturbed velocity vectors, Eqs. (5.29) and (5.30), to obtain the following condition

$$\frac{\partial \mathbf{r}}{\partial \boldsymbol{\alpha}} \frac{d\boldsymbol{\alpha}}{dt} = \mathbf{0} \tag{5.31}$$

Equation (5.31) must hold on the osculating orbit. This expression does not imply that $d\boldsymbol{\alpha}/dt = \mathbf{0}$ (we know that the orbital elements vary with time). Equation (5.31) states that multiplying the 3×6 matrix of partial derivatives $(\partial \mathbf{r}/\partial \boldsymbol{\alpha})$ by the 6×1 vector $d\boldsymbol{\alpha}/dt$ must produce a 3×1 null vector.

In a similar fashion, let us take the time derivative of Eq. (5.28)

$$\ddot{\mathbf{r}} = \frac{d\mathbf{v}}{dt} = \frac{\partial \mathbf{v}}{\partial t} + \frac{\partial \mathbf{v}}{\partial \boldsymbol{\alpha}} \frac{d\boldsymbol{\alpha}}{dt} \tag{5.32}$$

For two-body motion, $\boldsymbol{\alpha}$ is constant and Eq. (5.32) becomes

$$\text{Two-body motion:} \quad \ddot{\mathbf{r}} = \frac{d\mathbf{v}}{dt} = \frac{\partial \mathbf{v}}{\partial t} = -\frac{\mu}{r^3} \mathbf{r} \tag{5.33}$$

Substituting Eq. (5.33) for $\partial \mathbf{v}/\partial t$ in Eq. (5.32) and comparing the result with Eq. (5.26), we determine a second condition:

$$\frac{\partial \mathbf{v}}{\partial \boldsymbol{\alpha}} \frac{d\boldsymbol{\alpha}}{dt} = \nabla R \tag{5.34}$$

We should reiterate that our goal is to determine the six time derivatives of the orbital elements, $d\boldsymbol{\alpha}/dt$, as functions of the disturbing potential R. Equations (5.31) and (5.34) each provide three conditions, and therefore they must be solved simultaneously in order to determine the 6×1 vector $d\boldsymbol{\alpha}/dt$. We can combine these two equations with a few additional matrix manipulations. First, multiply Eq. (5.31) by the 6×3 matrix $[\partial \mathbf{v}/\partial \boldsymbol{\alpha}]^T$ to form six equations:

$$\left[\frac{\partial \mathbf{v}}{\partial \boldsymbol{\alpha}}\right]^T \frac{\partial \mathbf{r}}{\partial \boldsymbol{\alpha}} \frac{d\boldsymbol{\alpha}}{dt} = \mathbf{0} \tag{5.35}$$

The right-hand side of Eq. (5.35) is a 6×1 null vector. Next, multiply Eq. (5.34) by the 6×3 matrix $[\partial \mathbf{r}/\partial \boldsymbol{\alpha}]^T$ to form six additional equations:

$$\left[\frac{\partial \mathbf{r}}{\partial \boldsymbol{\alpha}}\right]^T \frac{\partial \mathbf{v}}{\partial \boldsymbol{\alpha}} \frac{d\boldsymbol{\alpha}}{dt} = \left[\frac{\partial \mathbf{r}}{\partial \boldsymbol{\alpha}}\right]^T \frac{\partial R}{\partial \mathbf{r}} \tag{5.36}$$

Note that we have expressed ∇R as $\partial R/\partial \mathbf{r}$ in Eq. (5.36). We can use the chain rule to express the right-hand side of Eq. (5.36) as $\partial R/\partial \boldsymbol{\alpha}$. Subtracting Eq. (5.35) from Eq. (5.36) yields

$$\mathbf{L}\frac{d\boldsymbol{\alpha}}{dt} = \frac{\partial R}{\partial \boldsymbol{\alpha}} \tag{5.37}$$

where the 6×6 *Lagrangian matrix* $\mathbf{L}$ is

$$\mathbf{L} = \left[\frac{\partial \mathbf{r}}{\partial \boldsymbol{\alpha}}\right]^T \frac{\partial \mathbf{v}}{\partial \boldsymbol{\alpha}} - \left[\frac{\partial \mathbf{v}}{\partial \boldsymbol{\alpha}}\right]^T \frac{\partial \mathbf{r}}{\partial \boldsymbol{\alpha}} \tag{5.38}$$

Finally, multiply both sides of Eq. (5.37) by the inverse of the Lagrangian matrix

$$\frac{d\boldsymbol{\alpha}}{dt} = \mathbf{L}^{-1}\frac{\partial R}{\partial \boldsymbol{\alpha}} \tag{5.39}$$

Equation (5.39) provides the means to determine each time derivative of the orbital elements. Recall that the 6×1 vector $d\boldsymbol{\alpha}/dt$ is

$$\frac{d\boldsymbol{\alpha}}{dt} = \begin{bmatrix} da/dt \\ de/dt \\ di/dt \\ d\Omega/dt \\ d\omega/dt \\ d\sigma/dt \end{bmatrix} \tag{5.40}$$

Battin [2; pp. 477–479] and Vallado [1; pp. 621–626] use the so-called Lagrange brackets to determine the ith row and jth column of the Lagrangian matrix $\mathbf{L}$

$$[\alpha_i, \alpha_j] = L_{ij} = \frac{\partial \mathbf{r}}{\partial \alpha_i} \cdot \frac{\partial \mathbf{v}}{\partial \alpha_j} - \frac{\partial \mathbf{r}}{\partial \alpha_j} \cdot \frac{\partial \mathbf{v}}{\partial \alpha_i} \tag{5.41}$$

The Lagrange brackets exhibit the properties $[\alpha_i, \alpha_i] = 0$ (i.e., diagonal elements of **L** are zero) and $[\alpha_i, \alpha_j] = -[\alpha_j, \alpha_i]$ (i.e., skew symmetry). Therefore, only 15 (out of 36) Lagrange brackets must be computed. The key to obtaining the Lagrange brackets (and ultimately Lagrange's variation of parameters) is the calculation of the partial derivatives of the ECI vectors **r** and **v** with respect to the six orbital elements, or $\partial\mathbf{r}/\partial\alpha_i$ and $\partial\mathbf{v}/\partial\alpha_i$. This step requires us to express position and velocity vectors in terms of orbital elements. Recall that in Section 3.5, we developed the transformation from orbital elements to the ECI state vector $(\mathbf{r},\mathbf{v})$. The transformation begins with the calculation of the position and velocity vectors in the perifocal (**PQW**) frame, Eqs. (3.18) and (3.23):

$$\mathbf{r}_{PQW} = \frac{a(1-e^2)}{1+e\cos\theta}\begin{bmatrix}\cos\theta\\\sin\theta\\0\end{bmatrix} \tag{5.42}$$

$$\mathbf{v}_{PQW} = \frac{\mu}{\sqrt{\mu a(1-e^2)}}\begin{bmatrix}-\sin\theta\\e+\cos\theta\\0\end{bmatrix} \tag{5.43}$$

Multiplying Eqs. (5.42) and (5.43) by the rotation matrix transforms the state vector to ECI coordinates

$$\mathbf{r} = \widetilde{\mathbf{R}}\mathbf{r}_{PQW} \tag{5.44}$$

$$\mathbf{v} = \widetilde{\mathbf{R}}\mathbf{v}_{PQW} \tag{5.45}$$

Using Eq. (3.39) for the 3×3 rotation matrix $\widetilde{\mathbf{R}}$, the ECI state vector is

$$\mathbf{r} = \frac{a(1-e^2)}{1+e\cos\theta}\begin{bmatrix}c_\Omega c_\omega - s_\Omega s_\omega c_i & -c_\Omega s_\omega - s_\Omega c_\omega c_i & s_\Omega s_i\\s_\Omega c_\omega + c_\Omega s_\omega c_i & -s_\Omega s_\omega + c_\Omega c_\omega c_i & -c_\Omega s_i\\s_\omega s_i & c_\omega s_i & c_i\end{bmatrix}\begin{bmatrix}\cos\theta\\\sin\theta\\0\end{bmatrix} \tag{5.46}$$

$$\mathbf{v} = \frac{\mu}{\sqrt{\mu a(1-e^2)}}\begin{bmatrix}c_\Omega c_\omega - s_\Omega s_\omega c_i & -c_\Omega s_\omega - s_\Omega c_\omega c_i & s_\Omega s_i\\s_\Omega c_\omega + c_\Omega s_\omega c_i & -s_\Omega s_\omega + c_\Omega c_\omega c_i & -c_\Omega s_i\\s_\omega s_i & c_\omega s_i & c_i\end{bmatrix}\begin{bmatrix}-\sin\theta\\e+\cos\theta\\0\end{bmatrix} \tag{5.47}$$

Recall that Eqs. (5.46) and (5.47) use a short-hand notation for sines and cosines of the three rotation angles i, Ω, and ω (e.g., $c_\Omega = \cos\Omega$ and $s_\Omega = \sin\Omega$).

We will not pursue the explicit calculation of the partial derivatives $\partial\mathbf{r}/\partial\alpha_i$ and $\partial\mathbf{v}/\partial\alpha_i$ any further. Although the partial-derivative process is tedious, it is fairly straightforward: perform the matrix-vector multiplication, Eqs. (5.46) and (5.47), to produce the 3×1 vectors **r** and **v** in terms of six orbital elements; compute the 15 Lagrange brackets using Eq. (5.41) and the appropriate partial derivatives; assemble the 6×6 Lagrangian matrix **L**; invert the Lagrangian matrix; multiply $\mathbf{L}^{-1}$ by the 6×1 disturbance vector $\partial R/\partial\boldsymbol{\alpha}$ to determine the variations of parameters $d\boldsymbol{\alpha}/dt$. The interested reader may consult Vallado [1] and Battin [2] for details of this process. The end result is *Lagrange's planetary equations*:

$$\frac{da}{dt} = \frac{2}{na} \frac{\partial R}{\partial \sigma} \tag{5.48}$$

$$\frac{de}{dt} = \frac{1-e^2}{na^2 e} \frac{\partial R}{\partial \sigma} - \frac{\sqrt{1-e^2}}{na^2 e} \frac{\partial R}{\partial \omega} \tag{5.49}$$

$$\frac{di}{dt} = \frac{1}{na^2 \sqrt{1-e^2}\sin i} \left[\cos i \frac{\partial R}{\partial \omega} - \frac{\partial R}{\partial \Omega} \right] \tag{5.50}$$

$$\frac{d\Omega}{dt} = \frac{1}{na^2 \sqrt{1-e^2}\sin i} \frac{\partial R}{\partial i} \tag{5.51}$$

$$\frac{d\omega}{dt} = \frac{\sqrt{1-e^2}}{na^2 e} \frac{\partial R}{\partial e} - \frac{\cos i}{na^2 \sqrt{1-e^2}\sin i} \frac{\partial R}{\partial i} \tag{5.52}$$

$$\frac{d\sigma}{dt} = -\frac{2}{na} \frac{\partial R}{\partial a} - \frac{1-e^2}{na^2 e} \frac{\partial R}{\partial e} \tag{5.53}$$

Equations (5.48)–(5.53) are Lagrange's variation of parameter equations, and they are in terms of the partial derivatives of the disturbance potential R. Recall that R must be a conservative disturbing function such as a non-spherical gravitational body or third-body gravity. Close inspection of Lagrange's equations shows that the disturbing potential R must be expressed in terms of the orbital elements so that we can compute the partial derivatives.

At this point, the reader may have lost sight of the fact that the original objective of this section is to develop a *general* perturbation method that results in *analytical expressions* for the effects of perturbations. Lagrange's variation of parameter equations (5.48)–(5.53) are highly nonlinear first-order differential equations – finding analytical solutions for each orbital element does not appear to be an easy task! The next subsection presents analytical expressions for the "mean" or "averaged" effect of Earth's oblateness.

5.3.2 Secular Perturbations due to Oblateness (J_2)

Recall that Example 5.1 illustrated the special perturbation method of numerically integrating the absolute acceleration of a satellite in the Cartesian (ECI) frame. The satellite's absolute acceleration is expressed as the gradient of the total geopotential function $U = U_{2b} + R$ [see Eq. (5.12)]. In Example 5.1, we only considered the zonal harmonic coefficient J_2 (oblateness). Hence, the disturbing function is

$$R(r,\phi') = -\frac{\mu}{r} J_2 \left(\frac{R_E}{r} \right)^2 P_2(\sin\phi') \tag{5.54}$$

Recall that P_2 is the second-order Legendre polynomial and ϕ' is geocentric latitude. Note that we can repeat Example 5.1 by numerically integrating the Lagrange variation of parameter equations (5.48)–(5.53) as long as we can compute the required partial derivatives of R with respect to the orbital elements. This approach would also be a special perturbation method but we would directly obtain time histories of the orbital elements, that is, $a(t)$, $e(t)$, and so on [remember that Example 5.1 required coordinate transformations from $(\mathbf{r},\mathbf{v})$ to the orbital elements]. Successfully completing this process would reproduce Figures 5.3a–e, the time histories of elements a, e, i, Ω, and ω. Recall that Figures 5.3a–c showed that semimajor axis, eccentricity, and inclination exhibited periodic changes due to J_2, but in all three cases the net changes were zero after an orbital revolution. Figures 5.3d and 5.3e showed that elements Ω and ω exhibited periodic variations about a "mean" or secular change that is linear with time.

It is possible to analytically derive the *secular* changes in the orbital elements due to Earth's oblateness (or J_2) effect. To begin, we must express the disturbing function R, Eq. (5.54), in terms of the orbital elements. The sine of the geocentric latitude is computed from spherical trigonometry

$$\sin\phi' = \sin(\omega + \theta)\sin i \tag{5.55}$$

Using this expression for $\sin\phi'$ in Eq. (5.15), the second-order Legendre polynomial is

$$P_2(\sin\phi') = \frac{1}{2}\left[3\sin^2(\omega + \theta)\sin^2 i - 1\right] \tag{5.56}$$

Substituting Eq. (5.56) into the oblateness disturbing function, Eq. (5.54) becomes

$$R = -\frac{\mu}{2r^3}J_2\left(\frac{R_E}{r}\right)^2\left[3\sin^2(\omega + \theta)\sin^2 i - 1\right] \tag{5.57}$$

Finally, we must substitute the trajectory equation

$$r = \frac{p}{1 + e\cos\theta} = \frac{a(1 - e^2)}{1 + e\cos\theta}$$

into Eq. (5.57) to obtain the disturbing function

$$R = -\frac{\mu}{2a^3(1 - e^2)^3}J_2 R_E^2(1 + e\cos\theta)^3\left[3\sin^2(\omega + \theta)\sin^2 i - 1\right] \tag{5.58}$$

Equation (5.58) presents R as a function of the orbital elements. We could take partial derivatives of R and use them in Lagrange's variation of parameter equations. However, this step would not result in analytical expressions because the first-order differential equations would remain highly nonlinear. Because we want the *secular* changes in the elements, we need to "average out" the "fast" variable that causes periodic fluctuations during an orbital revolution. The "fast" variable is the angular position of the satellite. The "mean" disturbance function is its average value over one orbit, that is,

$$\bar{R} = \frac{1}{2\pi}\int_0^{2\pi} R\,dM \tag{5.59}$$

where the over-bar indicates the "mean" or "average" value. Here the "averaging" integration is performed using mean anomaly M as the angular variable. Recall from Chapter 4 that the time-rate of mean anomaly is the mean motion, or

$$\frac{dM}{dt} = n \tag{5.60}$$

Therefore, the differential in the mean anomaly is

$$dM = n\,dt \tag{5.61}$$

The differential time dt can be found using the angular momentum relationship expressed as the product of radial position r and transverse velocity $v_\theta = r\dot\theta$

$$h = r^2\dot\theta = r^2\frac{d\theta}{dt} \tag{5.62}$$

Therefore, $dt = r^2 d\theta/h$, and Eq. (5.61) becomes

$$dM = \frac{nr^2}{h}d\theta \tag{5.63}$$

Substituting Eqs. (5.63) and (5.58) into the mean disturbance function (5.59) yields

$$\bar{R} = \frac{1}{2\pi}\int_0^{2\pi}\frac{-nr^2\mu}{2ha^3(1-e^2)^3}J_2 R_E^2(1+e\cos\theta)^3\left[3\sin^2(\omega+\theta)\sin^2 i - 1\right]d\theta \tag{5.64}$$

Finally, we substitute $h = \sqrt{\mu a(1-e^2)}$ and the trajectory equation for r in Eq. (5.64) and perform the integration to obtain

$$\bar{R} = \frac{n^2 J_2 R_E^2}{4(1-e^2)^{3/2}}(2-3\sin^2 i) \tag{5.65}$$

Equation (5.65) is the *mean* disturbance function where the periodic fluctuations (due to angular position θ) have been averaged out. If we use the partial derivatives of $\bar{R}$ in Lagrange's planetary equations (5.48)–(5.53), we get the *mean* or *averaged* rates for the orbital elements over one revolution. Because $\bar{R}$ is not a function of elements Ω, ω, or σ, the corresponding partial derivatives $\partial\bar{R}/\partial\Omega$, $\partial\bar{R}/\partial\omega$, and $\partial\bar{R}/\partial\sigma$ are zero. Consequently, Eqs. (5.48)–(5.50) show that the *mean rates* for semimajor axis, eccentricity, and inclination are zero:

$$\frac{d\bar{a}}{dt} = 0 \tag{5.66}$$

$$\frac{d\bar{e}}{dt} = 0 \tag{5.67}$$

$$\frac{d\bar{i}}{dt} = 0 \tag{5.68}$$

Equations (5.66)–(5.68) confirm the numerical results presented in Example 5.1: the orbital elements a, e, and i exhibit zero change when averaged over a single orbital revolution when J_2 is the only perturbation (again, see Figures 5.3a–c). Equations (5.66)–(5.68) are results we obtained from our *general* perturbation analysis: that is, the secular change in elements a, e, and i due to Earth oblateness is zero.

Note that Lagrange's planetary equations for the longitude of the ascending node and argument of perigee, Eqs. (5.51) and (5.52), contain the partial derivatives $\partial\bar{R}/\partial i$ and $\partial\bar{R}/\partial e$. Taking the partial derivative of Eq. (5.65) with respect to inclination yields

$$\frac{\partial\bar{R}}{\partial i} = \frac{-3n^2 J_2 R_E^2}{2(1-e^2)^{3/2}}\sin i\cos i \tag{5.69}$$

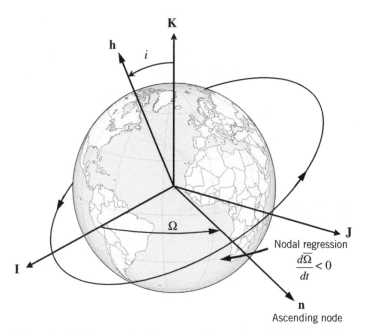

Figure 5.4 Nodal regression due to zonal harmonic J_2 (Earth oblateness).

Substituting Eq. (5.69) into Eq. (5.51), yields the *mean* rate in the longitude of the ascending node:

$$\frac{d\bar{\Omega}}{dt} = \frac{-3nJ_2}{2(1-e^2)^2}\left(\frac{R_E}{a}\right)^2 \cos i \tag{5.70}$$

Equation (5.70) is the secular rate of change for longitude of the ascending node. This average or mean rate is the linear "drift" shown by the sloping dashed line in Figure 5.3d. Equation (5.70) is sometimes called the *nodal regression* because the oblateness (J_2) effect causes the ascending node vector to *regress* or rotate from east to west for a direct orbit (i.e., inclination $i < 90°$). Figure 5.4 shows a direct Earth orbit where the zonal harmonic J_2 is causing the ascending node **n** to rotate westward. Of course, the nodal regression indicates that the orbital plane is rotating clockwise about the polar axis as viewed from above the North Pole.

Figure 5.5 shows the secular drift rate $d\bar{\Omega}/dt$ as computed by Eq. (5.70) for circular Earth orbits. Here the dimensionless zonal harmonic coefficient is $J_2 = 0.0010826267$ and the equatorial Earth radius is $R_E = 6,378.14$ km. The reader should note that the dimensions of Eq. (5.70) are radians per second because the ratio R_E/a is dimensionless and the mean motion n is an angular velocity in radians per second [mean motion n solely depends on semimajor axis a as shown by Eq. (5.23); do not confuse mean motion n with the magnitude of the ascending node vector **n**!]. The nodal regression presented in Figure 5.5 has been converted to units of degrees per day. Figure 5.5 and Eq. (5.70) show that $d\bar{\Omega}/dt$ is negative for direct orbits, zero for polar orbits ($i = 90°$), and positive for retrograde orbits ($i > 90°$). Furthermore, the nodal drift rate is significant for LEOs.

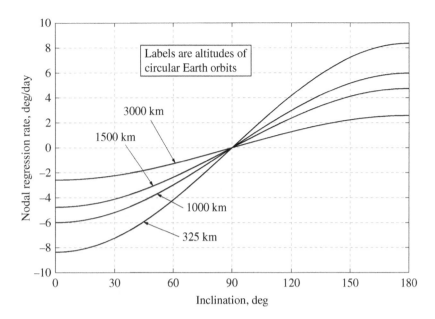

Figure 5.5 Nodal regression $d\bar{\Omega}/dt$ vs. inclination for circular Earth orbits.

For example, a 325-km altitude circular LEO with inclination $i = 28.5°$ will exhibit a nodal regression of −7.4 deg/day. This motion of the orbital plane must be taken into account when planning orbital transfers or orbital rendezvous maneuvers.

Because Earth oblateness has a net zero effect on a, e, and i, we can consider the nodal regression $d\bar{\Omega}/dt$ as constant for a given Earth orbit. Therefore, Eq. (5.70) can be integrated to yield

$$\bar{\Omega}(t) = \Omega_0 + \frac{d\bar{\Omega}}{dt}t \tag{5.71}$$

Equation (5.71) is the time history of the *mean* (or average) longitude of the ascending node. The initial value Ω_0 is the longitude of the ascending node at time $t = 0$. It should be clear that Eq. (5.71) only accounts for the secular drift in Ω and does not include the periodic variations during each orbital revolution. The following example illustrates nodal regression for an Earth orbit.

Example 5.2 The International Space Station (ISS) has the following orbital elements at time $t = 0$

> Semimajor axis $a = 6{,}790.6$ km
> Eccentricity $e = 0.0005$
> Inclination $i = 51.65°$
> Longitude of the ascending node $\Omega_0 = 295°$

Compute the longitude of the ascending node for the ISS 7 days after this epoch.

The ISS orbit is nearly circular at an altitude of about 412 km. Mean motion n is the critical value needed in the nodal-regression equation (5.70). Using Eq. (5.23), the mean motion is

$$n = \sqrt{\frac{\mu}{a^3}} = 0.001128 \text{ rad/s}$$

Using Eq. (5.70) with $J_2 = 0.0010826267$ and $R_E = 6{,}378.14$ km, we obtain the mean nodal regression

$$\frac{d\bar{\Omega}}{dt} = \frac{-3nJ_2}{2(1-e^2)^2}\left(\frac{R_E}{a}\right)^2 \cos i = -1.0029\left(10^{-6}\right) \text{ rad/s}$$

Converting the nodal regression to degrees per day yields $d\bar{\Omega}/dt = -4.965$ deg/day. Therefore, the longitude of the ascending node at time $t = 7$ days is

$$\bar{\Omega}(t) = \Omega_0 + \frac{d\bar{\Omega}}{dt}t = 295° + (-4.965 \text{ deg/day})(7 \text{ days}) = \boxed{260.245°}$$

The orbital plane of the ISS has rotated nearly 35° westward in 1 week. This simple example shows that two-body (Keplerian) motion does not hold for a LEO such as the ISS. The orbital plane is not fixed in inertial space; in reality the plane rotates westward because the Earth is not a perfect sphere.

The Earth-oblateness effect may be used advantageously to design a *sun-synchronous orbit* where the orbital plane rotates at the same rate as the Earth's rotation rate about the sun. Figure 5.6a shows a "top-down" view of the Earth's orbit about the sun. Earth's mean motion (or "average" angular velocity) is one revolution every year or $(360°)/(365.25 \text{ days})$ = 0.986 deg/day (of course, the Earth's heliocentric orbit is elliptical and hence its angular velocity slightly varies with time). Figure 5.6a also shows a top-down view of a nearly polar Earth orbit with a mean nodal rate $d\bar{\Omega}/dt$ that matches Earth's mean motion about the sun (because the Earth orbit is nearly polar it appears as an "edge" in Figure 5.6a). The Earth orbit shown in Figure 5.6a is a sun-synchronous orbit because its orbital plane rotates at the same rate as Earth's mean motion about the sun, that is, $d\bar{\Omega}/dt = 0.986$ deg/day. If the orbital plane is initially orientated so that it is perpendicular to the sun–Earth line (as in Figure 5.6a), then the Earth orbit will maintain this orientation with respect to the sun as the Earth follows its heliocentric path. Hence, a satellite in a sun-synchronous orbit shown in Figure 5.6a will always be in sunlight and will never experience an Earth eclipse. Figure 5.6b shows an Earth satellite in a polar orbit ($i = 90°$) where the J_2 zonal harmonic has a zero net effect on the orientation of the orbital plane (i.e., $d\bar{\Omega}/dt = 0$). An Earth satellite in a polar orbit may initially experience all-sunlight conditions (see time $t = 0$ in Figure 5.6b), but 3 months later nearly half of its orbit will be in the Earth's shadow (see time $t = 91$ days in Figure 5.6b).

Equation (5.70) shows that we have the freedom to select semimajor axis a, eccentricity e, and inclination i in order to establish a sun-synchronous orbit with a mean ascending node rate $d\bar{\Omega}/dt = 0.986$ deg/day. First, Eq. (5.70) and Figure 5.5 show that a sun-synchronous orbit must be slightly retrograde ($i > 90°$) so that the nodal rotation rate is positive (eastward). The following example involves computing the orbital elements for a sun-synchronous orbit.

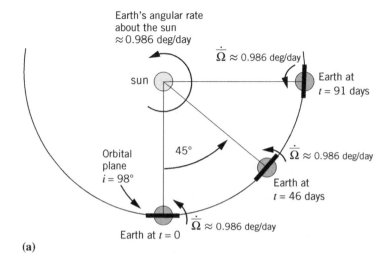

(a)

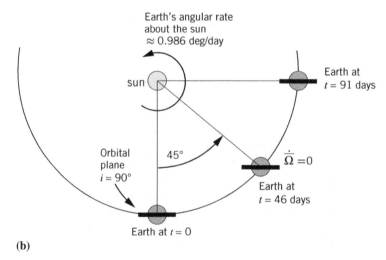

(b)

Figure 5.6 (a) Sun-synchronous orbit and (b) polar orbit with a stationary orbital plane.

Example 5.3 Determine the orbital inclination required for a circular sun-synchronous orbit. Plot inclination vs. orbital altitudes (up to 1,500 km) for circular sun-synchronous orbits.

We know that the mean rate of the longitude of the ascending node must match the Earth's mean motion about the sun (i.e., 360 deg/year = 0.9856 deg/day). Using Eq. (5.70), we have

$$\frac{d\bar{\Omega}}{dt} = \frac{-3nJ_2}{2(1-e^2)^2}\left(\frac{R_E}{a}\right)^2 \cos i = 0.9856 \text{ deg/day} = 1.991021\left(10^{-7}\right) \text{rad/s}$$

Because we are interested in circular orbits, we can set $e = 0$ and $a = r$ (radius). Substituting $n = \sqrt{\mu/a^3} = \sqrt{\mu/r^3}$ (for a circular orbit) and solving for inclination, we obtain

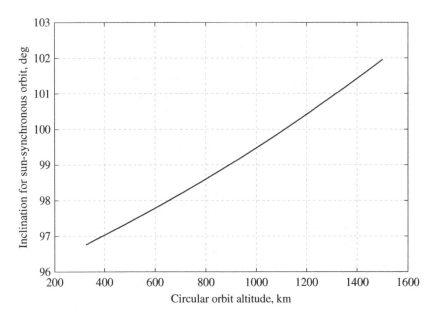

Figure 5.7 Inclination required for a circular sun-synchronous orbit (Example 5.3).

$$i = \cos^{-1}\left[\frac{-2\dot{\Omega}}{3J_2\,R_E^2}\sqrt{\frac{r^7}{\mu}}\right]$$

Using $J_2 = 0.0010826267$, $R_E = 6{,}378.14$ km, and $\dot{\Omega} = 1.991021(10^{-7})$ rad/s, we can determine the sun-synchronous orbital inclination for a given radius r. Figure 5.7 shows sun-synchronous orbit inclination for circular altitudes ranging from 325 km (LEO) to 1,500 km. We see that inclination must be retrograde for a sun-synchronous orbit (as expected) and that inclination increases with altitude. The sun-synchronous orbit shown in Figure 5.6a has inclination $i = 98°$, and therefore its circular orbital altitude must be about 654 km (see Figure 5.7).

The previous analysis has identified the secular change in longitude of the ascending node due to oblateness. Earth oblateness also causes a secular drift rate in argument of perigee. Equation (5.52) shows that the time-rate of ω depends on partial derivatives of the disturbance function with respect to eccentricity and inclination. Equation (5.69) provides the partial derivative $\partial\bar{R}/\partial i$. The partial derivative of Eq. (5.65) with respect to eccentricity is

$$\frac{\partial\bar{R}}{\partial e} = \frac{3n^2 J_2\,R_E^2 e}{4(1-e^2)^{5/2}}\left(2 - 3\sin^2 i\right) \tag{5.72}$$

Substituting Eqs. (5.69) and (5.72) into Eq. (5.52) yields the *mean* rate in the argument of perigee:

$$\frac{d\bar{\omega}}{dt} = \frac{3nJ_2}{4(1-e^2)^2}\left(\frac{R_E}{a}\right)^2\left(4 - 5\sin^2 i\right) \tag{5.73}$$

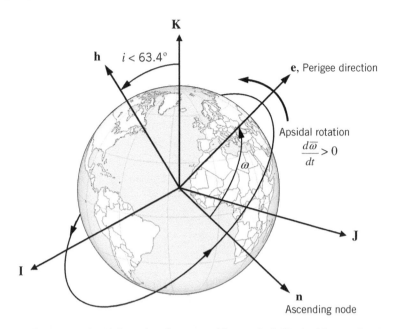

Figure 5.8 Apsidal rotation due to zonal harmonic J_2 (Earth oblateness).

Equation (5.73) is the secular rate of change for argument of perigee caused by Earth oblateness. This mean rate is the linear "drift" shown by the dashed line in Figure 5.3e (Example 5.1). Equation (5.73) is the *apsidal rotation* because the oblateness (J_2) effect causes the eccentricity vector **e** (or apse line) to rotate in the orbital plane. Integrating Eq. (5.73) yields the time history of the *mean* argument of perigee $\bar{\omega}$.

$$\bar{\omega}(t) = \omega_0 + \frac{d\bar{\omega}}{dt}t \tag{5.74}$$

Figure 5.8 shows a direct Earth orbit ($i < 90°$) where the zonal harmonic J_2 is causing the eccentricity vector **e** to rotate about the angular momentum vector **h** in the direction of orbital motion. Equation (5.73) shows that the mean argument of perigee rate is *positive* when the sine of the inclination satisfies the condition $\sin i < \sqrt{4/5}$. This condition is satisfied for direct orbits with inclination $i < 63.4°$ and retrograde orbits with inclination $i > 116.6°$. Because the direct elliptical Earth orbit shown in Figure 5.8 has inclination $i < 63.4°$, the secular change $d\bar{\omega}/dt$ is positive and vector **e** rotates counter-clockwise in the orbital plane.

Figure 5.9 shows the apsidal rotation rate, Eq. (5.73), as a function of inclination for elliptical Earth orbits. The four elliptical orbits presented in Fig. 5.9 have a common perigee altitude of 325 km. Figure 5.9 clearly shows that Earth oblateness causes positive apsidal rotation ($d\bar{\omega}/dt > 0$) for inclinations $i < 63.4°$ and $i > 116.6°$, and negative apsidal rotation when $63.4° < i < 116.6°$. Equation (5.73) and Figure 5.9 show that the apsidal rotation is maximized (and positive) for equatorial orbits ($i = 0°$ and $i = 180°$). Apsidal rotation is zero at $i = 63.4°$ (direct) and $i = 116.6°$ (retrograde) because both inclinations

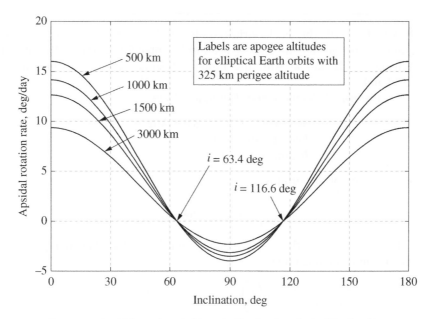

Figure 5.9 Apsidal rotation $d\bar{\omega}/dt$ vs. inclination for elliptical Earth orbits.

satisfy the condition $\sin i = \sqrt{4/5}$ (or $4 - 5\sin^2 i = 0$). Figure 5.10 shows the Molniya orbit used by Russia for communication satellites (see Section 3.4 in Chapter 3). This highly elliptical orbit has $i = 63.4°$ and argument of perigee $\omega = -90°$. Therefore, the apse line does not rotate within the orbital plane and apogee of a Molniya orbit remains at the highest possible latitude (63.4° N in this case).

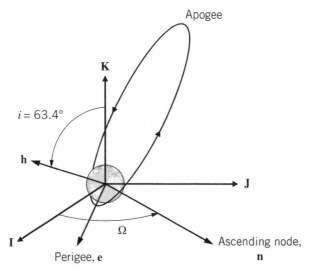

Figure 5.10 Molniya orbit.

Example 5.4   Compute the secular drift rates for the longitude of the ascending node and argument of perigee for the satellite orbit in Example 5.1. Compare the analytical secular change in Ω and ω with the numerical simulation results presented in Figures 5.3d and 5.3e.

Recall that the orbital elements from Example 5.1 are

$$\text{Semimajor axis } a = 8{,}059 \text{ km}$$
$$\text{Eccentricity } e = 0.15$$
$$\text{Inclination } i = 20°$$

The mean motion of this orbit is

$$n = \sqrt{\frac{\mu}{a^3}} = 0.000872664 \text{ rad/s}$$

Using this mean motion with $J_2 = 0.0010826267$ and $R_E = 6{,}378.14$ km in Eq. (5.70), the secular change in ascending node is

$$\frac{d\bar{\Omega}}{dt} = \frac{-3nJ_2}{2(1-e^2)^2}\left(\frac{R_E}{a}\right)^2 \cos i = -8.7297\left(10^{-7}\right) \text{rad/s}$$

Or, in degrees per day we have $\boxed{d\bar{\Omega}/dt = -4.32 \text{ deg/day}}$

Recall that Figure 5.3d (Example 5.1) shows the periodic and secular changes in the longitude of the ascending node resulting from numerical integration (i.e., special perturbation methods). The dashed line in Figure 5.3d is the secular drift in Ω. The longitude of the ascending node decreases approximately $-1.8°$ over 10 h (or 0.4167 days) so the approximate secular rate is $(-1.8°)/(0.4167 \text{ days}) = -4.32$ deg/day. Hence, an approximate linear fit through the numerically simulated response $\Omega(t)$ shows a good match with the analytically determined mean drift rate $d\bar{\Omega}/dt$.

Equation (5.73) gives us the secular change in argument of perigee

$$\frac{d\bar{\omega}}{dt} = \frac{3nJ_2}{4(1-e^2)^2}\left(\frac{R_E}{a}\right)^2 \left(4 - 5\sin^2 i\right) = 1.5863\left(10^{-6}\right) \text{rad/s}$$

Or, in degrees per day $\boxed{d\bar{\omega}/dt = 7.85 \text{ deg/day}}$

Figure 5.3e shows that argument of perigee increases by about $3.25°$ over 10 h or 7.8 deg/day. Again, the analytical secular change exhibits a good match with the approximate linear fit through the simulation results.

5.4   Gauss' Variation of Parameters

It is useful to summarize our perturbation analysis thus far. We began with a discussion of special perturbation methods where we numerically integrate the satellite's absolute acceleration (including the perturbing accelerations) in a Cartesian frame. Because the resulting time histories of $\mathbf{r}(t)$ and $\mathbf{v}(t)$ provide no insight to orbital variations, we must transform the state vector $(\mathbf{r},\mathbf{v})$ to orbital elements so that we may observe periodic and secular changes in the elements. Example 5.1 demonstrated this approach, where the total acceleration is the gradient of a geopotential function that includes a single zonal harmonic term (J_2) associated with Earth's oblateness. Next, we presented a general perturbation method where the goal is to derive analytical expressions for the non-Keplerian motion. We outlined Lagrange's variation of parameters which culminated with six first-order differential equations for the orbital elements. Here the time-rates of the elements are in terms of a disturbing function R, and therefore Lagrange's variation of parameters applies to *conservative* perturbations such as a non-spherical central body and third-body accelerations. We applied orbital averaging techniques to Lagrange's equations and developed analytical expressions for the secular (or *mean*) changes in the orbital elements due to oblateness. The zonal harmonic J_2 has a zero net effect on semimajor axis, eccentricity, and inclination but produces a secular change in the longitude of ascending node Ω and argument of perigee ω.

Gauss developed a form of the variation of parameters where the perturbing accelerations are expressed in a satellite-based coordinate frame that moves with the vehicle. Gauss' variation of parameters can handle conservative and non-conservative perturbations; the only constraint is that the perturbing accelerations must be expressed in terms of a satellite-fixed frame. The general form for Gauss' variation of parameters is

$$\frac{d\boldsymbol{\alpha}}{dt} = \mathbf{f}(\boldsymbol{\alpha}, \mathbf{a}_P, t) \tag{5.75}$$

Recall that $\boldsymbol{\alpha}$ is the 6×1 vector of the orbital elements and $\mathbf{a}_P$ is the 3×1 vector of perturbing accelerations in a convenient satellite-based frame. Although it is possible to derive all six variational equations, we will only present derivations for da/dt, de/dt, and di/dt. These three equations will be used in Chapter 9 when we analyze low-thrust orbit transfers, where the low-thrust propulsion force (divided by satellite mass) is treated as the perturbing acceleration $\mathbf{a}_P$.

We will derive Gauss' variation of parameters by using basic concepts from mechanics and by applying calculus to the orbital relationships developed in Chapter 2. Let us begin with the time-rate of semimajor axis, da/dt. Because semimajor axis is directly related to total energy, we start with an expression for *power* or the time-rate of energy:

$$\frac{d\xi}{dt} = \frac{\mathbf{F}_P \cdot \mathbf{v}}{m} = \mathbf{a}_P \cdot \mathbf{v} \tag{5.76}$$

Equation (5.76) is a familiar result from basic mechanics: the time-rate of energy is the dot product of the perturbing force vector $\mathbf{F}_P$ and the satellite's velocity $\mathbf{v}$ (the reader should note that if the perturbing force is zero, then we have Keplerian motion where energy is constant). Because ξ is total energy per unit mass, its time-rate is *specific* power where the perturbing acceleration vector is $\mathbf{a}_P = \mathbf{F}_P/m$. Now relate energy to semimajor axis

$$\xi = \frac{-\mu}{2a} \tag{5.77}$$

Using the chain rule and Eq. (5.77), the time-rate of energy is

$$\frac{d\xi}{dt} = \frac{d\xi}{da}\frac{da}{dt} = \frac{\mu}{2a^2}\frac{da}{dt} \tag{5.78}$$

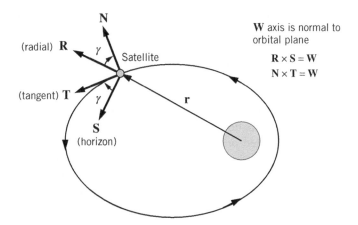

Figure 5.11 NTW and **RSW** satellite-based frames.

Substituting Eq. (5.76) for the left-hand side, Eq. (5.78) becomes

$$\frac{\mu}{2a^2}\frac{da}{dt} = \mathbf{a}_P \cdot \mathbf{v} \tag{5.79}$$

Therefore, we can derive the time-rate da/dt if we find an expression for the dot product of the perturbing acceleration $\mathbf{a}_P$ and velocity $\mathbf{v}$. Gauss used the orthogonal **RSW** coordinate frame where the **R** unit vector is along the radius vector, **S** is in the orbital plane and along the local horizon in the direction of motion, and **W** is along the angular momentum vector **h**. We will use the normal-tangent coordinate frame **NTW**. Figure 5.11 shows the **RSW** and **NTW** frames; these frames are fixed to the satellite. The **T** (tangent) unit vector is always tangent to the orbit in the direction of motion, the **N** axis is in the orbital plane and normal to the **T** axis (pointing away from the central body) and the common **W** axis is along **h**, or **N** × **T** = **W**. The **RSW** and **NTW** frames can be aligned by a rotation through the flight-path angle γ. For circular orbits, the **RSW** frame is always aligned with the **NTW** frame. Suppose we express the perturbing acceleration vector $\mathbf{a}_P$ in the **NTW** frame

$$\mathbf{a}_P = a_n\mathbf{N} + a_t\mathbf{T} + a_w\mathbf{W} \tag{5.80}$$

where a_n, a_t, and a_w are components along the orthogonal **NTW** axes. It should be clear that the velocity vector **v** has a single component when expressed in the **NTW** frame (i.e., $\mathbf{v} = v\mathbf{T}$). Hence, the dot product is

$$\mathbf{a}_P \cdot \mathbf{v} = a_t v \tag{5.81}$$

Substituting Eq. (5.81) into Eq. (5.79) and solving for the time-rate of semimajor axis yields

$$\frac{da}{dt} = \frac{2a^2 v}{\mu}a_t \tag{5.82}$$

Equation (5.82) is Gauss' variation of parameter equation for semimajor axis where the perturbing acceleration is expressed in the **NTW** frame. Component a_t is the sum of all perturbing force/mass vectors resolved into the local **T**-axis direction (tangent to the orbital path). Therefore, a_t may be due to conservative forces (e.g., gravity) or non-conservative forces (e.g., aerodynamic drag, solar radiation pressure, or thrust).

Next, we seek the time-rate de/dt. We begin with the orbital relationship for parameter p

$$p = \frac{h^2}{\mu} = a(1 - e^2) \tag{5.83}$$

Solving Eq. (5.83) for eccentricity yields

$$e = \sqrt{1 - \frac{h^2}{\mu a}} \tag{5.84}$$

Taking the time derivative of Eq. (5.84), we obtain

$$\frac{de}{dt} = \frac{-h}{\mu a \sqrt{1 - h^2/(\mu a)}} \frac{dh}{dt} + \frac{h^2}{2\mu a^2 \sqrt{1 - h^2/(\mu a)}} \frac{da}{dt} \tag{5.85}$$

We can substitute Eq. (5.84) for the common denominator term in Eq. (5.85) to produce

$$\frac{de}{dt} = \frac{h}{\mu a e} \left(-\frac{dh}{dt} + \frac{h}{2a} \frac{da}{dt} \right) \tag{5.86}$$

We can use Eq. (5.82) for the time-rate da/dt. In addition, we need the time rate of change of the *magnitude* of angular momentum dh/dt. From a basic dynamics course, we know that the time-rate of the angular momentum *vector* is the moment or torque. This is equal to the cross product of position $\mathbf{r}$ and applied force $\mathbf{F}$. Recall that $\mathbf{h}$ is the total angular momentum per unit mass, and therefore its time-rate is

$$\dot{\mathbf{h}} = \mathbf{r} \times \mathbf{a}_P \tag{5.87}$$

where $\mathbf{a}_P = \mathbf{F}_P/m$ is the perturbing acceleration (we already know that two-body gravity does not change angular momentum because the gravity force is aligned with $\mathbf{r}$). For now, let us express the perturbing acceleration $\mathbf{a}_P$ as components a_r, a_s, and a_w in the **RSW** frame (see Figure 5.11 for the **RSW** directions). It should be clear that radial perturbation a_r will not change angular momentum because it is aligned with radial position vector $\mathbf{r}$. A transverse perturbation a_s will increase the magnitude of the angular momentum vector. The orbit-normal perturbation a_w will cause the angular momentum vector $\mathbf{h}$ to rotate and change direction. Using these arguments, the time-rate of the *magnitude* of the angular momentum dh/dt is solely due to the in-plane perturbation along the **S** axis, or

$$\frac{dh}{dt} = r a_s \tag{5.88}$$

Because we want to develop the variation equations with perturbations expressed in the **NTW** frame, we can replace a_s in Eq. (5.88) with the in-plane perturbing accelerations a_n and a_t

$$\frac{dh}{dt} = r(a_t \cos\gamma - a_n \sin\gamma) \tag{5.89}$$

The reader should be able to easily identify the projections of the **T** and **N** components onto the **S** axis by reviewing Figure 5.11. Equation (5.89) shows that we need expressions for the cosine and sine of the flight-path angle γ. Equations (2.69) and (2.71) present the radial and transverse velocity components, $\dot{r}$ and $r\dot{\theta}$, in terms of e, h, and true anomaly θ

$$\dot{r} = \frac{\mu}{h}e\sin\theta \qquad (5.90)$$

$$r\dot{\theta} = \frac{\mu}{h}(1 + e\cos\theta) \qquad (5.91)$$

We know that $\sin\gamma = \dot{r}/v$ and $\cos\gamma = r\dot{\theta}/v$. Using Eqs. (5.90) and (5.91), we obtain the expressions for the sine and cosine of the flight-path angle

$$\sin\gamma = \frac{\mu e\sin\theta}{hv} \qquad (5.92a)$$

$$\cos\gamma = \frac{\mu(1 + e\cos\theta)}{hv} \qquad (5.92b)$$

Substituting Eqs. (5.92a) and (5.92b) into Eq. (5.89), we obtain

$$\frac{dh}{dt} = \frac{r\mu}{hv}[(1 + e\cos\theta)a_t - e\sin\theta a_n] \qquad (5.93)$$

Finally, substitute Eqs. (5.93) and (5.82) into Eq. (5.86) to yield an expression for the time-rate de/dt in terms of **NTW** perturbations a_n and a_t

$$\frac{de}{dt} = \frac{h}{\mu a e}\left\{\frac{-r\mu}{hv}[(1 + e\cos\theta)a_t - e\sin\theta a_n] + \frac{hav}{\mu}a_t\right\} \qquad (5.94)$$

The final steps involve substitutions of orbital relationships (such as the trajectory equation) and simplifications. These algebraic steps are omitted here. Equation (5.94) can be simplified to yield

$$\frac{de}{dt} = \frac{1}{v}\left[2(e + \cos\theta)a_t + \frac{r\sin\theta}{a}a_n\right] \qquad (5.95)$$

Equation (5.95) is Gauss' variational equation for eccentricity in terms of perturbing accelerations expressed in the satellite-based **NTW** frame. Only perturbations in the orbital plane cause eccentricity to change over time.

The variational equation for inclination can be obtained using calculus and geometrical methods. We start with the expression for the cosine of inclination, Eq. (3.8)

$$\cos i = \frac{\mathbf{K}\cdot\mathbf{h}}{h} \qquad (5.96)$$

where **K** is the unit vector along the Z axis of the ECI frame. Taking a time derivative yields

$$-\sin i\frac{di}{dt} = \frac{h(\mathbf{K}\cdot\dot{\mathbf{h}}) - \dot{h}(\mathbf{K}\cdot\mathbf{h})}{h^2} \qquad (5.97)$$

We may use Eq. (5.87) to compute the time-rate of vector **h** in terms of perturbation accelerations in the **RSW** frame

$$\dot{\mathbf{h}} = \mathbf{r} \times \mathbf{a}_P = \begin{vmatrix} \mathbf{R} & \mathbf{S} & \mathbf{W} \\ r & 0 & 0 \\ a_r & a_s & a_w \end{vmatrix} = -ra_w\mathbf{S} + ra_s\mathbf{W} \qquad (5.98)$$

We may use Eq. (5.96) to substitute $\mathbf{K} \cdot \mathbf{h} = h\cos i$ in Eq. (5.97). The dot product $\mathbf{K} \cdot \dot{\mathbf{h}}$ will involve the following dot products between $\mathbf{K}$ and unit vectors $\mathbf{S}$ and $\mathbf{W}$:

$$\mathbf{K} \cdot \mathbf{S} = \sin i \cos(\omega + \theta) \qquad (5.99)$$

$$\mathbf{K} \cdot \mathbf{W} = \cos i \qquad (5.100)$$

The angle $\omega + \theta$ (the *argument of latitude*; see Section 3.3) is measured in the orbital plane from the ascending node to the satellite. Making these substitutions (along with $\dot{h} = ra_s$), Eq. (5.97) becomes

$$-\sin i \frac{di}{dt} = \frac{h[-ra_w \sin i \cos(\omega + \theta) + ra_s \cos i] - ra_s h \cos i}{h^2} \qquad (5.101)$$

Canceling the two terms $hra_s \cos i$ in Eq. (5.101), we obtain

$$\frac{di}{dt} = \frac{r\cos(\omega + \theta)}{h} a_w \qquad (5.102)$$

Equation (5.102) is Gauss' variational equation for inclination. Only perturbations that are normal to the orbital plane (a_w) will change inclination.

We can follow the same basic procedures and derive the remaining Gauss variational equations. These steps will not be presented here. The interested reader may consult Vallado [1; pp. 633–636] or Bate *et al.* [3; pp. 402–406]. Gauss' variational equations in **NTW** coordinates are

$$\frac{da}{dt} = \frac{2a^2v}{\mu} a_t \qquad (5.103)$$

$$\frac{de}{dt} = \frac{1}{v}\left[2(e + \cos\theta)a_t + \frac{r\sin\theta}{a}a_n\right] \qquad (5.104)$$

$$\frac{di}{dt} = \frac{r\cos(\omega + \theta)}{h} a_w \qquad (5.105)$$

$$\frac{d\Omega}{dt} = \frac{r\sin(\omega + \theta)}{h\sin i} a_w \qquad (5.106)$$

$$\frac{d\omega}{dt} = \frac{1}{ev}\left[2\sin\theta a_t - \left(2e + \frac{r\cos\theta}{a}\right)a_n\right] - \frac{r\sin(\omega + \theta)\cos i}{h\sin i}a_w \qquad (5.107)$$

$$\frac{d\theta}{dt} = \frac{h}{r^2} - \frac{1}{ev}\left[2\sin\theta a_t - \left(2e + \frac{r\cos\theta}{a}\right)a_n\right] \qquad (5.108)$$

Note that, when all perturbing acceleration components vanish ($a_n = a_t = a_w = 0$), Gauss' variational equations show that the five elements (a, e, i, Ω, ω) remain constant while the time-rate of true anomaly is governed by conservation of angular momentum, or $h = r(r\dot{\theta}) = r^2\dot{\theta}$. It is also interesting to note that the orbit-normal perturbation a_w only affects the orientation of the orbital plane in three-dimensional space (i.e., orbital elements i, Ω, and ω).

Gauss' variation of parameters (5.103)–(5.108) may be used in special (numerical) or general (analytical) perturbation methods. Gauss' variational equations provide two distinct advantages for a special perturbation method: (1) numerical integration provides the time histories of the orbital elements without a coordinate transformation step; and (2) a relatively large time step may be used in the numerical integration process. Referring back to Example 5.1, we see that the special perturbation method was applied to a perturbed system in Cartesian coordinates [see Eqs. (5.18)–(5.20)], and therefore a coordinate transformation was required to obtain time histories of the orbital elements. In order to use Gauss' equations, we must provide the perturbing accelerations (third-body gravity, drag, thrust, etc.) as components in the satellite-based **NTW** frame. In Chapter 9 we will apply a general perturbation method to develop analytical solutions for low-thrust transfers where the small perturbing acceleration $\mathbf{a}_P$ is produced by an onboard propulsion system.

Finally, we should note that Gauss' variational equations possess singularities for equatorial orbits (Ω is not defined) and circular orbits (ω and θ are not defined). As inclination approaches zero, the time-rate $d\Omega/dt$ becomes infinite even if the perturbations are small. The same problem occurs for the time-rates $d\omega/dt$ and $d\theta/dt$ as eccentricity approaches zero. One solution is to use a non-singular set of elements that are nonlinear functions of the classical orbital elements. The non-singular element for angular position is measured from the inertial **I** axis to the satellite. Gauss' variational equations may be written in terms of non-singular elements (the so-called *equinoctial elements*). The interested reader may consult Battin [2; pp. 490–494] for the definition and use of non-singular orbital elements.

5.5 Perturbation Accelerations for Earth Satellites

The previous sections have presented special and general perturbation methods. The only perturbation that we have investigated in any detail has been the non-spherical Earth oblateness (J_2) effect. During our discussion of perturbations, we have noted that other perturbing forces (or accelerations) exist. In this section we will briefly describe the characteristics of orbital perturbations for an Earth satellite. This discussion will not delve into detailed calculations; instead the objective here is to quantify the effect of perturbations on "typical" Earth orbits such as LEO and GEO.

5.5.1 Non-Spherical Earth

The major effects of a non-spherical or oblate Earth (or any non-spherical central body) have been discussed in Section 5.3. We will provide a brief summary of the principle characteristics of non-Keplerian motion caused by a non-spherical gravity field. The Earth resembles a "flattened ellipsoid" with an additional "bulge" of mass at its equator. This equatorial bulge is modeled by a second-order spherical harmonic function with zonal coefficient J_2. The Earth-oblateness effect is a conservative perturbation that does not change the energy of the satellite's orbit. While the J_2 perturbation causes periodic variations in the orbital elements over each revolution (see Figures 5.3a–e), the secular or mean changes are zero for semimajor axis a, eccentricity e, and inclination i. The major

Table 5.1 Secular rates due to Earth oblateness (J_2).

Orbit	Semimajor axis, a (km)	Eccentricity, e	Inclination, i	$\dot{\Omega}_{J_2}$ (deg/day)	$\dot{\omega}_{J_2}$ (deg/day)
LEO	6,728	0.0	28.5°	−7.26	No perigee
GPS	26,558	0.0	55.0°	−0.04	No perigee
GTO	24,364	0.731	28.5°	−0.37	0.60

effect of Earth oblateness is the secular changes in the longitude of the ascending node Ω and argument of perigee ω. The secular "drift" time-rates for Ω and ω are repeated here:

$$\dot{\Omega}_{J_2} = \frac{-3nJ_2}{2(1-e^2)^2}\left(\frac{R_E}{a}\right)^2 \cos i \tag{5.109}$$

$$\dot{\omega}_{J_2} = \frac{3nJ_2}{4(1-e^2)^2}\left(\frac{R_E}{a}\right)^2 \left(4-5\sin^2 i\right) \tag{5.110}$$

Recall that the over-bar indicates the mean or average value and that for Earth, the second zonal harmonic coefficient is $J_2 = 0.0010826267$. Clearly, the secular drift rates $\dot{\Omega}_{J_2}$ and $\dot{\omega}_{J_2}$ diminish as the orbital energy (i.e., a) increases (remember that mean motion $n = \sqrt{\mu/a^3}$ also diminishes as a increases). Equations (5.109) and (5.110) show that the secular drift for Ω is zero for polar orbits ($i = 90°$) and the secular change in ω is zero at inclination $i = 63.4°$ and $116.6°$.

Table 5.1 presents the secular drift rates due to oblateness effects for a few selected Earth orbits. The LEO exhibits a significant nodal regression rate $\dot{\Omega}_{J_2}$ that causes the orbital plane to regress westward. Many spacecraft are launched into a LEO "parking orbit" before injection into a trajectory leading to their target orbit or target planet. Mission planners must carefully consider the nodal regression due to J_2 even if the spacecraft loiters for a few hours in LEO before its departure rocket burn. The second orbit in Table 5.1 is for the Global Positioning System (GPS). Note that the drift rate $\dot{\Omega}_{J_2}$ is very small for GPS satellites due to the large semimajor axis a and relatively large inclination i. The third entry in Table 5.1 is the geostationary transfer orbit (GTO), which is a highly elliptical orbit with a perigee altitude of 185 km and apogee altitude of 35,786 km. Thus, the GTO perigee is tangent to a 185-km circular LEO and its apogee is tangent to GEO. The argument of perigee for a GTO is zero (or 180°) so that its perigee direction **e** is in the equatorial plane and aligned with the line of nodes. The apogee of GTO *must* be in the equatorial plane so that it is tangent to its target, a circular geostationary equatorial orbit. During a 5-h transit (perigee to apogee), the GTO apse line will rotate more than 0.13° due to oblateness effects. Despite this small angular displacement, an unplanned 0.13° rotation in the major axis will cause a 96-km error north (or south) of the equatorial plane when the satellite reaches its apogee.

As a final note, the secular oblateness effect for *any* gravitational body can be quantified using Eqs. (5.109) and (5.110). All that is required is the zonal harmonic coefficient J_2

and equatorial radius for the body of interest. For example, our moon is less oblate than the Earth (the moon's zonal coefficient is $J_2 = 0.0002027$). A satellite in a 100-km altitude circular low-lunar orbit with an inclination of 10° will exhibit a nodal regression rate of $\dot{\Omega}_{J_2} = -1.18$ deg/day. Hence, the nodal regression for a satellite in LEO is six to seven times greater than the nodal regression for a satellite in a low-lunar orbit.

5.5.2 Third-Body Gravity

The moon and sun are the two obvious gravitational bodies that perturb Earth-orbiting satellites. Let us develop an expression for the perturbing acceleration due to a third gravitational body that is present in the two-body (Earth and satellite) system. Figure 5.12 shows a three-body system comprising the Earth, moon, and satellite where the respective (absolute) position vectors are $\mathbf{r}_E$, $\mathbf{r}_m$, and $\mathbf{r}_{sc}$. The relative position vectors shown in Figure 5.12 are

$$\text{moon-to-spacecraft:} \quad \mathbf{r}_{m\text{-}sc} = \mathbf{r}_{sc} - \mathbf{r}_m \tag{5.111}$$

$$\text{moon-to-Earth:} \quad \mathbf{r}_{m\text{-}E} = \mathbf{r}_E - \mathbf{r}_m \tag{5.112}$$

$$\text{Earth-to-spacecraft:} \quad \mathbf{r}_{E\text{-}sc} = \mathbf{r}_{sc} - \mathbf{r}_E \tag{5.113}$$

Applying Newton's second law and the law of gravitation to the spacecraft and Earth, we obtain

$$\text{Spacecraft:} \quad m\ddot{\mathbf{r}}_{sc} = -\frac{GM_E m}{r_{E\text{-}sc}^3}\mathbf{r}_{E\text{-}sc} - \frac{GM_m m}{r_{m\text{-}sc}^3}\mathbf{r}_{m\text{-}sc} \tag{5.114}$$

$$\text{Earth:} \quad M_E\ddot{\mathbf{r}}_E = \frac{GM_E m}{r_{E\text{-}sc}^3}\mathbf{r}_{E\text{-}sc} - \frac{GM_E M_m}{r_{m\text{-}E}^3}\mathbf{r}_{m\text{-}E} \tag{5.115}$$

The acceleration of the spacecraft relative to the Earth is the second time derivative of Eq. (5.113), which can be obtained by subtracting Eq. (5.115) from (5.114) (with the appropriate divisions by masses M_E and m, respectively); the result is

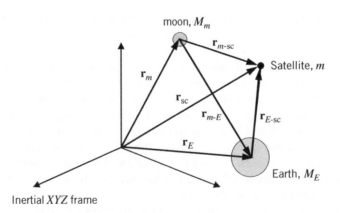

Inertial XYZ frame

Figure 5.12 Three-body system.

$$\ddot{\mathbf{r}}_{E\text{-sc}} = \ddot{\mathbf{r}}_{\text{sc}} - \ddot{\mathbf{r}}_E$$

$$= -\frac{GM_E}{r_{E\text{-sc}}^3}\mathbf{r}_{E\text{-sc}} - \frac{GM_m}{r_{m\text{-sc}}^3}\mathbf{r}_{m\text{-sc}} - \frac{Gm}{r_{E\text{-sc}}^3}\mathbf{r}_{E\text{-sc}} + \frac{GM_m}{r_{m\text{-}E}^3}\mathbf{r}_{m\text{-}E} \qquad (5.116)$$

Grouping like terms, we obtain

$$\ddot{\mathbf{r}}_{E\text{-sc}} = -\frac{G(M_E + m)}{r_{E\text{-sc}}^3}\mathbf{r}_{E\text{-sc}} - GM_m\left(\frac{\mathbf{r}_{m\text{-sc}}}{r_{m\text{-sc}}^3} - \frac{\mathbf{r}_{m\text{-}E}}{r_{m\text{-}E}^3}\right) \qquad (5.117)$$

Note that if we define $\mathbf{r} = \mathbf{r}_{E\text{-sc}}$ and use the approximation $G(M_E + m) \approx GM_E = \mu$ (Earth's gravitational parameter), then Eq. (5.117) matches Eq. (5.3). In this case, the first term on the right-hand side of Eq. (5.117) is central-body acceleration and the second term is the perturbation acceleration due to lunar gravity:

$$\text{Lunar:} \quad \mathbf{a}_m = -\mu_m\left(\frac{\mathbf{r}_{m\text{-sc}}}{r_{m\text{-sc}}^3} - \frac{\mathbf{r}_{m\text{-}E}}{r_{m\text{-}E}^3}\right) \qquad (5.118)$$

where $\mu_m = GM_m$ is the moon's gravitational parameter. Kaplan [4; pp. 357–358] calls this perturbing acceleration the *effective attraction* of the moon on the satellite. Closer inspection of Eq. (5.118) shows that the so-called effective attraction is the difference between the lunar gravity acting on the satellite and the lunar gravity acting on the Earth.

If we redraw Figure 5.12 with the sun as the third body and repeat the previous derivation, we obtain its perturbation acceleration acting on the satellite

$$\text{Solar:} \quad \mathbf{a}_s = -\mu_s\left(\frac{\mathbf{r}_{s\text{-sc}}}{r_{s\text{-sc}}^3} - \frac{\mathbf{r}_{s\text{-}E}}{r_{s\text{-}E}^3}\right) \qquad (5.119)$$

where $\mathbf{r}_{s\text{-sc}}$ is the sun-to-spacecraft position vector, $\mathbf{r}_{s\text{-}E}$ is the sun-to-Earth position vector, and μ_s is the sun's gravitational parameter. Equation (5.119) is the effective attraction of the sun on the satellite [4; p. 359]. Lunar and solar gravitational accelerations (5.118) and (5.119) may now be added as perturbation acceleration $\mathbf{a}_P$ to the central-body gravitational acceleration [see Eq. (5.3) or Cowell's formulation of the special perturbation method]. Table 5.2 presents the Earth, lunar, and solar gravitational acceleration magnitudes that act on an Earth-orbiting satellite. Equations (5.118) and (5.119) are used to compute the perturbing accelerations using the closest-approach distance between the third body and the satellite. Table 5.2 shows that the moon's perturbing acceleration is generally more than twice as large as the sun's perturbing acceleration for satellites in the vicinity of Earth. In both cases, however, the third-body gravitational acceleration is extremely small compared with central-body gravity.

Table 5.2 Third-body gravitational accelerations.

Orbit	Earth gravitational acceleration (m/s²)	Lunar gravitational acceleration (m/s²)	Solar gravitational acceleration (m/s²)
LEO[a]	8.938	$1.18(10^{-6})$	$5.29(10^{-7})$
GEO	0.224	$8.65(10^{-6})$	$3.34(10^{-6})$

[a] LEO: 300-km altitude circular orbit.

Lunar and solar gravity cause a secular drift rate in a satellite's ascending node and argument of perigee (we have already noted that third-body gravity is a conservative force that does not change the satellite's total energy). Larson and Wertz [5; p. 142] present expressions for the approximate secular drift in Ω and ω due to third-body gravity for nearly circular orbits

$$\text{Lunar: } \dot{\Omega}_m = \frac{-0.00338}{N_{rev}}\cos i \quad (\text{deg/day}) \tag{5.120}$$

$$\text{Solar: } \dot{\Omega}_s = \frac{-0.00154}{N_{rev}}\cos i \quad (\text{deg/day}) \tag{5.121}$$

$$\text{Lunar: } \dot{\omega}_m = \frac{0.00169}{N_{rev}}\left(4-5\sin^2 i\right) \quad (\text{deg/day}) \tag{5.122}$$

$$\text{Solar: } \dot{\omega}_s = \frac{0.00077}{N_{rev}}\left(4-5\sin^2 i\right) \quad (\text{deg/day}) \tag{5.123}$$

where N_{rev} is the number of orbital revolutions per day. Note that when $i = 63.4°$ the argument of perigee is unaltered by third-body gravity (just as in the case of Earth's oblateness). The reader should also note that the ratio of the lunar/solar drift-rate magnitudes is roughly 2.2, which is on the same order as the ratios of the respective gravitational accelerations presented in Table 5.2.

Example 5.5 Compute the secular drift rates in Ω and ω caused by lunar and solar gravity for the Earth-orbiting satellite in Example 5.1. Compare these gravity-induced secular changes with the nodal regression and apsidal rotation caused by Earth oblateness (J_2).

Recall that the orbital elements from Example 5.1 are

Semimajor axis a = 8,059 km
Eccentricity e = 0.15
Inclination i = 20°

The orbital period is

$$T_{period} = \frac{2\pi}{\sqrt{\mu}}a^{3/2} = 7{,}200\,\text{s} = 2\,\text{h}$$

Therefore, the number of revolutions per day is $N_{rev} = 12$.

Using Eqs. (5.120) and (5.121), the nodal drift rates due to lunar and solar gravity are

$$\text{Lunar: } \dot{\Omega}_m = \frac{-0.00338}{N_{rev}}\cos i = \boxed{-0.0002647 \text{ deg/day}}$$

$$\text{Solar: } \dot{\Omega}_s = \frac{-0.00154}{N_{rev}}\cos i = \boxed{-0.0001206 \text{ deg/day}}$$

Recall from Example 5.4 that the nodal regression due to Earth oblateness is $\dot{\Omega}_{J_2} = -4.32$ deg/day.

Using Eqs. (5.122) and (5.123), the apsidal rates due to lunar and solar gravity are

$$\text{Lunar: } \dot{\omega}_m = \frac{0.00169}{N_{rev}}\left(4-5\sin^2 i\right) = \boxed{0.0004810 \text{ deg/day}}$$

$$\text{Solar: } \dot{\bar{\omega}}_s = \frac{0.00077}{N_{\text{rev}}}\left(4 - 5\sin^2 i\right) = \boxed{0.0002191 \text{ deg/day}}$$

The apsidal rotation rate caused by Earth oblateness is $\dot{\bar{\omega}}_{J_2} = 7.85$ deg/day.

These calculations show that the nodal regression and apsidal rotation due to lunar-solar gravity are miniscule compared with the secular drift rates caused by Earth oblateness. For this particular low-Earth orbit, the moon causes the largest perturbation (apsidal rotation); however, even after 1 year in orbit, lunar gravity will increase the argument of perigee by less than 0.2°.

5.5.3 Atmospheric Drag

Satellites in low orbits (or orbits with a low perigee) will encounter particles of the upper atmosphere. This interaction is manifested as an aerodynamic drag force that can be calculated from the same basic equation used for airplane drag. Atmospheric drag acceleration is the drag force divided by the satellite's mass m

$$a_D = \frac{1}{2}\rho v_{\text{rel}}^2 \frac{SC_D}{m} \tag{5.124}$$

The drag force always opposes the satellite's velocity vector $\mathbf{v}_{\text{rel}}$ that is *relative* to the Earth's atmosphere

$$\mathbf{v}_{\text{rel}} = \mathbf{v} - \boldsymbol{\omega}_E \times \mathbf{r} \tag{5.125}$$

where $\mathbf{v}$ is the satellite's inertial velocity in the ECI frame, $\boldsymbol{\omega}_E$ is the angular velocity vector of the Earth (not to be confused with argument of perigee ω), and $\mathbf{r}$ is the satellite's ECI position vector. Equation (5.125) assumes that the atmosphere rotates with the Earth. The other terms in the drag acceleration equation (5.124) include atmospheric density ρ, satellite cross-sectional area S, and drag coefficient C_D. All three terms are difficult to determine accurately. We may model atmospheric density as an exponential function of altitude h

$$\rho = \rho_0 \exp\left[-\beta(h - h_0)\right] \tag{5.126}$$

where ρ_0 is the atmospheric density at reference altitude h_0, and β is the "inverse scale height." Larson and Wertz [5] present values of ρ_0 and β at discrete references altitudes h_0 ranging from 0 (sea level) to 1,500 km. Density of the upper atmosphere is strongly affected by the 11-year solar cycle. High solar flux will greatly increase the density of the upper atmosphere by a factor of 5 from its mean value. Low solar flux will reduce density. Larson and Wertz [5] present values of ρ_0 for three solar flux levels: "high," "mean," and "low." Table 5.3 provides atmospheric density values for mean solar activity at various altitudes above the Earth's surface [5].

Drag coefficient C_D and cross-sectional area S are difficult to determine because they depend on the satellite's orientation relative to the velocity vector. One solution is to group the terms C_D, S, and m into a single parameter called the *ballistic coefficient*

$$C_B = \frac{m}{SC_D} \tag{5.127}$$

Table 5.3 Density of the Earth's upper atmosphere [5].

Altitude (km)	Mean atmospheric density (kg/m³)
200	$2.53(10^{-10})$
350	$6.98(10^{-12})$
500	$4.89(10^{-13})$
650	$5.15(10^{-14})$
800	$9.63(10^{-15})$
1,000	$2.79(10^{-15})$

The ballistic coefficient has units of kilograms per meter squared. Typical values of C_B will be on the order of 10–100 kg/m².

Let us compute the drag acceleration for a representative Earth-orbiting satellite with ballistic coefficient $C_B = 85$ kg/m². We will assume that the satellite is in a circular equatorial orbit and therefore the relative velocity is $v_{rel} = \sqrt{\mu/r} - \omega_E r$ where $\omega_E = 7.292(10^{-5})$ rad/s (one revolution per sidereal day). Drag acceleration is computed using Eq. (5.124) for the circular altitudes and densities listed in Table 5.3 [the reader should note that v_{rel} must be expressed in base units of meters per second when using Eq. (5.124) so that the resulting acceleration is in meters per second squared]. Table 5.4 presents the inertial velocity $\sqrt{\mu/r}$, relative velocity v_{rel}, and corresponding drag acceleration a_D. Table 5.4 shows that while the atmospheric-relative velocity v_{rel} only varies by about 7%, the drag acceleration varies by 5 orders of magnitude for circular orbits between 200 and 1,000 km altitude. Comparing Tables 5.3 and 5.4, we see that drag acceleration is greater than third-body gravitational accelerations for LEOs with altitudes less than 350 km. However, we must remember that drag is a *non-conservative* force that always diminishes orbital energy while third-body gravity is a conservative force.

Table 5.4 Drag acceleration on Earth-orbiting satellite with $C_B = 85$ kg/m².

Circular altitude (km)	Inertial velocity (km/s)	Relative velocity (km/s)	Drag acceleration (m/s²)
200	7.7843	7.3046	$7.94(10^{-5})$
350	7.6971	7.2064	$2.13(10^{-6})$
500	7.6127	7.1111	$1.45(10^{-7})$
650	7.5310	7.0185	$1.49(10^{-8})$
800	7.4519	6.9284	$2.72(10^{-9})$
1,000	7.3502	6.8122	$7.62(10^{-10})$

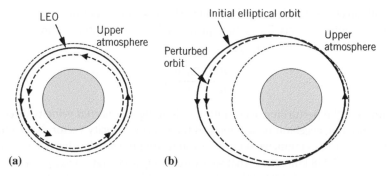

Figure 5.13 Atmospheric drag effects on an Earth-orbiting satellite: (a) "spiral in" from circular low-Earth orbit (LEO); and (b) eccentricity reduction of an elliptical orbit.

We end this brief subsection with a summary of the general effects of atmospheric drag on Earth-orbiting satellites. Because drag always acts in the opposite direction of the atmospheric-relative velocity $\mathbf{v}_{rel}$, it constantly reduces orbital energy. For a near-circular LEO, atmospheric drag will cause the satellite to slowly spiral inward (thus decreasing altitude and increasing atmospheric density) until it enters the dense atmosphere and either burns up or decelerates and eventually crashes into the Earth's surface. Figure 5.13a illustrates this scenario (of course, this figure is not to scale). Therefore, satellites operating in LEO must be periodically re-boosted using onboard propulsion in order to increase orbital altitude. Atmospheric drag (enhanced by higher than expected solar activity) caused the first US space station Skylab to slowly spiral inward from its 430-km orbit. After 6 years in orbit, Skylab entered the dense atmosphere and disintegrated due to extreme aerodynamic heating (some orbital debris crashed in Western Australia). Figure 5.13b shows an Earth-orbiting satellite in an elliptical orbit where its perigee is within the upper atmosphere. Passage through the upper atmosphere will reduce the satellite's speed and consequently reduce its apogee altitude as shown in Figure 5.13b. Over many orbital revolutions, the satellite's eccentricity is continually reduced until the orbit is nearly circular and entirely within the upper atmosphere; after this point the satellite follows the slow inward spiral presented in Figure 5.13a.

Example 5.6   The Space Shuttle is in a 300-km altitude circular orbit. The Space Shuttle has mass $m = 90{,}000$ kg and is oriented so that its maximum cross-sectional area is normal to the atmospheric-relative velocity vector $\mathbf{v}_{rel}$. Using $S = 367$ m^2 and $C_D = 2$, compute the drag acceleration and estimate the loss of altitude after 1 day in orbit.

First, let us compute the Shuttle's ballistic coefficient using Eq. (5.127)

$$C_B = \frac{m}{SC_D} = 122.62 \text{ kg/m}^2$$

Next, we will estimate the atmospheric-relative velocity as $v_{rel} = \sqrt{\mu/r} - \omega_E r$ even though the Shuttle's orbit is not equatorial. Using $r = 6{,}678$ km, we determine $v_{rel} = 7{,}239$ m/s. The mean atmospheric density at 300-km altitude is $\rho = 1.95(10^{-11})$ kg/m^3 [5]. Using Eq. (5.124), the drag acceleration is

$$a_D = \frac{1}{2}\rho v_{rel}^2 \frac{SC_D}{m} = \boxed{4.1667\left(10^{-6}\right) \text{m/s}^2}$$

Drag will cause the Shuttle's circular orbit to slowly shrink over time. Gauss' variational equation for semimajor axis, Eq. (5.103), provides the rate of altitude change because the orbit is circular (i.e., $a = r = h + R_E$):

$$\frac{da}{dt} = \frac{2a^2 v}{\mu} a_t$$

By definition, the perturbing acceleration a_t is along the *inertial* velocity direction (**T** axis), whereas drag acceleration a_D opposes the *relative* velocity vector. Neglecting this slight difference in direction, we set $a_t = -a_D$ and compute da/dt using $a = r = 6{,}678$ km and $v = \sqrt{\mu/r} = 7.7258$ km/s [note that we express drag acceleration as $a_D = 4.1667(10^{-9})$ km/s^2 so that we obtain da/dt in kilometers per second]. The time-rate of semimajor axis due to drag is

$$\frac{da}{dt} = \frac{-2a^2 v}{\mu} a_D = -7.2032\left(10^{-6}\right) \text{km/s}$$

This time-rate of semimajor axis is essentially the time-rate of radius (and the time-rate of altitude) because the inward spiraling orbit remains nearly circular. The change in semimajor axis after 1 day (86,400 s) is approximately

$$\Delta a \approx \frac{da}{dt}\Delta t = \left[-7.2032\left(10^{-6}\right)\text{km/s}\right](86{,}400\,\text{s}) = \boxed{-0.622 \text{ km}}$$

Hence, drag causes the Shuttle's orbital altitude to decrease by about 622 m after 1 day. This result is approximate because we assumed that the rate da/dt remained constant during $\Delta t = 1$ day (=86,400 s). This approximation is valid because orbital velocity and atmospheric density will change very little for a small change in altitude.

We can estimate the orbital lifetime of the Space Shuttle by numerically integrating Gauss' variational equation da/dt with drag as the sole perturbation. A simple Euler-integration scheme may be used:

$$a(t_i + \Delta t) = a(t_i) + \frac{da}{dt}\Delta t$$

where the Gauss variational equation (5.103) is used to update da/dt as semimajor axis decreases. Of course, the drag acceleration is computed at every time step using Eq. (5.124) with the updated values for v_{rel} and atmospheric density ρ (both are functions of orbital radius or a). Atmospheric density is calculated using the exponential model (5.126) with the reference values h_0 and ρ_0 defined in Larson and Wertz [5]. The Euler integration is carried out with a fixed time step $\Delta t = 0.25$ days (6 h) starting from the 300-km altitude circular orbit. Figure 5.14 shows the time history of the circular orbital altitude ($h = a - R_E$). Note that the altitude time-rate dh/dt (= da/dt) caused by drag is essentially constant for the first 30 days in orbit. The average altitude rate over the first 30 days is $(-25 \text{ km})/(30 \text{ days}) = -0.833$ km/day. The rate of altitude loss becomes larger as the Shuttle's orbit becomes lower and enters the denser atmosphere. Once the satellite reaches an altitude of about 150 km, the altitude rate increases dramatically (the integration time step must be reduced to less than 1 h). The satellite crashes into the Earth's surface about 8.5 h (approximately six orbital revolutions) after reaching $h = 150$ km. Figure 5.14 shows that the lifetime of a Space Shuttle orbit is about 62.3 days.

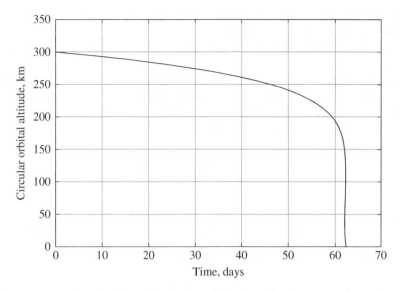

Figure 5.14 Circular orbit decay due to atmospheric drag: altitude vs. time (Example 5.6).

5.5.4 Solar Radiation Pressure

Satellites in sunlight are perturbed by a force from solar radiation pressure (SRP). The magnitude of the SRP force depends on the intensity of the solar energy (i.e., distance from the sun and solar activity), the satellite's area exposed to the sun, and the satellite's reflectivity characteristics. Like drag, solar radiation pressure is a non-conservative perturbation. However, because the SRP force acts in a direction opposite the satellite-to-sun vector, a perturbing SRP force can increase or decrease the energy of a satellite's orbit. Figure 5.15 shows a satellite initially in a circular orbit about the Earth. For simplicity, let us assume that the sun's rays are parallel to the orbital plane. When the satellite is near point A (right-hand side of the orbit in Figure 5.15), the SRP force will "push" in the same direction as the satellite's velocity vector and increase its speed and orbital energy. The perturbed orbit will have an apogee near point B in Figure 5.15. As the satellite transits point B, the SRP force is opposite the satellite's velocity vector causing deceleration (energy loss) and decreasing the perigee altitude near point A. Therefore, Figure 5.15 implies that over time, the SRP perturbation creates an elliptical orbit with continually increasing eccentricity but with essentially constant

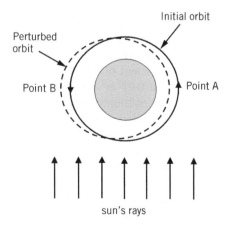

Figure 5.15 Solar-radiation pressure effects on a geocentric satellite.

semimajor axis (energy). This conclusion is a bit simplistic because the direction of the sun's rays will change as the Earth moves in its orbit. In addition, Earth eclipses (i.e., shadow conditions where SRP force is zero) must be included when analyzing the perturbed orbit.

The solar intensity (or solar flux) for the mean Earth–sun distance is $I_s = 1{,}361$ W/m^2. Note that solar intensity I_s is energy flux (power) per unit area. From basic mechanics, we know that power is the product of force and velocity. Dividing solar power per unit area by the speed of light [$c = 3(10^8)$ m/s] yields

$$P_{SRP} = \frac{I_s}{c} = 4.54(10^{-6})\,\text{N/m}^2 \tag{5.128}$$

Equation (5.128) is the solar radiation pressure P_{SRP} experienced by an Earth-orbiting satellite. Note that the units are pascals (or N/m^2) because solar intensity I_s has units of W/m^2 = N-m/(s-m^2) = N/(s-m); hence, dividing I_s by velocity (m/s) yields force per unit area. The perturbation acceleration from SRP is

$$a_{SRP} = \frac{P_{SRP}A_s C_R}{m} \tag{5.129}$$

where A_s is area of the satellite exposed to the sun and C_R is the reflectivity. It should be clear that the product $P_{SRP}A_s$ in Eq. (5.129) has units of force. Reflectivity $C_R = 0$ for a translucent body (i.e., light passes through the body), and hence there is no momentum transfer from the radiation and the SRP force is zero. Reflectivity $C_R = 1$ for a "black body" (i.e., all light is absorbed), and $C_R = 2$ for a "pure mirror" that reflects all radiation. Hence, a pure mirror doubles the momentum transfer (and force) when compared with a black body.

Equation (5.129) shows that the SRP perturbation acceleration is very difficult to accurately predict due to the uncertainty of the satellite's exposed area A_s and its reflectivity C_R. Let us attempt to quantify the SRP acceleration by considering a "typical" geocentric satellite with an area-to-mass ratio of 0.01 m^2/kg and reflectivity $C_R = 1.4$. Using Eq. (5.129) with $P_{SRP} = 4.54(10^{-6})$ N/m^2, we obtain $a_{SRP} = 6.36(10^{-8})$ m/s^2. We can compare the SRP and drag accelerations by interpolating the drag data in Table 5.4. Doing so, we find that a_{SRP} and a_D are equal for a 592-km altitude circular orbit (the reader should note that drag accelerations in Table 5.4 assumed a ballistic coefficient of 85 kg/m^2 which may not exactly match the assumed area-to-mass ratio of 0.01 m^2/kg used for the SRP acceleration calculation). Therefore, as a rough rule-of-thumb, drag will dominate SRP for circular orbits below 600 km, while SRP will dominate drag for altitudes above 600 km. Table 5.4 shows that drag acceleration is dramatically reduced to negligible values by modest increases in altitude above a 600-km LEO; SRP acceleration is essentially constant with orbital altitude for geocentric satellites that maintain a fixed orientation relative to the sun-pointing vector.

As a final demonstration, let us compute the magnitudes of the various perturbing accelerations on a geocentric satellite for a range of circular orbital altitudes. Third-body gravitational accelerations (moon and sun) are computed using Eqs. (5.118) and (5.119) with $r_{m\text{-}E} = 384{,}400$ km (mean moon–Earth distance), $r_{s\text{-}E} = 1.496(10^8)$ km (mean sun–Earth distance), and the appropriate gravitational parameters summarized in Table A.1 in Appendix A. We compute the maximum possible gravitational perturbations by assuming that the satellite is located on the line connecting the Earth and third body (moon or sun). Unlike calculating gravitational perturbations, computing drag and SRP accelerations require specific satellite characteristics. Here we will use a spherical satellite based on Sputnik 1 with mass $m = 83.6$ kg, diameter = 0.58 m, drag coefficient $C_D = 2$, and reflectivity $C_R = 1.7$. Drag acceleration is computed using Eqs. (5.124)–(5.126), where atmospheric density is determined by using ρ_0, h_0, and β values from Larson and Wertz [5] (recall that Table 5.3 presents density at representative altitudes). The perturbing acceleration from solar radiation pressure is computed using Eq. (5.129) with $P_{SRP} = 4.54(10^{-6})$ N/m^2. Figure 5.16 shows the various perturbing accelerations acting on the spherical satellite vs. circular orbital altitudes (note that Figure 5.16 is plotted on a log–log scale so that we can show a very wide range of perturbation accelerations and orbital altitudes). The Earth's gravitational acceleration μ/r^2 is also included in Figure 5.16 (of course, it is by far the dominant acceleration for orbits in the vicinity of Earth). Figure 5.16 clearly shows that drag acceleration is the dominant perturbation for circular orbits with altitudes between 100 and 300 km (i.e., the left-hand side of Figure 5.16). For altitudes between 350 and 400 km, drag acceleration and third-body accelerations (moon and sun) are approximately equal (note that the ratio between lunar and solar gravitational accelerations is nearly constant for orbits up to GEO). SRP acceleration is constant at $4.9(10^{-8})$ m/s^2 for all altitudes; drag and SRP accelerations are equal at an altitude of 530 km. Figure 5.16 also shows that the Earth and moon gravitational

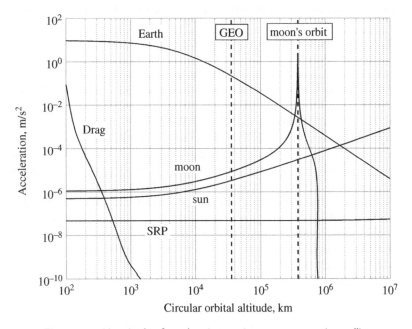

Figure 5.16 Magnitude of accelerations acting on a geocentric satellite.

accelerations are equal at two altitudes: roughly 340,000 km (in "front" of the moon) and 426,000 km ("behind" the moon). When the satellite is in the narrow altitude band near lunar orbit (between 340,000 km and 426,000 km), the moon is the dominant acceleration (again, we assume that the satellite is on the line connecting the Earth and moon). Figure 5.16 also shows that the Earth and sun gravitational accelerations are equal at an altitude of about 1.7 million km; beyond that distance the sun is the dominant acceleration. Extrapolating the Earth's gravitational acceleration in Figure 5.16 shows that it will be exceeded by SRP when the satellite is about 100 million km from Earth.

5.6 Circular Restricted Three-Body Problem

Some space mission scenarios involve a satellite moving in the vicinity of two gravitational bodies where their combined gravitational effect must be accounted for. One obvious example is an Earth–moon trajectory (such as the Apollo lunar missions), where the spacecraft's motion is simultaneously influenced by the Earth and lunar gravitational forces. We can imagine a point in a translunar trajectory where the magnitudes of the Earth and moon gravitational forces acting on the spacecraft are approximately equal (e.g., see Figure 5.16). Clearly, in this scenario we cannot treat either gravitational acceleration as a perturbation.

A spacecraft moving in Earth–moon space is an example of a general three-body problem. As previously mentioned, there is no closed-form solution of the N-body (or three-body) problem. However, we are able to develop some analytical expressions for the satellite's motion if we consider the *circular restricted three-body problem* (CR3BP). Two major assumptions are required for the CR3BP: (1) the two gravitational bodies move in circular orbits about their center of mass; and (2) the mass of the third body (the satellite) is negligible and does not influence the motion of the two primary bodies. While the second assumption seems reasonable, the first assumption causes some loss of accuracy for translunar trajectories (note that the moon's orbit about the Earth has an eccentricity of about 0.055). However, the CR3BP serves as a useful method for developing preliminary cislunar trajectories.

Figure 5.17 shows the geometry of the CR3BP (the Earth–moon system is used as an example throughout this section). The common origin of the inertial (non-rotating) Cartesian axes $(x_I y_I)$ and the rotating axes $(x_{rot} y_{rot})$ is located at the center of mass (c.m.) or barycenter of the Earth–moon system (Figure 5.17 is not to scale; the Earth–moon barycenter is actually below the Earth's surface). The rotating axis x_{rot} *always* points from the barycenter to the moon (the secondary gravitational body), and the rotating axis y_{rot} always points in a direction perpendicular to the Earth–moon line. Angular velocity of the $(x_{rot} y_{rot})$ frame is ω_r. Although Figure 5.17 only shows the x_{rot}-y_{rot} (or x_I-y_I) plane, both coordinate frames share a common z axis that points in a direction out of the page. The gravitational and centrifugal forces balance so that

$$\frac{GM_1 M_2}{D^2} = M_1 D_1 \omega_r^2 = M_2 D_2 \omega_r^2 \tag{5.130}$$

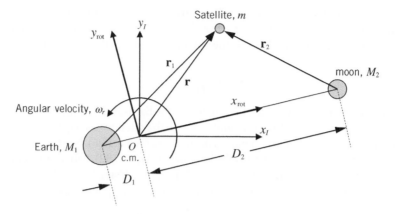

Figure 5.17 Geometry of the CR3BP for the Earth–moon system.

where G is the universal gravitational constant, M_1 and M_2 are the masses of the gravitational bodies, D_1 and D_2 are the distances of each body from the barycenter, and
$D = D_1 + D_2$ is the separation distance between the bodies. Manipulating Eq. (5.130)
using gravitational parameter $\mu_i = GM_i$, we obtain

$$\omega_r^2 = \frac{\mu_1 + \mu_2}{D^3} \tag{5.131}$$

We begin our analysis of the CR3BP with the absolute acceleration of the satellite [see
Eq. (C.16) in Appendix C]:

$$\ddot{\mathbf{r}} = \mathbf{a} = \mathbf{a}_O + \ddot{\mathbf{r}}_r + 2\boldsymbol{\omega}_r \times \dot{\mathbf{r}}_r + \dot{\boldsymbol{\omega}}_r \times \mathbf{r} + \boldsymbol{\omega}_r \times (\boldsymbol{\omega}_r \times \mathbf{r}) \tag{5.132}$$

where $\ddot{\mathbf{r}}$ is the satellite's acceleration relative to the inertial (fixed) axes $x_I y_I z_I$, $\mathbf{r}$ is the
position of the satellite from the barycenter origin O, $\ddot{\mathbf{r}}_r$ and $\dot{\mathbf{r}}_r$ are the satellite's acceleration and velocity relative to the rotating frame $x_{rot} y_{rot} z_{rot}$, and $\boldsymbol{\omega}_r$ is the constant angular velocity vector of the rotating frame. We can eliminate the acceleration of the
barycenter $\mathbf{a}_O$ and the term involving angular acceleration $\dot{\boldsymbol{\omega}}_r$ because both are zero.
Defining unit vectors $\mathbf{u}_x$, $\mathbf{u}_y$, and $\mathbf{u}_z$ along the *rotating* frame $x_{rot} y_{rot} z_{rot}$ allow us to express
the appropriate vectors as

$$\mathbf{r} = x\mathbf{u}_x + y\mathbf{u}_y + z\mathbf{u}_z \tag{5.133}$$

$$\dot{\mathbf{r}}_r = \dot{x}\mathbf{u}_x + \dot{y}\mathbf{u}_y + \dot{z}\mathbf{u}_z \tag{5.134}$$

$$\ddot{\mathbf{r}}_r = \ddot{x}\mathbf{u}_x + \ddot{y}\mathbf{u}_y + \ddot{z}\mathbf{u}_z \tag{5.135}$$

$$\boldsymbol{\omega}_r = \omega_r \mathbf{u}_z \tag{5.136}$$

Note that the time derivative of Eq. (5.133) does *not* produce Eq. (5.134), that is,
$d\mathbf{r}/dt \neq \dot{\mathbf{r}}_r$, because Eq. (5.134) does not account for the directional change of the rotating-frame unit vectors (see Appendix C for details). Carrying out the cross products for
the Coriolis and centrifugal terms in Eq. (5.132) yields

$$2\boldsymbol{\omega}_r \times \dot{\mathbf{r}}_r = -2\omega_r \dot{y}\mathbf{u}_x + 2\omega_r \dot{x}\mathbf{u}_y \tag{5.137}$$

$$\boldsymbol{\omega}_r \times (\boldsymbol{\omega}_r \times \mathbf{r}) = -\omega_r^2 x\mathbf{u}_x - \omega_r^2 y\mathbf{u}_y \tag{5.138}$$

In Section 5.2 we showed that the satellite's absolute acceleration due to gravity is equal to the gradient of a total potential function [see Eq. (5.5) for two-body motion]. For a system with two gravitational bodies, the total acceleration of the third body is

$$\ddot{\mathbf{r}} = \nabla\left(\frac{\mu_1}{r_1} + \frac{\mu_2}{r_2}\right) \tag{5.139}$$

where μ_1 and μ_2 are the gravitational parameters of the Earth and moon, respectively, and r_1 and r_2 are the distances from the Earth and moon to the satellite (see Figure 5.17). Remember that "del" is a vector operator of partial derivatives; see Eq. (5.6). Next, let us equate the satellite's absolute acceleration defined by kinematics, Eq. (5.132), with the absolute acceleration resulting from the gravitational forces, Eq. (5.139). Using Eqs. (5.135), (5.137), and (5.138), we may express the absolute acceleration in the component form:

$$\ddot{x} - 2\omega_r\dot{y} - \omega_r^2 x = \frac{\partial}{\partial x}\left(\frac{\mu_1}{r_1} + \frac{\mu_2}{r_2}\right) \tag{5.140}$$

$$\ddot{y} + 2\omega_r\dot{x} - \omega_r^2 y = \frac{\partial}{\partial y}\left(\frac{\mu_1}{r_1} + \frac{\mu_2}{r_2}\right) \tag{5.141}$$

$$\ddot{z} = \frac{\partial}{\partial z}\left(\frac{\mu_1}{r_1} + \frac{\mu_2}{r_2}\right) \tag{5.142}$$

Equations (5.140)–(5.142) are the governing equations for the satellite's motion in the CR3BP and are expressed in the rotating-frame coordinates.

5.6.1 Jacobi's Integral

Next, we will define an integral (constant) of the CR3BP. This integral will define allowable and forbidden regions for satellite motion. To start, let us multiply each acceleration component equation (such as the $\ddot{x}$ equation) by twice the corresponding velocity component ($2\dot{x}$ in this case) and add the three equations. The result is

$$2\ddot{x}\dot{x} + 2\ddot{y}\dot{y} + 2\ddot{z}\dot{z} - 2\omega_r^2 x\dot{x} - 2\omega_r^2 y\dot{y} = \frac{\partial}{\partial x}\left(\frac{\mu_1}{r_1} + \frac{\mu_2}{r_2}\right)2\dot{x} + \frac{\partial}{\partial y}\left(\frac{\mu_1}{r_1} + \frac{\mu_2}{r_2}\right)2\dot{y} + \frac{\partial}{\partial z}\left(\frac{\mu_1}{r_1} + \frac{\mu_2}{r_2}\right)2\dot{z} \tag{5.143}$$

Integrating Eq. (5.143) with respect to time, we obtain

$$\dot{x}^2 + \dot{y}^2 + \dot{z}^2 - \omega_r^2\left(x^2 + y^2\right) = \frac{2\mu_1}{r_1} + \frac{2\mu_2}{r_2} - C \tag{5.144}$$

It should be easy for the reader to verify that the time derivative of the left-hand side of Eq. (5.144) produces the left-hand side of Eq. (5.143). The time derivatives of the two right-hand side terms of Eq. (5.144) are determined by applying the chain rule:

$$\frac{d}{dt}\left(\frac{2\mu_1}{r_1}\right) = \frac{\partial}{\partial x}\left(\frac{2\mu_1}{r_1}\right)\frac{dx}{dt} + \frac{\partial}{\partial y}\left(\frac{2\mu_1}{r_1}\right)\frac{dy}{dt} + \frac{\partial}{\partial z}\left(\frac{2\mu_1}{r_1}\right)\frac{dz}{dt} \tag{5.145}$$

$$\frac{d}{dt}\left(\frac{2\mu_2}{r_2}\right) = \frac{\partial}{\partial x}\left(\frac{2\mu_2}{r_2}\right)\frac{dx}{dt} + \frac{\partial}{\partial y}\left(\frac{2\mu_2}{r_2}\right)\frac{dy}{dt} + \frac{\partial}{\partial z}\left(\frac{2\mu_2}{r_2}\right)\frac{dz}{dt} \tag{5.146}$$

Summing Eqs. (5.145) and (5.146) produces the right-hand side of Eq. (5.143).

The integration constant *C* in Eq. (5.144) is called *Jacobi's integral* or *Jacobi's constant*. Note that summing the first three terms on the left-hand side of Eq. (5.144) yields the *relative* velocity squared (i.e., $\dot{x}^2 + \dot{y}^2 + \dot{z}^2 = v_{rel}^2$). Using this definition, we can rearrange Eq. (5.144) to obtain

$$\frac{v_{rel}^2}{2} - \frac{\omega_r^2}{2}\left(x^2 + y^2\right) - \frac{\mu_1}{r_1} - \frac{\mu_2}{r_2} = -\frac{C}{2} \tag{5.147}$$

Equation (5.147) is similar to the two-body energy equation with a kinetic energy term $\left(v_{rel}^2/2\right)$ and two potential energy terms $(-\mu_1/r_1$ and $-\mu_2/r_2)$. Using this analogy, the term $-C/2$ is the total energy. We will show in the next subsection how the Jacobi integral *C* defines regions of allowable satellite motion in the CR3BP space.

5.6.2 Lagrangian Points

The next step is to characterize the CR3BP motion by investigating *equilibrium solutions* or (x,y,z) coordinates where the satellite has zero velocity and zero acceleration in the rotating frame. Therefore, we must develop explicit equations for the right-hand-side partial derivatives in Eqs. (5.140)–(5.142). The separation distances between the gravitational bodies and the satellite are

$$r_1 = \sqrt{\left(x + D_1\right)^2 + y^2 + z^2} \tag{5.148}$$

$$r_2 = \sqrt{\left(x - D_2\right)^2 + y^2 + z^2} \tag{5.149}$$

Using Eqs. (5.148) and (5.149) in Eqs. (5.140)–(5.142) and carrying out the partial derivatives yields

$$\ddot{x} - 2\omega_r\dot{y} - \omega_r^2 x = -\frac{\mu_1(x + D_1)}{r_1^3} - \frac{\mu_2(x - D_2)}{r_2^3} \tag{5.150}$$

$$\ddot{y} + 2\omega_r\dot{x} - \omega_r^2 y = -\frac{\mu_1 y}{r_1^3} - \frac{\mu_2 y}{r_2^3} \tag{5.151}$$

$$\ddot{z} = -\frac{\mu_1 z}{r_1^3} - \frac{\mu_2 z}{r_2^3} \tag{5.152}$$

It is useful at this point to rewrite the governing CR3BP equations (5.150)–(5.152) in terms of dimensionless variables that have been normalized by the sum of the gravitational parameters $(\mu_1 + \mu_2)$ and the constant distance between the gravitational bodies $(D = D_1 + D_2)$. First, let us define the mass ratio as

$$\tilde{\mu} = \frac{\mu_2}{\mu_1 + \mu_2} = \frac{GM_2}{GM_1 + GM_2} \tag{5.153}$$

Of course, $\tilde{\mu}$ is the mass of the (smaller) body M_2 normalized by the total system mass because *G* cancels out in Eq. (5.153). The mass ratio $\tilde{\mu}$ must be less than 0.5 because our convention is $M_2 < M_1$. The normalized mass of the primary body is $1 - \tilde{\mu}$ because the sum of the two normalized masses must equal unity. Balance between gravitational and centrifugal forces, Eq. (5.130), leads to the dimensionless distances

$$\widetilde{D}_1 = \frac{D_1}{D} = \widetilde{\mu} \quad \text{and} \quad \widetilde{D}_2 = \frac{D_2}{D} = 1 - \widetilde{\mu} \tag{5.154}$$

Finally, we note that the normalized angular rate $\widetilde{\omega}_r$ is unity. Using these results, we can express the CR3BP system dynamics in terms of dimensionless variables

$$\ddot{\widetilde{x}} - 2\dot{\widetilde{y}} - \widetilde{x} = -\frac{(1-\widetilde{\mu})(\widetilde{x}+\widetilde{\mu})}{\widetilde{r}_1^3} - \frac{\widetilde{\mu}(x-1+\widetilde{\mu})}{\widetilde{r}_2^3} \tag{5.155}$$

$$\ddot{\widetilde{y}} + 2\dot{\widetilde{x}} - \widetilde{y} = -\frac{(1-\widetilde{\mu})\widetilde{y}}{\widetilde{r}_1^3} - \frac{\widetilde{\mu}\widetilde{y}}{\widetilde{r}_2^3} \tag{5.156}$$

$$\ddot{\widetilde{z}} = -\frac{(1-\widetilde{\mu})\widetilde{z}}{\widetilde{r}_1^3} - \frac{\widetilde{\mu}\widetilde{z}}{\widetilde{r}_2^3} \tag{5.157}$$

where the tilde denotes dimensionless variables.

Now let us determine the equilibrium solutions of the dimensionless CR3BP dynamics (5.155)–(5.157). Setting all rotating-frame derivative terms to zero, we immediately see that the out-of-plane coordinate $\widetilde{z}$ must be zero for equilibrium. We will not present the equilibrium solution process (see Szebehely [6; pp. 131–138] for details). Five equilibrium solutions exist: three collinear points along the x_{rot} axis and two triangular points. These equilibrium solutions are called *Lagrangian points* or *libration points* (Euler discovered the three collinear points and Lagrange determined the two triangular points), and these points are typically labeled L_1–L_5. Figure 5.18 shows the five Lagrangian points in the rotating frame of the Earth–moon system. The five Lagrangian points exist because of the balance between gravitational and centrifugal forces. The three collinear Lagrangian points L_1, L_2, and L_3 lie on the Earth–moon line, and points L_4 and L_5 are located at vertices of equilateral triangles with length equal to D (i.e., the Earth–moon

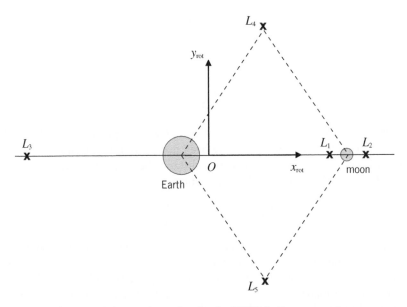

Figure 5.18 Lagrangian points for the CR3BP Earth–moon system.

Table 5.5 Locations of the Earth–moon Lagrangian points.

Lagrangian point	$\tilde{x}$	$\tilde{y}$
L_1	0.83692	0
L_2	1.15568	0
L_3	−1.00506	0
L_4	0.48785	0.86603
L_5	0.48785	−0.86603

separation distance). It is important for the reader to note that these Lagrangian points appear stationary when observed in the *rotating* x_{rot}–y_{rot} coordinate system. A satellite placed in any of the five Lagrangian points with zero *relative* velocity $(\dot{\tilde{x}} = \dot{\tilde{y}} = \dot{\tilde{z}} = 0)$ will appear to remain stationary in the rotating frame. However, a satellite at a Lagrangian point has non-zero *inertial* velocity when viewed from a non-rotating inertial frame.

Table 5.5 presents the locations of the Lagrangian points for the Earth–moon system in terms of dimensionless coordinates $\tilde{x}$ and $\tilde{y}$ (remember that $\tilde{z} = 0$ for all Lagrangian points). For the Earth–moon system, the mass ratio is $\tilde{\mu} = 0.01215$ and the dimensionless distances from the barycenter to the gravitational bodies are $\tilde{D}_1 = 0.01215$ (Earth) and $\tilde{D}_2 = 0.98785$ (moon). Using the mean Earth–moon distance $D = 384,400$ km, we see that the L_1 point is about 58,000 km from the center of the moon (the moon's radius is 1,738 km). The L_2 point is about 64,500 km from the moon's center on its "far side" when viewed from Earth.

Advanced analysis of the Lagrangian points would include checking the *stability* of these five equilibrium solutions. Stability may be evaluated by using a "standard" method from linear systems theory: (1) write the governing CR3BP dynamics (5.155)–(5.157) as a (nonlinear) system of six first-order state-variable equations; (2) linearize the system by computing the first-order partial derivatives of each state-variable equation; (3) compute the 6×6 state matrix by evaluating the partial derivatives at a Lagrangian point; and (4) compute the eigenvalues of the state matrix. We must perform the linearization process and corresponding eigenvalue calculation for each Lagrangian point. If any eigenvalue has a positive real part, then the linearized dynamics are unstable. This process shows that the three collinear Lagrangian points L_1, L_2, and L_3 are unstable. Therefore, a satellite that is slightly perturbed from an equilibrium condition at a collinear Lagrangian point will naturally diverge or drift away to another region of the CR3BP space. The two triangular Lagrangian points, L_4 and L_5, are stable. A satellite perturbed from the L_4 or L_5 point will continue to move in a region that is "close" to the triangular Lagrangian point.

Let us return to Eq. (5.144) and Jacobi's integral. Using dimensionless variables (e.g., $\tilde{\omega}_r = 1$), Eq. (5.144) becomes

$$\tilde{v}_{rel}^2 = \tilde{x}^2 + \tilde{y}^2 + \frac{2(1-\tilde{\mu})}{\tilde{r}_1} + \frac{2\tilde{\mu}}{\tilde{r}_2} - C \qquad (5.158)$$

where distances $\tilde{r}_1$ and $\tilde{r}_2$ are determined by using Eqs. (5.148) and (5.149) with the appropriate dimensionless variables. With the exception of $-C$, all terms in Eq. (5.158) are positive. We can use Eq. (5.158) and the Jacobi integral C to determine

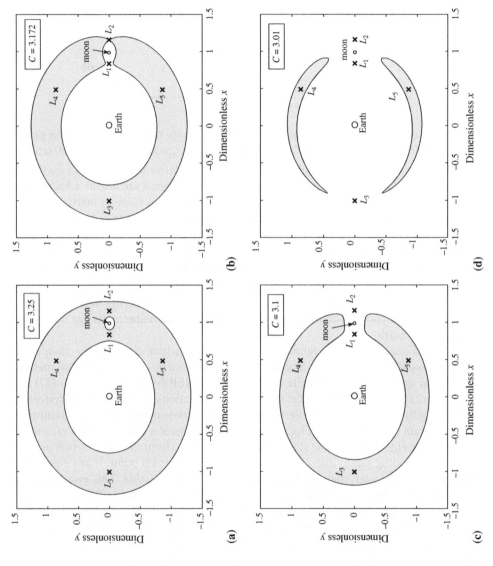

Figure 5.19 Allowable motion in CR3BP Earth–moon system for decreasing Jacobi constant C (motion is forbidden in the shaded regions).

a satellite's allowable and forbidden regions of motion in the three-body space. As an example, suppose the (dimensionless) Jacobi constant is $C = 3.25$ for a satellite in the Earth–moon system (the mass ratio is $\tilde{\mu} = 0.01215$ for the Earth–moon system). We can determine position coordinates $(\tilde{x}, \tilde{y})$ such that Eq. (5.158) is exactly equal to zero, that is, $\tilde{v}_{rel}^2 = 0$. Such coordinates would provide a locus of points that have *zero* relative velocity in the rotating frame. For the same constant $C = 3.25$, it is also possible to determine coordinates $(\tilde{x}, \tilde{y})$ such that Eq. (5.158) is *negative*, or $\tilde{v}_{rel}^2 < 0$. Such coordinates would define regions where satellite motion is infeasible or forbidden. Figure 5.19a illustrates this scenario with Jacobi constant $C = 3.25$. The shaded region in Fig. 5.19a represents $(\tilde{x}, \tilde{y})$ coordinates that cause Eq. (5.158) to be negative (or $\tilde{v}_{rel}^2 < 0$) and hence the shaded region is forbidden. When $C = 3.25$, the satellite may move anywhere outside the shaded region in Figure 5.19a; hence the satellite can move in a large circular region near Earth, a small circular region very close to the moon, and a large region very far from the Earth–moon system. The boundary of the shaded region is a locus of points with zero relative velocity. For this particular value of C, the satellite cannot reach any of the five Lagrangian points; nor can it travel from the Earth to the small region near the moon. As we decrease the Jacobi integral C (or, *increase* the "energy-like" constant $-C/2$), the forbidden region shrinks as seen in Figures 5.19a–d. Figure 5.19b shows that when $C = 3.172$, the collinear Lagrangian points L_1 and L_2 are accessible. Figure 5.19c shows that when C is further reduced to 3.1, a satellite may follow a trajectory from the Earth past the moon and reach regions that are well beyond the moon's orbit. Figure 5.19d shows that when $C = 3.01$, a satellite may move outside of the Earth–moon system along a path in the direction of the moon or in the direction of collinear Lagrangian point L_3. However, the triangular Lagrangian points remain inaccessible for this energy level. The reader should remember that *decreasing* the Jacobi integral C corresponds to *increasing* the total energy of the satellite's orbit.

Example 5.7 Consider a satellite that is located at the L_1 Lagrangian point in the CR3BP. Figure 5.20 shows the Earth–moon CR3BP where the origin of the rotating x_{rot}–y_{rot} system is located at the barycenter. An ECI coordinate system x_{ECI}–y_{ECI} is also shown in Figure 5.20, where the $+x_{ECI}$ and $+x_{rot}$ axes are aligned at this instant (the ECI frame does *not* rotate). Note that Figure 5.20 is not to scale. The non-rotating ECI frame and the rotating frame share the same $+z$ axis (i.e., the x_{ECI}–y_{ECI} plane is the plane containing the Earth and moon).

a) Determine the satellite's position vector in the non-rotating ECI coordinate system (in km) at this instant.

b) Determine the satellite's inertial velocity vector in the ECI frame (in km/s) at this instant.

c) Determine the Jacobi constant C for a satellite located at Lagrangian point L_1. Express C in the dimensionless system of units where D is the reference distance.

a) The position vector of the Lagrangian point L_1 in the *rotating* system is easy to determine because its coordinates are given in Table 5.5. The non-dimensional x_{rot} coordinate of L_1 is $\tilde{x} = 0.83692$ and its y_{rot} coordinate is $\tilde{y} = 0$. Thus, the position vector from the barycenter to the satellite at L_1 is

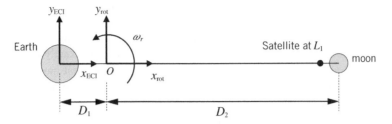

Figure 5.20 CR3BP for the Earth-moon system (Example 5.7).

$$\mathbf{r} = \begin{bmatrix} 0.83692 \\ 0 \\ 0 \end{bmatrix}$$

Recall that the normalizing distance unit is the mean Earth–moon separation, $D = 384,400$ km. Therefore, the satellite's dimensional position vector relative to the barycenter is

$$\mathbf{r} = D \begin{bmatrix} 0.83692 \\ 0 \\ 0 \end{bmatrix} = \begin{bmatrix} 321,712.05 \\ 0 \\ 0 \end{bmatrix} \text{ km}$$

The satellite's position vector in the Earth-centered frame is easy to compute because the $+x$ axes of coordinate systems x_{ECI}–y_{ECI} and x_{rot}–y_{rot} are aligned at this instant. We simply add the offset distance D_1 to the x_{rot} coordinate. Using the mass ratio $\tilde{\mu} = 0.01215$ and Eq. (5.154), we determine that the Earth–barycenter distance is $D_1 = D\tilde{\mu} = 4,670.46$ km and the Earth-to-satellite position vector is

$$\mathbf{r}_{ECI} = \begin{bmatrix} D_1 + D\tilde{x} \\ 0 \\ 0 \end{bmatrix} = \boxed{326,382.51\mathbf{I} \text{ km}}$$

Recall that $\mathbf{I}$ is a unit vector in the $+x_{ECI}$ direction.

b) The satellite's inertial velocity is determined using Eq. (C.12) in Appendix C

$$\mathbf{v}_{ECI} = \mathbf{v}_O + \mathbf{v}_{rot} + \boldsymbol{\omega}_r \times \mathbf{r}$$

where $\mathbf{v}_O$ is the inertial velocity of the barycenter (expressed in the ECI frame), $\mathbf{v}_{rot}$ is the satellite's velocity expressed in the rotating frame, and $\mathbf{r}$ is the satellite's position vector relative to the rotating frame. Because the barycenter moves in a circle with constant angular velocity ω_r (as seen by an observer at the origin of the ECI frame), the barycenter velocity is

$$\mathbf{v}_O = \omega_r D_1 \mathbf{J} = 0.0124\,\mathbf{J} \text{ km/s}$$

Recall that $\mathbf{J}$ is a unit vector in the $+y_{ECI}$ direction. The relative velocity in the rotating frame, $\mathbf{v}_{rot}$, is zero because the satellite is located at the Lagrangian L_1 point. The vector cross product is

$$\boldsymbol{\omega}_r \times \mathbf{r} = \omega_r D\tilde{x}\mathbf{J} = 0.8575 \ \mathbf{J} \ \text{km/s}$$

where the constant rotation rate is $\omega_r = 2.6653(10^{-6})$ rad/s and the magnitude of $\mathbf{r}$ is $D\tilde{x} = 321,712$ km. The Earth–moon angular velocity ω_r is computed using Eq. (5.131) with $\mu_1 = 3.986(10^5)$ km^3/s^2 (Earth) and $\mu_2 = 4,903$ km^3/s^2 (moon). Adding $\mathbf{v}_O$ and $\boldsymbol{\omega}_r \times \mathbf{r}$, we obtain the satellite's inertial velocity in the ECI frame:

$$\mathbf{v}_{ECI} = \mathbf{v}_O + \mathbf{v}_{rot} + \boldsymbol{\omega}_r \times \mathbf{r} = \boxed{0.8699 \ \mathbf{J} \ \text{km/s}}$$

c) The Jacobi constant can be computed using Eq. (5.158)

$$\tilde{v}_{rel}^2 = \tilde{x}^2 + \tilde{y}^2 + \frac{2(1-\tilde{\mu})}{\tilde{r}_1} + \frac{2\tilde{\mu}}{\tilde{r}_2} - C$$

Because the satellite is at L_1, it is has zero relative velocity, and therefore $\tilde{v}_{rel}^2 = 0$. Hence, the Jacobi constant for L_1 is

$$C = \tilde{x}^2 + \tilde{y}^2 + \frac{2(1-\tilde{\mu})}{\tilde{r}_1} + \frac{2\tilde{\mu}}{\tilde{r}_2}$$

We know that $\tilde{x} = 0.83692$ and $\tilde{y} = 0$. The dimensionless Earth–L_1 distance is $\tilde{r}_1 = \tilde{\mu} + \tilde{x} = 0.84907$, and the dimensionless moon–L_1 distance is $\tilde{r}_2 = 1 - \tilde{\mu} - \tilde{x} = 0.15093$ [see Eq. (5.154) for the dimensionless distances from the barycenter to each gravitational body]. Using these values, the Jacobi constant is

$$C = \tilde{x}^2 + \tilde{y}^2 + \frac{2(1-\tilde{\mu})}{\tilde{r}_1} + \frac{2\tilde{\mu}}{\tilde{r}_2} = \boxed{3.1883}$$

Compare this Jacobi constant with the values in Figures 5.19a and 5.19b. Figure 5.19a shows that when $C = 3.25$, the Lagrangian point L_1 is unreachable, whereas Figure 5.19b shows that motion near L_1 is possible for $C = 3.172$. Hence, $C = 3.1883$ is the *maximum* Jacobi constant that opens up the forbidden region and allows a satellite to reach the L_1 point.

Example 5.8   Figure 5.21 shows the Earth–moon CR3BP where the $+x$ axis of an ECI frame is aligned with the rotating x_{rot}–y_{rot} system at this instant. A satellite has the following position and velocity vectors in the ECI frame:

$$\mathbf{r}_{ECI} = -50,000 \ \mathbf{I} \ \text{km}, \quad \mathbf{v}_{ECI} = -3.675 \ \mathbf{J} \ \text{km/s}$$

Determine if this satellite can access the Lagrangian point L_1 and regions near the moon as shown in Figure 5.19b.

We can determine if L_1 and the lunar region is accessible by computing the Jacobi constant C. Equation (5.158) can be manipulated and solved for C

$$C = -\tilde{v}_{rel}^2 + \tilde{x}^2 + \tilde{y}^2 + \frac{2(1-\tilde{\mu})}{\tilde{r}_1} + \frac{2\tilde{\mu}}{\tilde{r}_2} \tag{5.159}$$

We need the satellite's position and velocity coordinates relative to the rotating x_{rot}–y_{rot} frame. The satellite's x_{rot} coordinate is

$$x = -D_1 - r_1 = -4,670.46 \ \text{km} - 50,000 \ \text{km} = -54,670.46 \ \text{km}$$

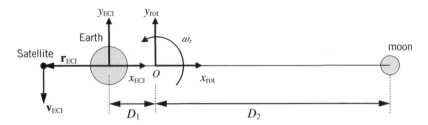

Figure 5.21 CR3BP for the Earth–moon system (Example 5.8).

where $r_1 = \|\mathbf{r}_{\text{ECI}}\|$ is the radial distance from the Earth center to the satellite. Normalizing this x_{rot} coordinate by using the Earth–moon distance D, we obtain

$$\tilde{x} = \frac{x}{D} = -0.1422$$

It is clear from Figure 5.21 that $\tilde{y} = 0$. The normalized Earth–satellite distance is

$$\tilde{r}_1 = \frac{r_1}{D} = 0.1301$$

The normalized moon–satellite distance is

$$\tilde{r}_2 = \frac{D_1 + D_2 + r_1}{D} = 1.1301$$

The satellite's inertial velocity in the ECI frame can be expressed as

$$\mathbf{v}_{\text{ECI}} = \mathbf{v}_O + \mathbf{v}_{\text{rot}} + \boldsymbol{\omega}_r \times \mathbf{r} = -3.675 \; \mathbf{J} \; \text{km/s}$$

where $\mathbf{v}_O$ is the inertial velocity of the barycenter ($=0.0124 \; \mathbf{J}$ km/s; see Example 5.7) and the cross-product term is

$$\boldsymbol{\omega}_r \times \mathbf{r} = \omega_r x \mathbf{J} = -0.1457 \; \mathbf{J} \; \text{km/s}$$

Therefore, the relative velocity vector expressed in the rotating frame is

$$\mathbf{v}_{\text{rot}} = \mathbf{v}_{\text{ECI}} - \mathbf{v}_O - \boldsymbol{\omega}_r \times \mathbf{r} = -3.5417 \; \mathbf{J} \; \text{km/s}$$

Hence, the magnitude of the relative velocity is $v_{\text{rel}} = 3.5417$ km/s. We must normalize the velocity using the reference speed, $D\omega_r = 1.0245$ km/s; the result is $\tilde{v}_{\text{rel}} = 3.4569$. Finally, we can substitute the dimensionless values for $\tilde{v}_{\text{rel}}, \tilde{x}, \tilde{y}, \tilde{r}_1, \tilde{r}_2$, and the mass ratio $\tilde{\mu} = 0.01215$ into Eq. (5.159) to obtain

$$C = 3.28$$

This Jacobi constant is greater than the value used to define the feasible and forbidden regions in Figure 5.19a (with $C = 3.25$), and therefore the satellite *cannot* transit from near-Earth space to regions surrounding L_1 and the moon. The Jacobi constant must be smaller so that the forbidden region near L_1 "opens up" and allows a satellite to move from near-Earth space to the vicinity of the moon (see Figures 5.19b and 5.19c). Increasing the satellite's relative velocity (or, increasing the magnitude of the inertial velocity $\mathbf{v}_{\text{ECI}}$ shown in Figure 5.21) will reduce C and allow the satellite to move in the forbidden regions shown in Figure 5.19a.

The Earth–moon system is not the only example of the restricted three-body problem. Five Lagrangian points also exist for the sun–Earth system. Several missions have sent spacecraft to the sun–Earth Lagrangian points. The International Sun-Earth Explorer-3 (ISEE-3) spacecraft (1978) was the first satellite to reach the sun–Earth L_1 point. Other spacecraft sent to sun–Earth L_1 include Solar and Heliospheric Observatory (SOHO, 1996), WIND (2004), and Lisa Pathfinder (2015). These L_1 spacecraft are orbiting solar observatories that measure the solar wind, solar activity, and gravitational waves. Spacecraft intended for solar system observations (such as the Planck and Herschel Space Observatories) operate near the sun–Earth L_2 point.

5.7 Summary

In this chapter, we make a distinction between *Keplerian* and *non-Keplerian* motion. Keplerian motion is governed by the two-body problem and consequently the orbital elements are constant. For non-Keplerian motion, the orbital elements vary with time. Recall that the two major assumptions that lead to Keplerian motion are: (1) the single gravitational body is spherically symmetric; and (2) no forces other than the central gravity act on the satellite. In this chapter, we investigated scenarios where both of these assumptions were violated. The reader should note that all real-world satellite orbits follow non-Keplerian motion. However, in many cases the two-body (Keplerian) solution accurately represents the satellite's motion.

In this chapter, we identified *perturbations* that cause deviations from the theoretical two-body motion. A non-spherical gravitational body, third-body gravity, atmospheric drag, and solar radiation pressure all cause orbital perturbations. Accounting for perturbations and obtaining a satellite's non-Keplerian motion is much more difficult than analyzing two-body motion. Two approaches exist: (1) *special perturbation methods* involve direct numerical integration of the perturbed two-body equations of motion; and (2) *general perturbation methods* develop analytical solutions by replacing the perturbations with a series expansion. We applied the general perturbation method to a non-spherical gravitational body to show that an oblate body causes the longitude of the ascending node (Ω) and argument of perigee (ω) to "drift" at a linear rate over time. This chapter also quantified the various perturbing forces that act on a satellite (the reader is encouraged to review Figure 5.16). Finally, we presented the CR3BP where two gravitational bodies simultaneously influence the satellite's motion. The "circular restricted" part of the CR3BP acronym arises from the assumption that the two gravitational bodies are moving in circular orbits about their common center of mass. While we cannot obtain closed-form solutions for the CR3BP, we can use the energy-like Jacobi integral to determine feasible and infeasible regions for satellite motion in the vicinity of two gravitational bodies.

References

1 Vallado, D.A., *Fundamentals of Astrodynamics and Applications*, 4th edn, Microcosm Press, Hawthorne, CA, 2013.

2 Battin, R.H., *An Introduction to the Mathematics and Methods of Astrodynamics, Revised Edition*, AIAA Education Series, Reston, VA, 1999.

3 Bate, R.R., Mueller, D.D., and White, J.E., *Fundamentals of Astrodynamics*, Dover, New York, 1971.

4 Kaplan, M.H., *Modern Spacecraft Dynamics and Control*, John Wiley & Sons, Inc., Hoboken, NJ, 1976.

5 Larson, W.J., and Wertz, J.R., *Space Mission Analysis and Design*, 3rd edn, Space Technology Library, El Segundo, CA, 2005.

6 Szebehely, V., *Theory of Orbits: The Restricted Problem of Three Bodies*, Academic Press, New York, 1967.

Problems

Conceptual Problems

Problems 5.1–5.4 involve special perturbation methods where a geocentric orbit is perturbed by Earth's oblateness (J_2). Each problem includes the time history plot of an orbital element. Use the special perturbations result to determine the requested orbital characteristic.

5.1 Figure P5.1 presents the time history of the longitude of the ascending node for a geocentric orbit with eccentricity $e = 0.183$ and inclination $i = 30°$. Estimate the semimajor axis a.

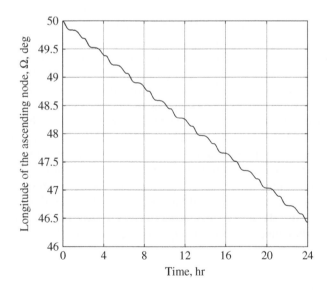

Figure P5.1

5.2 Figure P5.2 shows the time history of the longitude of the ascending node for a circular geocentric orbit with radius $r = 7{,}178$ km. Estimate the inclination.

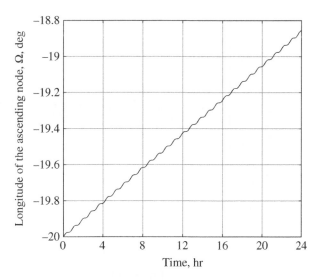

Figure P5.2

5.3 Figure P5.3 shows the time history of the argument of perigee for a geocentric orbit with eccentricity $e = 0.3$ and inclination $i = 10°$. Estimate the semimajor axis a.

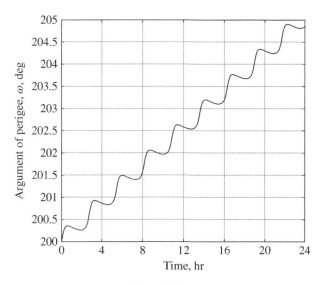

Figure P5.3

5.4 Figure P5.4 shows the time history of the argument of perigee for a geocentric orbit with eccentricity $e = 0.2$ and semimajor axis $a = 9,500$ km. Estimate the inclination.

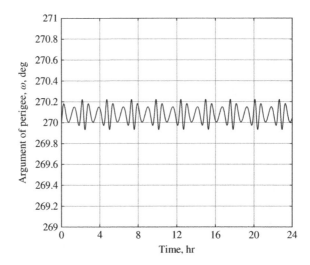

Figure P5.4

5.5 Carry out the partial derivatives (the "del" operation) of the total geopotential function for an oblate Earth, Eq. (5.16), and verify the absolute acceleration components presented by Eqs. (5.18)–(5.20).

5.6 A satellite is in a circular, equatorial low-Earth orbit (LEO) with an altitude of 185 km. The atmospheric density is $\rho = 4.65(10^{-10})$ kg/m^3 at this altitude. After 3 h, the satellite's orbital radius has decreased by 2.62 km. Estimate the satellite's ballistic coefficient C_B.

5.7 The special perturbation method is used to obtain the orbital characteristics of a low-altitude geocentric orbit. Figure P5.7 shows the satellite's altitude history with

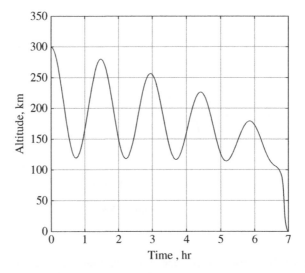

Figure P5.7

time. Estimate the satellite's semimajor axis and eccentricity at time $t = 0$. Using Figure P5.7, describe the progression of the satellite's energy (or semimajor axis) and eccentricity. Is a perturbation force acting on this satellite? If so, identify the perturbing force and describe how it is affecting the satellite's orbit.

5.8 Derive Gauss' variation of parameter equation for semimajor axis, da/dt, where the perturbing acceleration $\mathbf{a}_P$ is expressed in the **RSW** frame, i.e., $\mathbf{a}_P = a_r\mathbf{R} + a_s\mathbf{S} + a_w\mathbf{W}$ (see Figure 5.11).

5.9 Derive Gauss' variation of parameter equation for eccentricity, de/dt, where the perturbing acceleration $\mathbf{a}_P$ is expressed in the **RSW** frame, i.e., $\mathbf{a}_P = a_r\mathbf{R} + a_s\mathbf{S} + a_w\mathbf{W}$ (see Figure 5.11).

5.10 Derive Gauss' variation of parameter equation for parameter, dp/dt, where the perturbing acceleration $\mathbf{a}_P$ is expressed in the **NTW** frame.

5.11 Derive Gauss' variation of parameter equation for perigee radius, dr_p/dt, where the perturbing acceleration $\mathbf{a}_P$ is expressed in the **NTW** frame.

5.12 Derive Gauss' variation of parameter equation for apogee radius, dr_a/dt, where the perturbing acceleration $\mathbf{a}_P$ is expressed in the **NTW** frame.

5.13 A geocentric satellite is in an elliptical orbit with a perigee altitude of 500 km and inclination of 28.5°. Compute its apogee altitude such that the secular drift in argument of perigee is 5 deg/day.

Problems 5.14 and 5.15 involve the geocentric satellite shown in Figure P5.14. The satellite is in a highly inclined circular orbit such that its orbital plane is normal to Earth–sun direction on the June solstice.

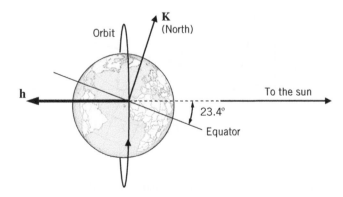

Figure P5.14

5.14 Suppose we desire a *circular* sun-synchronous orbit for the orbital geometry shown in Figure P5.14. Determine the orbital altitude for a sun-synchronous orbit.

5.15 Suppose that the (constant) perturbing acceleration from solar-radiation pressure is $6.2(10^{-8})$ m/s^2. Use Gauss' variation of parameter equations to show that solar radiation pressure does not cause a secular drift in the longitude of the ascending node Ω. [Hint: express the appropriate Gauss variational differential equation with argument of latitude u as the independent variable (instead of time) and integrate this transformed Gauss equation over one orbital revolution.]

5.16 Determine the Jacobi constant C for a satellite that is located at the Lagrangian point L_4 (use dimensionless variables).

5.17 Determine the inertial position and velocity vectors (in km and km/s) for a satellite that is located at the Lagrangian point L_4. Use the non-rotating, Earth-centered inertial (ECI) frame x_{ECI}–y_{ECI} that is defined at epoch t_0 as shown in Figure P5.17. In addition, determine the satellite's ECI position and velocity vectors at epoch $t_0 + 24$ h.

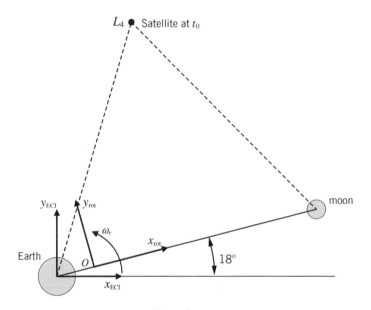

Figure P5.17

Problems 5.18–5.20 involve the circular restricted three-body problem (CR3BP) with the Earth and moon as the two primary gravitational bodies. Figure P5.18 depicts an instant in the CR3BP where the $+x_{ECI}$ axis of an Earth-centered inertial (ECI) frame is aligned with the $+x_{rot}$ axis of a rotating frame with its origin at the Earth–moon barycenter. The x_{ECI}–y_{ECI} plane coincides with the plane containing the Earth and moon.

5.18 A satellite has the following position and velocity vectors in the ECI frame:

$$\mathbf{r}_{ECI} = -6,578\ \mathbf{I}\ \text{km}, \quad \mathbf{v}_{ECI} = -10.7\ \mathbf{J}\ \text{km/s}$$

Determine if this satellite can access the Lagrangian points L_1 and L_2 and regions near the moon. Justify your answer.

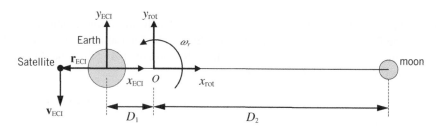

Figure P5.18

5.19 Repeat Problem 5.18 for a satellite with the following position and velocity vectors in the ECI frame:

$$\mathbf{r}_{ECI} = -6,578\ \mathbf{I}\ \text{km}, \quad \mathbf{v}_{ECI} = -10.89\ \mathbf{J}\ \text{km/s}$$

5.20 A GEO satellite has the following position and velocity vectors in the ECI frame:

$$\text{GEO:} \quad \mathbf{r}_{ECI} = -42,164\ \mathbf{I}\ \text{km}, \quad \mathbf{v}_{ECI} = -3.074665\ \mathbf{J}\ \text{km/s}$$

A second satellite is in an elliptical orbit with the ECI state vector:

$$\text{Elliptical orbit:} \quad \mathbf{r}_{ECI} = -6,746.24\mathbf{I}\ \text{km}, \quad \mathbf{v}_{ECI} = -10.426695\ \mathbf{J}\ \text{km/s}$$

a) Show that both orbits have the same two-body energy relative to Earth.
b) Which satellite orbit has the "largest" region for allowable motion in the CR3BP? (Or, which orbit has the smallest forbidden region in the CR3BP?) Justify your answer.

MATLAB Problem

5.21 Write an M-file that can compute a satellite's position and velocity vectors relative to the rotating coordinates that are used for the circular restricted three-body problem (CR3BP) for the Earth–moon system. The inputs to the M-file are the satellite's inertial position and velocity vectors, $\mathbf{r}_{ECI}$ and $\mathbf{v}_{ECI}$, expressed in an Earth-centered inertial frame (in km and km/s, respectively). Assume that the $+x_{ECI}$ axis of the ECI frame is aligned with the $+x_{rot}$ axis, and that both frames share a common $+z$ axis. The M-file should also compute the Jacobi constant C in terms of the dimensionless variables. Test your M-file by using the data contained in Examples 5.7 and 5.8.

Mission Applications

5.22 The Tropical Rainfall Measuring Mission (TRMM) satellite operated in low-Earth orbit from late 1997 to early 2015. The TRMM orbit was essentially circular at an altitude of 175 km and inclination of 34.9°. Suppose that the TRMM's longitude of the ascending node is $\Omega_0 = 78°$ at epoch t_0. Compute the longitude of the ascending node for the TRMM 10 days after this epoch.

5.23 Envisat ("Environmental Satellite") was an Earth-observing satellite deployed by the European Space Agency (ESA) in 2002. The Envisat orbit was essentially circular at an altitude of 773 km with an inclination of 98.4°. Show that Envisat operated in a sun-synchronous orbit.

5.24 The orbital elements of the Hubble Space Telescope (HST) at a particular epoch are

$$a = 6,922.3 \text{ km}, \quad e = 0.001143, \quad i = 50.75°$$
$$\Omega = 193.89°, \quad \omega = 85.41°, \theta = 10.23°$$

Determine the secular drift rates for Ω and ω (in deg/day) caused by Earth's oblateness.

5.25 The Cosmos 1687 was a mid-1980s Russian military communication satellite. The Cosmos 1687 followed an elliptical orbit with perigee and apogee altitudes of 6,566 and 33,913 km, respectively, with an inclination of 69.7°. Determine the secular drift rates for Ω and ω (in deg/day) caused by Earth's oblateness.

5.26 The Global Positioning System (GPS) is a constellation of satellites that occupy circular orbits at an altitude of 20,180 km with an inclination of 55°. Determine the secular drift rates for GPS satellites (in deg/day) caused by Earth's oblateness, lunar gravity, and solar gravity.

5.27 The Chandra X-ray Observatory (CXO) is an Earth-orbiting satellite with perigee and apogee altitudes of 14,308 and 134,528 km, respectively, and an inclination of 76.72°. Determine the secular drift rates for the CXO (in deg/day) caused by Earth's oblateness.

5.28 An upper-stage rocket burn inserts a communication satellite into in a geostationary transfer orbit (GTO) with a perigee altitude of 185 km, apogee altitude of 35,786 km, inclination $i = 28.5°$, longitude of ascending node $\Omega = 300°$, and argument of perigee $\omega = 180°$. At epoch t_0, the satellite is at perigee. A malfunction in the satellite's health-monitoring system causes the GEO-insertion rocket burn (at apogee) to be cancelled and consequently the satellite completes one orbital revolution in GTO and returns to perigee.
   a) Estimate the longitude of the ascending node Ω and argument of perigee ω after one orbital revolution.
   b) Determine the time-rate of energy change $d\xi/dt$ at perigee passage. Assume that the satellite's ballistic coefficient is $C_B = 85$ kg/m^2 and that the atmospheric density is $\rho = 4.65(10^{-10})$ kg/m^3 at 185-km altitude.

c) Estimate the total energy dissipated by drag for a single perigee passage. [Hint: assume that drag is only significant for flight at altitudes below 200 km and use an average atmospheric density and average velocity for altitudes between 200 and 185 km. Use Table 5.3 to determine the atmospheric density at 200 km.]

5.29 A lunar probe has the following position and velocity vectors as expressed in a *rotating* coordinate system with its origin at the Earth–moon barycenter:

$$\mathbf{r} = \begin{bmatrix} 146{,}810 \\ 73{,}045 \\ 0 \end{bmatrix} \text{km}, \quad \mathbf{v}_{\text{rot}} = \begin{bmatrix} 1.3137 \\ 0.3808 \\ 0 \end{bmatrix} \text{km/s}$$

Show that this lunar probe is capable of reaching the vicinity of the moon including the Lagrangian points L_1 and L_2. Can this satellite possibly reach the Lagrangian point L_3?

6

Rocket Performance

6.1 Introduction

Chapters 2–5 presented the fundamentals of satellite motion in an orbit. However, we have not yet discussed how to transport satellites from the surface of the Earth to orbit. Clearly, some form of rocket propulsion launches a satellite and delivers it to orbit. The next four chapters will discuss orbital maneuvers (i.e., changing a satellite's orbit) and interplanetary trajectories. Again, rocket propulsion is required to perform a desired orbital maneuver or initiate a trajectory to a planet or the moon. We will see in Chapters 7–10 that orbital maneuvers and orbital transfers involve changing the satellite's velocity vector.

This chapter presents the fundamental concepts of rocket propulsion and rocket performance. We do not present an in-depth treatment of rocket propulsion here; the interested reader may consult Sutton and Biblarz [1] and Hill and Peterson [2] for a detailed discussion of this topic. Instead, the primary objectives of this chapter are: (1) to present rocket performance metrics that aerospace engineers can use to link orbital maneuvers with propellant mass requirements; and (2) to present the important issues associated with launching a satellite into orbit.

6.2 Rocket Propulsion Fundamentals

Liquid-propellant engines and/or solid-rocket motors provide the thrust for launch vehicles. Rocket thrust is the result of a chemical reaction in the combustion chamber between the oxidizer and fuel in the propellant. Hot gases from the reaction are ejected out the nozzle of the rocket at a high velocity to impart a thrust force on the rocket. Using a control-volume approach, we can characterize the rocket thrust force as

$$T = \dot{m}v_e + (p_e - p_a)A_e \qquad (6.1)$$

where v_e is the exhaust velocity of the gas relative to the nozzle exit, $\dot{m}$ is the mass-flow rate of the exhaust gas (positive for outflow), p_e is the exhaust pressure at the nozzle exit, p_a is the ambient atmospheric pressure, and A_e is the area of the nozzle exit. The first term on the right-hand side in Eq. (6.1), $\dot{m}v_e$, is often labeled as the "momentum thrust" whereas the second term is called the "pressure thrust." Although the momentum thrust

Space Flight Dynamics, First Edition. Craig A. Kluever.
© 2018 John Wiley & Sons Ltd. Published 2018 by John Wiley & Sons Ltd.
Companion website: www.wiley.com/go/Kluever/spaceflightmechanics

is the dominant term in Eq. (6.1), the pressure imbalance term is not negligible for launch vehicles. For example, the Space Shuttle main engines experienced more than 18% thrust degradation at sea level when compared with thrust at vacuum conditions.

It is convenient to divide the total thrust in Eq. (6.1) by mass-flow rate in order to define the *effective exhaust velocity*

$$v_{eff} = v_e + \frac{(p_e - p_a)A_e}{\dot{m}} \tag{6.2}$$

Of course, we can concisely express the thrust as $T = \dot{m}v_{eff}$ using the effective exhaust velocity. *Specific impulse* is an important metric of rocket performance

$$I_{sp} = \frac{T}{\dot{m}g_0} \tag{6.3}$$

where $g_0 = 9.80665 \text{ m/s}^2$ is Earth's "standard gravitational acceleration near sea level." We compute specific impulse by dividing the impulse of a rocket (i.e., the time integral of thrust) by the total propellant weight consumed during the burn. Therefore, I_{sp} has units of time. Another definition of specific impulse is the time duration that a given quantity of propellant can generate a thrust equal to the propellant's initial weight. Substituting $T = \dot{m}v_{eff}$ into Eq. (6.3) yields the following expression for effective exhaust velocity

$$v_{eff} = g_0 I_{sp} \tag{6.4}$$

Hence, we may think of effective exhaust velocity and specific impulse as the same metric for rocket performance. We shall soon see that I_{sp} is the most important metric for characterizing the efficiency of a rocket engine and propellant combination.

6.3 The Rocket Equation

Let us assume that a rocket is operating in "field-free space" where no forces other than the rocket thrust T act on the vehicle. Hence, we may write Newton's second law as

$$m\dot{v} = T \tag{6.5}$$

where v is the velocity of the space vehicle. Vehicle acceleration is

$$\dot{v} = \frac{T}{m} = \frac{\dot{m}v_{eff}}{m} \tag{6.6}$$

Replacing the over-dot notation (e.g., $\dot{v} = dv/dt$), Eq. (6.6) becomes

$$\frac{dv}{dt} = \frac{-v_{eff}}{m}\frac{dm}{dt} \tag{6.7}$$

We inserted a minus sign in Eq. (6.7) because the outflow parameter $\dot{m}$ (positive by convention) has been replaced by the time derivative of mass (note that dm/dt is negative because propellant mass is being exhausted out the rocket nozzle). After canceling the dt terms and substituting $v_{eff} = g_0 I_{sp}$ into Eq. (6.7), we obtain

$$dv = \frac{-g_0 I_{sp}}{m}dm \tag{6.8}$$

which can be integrated to yield

$$\Delta v = -g_0 I_{sp} \ln\left(\frac{m_f}{m_0}\right) \tag{6.9}$$

where m_0 and m_f are the initial and final masses of the space vehicle before and after the propulsive burn, respectively. We can eliminate the minus sign in Eq. (6.9) by inverting the natural logarithm argument

$$\Delta v = g_0 I_{sp} \ln\left(\frac{m_0}{m_f}\right) \tag{6.10}$$

Equation (6.10) is often called the "rocket equation" or "ideal rocket equation." It can be used to compute the idealized (or maximum) velocity increment Δv that can be achieved by a given rocket engine-propellant combination (I_{sp}) and propellant mass $m_p = m_0 - m_f$. In Chapters 7–10, we will show how to compute the required velocity increment Δv for a desired orbital maneuver. The rocket equation (6.10) shows that for fixed initial and final masses, m_0 and m_f, the propulsive velocity increment Δv is maximized by using an engine-propellant combination with the highest possible specific impulse I_{sp}.

It is useful to modify the rocket equation so that we can determine the mass ratio m_0/m_f for a given Δv. Applying the exponential function to Eq. (6.10), we obtain

$$\frac{m_0}{m_f} = \exp\left(\frac{\Delta v}{g_0 I_{sp}}\right) \tag{6.11}$$

The mass ratio m_0/m_f is greater than unity for any nonzero Δv. It is also useful to modify Eq. (6.11) so that we can determine the propellant mass m_p required to impart a desired Δv. Substituting $m_f = m_0 - m_p$ into Eq. (6.11) yields

$$\frac{m_0}{m_0 - m_p} = \exp\left(\frac{\Delta v}{g_0 I_{sp}}\right) \tag{6.12}$$

Solving Eq. (6.12) for propellant mass, we obtain

$$m_p = m_0 \left[1 - \exp\left(\frac{-\Delta v}{g_0 I_{sp}}\right)\right] \tag{6.13}$$

Equation (6.13) shows that increasing I_{sp} will decrease the propellant mass required for a given velocity increment Δv. Table 6.1 presents the specific impulses for a variety of rocket-propellant combinations. In general, cryogenic liquid propellants with low molecular weight (e.g., liquid oxygen, LO_2, and liquid hydrogen, LH_2) are much more efficient than solid propellants. Figure 6.1 presents the final-to-initial mass fraction m_f/m_0 after a propulsive burn as computed by the reciprocal of Eq. (6.11). Note that the velocity increment Δv in Figure 6.1 ranges from zero (no burn) to 3.5 km/s; a velocity change at this upper value will result in a hyperbolic escape trajectory from low-Earth orbit (LEO). Figure 6.1 shows that the liquid-propellant J-2 stage delivers a much higher mass fraction when compared with the solid-propellant Star 48 stage. The Apollo lunar missions utilized the Rocketdyne J-2 stage to attain the velocity change to depart LEO and follow a high-energy trajectory to the moon. Figure 6.1 shows that the mass fraction after the translunar burn ($\Delta v \approx 3.05$ km/s) was about 0.48; hence more than half of the

Table 6.1 Rocket-propellant combinations and specific impulse.

Rocket	Propellant type	Fuel	Oxidizer	I_{sp} (s)
Space Shuttle main engine	Liquid	LH_2	LO_2	452
Rocketdyne J-2	Liquid	LH_2	LO_2	421
Aerojet Rocketdyne R-4D	Liquid	CH_6N_2	N_2O_4	311
Orbital ATK Star 48	Solid	HTPB/Al[a]	AP[b]	287
Shuttle solid rocket booster	Solid	Al	AP[b]	268

[a] Hydroxyl-terminated polybutadiene/aluminum.
[b] Ammonium perchlorate.

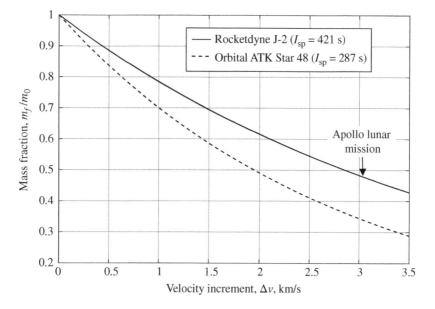

Figure 6.1 Final-to-initial mass fraction m_f/m_0 vs. velocity increment Δv.

total spacecraft mass in the low-Earth "parking orbit" consisted of propellant mass required to achieve the trajectory to the moon.

We cannot overemphasize the utility of either form of the rocket equation. Estimating the propellant mass is a principal objective of space mission design because it contributes to the total spacecraft mass, which in turn determines the launch-vehicle selection process. The overall mission costs are heavily influenced by the launch costs. Furthermore, reducing a space vehicle's onboard propellant mass may result in an increase in payload mass (e.g., additional scientific instruments).

Example 6.1 In July 1999, the Inertial Upper Stage (IUS) booster rocket transferred the Chandra X-ray Observatory (CXO) from a 300-km altitude circular LEO to an elliptical

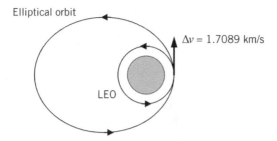

Elliptical orbit

$\Delta v = 1.7089$ km/s

LEO

Figure 6.2 CXO orbit transfer (Example 6.1).

orbit with an apogee altitude of 13,200 km (see Figure 6.2). The IUS provided a velocity increment of $\Delta v = 1.7089$ km/s as shown in Figure 6.2. The initial vehicle mass in LEO was 12,500 kg (CXO + IUS), and the solid-propellant IUS has a specific impulse of 296 s. Determine the propellant mass required for the orbit transfer.

Using Eq. (6.13) with $I_{sp} = 296$ s, $m_0 = 12,500$ kg, $\Delta v = 1,708.9$ m/s, and $g_0 = 9.80665$ m/s^2, yields the IUS propellant mass:

$$m_p = m_0\left[1-\exp\left(\frac{-\Delta v}{g_0 I_{sp}}\right)\right] = \boxed{5,562 \text{ kg}}$$

Note that Δv and effective exhaust speed $g_0 I_{sp}$ must have consistent units in the rocket equation (we used meters per second in this example).

It turns out that the propulsive maneuver presented in this example did not establish the final target orbit for the CXO. A sequence of additional propulsive burns was needed to place the CXO in its highly elliptical orbit with semimajor axis $a = 80,796$ km and eccentricity $e = 0.744$ (recall that Example 2.3 analyzes the final CXO orbit).

Example 6.2   Consider a 2,500 kg satellite in a 300-km altitude circular LEO as shown in Figure 6.3a. Compute the propellant mass required for a propulsive burn that increases the satellite's velocity from circular speed to the local escape speed (Figure 6.3b). Determine the propellant mass for a liquid-propellant stage with $I_{sp} = 325$ s and a solid-propellant stage with $I_{sp} = 280$ s.

We will use the propellant-mass version of the rocket equation, that is, Eq. (6.13). Therefore, we need to compute the incremental change in velocity Δv between the circular LEO (Figure 6.3a) and the parabolic escape trajectory shown (Figure 6.3b). Circular orbital speed is

$$v_{LEO} = \sqrt{\frac{\mu}{r_{LEO}}} = 7.726 \text{ km/s}$$

where $r_{LEO} = R_E + 300 = 6,678$ km is the radius of LEO. Local escape speed for a parabolic trajectory [see Eq. (2.88) in Section 2.6] is

$$v_{esc} = \sqrt{\frac{2\mu}{r_{LEO}}} = 10.926 \text{ km/s}$$

Figure 6.3b shows that the escape parabola is tangent to the circular orbit and hence they share the common radius r_{LEO}. The velocity increment required for the escape trajectory

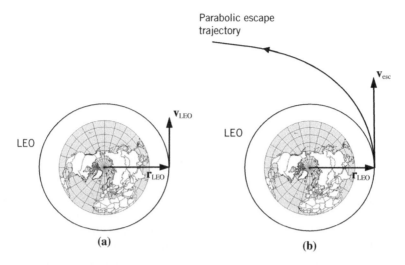

Figure 6.3 (a) Circular low-Earth orbit (LEO) and (b) escape trajectory (Example 6.2).

is $\Delta v = v_{\text{esc}} - v_{\text{LEO}} = 3.2$ km/s. Using Eq. (6.13) with $I_{\text{sp}} = 325$ s (liquid propellant), $m_0 = 2{,}500$ kg, and $g_0 = 9.80665$ m/s^2, yields the propellant mass:

$$\text{Liquid propellant:} \quad m_p = m_0 \left[1 - \exp\left(\frac{-\Delta v}{g_0 I_{\text{sp}}} \right) \right] = \boxed{1{,}584.0\,\text{kg}}$$

The reader should take care to express Δv and effective exhaust speed $g_0 I_{\text{sp}}$ in consistent units (km/s or m/s). Repeating this calculation using the solid-propellant rocket ($I_{\text{sp}} = 280$ s), we obtain

$$\text{Solid propellant:} \quad m_p = m_0 \left[1 - \exp\left(\frac{-\Delta v}{g_0 I_{\text{sp}}} \right) \right] = \boxed{1{,}720.5\,\text{kg}}$$

Therefore, the solid-propellant stage requires 136.5 kg additional propellant to perform the same velocity change as the liquid-propellant stage in this example. The reader should also note that in either case the propellant mass is a significant percentage of the initial mass of the satellite before the burn. For the more efficient liquid-propellant stage,

$$\text{Liquid propellant:} \quad m_p / m_0 = 0.634$$

Over 63% of the initial mass in LEO must be liquid propellant.

Example 6.3 In Example 6.2, the velocity increment required for an escape trajectory from LEO was computed to be $\Delta v = 3.2$ km/s. The propellant mass for the solid-motor stage was determined to be $m_p = 1{,}720.5$ kg (recall that the initial satellite mass is 2,500 kg and the specific impulse is $I_{\text{sp}} = 280$ s). Estimate the burn time for the solid-propellant stage if the engine can produce 67,000 N of thrust.

We use Eq. (6.3) to compute mass-flow rate from thrust and specific impulse

$$\dot{m} = \frac{T}{g_0 I_{sp}} = 24.4 \text{ kg/s}$$

The burn time is the total propellant mass divided by the mass-flow rate

$$t_{burn} = \frac{m_p}{\dot{m}} = \boxed{70.511 \text{ s}}$$

6.4 Launch Trajectories

A launch (or ascent) trajectory is designed to deliver a specified payload mass to a desired orbit at engine cut-off using the least amount of propellant. It is difficult to analyze launch trajectories due to the inclusion of aerodynamic and thrust forces along with gravity forces. Recall that obtaining an analytical solution for orbital motion is possible only for the case where a central-force (conservative) gravity field is the only force acting on the satellite. Accurately determining a launch trajectory relies on numerical integration of the ascent equations of motion

Figure 6.4 shows a free-body diagram of a launch vehicle during its planar ascent trajectory over a spherical Earth. Three forces affect the vehicle's motion: gravitational force mg acting from the vehicle to the center of the Earth, aerodynamic force (resolved into lift L and drag D), and thrust force T. By definition, lift L is the aerodynamic force component normal to the flight path and drag D is tangent to the flight path (opposite the velocity vector). We assume that thrust T always points along the longitudinal axis of the vehicle, and therefore it can be steered by changing the angle-of-attack α as shown in Figure 6.4. Because we are treating the launch vehicle as a point mass moving in a vertical plane, we do not develop a dynamical equation for angle-of-attack; instead we assume

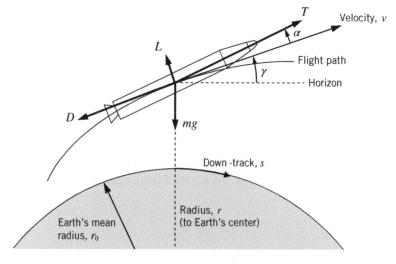

Figure 6.4 Launch trajectory in a vertical plane.

that the "steering program" $\alpha(t)$ is a known function of time. Attitude dynamics involve the rotational motion of a space vehicle about its center of mass and are treated in Chapters 12 and 13.

Our goal is to derive the equations of motion for the ascent trajectory. For simplicity, we will assume that the Earth (or planet) does not rotate. Thus, the vehicle's Earth-relative velocity v is also the inertial velocity. It is convenient to express accelerations in a local rotating frame that moves with the vehicle. The absolute acceleration of the launch vehicle in a rotating coordinate frame is

$$\mathbf{a} = \dot{v}\mathbf{u}_t + \left(v\dot{\gamma} - \frac{v^2 \cos\gamma}{r} \right) \mathbf{u}_n \tag{6.14}$$

where $\mathbf{u}_n$ and $\mathbf{u}_t$ are unit vectors that move with the vehicle and remain normal and tangential to the flight path (see Section C.4 in Appendix C for the full derivation of this equation). Radius r (from the Earth's center) and flight-path angle γ are the same parameters used to describe orbital motion in Chapters 2–5. Applying Newton's second law and the summation of forces in tangential and normal directions (see Figure 6.4) yields

$$m\dot{v} = T\cos\alpha - D - mg\sin\gamma \tag{6.15}$$

$$m\left(v\dot{\gamma} - \frac{v^2 \cos\gamma}{r} \right) = T\sin\alpha + L - mg\cos\gamma \tag{6.16}$$

Note we placed that the absolute acceleration terms on the left-hand sides of Eqs. (6.15) and (6.16) and the external forces on the right-hand sides. Equations (6.15) and (6.16) do not account for the Coriolis and centripetal accelerations caused by the Earth's rotation. We may divide all terms in Eqs. (6.15) and (6.16) by mass m in order to obtain accelerations and rearrange the equations so that the first-order derivative terms are on the left-hand sides:

$$\dot{v} = \frac{T}{m}\cos\alpha - \frac{D}{m} - g\sin\gamma \tag{6.17}$$

$$v\dot{\gamma} = \frac{T}{m}\sin\alpha + \frac{L}{m} - \left(g - \frac{v^2}{r} \right)\cos\gamma \tag{6.18}$$

In order to evaluate all right-hand side terms, we need expressions for Earth's gravitational acceleration, g, and the lift and drag forces, L and D. Gravitational acceleration has an inverse-square relationship with radius

$$g(r) = g_0 \left(\frac{r_0}{r} \right)^2 \tag{6.19}$$

where r_0 is the "mean radius" of the Earth that corresponds to the standard gravity; that is, $g_0 = \mu/r_0^2$ (note that the mean radius r_0 is not the Earth's equatorial radius). The lift and drag forces are

$$L = \frac{1}{2}\rho v^2 S C_L \tag{6.20}$$

$$D = \frac{1}{2}\rho v^2 S C_D \tag{6.21}$$

where ρ is the atmospheric density, S is a vehicle reference area, and C_L and C_D are the aerodynamic lift and drag coefficients, respectively. Because we use inertial velocity v in the aerodynamic force calculations, we have assumed that the Earth's atmosphere is stationary.

We need three additional equations to fully describe the state of the launch vehicle:

$$\dot{r} = v\sin\gamma \tag{6.22}$$

$$\dot{s} = \frac{r_0}{r}v\cos\gamma \tag{6.23}$$

$$\dot{m} = -\frac{T}{g_0 I_{sp}} \tag{6.24}$$

Equations (6.22) and (6.23) are kinematic equations that define the radial velocity and down-track velocity projected along the Earth's surface (Figure 6.4). Equation (6.24) defines the vehicle's decreasing mass due to burning and exhausting the propellant.

Equations (6.17), (6.18), and (6.22)–(6.24) are the equations of motion for the launch trajectory [of course we also need Eqs. (6.19)–(6.21) to fully define all terms in the dynamical equations]. These ordinary differential equations (ODEs) are highly nonlinear and hence there is no analytical (closed-form) solution. We must use a computer to numerically integrate the ODEs given the initial conditions, the thrust-steering program $\alpha(t)$, and appropriate functions that define atmospheric density ρ and aerodynamic coefficients C_L and C_D.

In order to gain some insight into the launch trajectory, we can analyze the losses due to atmospheric drag and the gravitational force. Consider the integral of Eq. (6.17), or integral of the acceleration along the flight path:

$$\int \dot{v}\,dt = \int \frac{T}{m}\cos\alpha\,dt - \int \frac{D}{m}\,dt - \int g\sin\gamma\,dt \tag{6.25}$$

The left-hand-side integral is the vehicle's *actual* velocity increment (or, change in magnitude of velocity vector **v**) because it is the time integral of absolute acceleration along the flight path. Let us denote the left-hand-side integral as Δv_{actual}. We can express all acceleration integrals in Eq. (6.25) as velocity increments

$$\Delta v_{actual} = \Delta v_{thrust} - \Delta v_{drag} - \Delta v_{grav} \tag{6.26}$$

where Δv_{thrust} is the integral of tangential thrust acceleration $T\cos\alpha/m$, Δv_{drag} is the integral of drag acceleration D/m, and Δv_{grav} is the integral of the tangential gravitational acceleration $g\sin\gamma$, respectively. If we consider the special case of tangential thrust (T is aligned with the velocity vector; i.e., $\alpha = 0$), then Δv_{thrust} is the *ideal* velocity increment defined by the rocket equation (6.10). The reader can see this fact by reviewing the derivation of the ideal rocket equation, Eqs. (6.5)–(6.10), where we assumed that tangential thrust is the only force acting on the vehicle. It should be clear to the reader that Eq. (6.10) determines the ideal or maximum theoretical velocity change that can be produced by a propulsive force aligned with the velocity vector. Thus, for the case of tangential thrust (or $\alpha = 0$), we can rewrite Eq. (6.26) as

$$\Delta v_{actual} = \Delta v_{ideal} - \Delta v_{drag} - \Delta v_{grav} \tag{6.27}$$

where the *ideal* velocity increment is computed using Eq. (6.10)

$$\Delta v_{ideal} = g_0 I_{sp} \ln \left(\frac{m_0}{m_f} \right) \tag{6.28}$$

The two energy-loss integrals that diminish the ideal velocity increment in Eq. (6.27) are called the *drag loss* and *gravity loss* and are defined as

$$\Delta v_{drag} = \int \frac{D}{m} dt \tag{6.29}$$

$$\Delta v_{grav} = \int g \sin \gamma \, dt \tag{6.30}$$

In general, we must compute both losses using numerical integration because there are no analytical solutions for the drag acceleration or flight-path angle histories during the ascent trajectory. Ascending vertically through the dense atmosphere as slowly as possible minimizes Δv_{drag} because aerodynamic drag is proportional to the square of the vehicle's velocity; see Eq. (6.21). Reaching horizontal flight ($\sin \gamma = 0$) as rapidly as possible, on the other hand, minimizes Δv_{grav}. However, a launch vehicle that accelerates along a long, shallow climb at relatively low altitudes will experience severe drag losses. Launch trajectories exhibit a compromise between these two conflicting flight programs.

Figure 6.5 shows the ascent profile of the Saturn V rocket used for the Apollo moon landings and the launch of the Skylab space station (1969–1973). We see that soon after the initial vertical ascent, the Saturn V pitched downward and the flight path became more horizontal (note the scale of the axes in Figure 6.5). When the launch vehicle reaches the first-stage engine cut-off altitude of approximately 65 km, the atmospheric density is greatly diminished, and hence drag is negligible (atmospheric density is reduced to less than 1% of its maximum sea-level value at an altitude of 32.5 km). Figure 6.5 shows that the majority of the second-stage flight and all of the third-stage flight is nearly horizontal as the Saturn V accelerates to orbital speed. Rocket staging

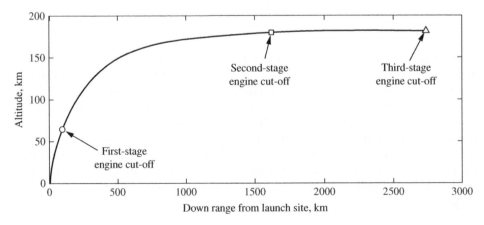

Figure 6.5 Altitude vs. down range for Saturn V launch trajectory.

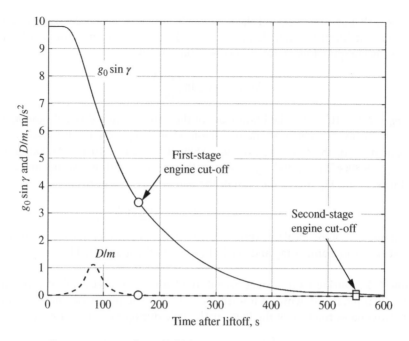

Figure 6.6 Energy-loss accelerations along the Saturn V flight path.

(discarding empty engine stages during flight) greatly increases the payload capability of launch vehicles. We will discuss staging in Section 6.5.

Figure 6.6 shows the acceleration losses along the flight path for the Saturn V booster: gravitational acceleration component tangent to the flight path, $g_0 \sin \gamma$, and drag acceleration, D/m. The gravity and drag losses, Δv_{grav} and Δv_{drag}, are the respective time integrals or areas under each curve in Figure 6.6. We see that $g_0 \sin \gamma \approx g_0$ for the first 30 s due to the initial vertical ascent ($\gamma = 90°$); for $t > 30$ s the Saturn V pitches down toward the horizon and hence the gravity component along the flight path diminishes. For the latter half of second-stage flight (and third-stage flight), the gravity-loss contribution is negligible because the flight path is essentially horizontal (see Figure 6.5). Drag loss is negligible for the first 30 s of vertical ascent due to the relatively low velocity and peaks at about 1.1 m/s^2 at $t = 80$ s due to supersonic flight through the dense atmosphere. Drag loss becomes small at first-stage engine cut-off because the atmosphere is relatively thin at an altitude of 65 km. Figure 6.6 shows that the gravity loss is far more significant than the drag loss for the Saturn V launch vehicle; the following examples will quantify the losses.

Example 6.4 The fully loaded mass of the Saturn V at liftoff is 2.899(10^6) kg and the total propellant mass of the first stage is 2.074(10^6) kg. At 161 s after liftoff, the first stage is depleted and shut down. The launch vehicle is at an altitude of 65 km and Earth-relative velocity $v_1 = 2,372$ m/s at first-stage engine cut-off. Assume that the average specific impulse for the first stage is 290 s. Estimate the sum of the drag and gravity losses experienced by the first stage of the Saturn V launch profile.

We know that at engine cut-off the *actual* velocity increment is $\Delta v_{\text{actual}} = v_1 = 2{,}372$ m/s because at liftoff the vehicle has zero Earth-relative velocity. The *ideal* velocity increment is computed using the rocket equation (6.10)

$$\Delta v_{\text{ideal}} = g_0 I_{\text{sp}} \ln\left(\frac{m_0}{m_f}\right)$$

where $m_0 = 2.899(10^6)$ kg is the liftoff mass of the entire Saturn V vehicle and m_f is the mass of the Saturn V vehicle at engine cut-off. Hence, m_f is the liftoff mass minus the propellant mass of the first stage, or $m_f = 2.899(10^6) - 2.074(10^6) = 825{,}000$ kg. Using m_0, m_f, $g_0 = 9.80665$ m/s^2, and $I_{\text{sp}} = 290$ s in the rocket equation, we obtain $\Delta v_{\text{ideal}} = 3{,}574.1$ m/s. Using Eq. (6.27), the sum of the two energy-loss terms is

$$\Delta v_{\text{drag}} + \Delta v_{\text{grav}} = \Delta v_{\text{ideal}} - \Delta v_{\text{actual}} = \boxed{1{,}202.1 \text{ m/s}}$$

Hence about 34% of the ideal (maximum) theoretical velocity increment from thrust has been lost to aerodynamic drag and vertical ascent in the gravity field.

Example 6.5 During its first-stage ascent, the Saturn V rocket climbed vertically for 30 s and then pitched downward toward the horizon. The piecewise linear function of time approximates the sine of the flight-path angle during first-stage ascent:

$$\sin\gamma = \begin{cases} 1 & \text{for } 0 \le t \le 30\,\text{s} \\ 1.15 - 0.005t & \text{for } 30 < t \le 161\,\text{s} \end{cases}$$

Estimate the total gravity loss at first-stage engine cut-off, $t = 161$ s.

Note that at liftoff $(t = 0)$, the flight-path angle is $\gamma(0) = \sin^{-1}(1) = 90°$ (vertical flight, as expected), and that at first-stage cut-off, $\gamma(161) = \sin^{-1}(0.345) = 20.2°$ (nearly horizontal flight). Equation (6.30) defines the gravity loss, which we separate into two integrals

$$\Delta v_{\text{grav}} = \int_0^{161} g\sin\gamma\, dt = \int_0^{30} g\, dt + \int_{30}^{161} g(1.15 - 0.005t)\, dt$$

Assuming constant gravitational acceleration $g = g_0$, we obtain the gravity loss

$$\Delta v_{\text{grav}} = g_0 t\big|_0^{30} + \left(1.15 g_0 t - 0.0025 g_0 t^2\right)\big|_{30}^{161} = \boxed{1{,}158.1 \text{ m/s}}$$

As an aside, we can use the gravity loss with the results from Example 6.4 to estimate the drag loss

$$\Delta v_{\text{drag}} = \Delta v_{\text{ideal}} - \Delta v_{\text{actual}} - \Delta v_{\text{grav}} = \boxed{44 \text{ m/s}}$$

This example shows that the gravity loss is much more significant than the drag loss for the first stage of the Saturn V launch vehicle. The relatively low drag loss is due to the Saturn V's low thrust-to-weight ratio and subsequent "slow" acceleration through the dense atmosphere. It turns out that Δv_{drag} estimated here is essentially the total drag loss for the entire launch trajectory because atmospheric density is relatively thin for flight above first-stage engine cut-off. Furthermore, gravity loss contributions after first-stage cut-off will diminish as the vehicle continues to pitch down from $\gamma(161) = 20.2°$ toward nearly horizontal flight. The next example demonstrates the diminished gravity loss for the nearly 6.4-min second-stage trajectory.

Example 6.6 At the start of the second stage of the Saturn V rocket ($t = 164$ s after liftoff), the vehicle has an Earth-relative velocity $v_1 = 2{,}372$ m/s and its total mass is 659,200 kg. The total propellant mass for the second stage is 444,100 kg and the specific impulse of the Rocketdyne J-2 engine is 421 s in vacuum conditions. An exponential function is used to approximate the sine of the flight-path angle during second-stage flight:

$$\sin\gamma = 1.54e^{-t/109} \text{ for } 164 \leq t \leq 550\,\text{s}$$

Figure 6.6 shows that $g_0 \sin\gamma$ approximately follows an exponential decay for $164 \leq t \leq 550$ s. Estimate the Earth-relative velocity of the launch vehicle at time $t = 550$ s (second-stage engine cut-off).

The ideal velocity increment can be computed using the rocket equation (6.10)

$$\Delta v_{ideal} = g_0 I_{sp} \ln\left(\frac{m_0}{m_f}\right)$$

where $m_0 = 659{,}200$ kg and m_f is the mass of the Saturn V at second-stage cut-off; $m_f = 659{,}200 - 444{,}100 = 215{,}100$ kg. Using m_0, m_f, $g_0 = 9.80665$ m/s^2, and $I_{sp} = 421$ s in the rocket equation, we obtain $\Delta v_{ideal} = 4{,}623.7$ m/s.

We use Eq. (6.30) to compute the gravity loss for second-stage flight

$$\Delta v_{grav} = \int_{164}^{550} g\sin\gamma\, dt = g_0 \int_{164}^{550} 1.54e^{-t/109}\, dt = -g_0(109)(1.54)e^{-t/109}\Big|_{164}^{550} = 355.0\,\text{m/s}$$

where we have assumed constant gravitational acceleration $g = g_0$. The actual velocity increment at second-stage engine cut-off is

$$\Delta v_{actual} = \Delta v_{ideal} - \Delta v_{grav} = 4{,}268.7\,\text{m/s}$$

Because atmospheric density is very small during second-stage flight, we have neglected drag losses. Hence, the Earth-relative velocity at the end of second stage is

$$v_2 = v_1 + \Delta v_{actual} = 2{,}372 + 4{,}268.7 = \boxed{6{,}640.7\,\text{m/s}}$$

As an aside, we can compute the *inertial* velocity of the launch vehicle relative to a non-rotating geocentric frame. To do so, we must add the inertial velocity of the rotating Earth to the vehicle's Earth-relative velocity v_2 [recall that the launch trajectory equations of motion (6.17) and (6.18) assumed a non-rotating planet]. Of course, a stationary vehicle on the surface of the Earth has inertial velocity equal to the product of the Earth's angular velocity and the distance normal to the spin axis. For launch from Kennedy Space Center (latitude $\phi = 28.5°$), the inertial velocity of the Earth's surface is $\omega_E R_E \cos\phi = 408.7$ m/s where $\omega_E = 7.292(10^{-5})$ rad/s is the Earth's spin rate and R_E is the radius of the Earth. At second-stage cut-off, the Saturn V is at an altitude of approximately 180 km and hence $r_2 = 6{,}558$ km (assuming a spherical Earth). Therefore, the inertial velocity of the launch vehicle at second-stage cut-off is

$$v_{2,\text{fix}} = v_{2,\text{rot}} + \omega_E r_2 \cos\phi = 6{,}640.7 + 420.3 = 7{,}061 \text{ m/s}$$

where we have assumed that the Saturn V is launched due east and remains at a latitude of 28.5°, and has horizontal velocity at second-stage engine cut-off. Note that circular orbital speed at 180-km altitude is $\sqrt{\mu/r_2} = 7{,}796$ m/s. A third stage of the Saturn V launch vehicle, consisting of a single J-2 engine, burned for about 2.5 min to provide the extra velocity increment to achieve circular orbital speed.

Figures 6.5 and 6.6 show that the launch profile usually begins with a vertical ascent through the dense layers of the atmosphere, followed by a rotation of the velocity vector toward the horizon. One way to begin the downward pitch motion is to produce a negative side force (i.e., $\alpha < 0$; see Figure 6.4) by slightly tilting the rocket from its vertical path. Equation (6.18) shows that a lateral thrust force, $T \sin \alpha$, will create a negative flight-path angle rate, $\dot{\gamma} < 0$, when $\alpha < 0$. Thrust is then realigned with the velocity vector ($\alpha = 0$) in order to maximize the rate of energy gain and reduce the side forces on a slender rocket. This maneuver initiates the so-called *gravity turn* where the gravitational acceleration normal to the flight path, $g \cos \gamma$, causes the rocket to pitch downward toward the horizon. The rotational rate of the gravity turn can be determined from Eq. (6.18) with $L = 0$ and $\alpha = 0$ (i.e., zero thrust and aerodynamic force components normal to the flight path)

$$\dot{\gamma} \cong \frac{-g \cos\gamma}{v} \tag{6.31}$$

Note that we have neglected the "centrifugal acceleration" term $v^2 \cos \gamma / r$ in Eq. (6.18) because it is relatively small compared with gravity during the initial launch trajectory. Table 6.2 presents approximate flight-path angle and velocity values for the first-stage ascent phase of the Saturn V booster. Column four in Table 6.2 displays the gravity turn rate as computed by Eq. (6.31). While the Saturn V did not utilize a pure gravity turn during the initial ascent phase, Table 6.2 shows that gravitational acceleration is capable of providing a gentle pitch-down maneuver without the need to impart aerodynamic or thrust side forces.

Table 6.2 Gravity turn for a Saturn V launch trajectory.

Time from liftoff (s)	Flight-path angle, γ	Velocity, v (m/s)	Gravity turn rate, $\dot{\gamma}$ (deg/s)
30	89.7°	89	−0.03
45	82.7°	162	−0.44
60	70.1°	264	−0.72
90	47.8°	591	−0.64
120	34.3°	1,178	−0.39
150	25.1°	1,998	−0.26

6.5 Staging

It is not possible to inject a satellite into LEO with a single rocket stage using the current level of technology. Single-stage rockets exhibit limited performance because the rocket engines must accelerate the entire structural weight from the surface of the Earth to orbital velocity.

To demonstrate the performance limitation of a single-stage rocket, we begin with analysis of its mass components:

$$m_0 = m_p + m_{st} + m_{PL} \tag{6.32}$$

where the liftoff mass m_0 consists of the total propellant mass m_p, the total structural mass m_{st}, and the payload mass m_{PL}. Structural mass m_{st} is the summation of all components that are not propellant or payload: tanks, engines, pumps, plumbing, support structure, electrical cables, and so on. It is useful to define non-dimensional mass ratios. To begin, we define the *payload ratio* λ as

$$\lambda = \frac{m_{PL}}{m_p + m_{st}} \tag{6.33}$$

The *structural coefficient* ε is

$$\varepsilon = \frac{m_{st}}{m_p + m_{st}} \tag{6.34}$$

Note that the "gross mass" of the rocket stage, $m_p + m_{st}$, is the denominator in both definitions. Clearly, we desire a large payload ratio λ. The *mass ratio* $\mathcal{R}$ is the liftoff mass divided by the final or burnout mass, $m_f = m_{st} + m_{PL}$

$$\mathcal{R} = \frac{m_0}{m_f} = \frac{m_0}{m_{st} + m_{PL}} \tag{6.35}$$

Using these non-dimensional terms in Eq. (6.32), we can express the mass ratio as

$$\mathcal{R} = \frac{1 + \lambda}{\varepsilon + \lambda} \tag{6.36}$$

Note that the mass ratio (6.35) is equal to Eq. (6.11), a variant of the ideal rocket equation, and hence we can write

$$\mathcal{R} = \exp\left(\frac{\Delta v_{ideal}}{g_0 I_{sp}}\right) = \frac{1 + \lambda}{\varepsilon + \lambda} \tag{6.37}$$

Next, we can characterize $\mathcal{R}$ by defining a particular target orbit (Δv_{ideal}) and rocket-propellant combination (I_{sp}). First, recall that we derived the ideal rocket equation by assuming that thrust is the sole force acting on the rocket (i.e., zero drag and gravity forces). Hence, the idealized (or theoretical maximum) velocity increment Δv_{ideal} used in Eq. (6.37) is computed by adding the drag and gravity losses to the *actual* velocity change to achieve the desired orbit. Using Eq. (6.27), we obtain

$$\Delta v_{ideal} = \Delta v_{actual} + \Delta v_{drag} + \Delta v_{grav} \tag{6.38}$$

For insertion into a 180-km circular LEO, the inertial velocity must be circular orbital speed, or $v_{LEO} = 7,796$ m/s. However, an eastward launch from Cape Canaveral (latitude $\phi = 28.5°$) will provide an additional 420 m/s due to Earth's rotation (see Example 6.6), and hence the *actual* velocity increment from liftoff to LEO is $\Delta v_{actual} = 7,796 - 420 = 7,376$ m/s. Assuming "typical" drag and gravity losses (e.g., $\Delta v_{drag} = 120$ m/s and

Table 6.3 Specific impulse and structural coefficient.

Stage	Propellant	Vacuum specific impulse, I_{sp} (s)	Structural coefficient, ε
Falcon 9	Kerosene/LO$_2$	311	0.06
Atlas V	Kerosene/LO$_2$	338	0.07
Delta IV	LH$_2$/LO$_2$	409	0.12

$\Delta v_{grav} = 1{,}600$ m/s), the ideal velocity increment required for LEO insertion is $\Delta v_{ideal} = 7{,}376 + 120 + 1{,}600 = 9{,}096$ m/s.

Now we can use Eq. (6.37) to determine mass ratio $\mathcal{R}$ for the launch-to-LEO $\Delta v_{ideal} = 9{,}096$ m/s and the rocket's specific impulse I_{sp}. However, the launch vehicle's structural coefficient ε will show a correlation with I_{sp}. Table 6.3 presents the specific impulses along with estimates of the structural coefficients for three existing liquid-propulsion stages. Using a low-weight fuel such as LH$_2$ will increase I_{sp}, but the corresponding low density will increase the size of the tank and consequently the structural coefficient ε (see the Delta IV data in Table 6.3).

To complete this simple analysis, let us compute the payload ratio λ for a fictitious single-stage vehicle based on the information in Table 6.3. We have established that the ideal velocity increment for launch-to-LEO insertion is $\Delta v_{ideal} = 9{,}096$ m/s. Solving Eq. (6.37) for payload ratio yields

$$\lambda = \frac{1 - \mathcal{R}\varepsilon}{\mathcal{R} - 1} \tag{6.39}$$

Table 6.4 presents the mass and payload ratios for a single-stage ascent to LEO. Mass ratio $\mathcal{R}$ dramatically decreases as I_{sp} increases (recall that $m_f/m_0 = 1/\mathcal{R}$) and hence the more efficient stages deliver greater mass to LEO as expected. However, the payload ratio λ is negative for all three cases indicating the infeasibility of a single-stage launch to LEO. It is interesting that the stage that uses the most efficient fuel (LH$_2$, $I_{sp} = 409$ s) exhibits the worst payload performance. While low-weight LH$_2$ provides the highest possible exhaust velocity (and smallest mass ratio $\mathcal{R}$), its use requires a very large tank (due to its low density), which in turn increases the structural coefficient.

All launch vehicles use multiple rocket stages to reach their desired target orbit. The main advantage of a multiple-stage rocket is that after propellant is burned and ejected, the structural "dead weight" of the empty stage is discarded thus making the entire rocket

Table 6.4 Single-stage to low-Earth orbit performance.[a]

Stage	Vacuum specific impulse, I_{sp} (s)	Structural coefficient, ε	Mass ratio, $\mathcal{R}$	Payload ratio, λ
Falcon 9	311	0.06	19.736	−0.010
Atlas V	338	0.07	15.552	−0.006
Delta IV	409	0.12	9.658	−0.018

[a] Ideal $\Delta v_{ideal} = 9{,}096$ m/s.

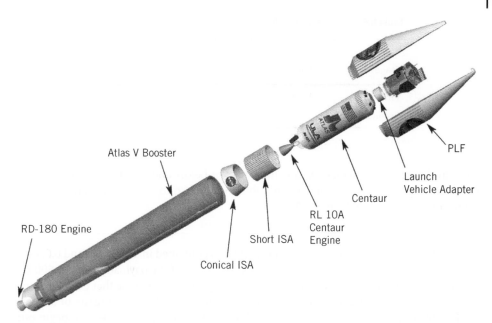

Figure 6.7 Atlas V 401 launch vehicle. *Source*: Reproduced with permission of United Launch Alliance, LLC.

Table 6.5 Atlas V 401 launch vehicle.

Stage	Engine	Specific impulse, I_{sp} (s)	Structural mass, m_{st} (kg)	Propellant mass, m_p (kg)
First	RD-180	315	25,600[a]	284,000
Second	RL-10	451	2,200	20,800

[a] Includes interstage adapter mass and payload fairing mass.

more efficient. A secondary advantage of staging is that smaller and/or fewer rocket engines can be employed for the latter stages because they are used to accelerate a less massive vehicle. A very large initial thrust force is required for liftoff, and hence vehicle acceleration for a single-stage vehicle may become excessive as the majority of the propellant mass is exhausted.

Figure 6.7 shows an exploded view of the two-stage Atlas V 401 launch vehicle. The first stage, a single RD-180 engine, burns for roughly 250 s. The Centaur second stage uses a single RL-10 engine with variable burn time. The payload fairing (PLF) is a shroud that protects the satellite (payload) against aerodynamic forces and thermal loads during flight through the atmosphere. Table 6.5 presents the approximate characteristics of the Atlas V 401 launch vehicle, while Table 6.6 presents the approximate gravity and drag losses for each stage. We may employ the rocket equation (6.10) to obtain the *idealized* velocity increment for each stage of a particular launch vehicle. However, we must

Table 6.6 Gravity and drag losses for the Atlas V 401 launch vehicle.

Stage	Gravity loss, Δv_{grav} (m/s)	Drag loss, Δv_{drag} (m/s)
First	1,300	120
Second	500	0

carefully define the initial mass m_0 and final mass m_f before and after the propulsive burn. The following example uses the data in Tables 6.5 and 6.6 to analyze the launch performance of the two-stage Atlas V 401.

Example 6.7　Using the Atlas V 401 specifications contained in Tables 6.5 and 6.6, estimate the final inertial velocity of the separated spacecraft if the payload mass is 9,800 kg.

　The first stage of the Atlas V 401 consists of a single RD-180 engine that uses kerosene and liquid oxygen as the propellant. The Centaur (or second) stage consists of a single RL-10 engine that burns a mixture of liquid hydrogen and liquid oxygen. To begin our calculations, let us use the rocket equation (6.10) to determine the ideal velocity increment for the first stage:

$$\Delta v_{ideal,1} = g_0 I_{sp,1} \ln\left(\frac{m_{0,1}}{m_{f,1}}\right)$$

where $m_{0,1}$ is the liftoff mass of the Atlas V, $m_{f,1}$ is the total mass of the Atlas V at first-stage engine cut-off, and $I_{sp,1} = 315$ s is the specific impulse of the first stage. Liftoff mass is the sum of all propellant masses, structural masses, and the payload mass. Using the data in Table 6.5 and a payload mass of 9,800 kg, we obtain

$$m_{0,1} = m_{p,1} + m_{st,1} + m_{p,2} + m_{st,2} + m_{PL} = 342,400\,\text{kg}$$

Total mass at first-stage burnout is the liftoff mass minus the first-stage propellant mass, or $m_{f,1} = 342,400 - 284,000 = 58,400$ kg. Hence, the first-stage ideal velocity increment is $\Delta v_{ideal,1} = 5,464$ m/s. Subtracting the first-stage gravity and drag losses (Table 6.6), yields the actual velocity increment at the end of the first stage

$$\Delta v_{actual,1} = \Delta v_{ideal,1} - \Delta v_{grav,1} - \Delta v_{drag,1} = 4,044\,\text{m/s}$$

Next, we apply the same steps to the second stage. The ideal velocity increment is

$$\Delta v_{ideal,2} = g_0 I_{sp,2} \ln\left(\frac{m_{0,2}}{m_{f,2}}\right)$$

Recall that the structural mass of the first stage has been discarded and consequently the initial mass at second-stage engine start is $m_{0,2} = m_{p,2} + m_{st,2} + m_{PL} = 32,800$ kg. The total mass at second-stage engine cut-off is $m_{f,2} = m_{0,2} - m_{p,2} = 12,000$ kg. Using $I_{sp,2} = 451$ s, the ideal velocity increment from the second stage is $\Delta v_{ideal,2} = 4,447$ m/s. After subtracting the 500 m/s gravity loss (there is no drag loss for second stage), the actual velocity increment is $\Delta v_{actual,2} = 3,947$ m/s. Finally, the spacecraft's Earth-relative velocity at the end of the second stage is the sum of the two velocity increments:

$$v_2 = \Delta v_{actual,1} + \Delta v_{actual,2} = 4,044 + 3,947 = 7,991 \text{ m/s}$$

We may assume that the spacecraft has a horizontal flight path (i.e., $\gamma = 0$) at second-stage burnout. Recall that our analysis of the flight mechanics of the ascent trajectory in Section 6.4 assumed a non-rotating Earth. In reality, of course, $v_{2,rot} = 7,991$ m/s is the spacecraft's Earth-relative velocity and we must add the velocity of the rotating frame in order to estimate the inertial velocity of the spacecraft:

$$v_{2,fix} = v_{2,rot} + \omega_E r_2 \cos\phi = 7,991 + 422 = \boxed{8,413 \text{ m/s}}$$

where Earth's spin rate is $\omega_E = 7.292(10^{-5})$ rad/s, $r_2 = 6,578$ km (i.e., 200-km altitude LEO), and latitude $\phi = 28.5°$ (launch from Cape Canaveral). Velocity for a 200-km altitude circular orbit is 7,784 m/s, and hence the estimated inertial velocity at second-stage burnout exceeds the LEO target speed by more than 600 m/s. One possible explanation for the overly optimistic performance is that our analysis assumes all propellant mass is consumed during both burns.

6.6 Launch Vehicle Performance

As noted in the previous section, determining a launch trajectory requires numerical integration of the equations of motion with accurate models for atmospheric density, subsonic and supersonic aerodynamic coefficients, and rocket thrust. Furthermore, maximizing the payload mass delivered to a desired orbit requires trajectory optimization methods that are beyond the scope of this textbook. Fortunately, launch vehicle suppliers provide a "payload user's guide" document that presents launch vehicle performance as a function of the orbital target. The launch performance is often times presented as a plot of payload mass vs. an orbital parameter such as circular orbit altitude. Thus, the launch vehicle supplier has accounted for all of the aforementioned complexities associated with determining a launch trajectory. References [3] and [4] are the "payload user's guides" for the Atlas and Delta families of launch vehicles, respectively.

Figure 6.8 shows the payload mass in circular LEO (inclination $i = 28.5°$) for two versions of the 500-series Atlas V launch vehicle (the second digit indicates the number of strap-on solid-propellant rockets that augment thrust at liftoff). Payload mass is typically defined by launch vehicle suppliers as the total mass of the spacecraft plus the mass required for the mechanical interface between the spacecraft and launch vehicle. Figure 6.8 shows that a single upper-stage burn is used for orbits with altitudes below 500 km, and a two-burn maneuver is used for circular altitudes greater than 500 km. Clearly, the payload mass decreases with LEO altitude because additional propellant mass is required to achieve the demand in additional orbital energy. Launch performance curves (similar to Figure 6.8) are used by satellite builders to determine the best match between existing launch vehicles and a desired spacecraft mass and target orbit.

Launch vehicles are also used to send a spacecraft on an escape trajectory that eventually leads to an interplanetary target. Recall from Chapter 2 that a satellite leaves a

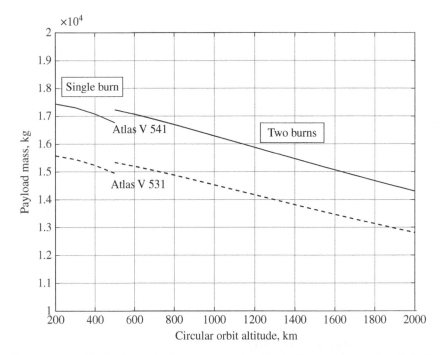

Figure 6.8 Low-Earth orbit payload mass vs. circular altitude. *Source:* Adapted from Ref. [3].

gravitational body along a hyperbolic trajectory with positive specific energy. Further-more, when the satellite has reached a very large distance from the gravitational body, the potential energy is negligible, that is, $-\mu/r_\infty \approx 0$ (recall that r_∞ is the radial distance "at infinity"). Therefore, total energy of a hyperbolic trajectory equals kinetic energy "at infinity"

$$\xi = \frac{v_\infty^2}{2} \tag{6.40}$$

Hence, the hyperbolic excess velocity v_∞ defines the energy of the departure hyperbola. Interplanetary launch performance is usually presented as a function of the so-called "launch energy" C_3 that is defined as the square of hyperbolic excess velocity

$$C_3 = v_\infty^2 = 2\xi \tag{6.41}$$

Figure 6.9 shows the payload mass capability of the Atlas V 551 and Delta IV Heavy vehicles for interplanetary missions [5]. Launch energy C_3 is the independent variable in Figure 6.9, and it defines the energy of the escape hyperbola. Note that the payload mass is greatest when $C_3 = 0$ for a parabolic trajectory with zero excess velocity at infinity (of course, a parabolic escape from Earth will require an additional propulsive burn to reach an interplanetary target). We will see in Chapter 10 that specifying a particular interplan-etary target (such as Mars or Venus) and departure and arrival dates determines the hyperbolic escape trajectory and hence the required C_3. Mission planners use launch-energy performance curves (similar to Figure 6.9) to estimate the total spacecraft mass as it leaves Earth's gravitational field.

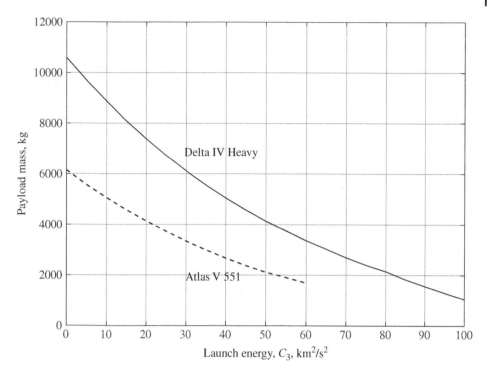

Figure 6.9 Interplanetary payload mass vs. launch energy C_3. *Source*: Ref. [5]. Courtesy of NASA.

6.7 Impulsive Maneuvers

This chapter has focused on the relationships between propulsive maneuvers and space-craft mass (i.e., the propellant mass) as well as launch trajectories. For some on-orbit propulsive maneuvers, the burn time is relatively short compared with the orbital period and consequently we can treat the burn as an impulsive maneuver (of course, the long-duration burn of a launch trajectory does not fit this category). We can develop an impulsive-thrust approximation by considering again Eq. (6.6) where thrust T is the only external force acting on the satellite:

$$\dot{v} = \frac{T}{m} \tag{6.42}$$

Next, we approximate the vehicle acceleration using $\dot{v} = \Delta v / \Delta t$ and solve Eq. (6.42) for finite burn time Δt

$$\Delta t = \frac{\Delta v}{T/m} \tag{6.43}$$

Equation (6.43) shows that a finite velocity change Δv will require a relatively short burn time if the thrust acceleration T/m is large. Equation (6.43) also shows that burn time Δt shrinks to an infinitesimal value as thrust acceleration becomes infinite. This limiting case is *impulsive thrust* or an *impulsive maneuver*. In this scenario, a very "large" thrust force instantaneously produces the desired velocity change Δv without any change in orbital position.

Let us consider the everyday example of a golfer driving a golf ball. The golf club imparts a very large force on the golf ball such that it leaves the tee at a speed of 150 mph (67 m/s). Therefore, we may treat a golfer's drive as an impulse: the ball's velocity changes from zero (at rest on the tee) to 67 m/s in a very short time with essentially no change in position of the ball.

As a space flight dynamics example, consider again the scenario in Examples 6.2 and 6.3 where a solid-propellant rocket delivers a force that changes the vehicle's velocity to achieve a parabolic escape trajectory from LEO. Referring back to Example 6.3, we see that a 70.5 s (1.2 min) burn will achieve the desired velocity increment $\Delta v = 3{,}200$ m/s. Because the orbital period of LEO is 90.5 min, the burn arc is about 1/90 of an orbital revolution or about 4°. Hence, there is little change in the spacecraft's position during the burn (the change in orbital position and burn time are small relative to the large time and distance scales associated with space flight). The short burn time supports the impulsive-maneuver approximation. Note that the initial thrust acceleration for Example 6.3 is $T/m_0 = 67{,}000$ N/2,500 kg $= 26.8$ m/s$^2 = 2.73\,g_0$. This relatively large thrust acceleration results in a short burn time. Reducing the thrust acceleration increases the burn time, and consequently the propulsive maneuver can no longer be approximated as an impulse.

Employing impulsive maneuvers greatly simplifies the analysis of orbital transfers, as we shall see in Chapters 7, 8, and 10. For impulsive maneuvers, the important performance metric is the velocity increment, Δv, and it is readily computed from knowledge of the initial orbit and transfer orbit. Once Δv is known, we can easily determine the propellant mass by using the ideal rocket equation and the propulsion system characteristics such as specific impulse.

6.8 Summary

This chapter has provided a brief introduction to the fundamentals of rocket propulsion and rocket performance. The primary outcome of this chapter is the so-called "rocket equation" that links the incremental change in velocity (Δv) with the corresponding propellant mass. It is important to reiterate that the rocket equation was derived for a vehicle in "field-free space" where propulsive thrust is the only external force acting on the vehicle. Consequently, we may use the rocket equation to determine the theoretical or *ideal* velocity increment Δv achieved by burning a given amount of propellant. The ideal Δv always represents the maximum possible velocity increment produced by rocket propulsion because other forces (such as aerodynamic drag and gravity) are ignored. For launch trajectories originating at the Earth's surface, drag and gravity losses must be included in order to estimate the vehicle's *actual* change in velocity after a propulsive maneuver. However, when an orbiting satellite fires an onboard rocket with a sufficiently "high" thrust-to-mass ratio, the burn time is relatively short and consequently the satellite's velocity changes rapidly with little change in orbital position. This scenario is approximated by an impulsive maneuver with an instantaneous change in velocity. In this case, the ideal velocity increment predicted by the rocket equation exhibits a good match with the vehicle's actual change in velocity.

References

1 Sutton, G.P. and Biblarz, O., *Rocket Propulsion Elements*, 8th edn, John Wiley & Sons, Inc., Hoboken, NJ, 2010.
2 Hill, P.G., and Peterson, C.R., *Mechanics and Thermodynamics of Propulsion*, 2nd edn, Addison-Wesley, Reading, MA, 1992.
3 Atlas V Launch Services User's Guide, United Launch Alliance, March 2010, http://www.ulalaunch.com/uploads/docs/AtlasVUsersGuide2010.pdf. Accessed August 17, 2017.
4 Delta IV Launch Services User's Guide, United Launch Alliance, June 2013, http://www.ulalaunch.com/uploads/docs/Launch_Vehicles/Delta_IV_Users_Guide_June_2013.pdf. Accessed August 17, 2017.
5 NASA Launch Services Program, July 2015, https://elvperf.ksc.nasa.gov/Pages/Query.aspx. Accessed August 17, 2017.

Problems

Conceptual Problems

6.1 A rocket engine and propellant combination has a specific impulse of 410 s. Determine the mass-flow rate required to obtain a thrust of 432,500 N.

6.2 A single Space Shuttle Main Engine (SSME) has the following characteristics:

Thrust at sea level, $T_{SL} = 1{,}859.4$ kN
Specific impulse at sea level is 366 s
Nozzle exit area, $A_e = 4.168$ m^2
Nozzle exit pressure, $p_e = 11{,}600$ Pa

Determine the specific impulse of the SSME at an altitude of about 6 km where the ambient atmospheric pressure is $p_a = 46{,}563$ Pa. Atmospheric pressure at sea level is 101,325 Pa.

6.3 A launch vehicle is currently at an altitude of 6.65 km with velocity $v = 280$ m/s. The launch vehicle is using a gravity turn to create a pitch-down maneuver with angular velocity $\dot{\gamma} = -0.735$ deg/s. Determine the launch vehicle's current flight-path angle.

6.4 Consider a rocket stage operating in "field-free space" where the rocket thrust is the only force. If the rocket's effective exhaust velocity is 3,140 m/s and the initial total vehicle mass is 4,500 kg, determine the propellant mass required to increase the vehicle's velocity by 2,600 m/s.

6.5 Using the rocket stage data in Problem 6.4, determine the payload ratio if its structural coefficient is 0.097 and the desired velocity increment is 2,600 m/s.

6.6 Consider a single-stage-to-orbit (SSTO) launch vehicle. Suppose the *ideal* velocity increment to achieve low-Earth orbit (LEO) is $\Delta v_{ideal} = 9{,}100$ m/s (this ideal increment factors in gravity and drag losses and the small velocity boost from Earth-rotation effects). If a SSTO vehicle could achieve a structural coefficient of 0.08 and a specific impulse of 375 s, determine the vehicle's total liftoff mass m_0 required to deliver a payload mass of 1,000 kg to LEO.

6.7 Consider the SSTO launch vehicle scenario described in Problem 6.6 (i.e., $m_{PL} = 1{,}000$ kg and $\Delta v_{ideal} = 9{,}100$ m/s). The specific impulse is a function of structural coefficient ε:

$$I_{sp} = -6{,}361\varepsilon^2 + 2{,}751\varepsilon + 195.6 \ (\text{in s})$$

Recall that the structural coefficient generally increases as specific impulse increases. Plot the total liftoff mass m_0 vs. structural coefficient for $0.05 \le \varepsilon \le 0.1$ and determine the structural coefficient that minimizes m_0 (note that the given specific impulse function is overly optimistic; e.g., $I_{sp} = 357$ s for $\varepsilon = 0.07$ which is significantly better than the specific impulse values listed in Table 6.3. In addition, I_{sp} in Table 6.3 are values for vacuum conditions whereas the I_{sp} values used for a SSTO vehicle must be "averaged" in order to account for both atmospheric and vacuum flight).

6.8 A satellite is in a 200-km altitude LEO and its total initial mass is 3,000 kg. The satellite needs to fire its onboard rocket engine to increase its inertial velocity to 8.6 km/s. If its onboard rocket engine has a thrust of 900 N, can we approximate this single rocket burn as an impulsive maneuver? Explain your answer.

Mission Applications

Problems 6.9–6.14 involve the Atlas V launch vehicle. Its characteristics are tabulated as follows:

Stage	Propellant mass (kg)	Structural mass (kg)	I_{sp} (s)	Total burn time (s)
First (RD-180)	284,000	21,000	315	253
Second (Centaur RL-10)	20,800	2,200	451	927
Solid rocket booster	41,000	5,700	279	94

The structural masses for the first and second stages accounts for the engines, pumps, tanks, and so on, and do not include the interstage adapter mass and payload fairing mass (see Figure 6.7). The total interstage adapter mass is 2,500 kg and the payload fairing mass is 3,600 kg. Note that the solid rocket booster (SRB) characteristics in the above table pertain to a single SRB.

In addition, use Table 6.6 for the gravity and drag losses for the Atlas V launch vehicle.

6.9 First, consider the Atlas V 501 which does *not* use any strap-on SRBs (hence the "0" second digit in 501). Compute the *ideal* and *actual* velocity increments (Δv) at the completion of the first-stage burn if the payload mass is 8,000 kg.

6.10 Next, let us consider the Atlas V 531 which uses three strap-on SRBs (hence the "3" second digit in 531). Compute the *ideal* velocity increment at the completion of the SRB burn (the SRBs are ignited at liftoff). You cannot estimate the *actual* Δv because the first-stage burn is not yet completed. In addition, compute the *ideal* and *actual* velocity increments at the completion of the first-stage burn. The payload mass is 8,000 kg.

6.11 Finally, consider the Atlas V 551 which uses five strap-on SRBs (hence the "5" second digit in 551). Compute the *ideal* and *actual* velocity increments at the completion of the first-stage burn (the SRBs are ignited at liftoff). The payload mass is 8,000 kg.

6.12 Now consider the final velocity of the spacecraft (payload mass) after all rocket stages have been exhausted. First, let us analyze the Atlas V 501 (no SRBs). Determine the final inertial velocity of the spacecraft for a payload mass of 8,000 kg. Assume that the Earth's rotation adds 422 m/s directly to the velocity increments.

6.13 Determine the final inertial spacecraft velocity delivered by the Atlas V 551 (five SRBs). Use a payload mass of 4,000 kg and assume that the Earth's rotation adds 422 m/s directly to the velocity increments. Assuming that the final altitude is 6,563 km at second-stage burnout (i.e., 185-km altitude) compute the launch energy C_3 and compare your result to the interplanetary payload performance predicted in Figure 6.9.

6.14 Repeat Problem 6.13 and determine the final inertial spacecraft velocity and launch energy C_3 delivered by the Atlas V 531 for a payload mass of 4,000 kg.

Problems 6.15–6.17 involve the S-IVB which was the third stage of the Saturn V rocket used for the Apollo missions. It was burned twice; the first burn completed the launch phase and placed the payload in a 185-km altitude circular LEO. The second burn was "translunar injection" (TLI) which sent the payload on a highly eccentric orbit to the moon. The S-IVB could be throttled to vary the thrust and mass-flow rate. The characteristics of these two burns are summarized in the following table:

Event	Mass-flow rate (kg/s)	Burn time (s)
LEO insertion	218.1	147
Translunar injection	204.4	347

The specific impulse of the J-2 engine used by the S-IVB is $I_{sp} = 421$ s. Flight-path angle is essentially zero during both burns of the S-IVB stage.

6.15 Compute the velocity increment Δv_{LEO} for the LEO insertion burn if the total mass in LEO *after* the burn is 132,000 kg [total mass in LEO includes the S-IVB stage and the payload mass consisting of the command/service module (CSM), and lunar module (LM)].

6.16 Compute the velocity increment Δv_{TLI} for translunar injection (TLI). The total mass before the TLI burn is 132,000 kg.

6.17 If the payload mass for TLI (i.e., the CSM and LM) is $m_{PL} = 44,000$ kg, estimate the structural coefficient and payload ratio of the S-IVB stage.

Problems 6.18 and 6.19 involve the LM ascent stage which was used to transport the Apollo astronauts from the surface of the moon to the orbiting CSM. The characteristics of the LM are summarized as follows:

Total mass of the LM ascent stage on the moon's surface is 4,980 kg
LM specific impulse $I_{sp} = 311$ s
LM thrust $T = 15,520$ N

6.18 Compute the vertical velocity of the LM 10 s after igniting the ascent engine. Assume vertical flight for the first 10 s of the ascent phase and assume constant lunar gravitational acceleration $g_m = 1.625$ m/s^2.

6.19 When the LM completed the ascent phase, its mass was 2,713 kg. The actual velocity change in the LM after the ascent burn was 1.687 km/s. Compute the gravity loss during the lunar ascent phase. Assume constant lunar gravitational acceleration $g_m = 1.625$ m/s^2.

Problems 6.20 and 6.21 involve the Pegasus, which is an air-launched system dropped from the underside of a Lockheed L-1011 at 12 km. The Pegasus consists of three solid rocket stages where first-stage flight is aided by lift from a fixed wing.

6.20 At second-stage ignition, the Pegasus has mass $m = 5,652$ kg and is at an altitude of 70 km, inertial velocity of 2.5 km/s, and flight-path angle of 38°. The second stage burns for 71 s and at burnout the vehicle is at an altitude of 192 km and flight-path angle of 26°. The Orion 50 XL second stage has specific impulse $I_{sp} = 290$ s and propellant mass $m_p = 3,915$ kg. Estimate the inertial velocity of the Pegasus after second-stage burnout. Second-stage flight is essentially at vacuum conditions. Assume that $\sin \gamma$ is a linear function of time during the second-stage burn.

6.21 At third-stage ignition, the Pegasus is at an altitude of 736 km, inertial velocity of 4.566 km/s, and flight-path angle of 2.2°. Its total mass at third-stage ignition is 1,198 kg. The Pegasus' target orbit after the third-stage burn is a 741-km circular orbit. The Orion 38 third stage has specific impulse $I_{sp} = 290$ s, structural mass $m_{st} = 110$ kg, and its burn time is 65 s. Determine the third-stage propellant mass and the payload mass delivered to the target orbit. Assume that $\sin \gamma$ is a linear function of time during the third-stage burn.

6.22 The Ares I or Crew Launch Vehicle (CLV) was designed for human space flight missions after the retirement of the US Space Shuttle in 2011. The second stage of the Ares I was powered by a single J-2X engine with thrust $T = 1,307,000$ N and specific impulse $I_{sp} = 421$ s. Suppose that the Ares I is currently at an altitude of 130 km with inertial velocity $v = 6.740$ km/s and flight-path angle $\gamma = 1.8°$. Its mass is 267,100 kg at this instant.

a) If the time rate of the flight-path angle is zero at this instant, compute the pitch attitude thrust-steering angle α. Neglect atmospheric forces at this altitude.

b) The CLV accelerates to an inertial velocity of 7.220 km/s. How much propellant mass was required for this acceleration? Assume that the flight-path angle remains constant during this flight phase.

6.23 An oxygen tank explosion on the service module during the translunar trajectory ended the lunar-landing flight plan for Apollo 13. Consequently, the Apollo 13 mission used a "free-return" (coasting) trajectory around the moon to return the astronauts to Earth. In order to shorten the return trajectory, the astronauts fired the LM descent-stage engine to provide a velocity increment Δv of 262 m/s. Estimate the burn time for this maneuver during the return trajectory. Use the following parameters: total initial spacecraft mass before the burn is 47,200 kg, LM descent-stage engine thrust $T = 45,040$ N, LM descent-stage $I_{sp} = 311$ s.

6.24 The Payload Assist Module (PAM)-D is a solid rocket upper stage with thrust $T = 66,440$ N and specific impulse $I_{sp} = 293$ s. The PAM-D stage is mated to a communication satellite in circular LEO and the total mass is 2,210 kg. Compute the velocity increment Δv achieved by the PAM-D if the total burn time is 55 s.

6.25 The Centaur upper stage uses an RL-10 rocket engine with liquid oxygen and liquid hydrogen ($I_{sp} = 451$ s). The Centaur structural mass is 2,200 kg and its structural coefficient is 0.096. The Centaur stage has the capability to impart a total ideal velocity increment of $\Delta v = 6,120$ m/s. Determine the payload mass and payload mass ratio associated with the Centaur stage.

7

Impulsive Orbital Maneuvers

7.1 Introduction

Chapters 2–5 dealt with satellite position, velocity, and flight time in an orbit, while Chapter 6 linked a propulsive rocket burn to changes in orbital velocity and mass. This chapter involves altering a satellite's orbit by applying a propulsive thrust force. We assume that the thrust-to-mass ratio is sufficiently high so that the propulsive maneuver is approximated by an *impulse* with an instantaneous change in orbital velocity and no change in position (the reader should consider reviewing impulsive maneuvers in Section 6.7).

This chapter focuses on orbital maneuvers: transporting a satellite from one orbit to a target orbit. Often a launch vehicle delivers a satellite to an intermediate (lower-energy) orbit and a subsequent engine stage transfers the satellite to its intended orbit for operation. We have seen in previous chapters that the geostationary-equatorial orbit (GEO) is a circular orbit with a period that matches Earth's rotation rate making it a desirable destination for communication and weather satellites. Because GEO altitude is 35,786 km (over 5.6 Earth radii), an orbital maneuver is required to transfer a satellite from a lower-energy orbit to GEO. This scenario is only one example of an orbital maneuver; other examples include small-scale orbit corrections, changing the orbital inclination, and rendezvous with an orbiting satellite such as a space station.

Figure 7.1 shows a schematic diagram of a two-impulse, coplanar orbit transfer between inner and outer circular orbits. The first impulse at point A changes the circular velocity vector so that the satellite follows an elliptical orbit (the dashed path); the second impulse at point B changes the elliptical velocity vector so that the satellite enters the desired outer circular orbit (Orbit 2). Because the durations of the two propulsive "burns" are relatively short, we may model the velocity changes at A and B as impulsive maneuvers without change in orbital position. In practice, firing an onboard chemical rocket twice produces the orbit transfer shown in Figure 7.1. It is important for the reader to note that the satellite follows a "coasting arc" between points A and B where gravity is the only force acting on the satellite, and hence the dashed path in Figure 7.1 is an elliptical orbit. Note also that if the onboard engine failed to fire at point B, the satellite would continue its elliptical orbit and eventually return to point A. Computing the velocity increments (Δv) required at orbit-intersection points A and B is the focus of this chapter.

Space Flight Dynamics, First Edition. Craig A. Kluever.
© 2018 John Wiley & Sons Ltd. Published 2018 by John Wiley & Sons Ltd.
Companion website: www.wiley.com/go/Kluever/spaceflightmechanics

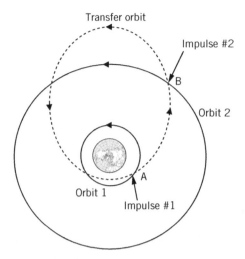

Figure 7.1 Two-impulse, coplanar orbit transfer.

7.2 Orbit Shaping

Before we present transfers between two orbits, we will discuss *orbit shaping*, where firing a rocket changes orbital elements such as semimajor axis and eccentricity. The reader should keep in mind that the rocket burn is treated as an *impulsive* maneuver that results in an instantaneous change in velocity, or Δv. Our basic objective in orbit-shaping problems is to determine the velocity increment Δv given the satellite's current orbit and the desired (or target) orbit. Once we determine Δv, we may compute the associated propellant mass using the rocket equation and the characteristics of the rocket engine (i.e., specific impulse or effective exhaust velocity). The following examples demonstrate orbit-shaping maneuvers.

Example 7.1 A geocentric satellite is in an elliptical orbit with semimajor axis $a = 8,500$ km and eccentricity $e = 0.15$. Determine the impulsive Δv required to create a circular orbit with a radius equal to the apogee radius of the elliptical orbit.

Figure 7.2 shows the initial elliptical orbit. We can compute perigee and apogee radii of the initial ellipse from a and e:

$$\text{Perigee:} \quad r_p = a(1-e) = (8,500 \text{ km})(0.85) = 7,225 \text{ km}$$
$$\text{Apogee:} \quad r_a = a(1+e) = (8,500 \text{ km})(1.15) = 9,775 \text{ km}$$

The target circular orbit has radius $r = r_a$, and therefore the circle is tangent to the ellipse at apogee as shown in Figure 7.2. Using the satellite's apogee state in the energy equations (2.29) and (2.63) yields

$$\xi = \frac{v_a^2}{2} - \frac{\mu}{r_a} = -\frac{\mu}{2a}$$

Solving for apogee velocity, we obtain

$$v_a = \sqrt{\frac{-\mu}{a} + \frac{2\mu}{r_a}} = 5.8873 \text{ km/s}$$

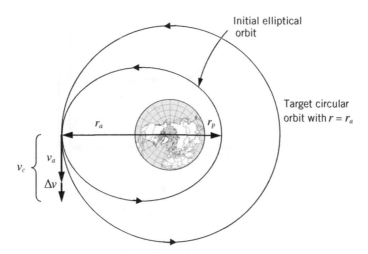

Figure 7.2 Circularization Δv at apogee (Example 7.1).

The circular velocity for the target orbit is

$$v_c = \sqrt{\frac{\mu}{r_a}} = 6.3857 \text{ km/s}$$

Circular speed is greater than the apogee velocity. The required velocity increment is

$$\Delta v = v_c - v_a = \boxed{0.4983 \text{ km/s}}$$

A collinear summation of the vectors is shown in Figure 7.2: apogee velocity v_a must be increased by an increment Δv to produce the circular velocity v_c. Because the velocity change Δv is impulsive, the radial position r_a does not change during the maneuver.

This type of orbit-shaping maneuver is commonly called a *perigee-raising* (or *periapsis-raising*) maneuver. In an operational setting, the thrust vector is aligned with the apogee velocity v_a for a so-called "tangent burn." In this example, the tangent Δv impulse at apogee increases the orbital energy and perigee radius so that the new orbit is circular. The reader should note that for this example, an impulsive Δv greater than 0.4983 km/s raises the initial perigee radius beyond the radius r_a such that it becomes the apogee of the new elliptical orbit (in which case the initial apogee radius will become the perigee radius of the new ellipse).

Example 7.2 Consider again a geocentric satellite in an elliptical orbit with semimajor axis a = 8,500 km and eccentricity e = 0.15. Determine the impulsive Δv required to create a circular orbit with a radius equal to the perigee radius of the elliptical orbit.

Figure 7.3 shows the initial elliptical orbit and the target circular orbit. Because the radius of the target circular orbit is the initial *perigee* radius, we must apply the Δv impulse at perigee as shown in Figure 7.3. We may determine the initial perigee velocity and the target circular velocity using perigee radius r_p = 7,225 km (see Example 7.1 for the calculation of perigee radius):

$$\text{Perigee velocity: } v_p = \sqrt{\frac{-\mu}{a} + \frac{2\mu}{r_p}} = 7.9652 \text{ km/s}$$

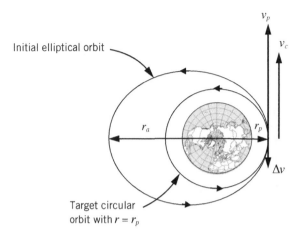

Figure 7.3 Circularization Δv at perigee (Example 7.2).

$$\text{Circular velocity:} \quad v_c = \sqrt{\frac{\mu}{r_p}} = 7.4276 \text{ km/s}$$

Because the target circular velocity is less than the perigee velocity, the impulsive Δv is applied in the *opposite* direction of v_p in order to reduce the speed (see Figure 7.3). The magnitude of the required velocity increment is

$$\Delta v = v_p - v_c = \boxed{0.5376 \text{ km/s}}$$

This orbit-shaping maneuver is commonly called an *apogee-lowering* (or *apoapsis-lowering*) maneuver. In practice, the thrust vector from the rocket burn is aligned in the opposite direction of the perigee velocity v_p so that the resulting Δv impulse decreases the orbital energy and lowers the apogee radius.

Example 7.3 Determine the propellant mass and burn time for the impulsive Δv maneuver from Example 7.1. Assume that the initial spacecraft mass is $m_0 = 2{,}000$ kg and that the onboard rocket engine has a thrust magnitude $T = 6{,}000$ N and specific impulse $I_{sp} = 320$ s.

Using Eq. (6.13) with $I_{sp} = 320$ s, $m_0 = 2{,}000$ kg, $\Delta v = 498.3$ m/s, and $g_0 = 9.80665$ m/s^2, we obtain the propellant mass:

$$m_p = m_0 \left[1 - \exp\left(\frac{-\Delta v}{g_0 I_{sp}} \right) \right] = \boxed{293.6 \text{ kg}}$$

Note that we expressed Δv in units of meters per second in order to be consistent with the effective exhaust speed $g_0 I_{sp}$.

We need the engine mass-flow rate in order to determine the burn time. Using Eq. (6.3), we obtain

$$\dot{m} = \frac{T}{g_0 I_{sp}} = 1.9120 \text{ kg/s}$$

The burn time is

$$t_{\text{burn}} = \frac{m_p}{\dot{m}} = 153.6 \text{ s} = \boxed{2.56 \text{ min}}$$

We could compute the propellant mass and burn time for the apogee-lowering maneuver (Example 7.2) in the same manner using the appropriate Δv.

7.3 Hohmann Transfer

We begin our discussion of orbit transfers with one of the most basic and simplest orbital maneuver: the minimum-energy transfer between circular coplanar orbits. This orbit transfer problem, first solved by Walter Hohmann in 1925 [1], is called the *Hohmann transfer*. Figure 7.4 shows that the Hohmann transfer is an elliptical orbit tangent to both the inner and outer circular orbits. The semimajor axis of the Hohmann transfer is one-half of the sum of the radii of the two circular orbits:

$$a_t = \frac{1}{2}(r_1 + r_2) \tag{7.1}$$

We will use subscript t to denote characteristics of the transfer orbit. Knowledge of semimajor axis allows computation of the energy of the Hohmann transfer ellipse:

$$\xi_t = \frac{-\mu}{2a_t} \tag{7.2}$$

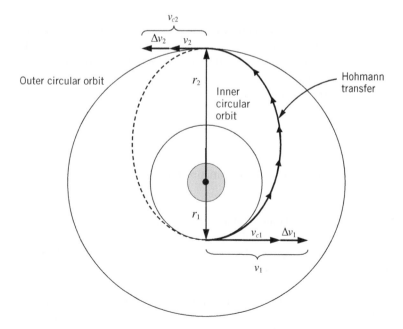

Figure 7.4 Hohmann transfer.

Finally, we can compute the required velocities at each apse of the Hohmann transfer ellipse using the energy equation (2.29) and Eq. (7.2):

$$\text{Periapsis:} \quad v_1 = \sqrt{2\left(\xi_t + \frac{\mu}{r_1}\right)} = \sqrt{\frac{-\mu}{a_t} + \frac{2\mu}{r_1}} \tag{7.3}$$

$$\text{Apoapsis:} \quad v_2 = \sqrt{2\left(\xi_t + \frac{\mu}{r_2}\right)} = \sqrt{\frac{-\mu}{a_t} + \frac{2\mu}{r_2}} \tag{7.4}$$

Let us identify the circular and Hohmann-transfer velocities in Figure 7.4. The satellite's initial circular velocity, $v_{c1} = \sqrt{\mu/r_1}$, is shown as an arrow tangent to the inner orbit. Because the Hohmann ellipse is tangent to both circles, the periapsis and apoapsis velocities are collinear with the inner and outer circular velocities. The Hohmann-transfer periapsis velocity v_1 is greater than circular velocity v_{c1} because the semimajor axis (energy) of the transfer ellipse is greater than the semimajor axis of the inner circle. Figure 7.4 shows that the Hohmann periapsis velocity is $v_1 = v_{c1} + \Delta v_1$. The first velocity increment is

$$\Delta v_1 = v_1 - v_{c1} \tag{7.5}$$

Firing a rocket engine aligned with the inner circular velocity vector produces this impulsive Δv_1 and establishes the Hohmann-transfer ellipse. When the satellite coasts for one-half of the elliptical period, it reaches the apoapsis of the Hohmann transfer, radius r_2. Figure 7.4 shows that the satellite arrives at apoapsis with velocity v_2 [as determined by Eq. (7.4)], and this velocity is less than the circular velocity of the outer orbit, $v_{c2} = \sqrt{\mu/r_2}$. Therefore, a second tangential engine burn must be employed to impart velocity increment Δv_2. The second velocity increment is

$$\Delta v_2 = v_{c2} - v_2 \tag{7.6}$$

The sum of the two impulsive velocity increments, $\Delta v = \Delta v_1 + \Delta v_2$, can be used in the ideal rocket equation (6.13) to determine the propellant mass required for the Hohmann transfer (of course, we must also know the initial mass of the satellite and the specific impulse of the rocket engine).

The flight time on the Hohmann-transfer ellipse is simply one-half the elliptical period [see Eq. (2.80)]

$$t_f = \frac{\pi}{\sqrt{\mu}} a_t^{3/2} \tag{7.7}$$

The following example illustrates a Hohmann transfer.

Example 7.4 A Delta IV launch vehicle delivers a Global Positioning System (GPS) satellite to a 185-km altitude circular low-Earth orbit (LEO). Determine the two Δv impulses required for a coplanar Hohmann transfer to a 20,180-km altitude GPS circular orbit. In addition, compute the coast time for the Hohmann transfer to the target GPS orbit.

Figure 7.5 shows the geometry of the Hohmann transfer where $r_1 = 185$ km $+ R_E = 6,563$ km (inner LEO) and $r_2 = 20,180$ km $+ R_E = 26,558$ km (GPS orbit). Recall that R_E is the Earth's radius. Using Eq. (7.1), we find that the semimajor axis of the Hohmann

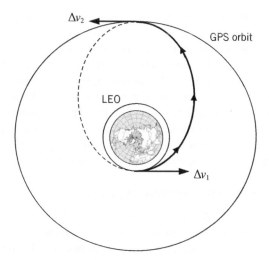

Figure 7.5 Hohmann transfer from low-Earth orbit (LEO) to Global Positioning System (GPS) orbit (Example 7.4).

transfer is $a_t = (r_1 + r_2)/2 = 16{,}560.5$ km. The apse velocities on the Hohmann transfer are found using Eqs. (7.3) and (7.4)

$$\text{Hohmann perigee: } v_1 = \sqrt{\frac{-\mu}{a_t} + \frac{2\mu}{r_1}} = 9.869 \text{ km/s}$$

$$\text{Hohmann apogee: } v_2 = \sqrt{\frac{-\mu}{a_t} + \frac{2\mu}{r_2}} = 2.439 \text{ km/s}$$

The inner and outer circular velocities are

$$\text{LEO: } v_{c1} = \sqrt{\frac{\mu}{r_1}} = 7.793 \text{ km/s}$$

$$\text{GPS orbit: } v_{c2} = \sqrt{\frac{\mu}{r_2}} = 3.874 \text{ km/s}$$

Finally, the two impulses are determined by using Eqs. (7.5) and (7.6)

$$\text{Impulse in LEO: } \Delta v_1 = v_1 - v_{c1} = \boxed{2.076 \text{ km/s}}$$

$$\text{Impulse at apogee: } \Delta v_2 = v_{c2} - v_2 = \boxed{1.435 \text{ km/s}}$$

The total velocity increment for the Hohmann transfer is $\Delta v_1 + \Delta v_2 = 3.511$ km/s. We use Eq. (7.7) to determine the flight time on the Hohmann transfer:

$$t_f = \frac{\pi}{\sqrt{\mu}} a_t^{3/2} = 10{,}604.5 \text{ s} = 2.95 \text{ h}$$

Hence, the flight time between the two rocket burns is nearly 3 h. Both burns are aligned with the local horizon and collinear with the satellite's velocity vector.

The Hohmann transfer is actually the most energy-efficient coplanar transfer for a specific range of outer-to-inner circular radii ratio. A three-burn bi-elliptic transfer, consisting of two successive Hohmann transfers, may be more efficient than the standard two-burn Hohmann transfer. The second burn of the bi-elliptic transfer is actually outside the outer (target) circular orbit. If the ratio $R = r_2/r_1$ is less than 11.94, then the two-burn Hohmann transfer is the most efficient transfer. If $R > 11.94$, then the bi-elliptic transfer *may* be the most efficient transfer. The bi-elliptic transfer may be unrealistic in some cases due to an extremely large intermediate orbit with a prohibitively long transfer time. The interested reader can consult Vallado [2; pp. 328–330] for additional details and criteria for when the bi-elliptic transfer is superior to the Hohmann transfer.

7.3.1 Coplanar Transfer with Tangential Impulses

The Hohmann transfer is an ellipse that is tangent to both the inner and outer circular orbits and hence it utilizes two tangential impulses. We can envision other coplanar orbit transfers that use tangential impulses. Figure 7.6 shows an orbit transfer between an inner elliptical orbit and a target outer circular orbit. If we compare Figure 7.6 with the Hohmann transfer (Figure 7.4), we see that the only difference is that the first tangential impulse Δv_1 is applied at the periapsis of the inner elliptical orbit (r_{p1}). The semi-major axis and energy of the transfer orbit are determined using Eqs. (7.1) and (7.2) with initial periapsis radius r_{p1} replacing the initial circular radius r_1:

$$a_t = \frac{1}{2}\left(r_{p1} + r_2\right) \tag{7.8}$$

$$\xi_t = \frac{-\mu}{2a_t} \tag{7.9}$$

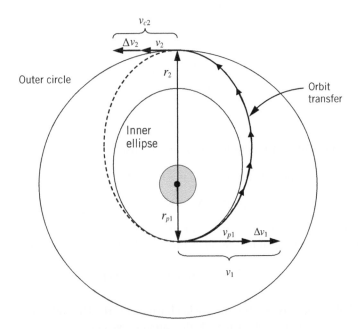

Figure 7.6 Orbit transfer using two tangential impulses.

The velocity required at periapsis of the transfer ellipse is computed using Eq. (7.3) with radius r_1 replaced by periapsis radius r_{p1}

$$\text{Periapsis:} \quad v_1 = \sqrt{2\left(\xi_t + \frac{\mu}{r_{p1}}\right)} = \sqrt{\frac{-\mu}{a_t} + \frac{2\mu}{r_{p1}}} \tag{7.10}$$

The first tangential impulse is

$$\Delta v_1 = v_1 - v_{p1} \tag{7.11}$$

where v_{p1} is the periapsis velocity of the inner elliptical orbit. The second tangential impulse (at apoapsis) is the difference between the outer circular velocity v_{c2} and the apoapsis velocity of the transfer ellipse computed via Eq. (7.4). The following example illustrates this type of orbit transfer.

Example 7.5 In July 2001, the European Space Agency (ESA) launched the ARTEMIS spacecraft on an Ariane 5 booster rocket. Because ARTEMIS was a communication satellite, its ultimate destination was GEO. Due to a partial failure of the Ariane booster, the ARTEMIS spacecraft reached a sub-nominal elliptical orbit with perigee and apogee altitudes of 580 and 17,350 km, respectively. The Ariane's target was a geostationary transfer orbit (GTO) with an apogee altitude of 35,786 km (i.e., GEO altitude). Mission operators decided to use the onboard chemical-propulsion rocket to transfer the ARTEMIS spacecraft from the sub-nominal elliptical orbit to a circular orbit with an altitude of 31,000 km (see Figure 7.7). ARTEMIS' onboard electric-propulsion stage performed the remaining orbit transfer from the 31,000-km altitude circle to GEO (we will analyze low-thrust transfers in Chapter 9).

Determine the two tangential Δv impulses required for the coplanar orbit transfer. In addition, compute the burn time for the first Δv impulse if the initial mass of the ARTEMIS satellite is 3,100 kg and the onboard chemical rocket provides 400 N of thrust with a specific impulse of 318 s.

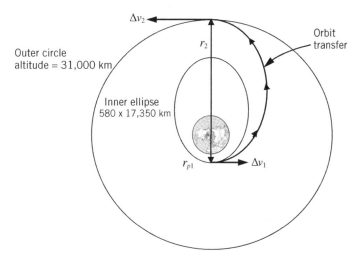

Figure 7.7 ARTEMIS coplanar orbit transfer (Example 7.5).

Figure 7.7 shows the geometry of the transfer where $r_{p1} = 580$ km $+ R_E = 6,958$ km (perigee of inner ellipse) and $r_2 = 31,000$ km $+ R_E = 37,378$ km (outer circular orbit). We use Eq. (7.8) to determine the semimajor axis of the transfer: $a_t = (r_{p1} + r_2)/2 = 22,168$ km. The apse velocities on the transfer orbit are found using Eqs. (7.10) and (7.4)

$$\text{Transfer perigee:} \quad v_1 = \sqrt{\frac{-\mu}{a_t} + \frac{2\mu}{r_{p1}}} = 9.8281 \text{ km/s}$$

$$\text{Transfer apogee:} \quad v_2 = \sqrt{\frac{-\mu}{a_t} + \frac{2\mu}{r_2}} = 1.8295 \text{ km/s}$$

Next, we need the perigee velocity of the initial inner elliptical orbit. The apogee radius of the inner ellipse is $r_{a1} = 17,350$ km $+ R_E = 23,728$ km. Hence, the semimajor axis of the inner ellipse is $a_1 = (r_{p1} + r_{a1})/2 = 15,343$ km. Perigee velocity on the inner ellipse is computed using Eq. (7.10) with initial semimajor axis a_1

$$\text{Inner ellipse perigee:} \quad v_{p1} = \sqrt{\frac{-\mu}{a_1} + \frac{2\mu}{r_{p1}}} = 9.4124 \text{ km/s}$$

Circular velocity of the outer orbit is

$$\text{Outer circular orbit:} \quad v_{c2} = \sqrt{\frac{\mu}{r_2}} = 3.2656 \text{ km/s}$$

Finally, the two tangential impulses are

$$\text{Impulse at perigee:} \quad \Delta v_1 = v_1 - v_{p1} = \boxed{0.4157 \text{ km/s}}$$

$$\text{Impulse at apogee:} \quad \Delta v_2 = v_{c2} - v_2 = \boxed{1.4361 \text{ km/s}}$$

Hence, the total velocity increment for the transfer is $\Delta v_1 + \Delta v_2 = 1.8518$ km/s.

To compute the burn time for the first impulse, we need the engine mass-flow rate and the propellant mass. Using Eq. (6.3) with thrust $T = 400$ N and $I_{sp} = 318$ s, the mass-flow rate is

$$\dot{m} = \frac{T}{g_0 I_{sp}} = 0.1287 \text{ kg/s}$$

The rocket equation (6.13) provides the propellant mass for the first burn:

$$m_{p1} = m_0 \left[1 - \exp\left(\frac{-\Delta v_1}{g_0 I_{sp}} \right) \right] = 386.88 \text{ kg}$$

Therefore, the burn time is

$$t_{\text{burn1}} = \frac{m_{p1}}{\dot{m}} = 3,016.2 \text{ s} = 50.3 \text{ min}$$

This burn time is too long to satisfy our assumption of an impulsive rocket burn. We can use Kepler's equation to show that the burn-arc angle $\Delta\theta = 122°$ (i.e., one-third an orbital revolution) corresponds to a 50-min transit from perigee to the end of the burn. The problem here is that the initial thrust acceleration $T/m_0 = (400$ N$)/(3,100$ kg$) = 0.129$ m/s^2 is too small to complete the first required velocity change ($\Delta v_1 = 416$ m/s) as an "impulse." To circumvent this issue, the mission operators divided the first tangential Δv_1 burn into five successive (smaller) perigee burns. Each perigee burn raised apogee; the fifth and final perigee burn established the elliptical transfer orbit shown in Figure 7.7. For the same reason, the second tangential impulse ($\Delta v_2 = 1.436$ km/s) was divided into three successive apogee burns to establish the circular orbit with an altitude of 31,000 km.

Example 7.5 and Figure 7.7 present an orbit transfer between an inner elliptical orbit and an outer circular orbit with two tangential impulses. However, Figure 7.7 does not represent the only feasible coplanar transfer that uses two tangential impulsive burns. Figure 7.8 shows another option: the first impulse Δv_1 occurs at *apogee* of the inner ellipse, and the second impulse Δv_2 establishes the desired circular orbit. Which option provides the smallest total Δv? Using the ARTEMIS orbit-transfer scenario presented in Example 7.5, we find that the first impulse (applied at apogee) is $\Delta v_1 = 1.7732$ km/s, and the second impulse is $\Delta v_2 = 0.3878$ km/s. The total velocity increment for this second option is 2.1610 km/s, which is nearly 17% greater than the total Δv for the transfer depicted in Figure 7.7. Clearly, this result shows why the ARTEMIS mission operators selected the orbit-transfer strategy illustrated by Example 7.5 and Figure 7.7 (i.e., initial tangential burn at perigee) to achieve a 31,000-km altitude circular orbit.

The previous numerical comparison suggests that it is more efficient in terms of Δv (and therefore propellant mass) to perform impulsive maneuvers at periapsis. Another way to pose this problem is to consider a satellite with a fixed propellant mass or fixed Δv. Suppose we want to increase the orbital energy for a given Δv impulse – where is the optimal location in the orbit for the rocket engine burn? The answer can be determined by manipulating the energy equation

$$\xi = \frac{v^2}{2} - \frac{\mu}{r} \tag{7.12}$$

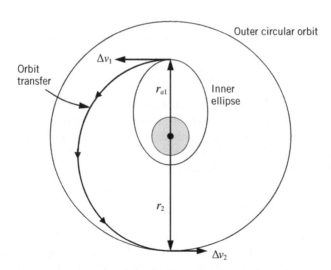

Figure 7.8 Feasible coplanar orbit transfer with poor Δv performance.

Taking differentials of Eq. (7.12) yields

$$d\xi = v\,dv + \frac{\mu}{r^2}dr \tag{7.13}$$

For an impulsive velocity change, there is no change in position (i.e., $dr = 0$). Replacing the differentials $d\xi$ and dv in Eq. (7.13) with incremental values, we obtain

$$\Delta\xi = v\Delta v \tag{7.14}$$

Equation (7.14) clearly shows that for a fixed Δv, we can achieve the largest energy increase by applying the impulse where velocity v is maximum, that is, at periapsis. Of course, aligning the impulsive Δv with the periapsis velocity (i.e., a tangent burn) maximizes the change in kinetic energy. This discussion is an example of the so-called *Oberth effect* named after the German physicist Hermann Oberth.

7.4   General Coplanar Transfer

Figure 7.1 shows that it is possible to perform an orbit transfer between coplanar circular orbits if the periapsis of the transfer orbit is less than or equal to the radius of the inner circle, and the apoapsis of the transfer ellipse is greater than or equal to the radius of the outer circle. Figure 7.9a is a more detailed depiction of the general coplanar orbit transfer presented in Figure 7.1. Figure 7.9a shows the circular and elliptical-orbit velocity vectors at the two intersections where the impulsive maneuvers occur. For a general coplanar orbit transfer, the velocity vectors on the circular and elliptical transfer orbits will not necessarily be collinear. Therefore, vector addition and geometry is required to calculate the magnitude and orientation of the impulsive vector $\Delta\mathbf{v}$. To show this, consider the first velocity impulse in Figure 7.9b:

$$\mathbf{v}_{c1} + \Delta\mathbf{v}_1 = \mathbf{v}_1 \quad \text{or} \quad \Delta\mathbf{v}_1 = \mathbf{v}_1 - \mathbf{v}_{c1} \tag{7.15}$$

Perhaps the easiest way to compute the magnitude of $\Delta\mathbf{v}_1$ is by using the law of cosines:

$$\Delta v_1 = \sqrt{v_1^2 + v_{c1}^2 - 2v_1 v_{c1}\cos\gamma_1} \tag{7.16}$$

where γ_1 is the flight-path angle of the elliptical-orbit velocity $\mathbf{v}_1$ (see Figure 7.9b). In general, the angle between the two velocity vectors at the impulse maneuver point is the difference in their respective flight-path angles; in this case, however, the initial orbit is circular and hence the flight-path angle for $\mathbf{v}_{c1}$ is zero. Law of cosines also determines the magnitude of the second impulse $\Delta\mathbf{v}_2$

$$\Delta v_2 = \sqrt{v_2^2 + v_{c2}^2 - 2v_2 v_{c2}\cos\gamma_2} \tag{7.17}$$

Knowledge of the Δv magnitude allows calculation of the propellant mass required for the orbit transfer (of course, we also need initial satellite mass and specific impulse of the propulsion system). From an operational viewpoint, we also need to determine the *direction* of the impulsive maneuver (or, the direction of the thrust vector). Figure 7.9b (a close-up view of the first impulse) shows that ϕ_1 is the elevation angle of vector

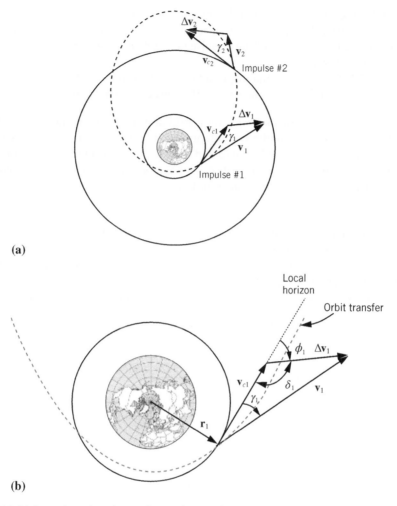

Figure 7.9 (a) General two-impulse, coplanar orbit transfer, and (b) close-up view of the first impulsive maneuver.

$\Delta \mathbf{v}_1$ with respect to the local horizon. Angle ϕ_1 is the supplementary angle of angle δ_1 in Figure 7.9b

$$\phi_1 = \pi - \delta_1 \tag{7.18}$$

where angle δ_1 can be computed using the law of sines:

$$\frac{\sin\delta_1}{v_1} = \frac{\sin\gamma_1}{\Delta v_1} \tag{7.19}$$

Firing the onboard rocket at elevation angle ϕ_1 performs the first impulsive burn.

In summary, we can determine the magnitude and direction of the impulsive maneuver by using the following steps:

1) Given the semimajor axis (or energy) of the desired transfer orbit, compute the velocity magnitude v_1 required at the impulse location (radius r_1) using the energy equation (2.29).

2) Given the angular momentum of the transfer orbit, compute the flight-path angle γ_1 at radius r_1 using Eq. (2.22). Additional information about the transfer (such as a sketch or the true anomaly of the impulse point) determines if the flight-path angle is positive or negative.

3) Use Eq. (7.16) to calculate the magnitude of the first impulse Δv_1.

4) Use Eqs. (7.18) and (7.19) to determine the elevation angle ϕ_1 of the first impulse.

The reader should note that knowing the energy and angular momentum of the transfer orbit is an essential part of the solution process. Recall from Chapter 2 that energy ξ and angular momentum h can be computed from a combination of orbital parameters such as semimajor axis a and eccentricity e, or periapsis and apoapsis radii. The following example illustrates a general coplanar orbit transfer.

Example 7.6 Figure 7.10 shows a two-impulse Earth-orbit transfer where the first impulse occurs after perigee passage on the transfer ellipse. The inner and outer circular orbits have radii $r_1 = 2.5R_E$ and $r_2 = 6R_E$, respectively where R_E is the radius of the Earth. The transfer orbit has a perigee radius of $1.9R_E$ and an apogee radius of $8.5R_E$. Determine (a) the magnitude and direction of the first impulse, and (b) the time of flight on the transfer ellipse.

a) We begin by computing the orbital characteristics of the transfer ellipse from the given apse radii:

$$\text{Semimajor axis:} \quad a_t = \frac{r_p + r_a}{2} = \frac{(1.9 + 8.5)R_E}{2} = 5.2R_E = 33{,}166 \text{ km}$$

$$\text{Eccentricity:} \quad e_t = \frac{r_a - r_p}{r_a + r_p} = \frac{(8.5 - 1.9)R_E}{(8.5 + 1.9)R_E} = 0.6346$$

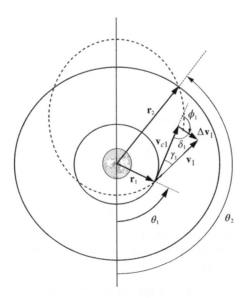

Figure 7.10 General two-impulse, coplanar orbit transfer (Example 7.6).

Parameter: $p_t = a_t(1 - e_t^2) = 19{,}809$ km

Angular momentum: $h_t = \sqrt{p_t \mu} = 88{,}858$ km^2/s

Using the energy equation and Eq. (2.22), we find the velocity and flight-path angle on the transfer ellipse at radius r_1

$$v_1 = \sqrt{\frac{-\mu}{a_t} + \frac{2\mu}{r_1}} = 6.163 \text{ km/s}$$

$$\gamma_1 = \cos^{-1}\left(\frac{h_t}{r_1 v_1}\right) = 25.27°$$

Note that flight-path angle γ_1 is positive because the first impulse occurs after perigee passage (see Figure 7.10). Finally, we use Eq. (7.16) to determine the magnitude of the first impulse

$$\Delta v_1 = \sqrt{v_1^2 + v_{c1}^2 - 2 v_1 v_{c1} \cos\gamma_1} = \boxed{2.693 \text{ km/s}}$$

where the circular speed for the inner orbit is $v_{c1} = \sqrt{\mu/r_1} = 5.000$ km/s. The angle δ_1 (see Figure 7.10) is determined from the law of sines

$$\delta_1 = \sin^{-1}\left(\frac{v_1}{\Delta v_1}\sin\gamma_1\right) = 77.72°$$

The elevation angle of impulse vector $\Delta \mathbf{v}_1$ is the supplementary angle:

$$\phi_1 = 180° - \delta_1 = \boxed{102.28°}$$

b) In order to determine the time of flight between the two impulses, we need to compute the true anomaly for each impulse. Begin by using the trajectory equation (2.45) to express the radial positions at each impulse:

$$\text{First impulse: } r_1 = \frac{p_t}{1 + e_t \cos\theta_1}$$

$$\text{Second impulse: } r_2 = \frac{p_t}{1 + e_t \cos\theta_2}$$

Solving for true anomaly, we obtain

$$\theta_1 = \cos^{-1}\left[\frac{1}{e_t}\left(\frac{p_t}{r_1} - 1\right)\right] = 67.55° \quad \text{and} \quad \theta_2 = \cos^{-1}\left[\frac{1}{e_t}\left(\frac{p_t}{r_2} - 1\right)\right] = 139.47°$$

Time of flight between impulses is determined by using Eq. (4.21)

$$t_2 - t_1 = \frac{1}{n}(M_2 - M_1)$$

where the two mean anomalies are

$$M_1 = E_1 - e_t \sin E_1 = 0.2477 \text{ rad} \quad \text{and} \quad M_2 = E_2 - e_t \sin E_2 = 1.200 \text{ rad}$$

[we use Eq. (4.33) to determine eccentric anomaly E from true anomaly θ]. Using mean motion $n = \sqrt{\mu/a_t^3} = 1.0453(10^{-4})$ rad/s, the transfer flight time is

$$t_2 - t_1 = 9,110 \text{ s} = \boxed{2.53 \text{ h}}$$

For comparison, the flight time for a Hohmann transfer between the circular orbits [i.e., semimajor axis $a_H = (r_1 + r_2)/2 = 27,107$ km] is 6.17 h and the first Hohmann-transfer impulse is 0.941 km/s. Therefore, the orbit transfer depicted in Figure 7.10 is a "fast transfer" between the circular orbits at the cost of much greater Δv or propellant mass.

7.5 Inclination-Change Maneuver

A velocity increment with a component normal to the plane of the orbit will change the orientation of the orbital plane. In general, a $\Delta \mathbf{v}$ component normal to the orbital plane will change the longitude of the ascending node Ω and inclination i. In this section, we will only focus on impulsive maneuvers that change orbital inclination. Figure 7.11 shows a plane-change maneuver between an inclined circular orbit (Orbit 1) and an equatorial orbit with the same circular radius (Orbit 2). To change inclination only, the velocity increment ($\Delta \mathbf{v}$) *must* be applied to Orbit 1 at the nodal crossing (i.e., the equatorial plane crossing). Figure 7.11a shows the plane-change impulse applied at the ascending node. Note that the orbit-normal component of $\Delta \mathbf{v}$ is in the opposite direction as the angular momentum vector $\mathbf{h}_1$ for Orbit 1. Of course, we can perform the inclination change one-half of a revolution later (or earlier) at the *descending* node; in this case the normal component of $\Delta \mathbf{v}$ will be in the same direction as $\mathbf{h}_1$. In either scenario, the vector addition shown in Figure 7.11b forms an isosceles triangle because Orbits 1 and 2 have the same circular speeds (i.e., $v_{c1} = v_{c2}$). Thus, the impulse $\Delta \mathbf{v}$ in Figure 7.11a has rotated velocity vector $\mathbf{v}_{c1}$ "downward" (south) to the equatorial plane without changing its magnitude. In addition, Figure 7.11a shows that the impulse $\Delta \mathbf{v}$ has rotated the angular momentum $\mathbf{h}_1$

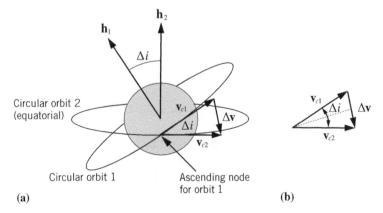

Figure 7.11 (a) Inclination change at the ascending node, and (b) vector diagram for the inclination change.

through angle Δi to $\mathbf{h}_2$ without changing its magnitude. We can determine the magnitude of the impulse by observing the "top" right triangle that is half of the isosceles triangle in Figure 7.11b:

$$\sin\frac{\Delta i}{2} = \frac{\Delta v/2}{v_{c1}} \tag{7.20}$$

Substituting $v = v_{c1} = v_{c2}$ in Eq. (7.20), the magnitude of the plane-change velocity increment is

$$\Delta v = 2v\sin\frac{\Delta i}{2} \tag{7.21}$$

Equation (7.21) is the velocity increment required for changing the orbital inclination of a circular orbit. The reader should note that Δi is the *magnitude* of the inclination change. A pure inclination-change maneuver *must* occur at a nodal crossing. However, an inclination-change maneuver does not necessarily have to result in an equatorial orbit as illustrated in Figure 7.11.

Determining the inclination-change Δv for an *elliptical* orbit requires an additional term in Eq. (7.21). Recall that we can define inclination as the angle measured from the equatorial plane to the projection of the velocity vector onto the horizontal plane at a nodal crossing (see Figure 3.12). Therefore, for an elliptical orbit, the two equal velocity vector "legs" of the isosceles triangle in Figure 7.11b must be the *horizontal* velocity component, or $v\cos\gamma$. Replacing the circular velocity vectors in Figure 7.11b with the horizontal velocity components, we obtain

$$\Delta v = 2v\cos\gamma\sin\frac{\Delta i}{2} \tag{7.22}$$

Equation (7.22) is the velocity increment required for changing the orbital inclination of an elliptical orbit [actually, Eq. (7.22) is the *general* equation for the inclination-change Δv because it also holds for circular orbits]. The reader should note that we must compute velocity v and flight-path angle γ at the nodal crossing. Furthermore, because velocity changes along an elliptical orbit, we must check both the ascending and descending nodal crossings and perform the inclination change at the crossing where horizontal velocity $v\cos\gamma$ is smallest. The following examples illustrate orbital maneuvers with inclination changes.

Example 7.7 A GPS satellite is in a circular orbit with an altitude of 20,180 km and inclination of 41°. Determine the Δv impulse required for a plane change to a target inclination of 55°.

We use Eq. (7.21) to determine the single velocity impulse

$$\Delta v = 2v\sin\frac{\Delta i}{2}$$

where $\Delta i = 55° - 41° = 14°$. The circular velocity of the GPS orbit is $v = \sqrt{\mu/r} = 3.874$ km/s (see Example 7.4). Therefore, the velocity increment for the plane change is

$$\Delta v = 2(3.874)\sin(7°) = \boxed{0.944 \text{ km/s}}$$

Example 7.8   A more realistic scenario for the GPS satellite orbit insertion is to perform the plane change at apogee of the Hohmann-transfer ellipse (see Example 7.4
and Figure 7.5). Assume that the GPS satellite is on a Hohmann-transfer ellipse with
a perigee altitude of 185 km, an apogee altitude of 20,180 km, an inclination of 41°,
and argument of perigee of 180°. Compute the velocity impulse at apogee that increases
the inclination to 55° but does not change any other orbital elements.

We can perform a pure inclination-change maneuver at apogee because the apse line
lies in the equatorial plane and apogee coincides with the ascending node (note that argument of perigee is 180°, i.e., perigee direction is collinear with the *descending* node).
Because the satellite is in the elliptical Hohmann-transfer orbit, we must use
Eq. (7.22) to determine the velocity impulse

$$\Delta v = 2v_2 \cos\gamma_2 \sin\frac{\Delta i}{2}$$

where $\Delta i = 55° - 41° = 14°$. In this scenario, the velocity v_2 is the orbital speed at apogee
of the Hohmann-transfer ellipse in Figure 7.5. Flight-path angle γ_2 is zero at apogee. We
must compute apogee speed using the energy (semimajor axis) of the Hohmann-transfer
ellipse (see Example 7.4). First, we find that perigee radius is $r_1 = 185$ km $+ R_E = 6{,}563$ km
and apogee radius is $r_2 = 20{,}180$ km $+ R_E = 26{,}558$ km (GPS orbit). Thus, the semimajor
axis of the Hohmann transfer is $a_t = (r_1 + r_2)/2 = 16{,}560.5$ km. The apogee velocity on
the Hohmann transfer is

$$\text{Hohmann apogee:}\quad v_2 = \sqrt{\frac{-\mu}{a_t} + \frac{2\mu}{r_2}} = 2.439 \text{ km/s}$$

Using the Hohmann apogee speed v_2 in Eq. (7.22) yields the plane-change impulse:

$$\Delta v = 2(2.439) \cos(0°) \sin(7°) = \boxed{0.594 \text{ km/s}}$$

Examples 7.7 and 7.8 both involve a 14° pure inclination-change maneuver performed
at radius $r_2 = 26{,}558$ km (GPS orbit). The plane-change Δv is lower in this example
because the maneuver occurs at apogee of the Hohmann transfer where orbital velocity
(2.439 km/s) is less than the GPS circular velocity (3.874 km/s).

Example 7.9   Russia launches a satellite intended for a Molniya orbit with targeted
orbital elements:

$$a = 26{,}565 \text{ km}$$
$$e = 0.7411$$
$$i = 63.4°$$
$$\Omega = 50°$$
$$\omega = -90°$$

Figure 7.12 shows the satellite's orbit after the launch phase. The orbit-insertion phase
achieved four of the five Molniya orbital elements: a, e, Ω, and ω. However, the orbital
inclination after the launch phase is $i_0 = 60.2°$. Determine the impulsive Δv required to
correct the satellite's inclination.

Because we only need to change inclination, the plane-change impulse occurs at
the nodal crossing. In order to use Eq. (7.22) to compute Δv, we need the velocity

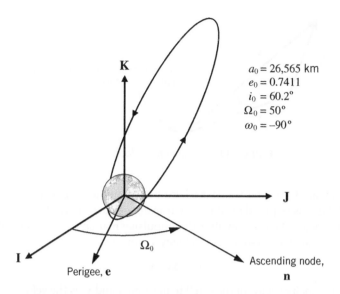

$$a_0 = 26{,}565 \text{ km}$$
$$e_0 = 0.7411$$
$$i_0 = 60.2°$$
$$\Omega_0 = 50°$$
$$\omega_0 = -90°$$

Figure 7.12 Molniya orbit after the launch phase (Example 7.9).

and flight-path angle of the satellite on the elliptical orbit at the nodal crossing. The satellite crosses the ascending node at true anomaly $\theta = 90°$, and the corresponding radius is determined from the trajectory equation (2.45):

$$r = \frac{p}{1 + e\cos\theta} = \frac{a(1 - e^2)}{1 + e\cos\theta} = 11{,}974.7 \text{ km}$$

The radial distance is equal to the parameter p when true anomaly is 90°. The velocity at the ascending node is determined using the energy equation:

$$v = \sqrt{\frac{-\mu}{a} + \frac{2\mu}{r}} = 7.181 \text{ km/s}$$

The flight-path angle at this point is

$$\gamma = \cos^{-1}\left(\frac{h}{rv}\right) = \cos^{-1}\left(\frac{\sqrt{\mu p}}{rv}\right) = 36.54°$$

Finally, the inclination-change impulse is computing using Eq. (7.22) with $\Delta i = 63.4° - 60.2° = 3.2°$:

$$\Delta v = 2v\cos\gamma\sin\frac{\Delta i}{2} = 2(7.181)\cos(36.54°)\sin(1.6°) = \boxed{0.322 \text{ km/s}}$$

Because the Molniya satellite's speed is relatively large at the nodal crossing (7.2 km/s), even a small inclination change requires a significant impulse.

7.6 Three-Dimensional Orbit Transfer

The previous sections have involved coplanar orbit transfers that change energy and plane-change maneuvers that only change inclination. It is possible to perform a

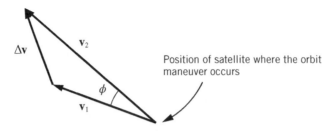

Figure 7.13 Change in the velocity vector.

three-dimensional (3-D) orbit transfer where the $\Delta\mathbf{v}$ impulse changes both energy and inclination. All that is required for the orbit-transfer calculation is a basic vector diagram of the velocity vectors before and after the impulsive maneuver. For example, consider Figure 7.13 and the associated simple velocity vector addition:

$$\mathbf{v}_1 + \Delta\mathbf{v} = \mathbf{v}_2 \tag{7.23}$$

where $\mathbf{v}_1$ is the velocity vector of the satellite in Orbit 1 and $\mathbf{v}_2$ is the velocity vector associated with Orbit 2. Of course, both velocity vectors share a common orbital position vector $\mathbf{r}$ where the impulsive maneuver occurs. One way to interpret Figure 7.13 is to view the satellite from the central body along "line-of-sight" position vector $\mathbf{r}$. The law of cosines determines the magnitude of the velocity increment

$$\Delta v = \sqrt{v_1^2 + v_2^2 - 2v_1 v_2 \cos\phi} \tag{7.24}$$

where ϕ is the angle between velocity vectors $\mathbf{v}_1$ and $\mathbf{v}_2$ (see Figure 7.13). Note that in the special case where the impulse occurs at the ascending (or descending) node, the angle ϕ in Figure 7.13 would represent a change in inclination between Orbits 1 and 2. In this case, Figure 7.13 illustrates an orbital maneuver where $\Delta\mathbf{v}$ increases the orbital energy (because magnitude $v_2 > v_1$) and increases the orbital inclination from i_1 to $i_2 = i_1 + \phi$. We will call this type of maneuver where a single impulse alters both the energy and plane a "3-D orbit transfer." The following example illustrates a 3-D orbit maneuver.

Example 7.10   Examples 7.4, 7.7, and 7.8 have illustrated a coplanar Hohmann transfer and inclination-change maneuvers for a GPS satellite. This example will demonstrate a 3-D orbit maneuver to the desired GPS orbit and compare the 3-D transfer performance with a "piecewise" approach where various orbit changes occur individually and sequentially. In all orbit-transfer cases here, the satellite begins in a circular 185-km altitude LEO with inclination $i_1 = 41°$. The GPS target orbit is a circular orbit with an altitude of 20,180 km and inclination $i_2 = 55°$. Determine the total Δv for the following orbit-transfer scenarios:

a) Perform a coplanar Hohmann transfer to the circular GPS orbit, and then perform a pure inclination change (i.e., three impulsive maneuvers).
b) Perform the first Hohmann-transfer Δv to raise apogee to the GPS orbit, perform a pure inclination change at apogee, and then perform a coplanar maneuver at apogee (one revolution later) to establish the circular GPS orbit (i.e., three impulsive maneuvers).

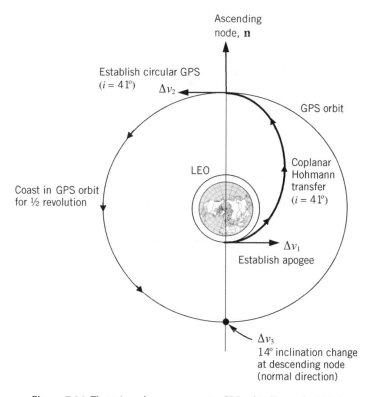

Figure 7.14 Three-impulse maneuver to GPS orbit (Example 7.10a).

c) Perform the first Hohmann-transfer Δv to raise apogee to the GPS orbit, and then perform a 3-D maneuver at apogee to change the plane and circularize the orbit (i.e., two impulsive maneuvers).

a) Figure 7.14 illustrates the three Δv maneuvers for Case (a). The first two impulses perform a coplanar Hohmann transfer (see Example 7.4 for details):

Coplanar impulse in LEO: $\Delta v_1 = 2.076$ km/s (establish apogee at GPS orbit)

Coplanar impulse at apogee: $\Delta v_2 = 1.435$ km/s (establish circular GPS orbit)

The third Δv is a pure 14° inclination change performed at the next nodal crossing (the descending node), one-half circular orbit revolution from the apogee burn (the Δv_3 vector is shown as a "dot" in Figure 7.14 because it is primarily normal to the orbit plane). We determined this inclination-change impulse in Example 7.7:

Plane-change impulse at descending node: $\Delta v_3 = 0.944$ km/s

Therefore, the total velocity increment for Case (a) is $\boxed{\Delta v = 4.455 \text{ km/s}}$

b) Figure 7.15 illustrates the three Δv maneuvers for Case (b). The first impulse establishes the coplanar Hohmann transfer [same as Case (a)]:

Coplanar impulse in LEO: $\Delta v_1 = 2.076$ km/s (establish apogee at GPS orbit)

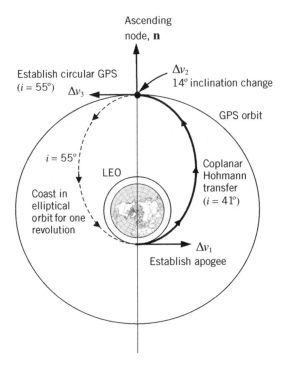

Figure 7.15 Three-impulse maneuver to GPS orbit (Example 7.10b).

The second impulse at apogee performs a pure 14° inclination change without altering the orbital energy (see Example 7.8 for details):

Inclination-change impulse at ascending node : $\Delta v_2 = 0.594$ km/s

("dot" in Figure 7.15)

Finally, the third impulse occurs at apogee after one full coasting revolution on the elliptical orbit. This third Δv establishes the circular GPS orbit, and hence it is equivalent to the second coplanar Hohmann-transfer impulse from Example 7.4:

Coplanar impulse at apogee: $\Delta v_3 = 1.435$ km/s (establish circular GPS orbit)

Therefore, the total velocity increment for Case (b) is $\boxed{\Delta v = 4.105 \text{ km/s}}$

c) Figure 7.16 shows the two Δv maneuvers for Case (c). The first impulse establishes the coplanar Hohmann transfer [same as Cases (a) and (b)]:

Coplanar impulse in LEO: $\Delta v_1 = 2.076$ km/s (establish apogee at GPS orbit)

The second impulse at apogee is a combined maneuver that changes inclination and energy to create the desired 55° inclined circular GPS orbit. A vector diagram of the 3-D maneuver is to the right of the orbit transfer shown in Figure 7.16. This vector diagram represents a view from the Earth to the satellite along the ascending node direction. Hence, v_2 (the satellite's velocity vector at apogee) has an angle of 41° (relative to the equatorial plane) at the nodal crossing, and the target GPS velocity

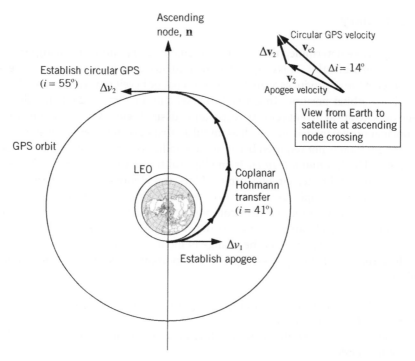

Figure 7.16 Two-impulse, three-dimensional maneuver to GPS orbit (Example 7.10c).

vector v_{c2} has an angle of 55° (equal to the target inclination). Therefore, the angle between the two velocity vectors is the inclination change $\Delta i = 14°$. From the previous examples, we know that the magnitudes of these velocities are $v_2 = 2.439$ km/s and $v_{c2} = 3.874$ km/s. The law of cosines determines the magnitude of the combined maneuver

$$\text{3-D impulsive maneuver:} \quad \Delta v_2 = \sqrt{v_2^2 + v_{c2}^2 - 2v_2 v_{c2} \cos\Delta i} = 1.619 \text{ km/s}$$

Therefore, the total velocity increment for Case (c) is $\boxed{\Delta v = 3.695 \text{ km/s}}$

Clearly, the combined 3-D maneuver [Case (c)] is the best option for inserting a GPS satellite into its proper orbit because it requires the lowest total Δv.

In practice, the orbit-transfer scenario of Case (c) delivers a GPS satellite to its desired orbit. First, the launch vehicle inserts the satellite into a circular low-Earth orbit (LEO). Next, the upper-stage engine fires at the nodal crossing to establish apogee at the GPS target altitude. After a 3-h (half-revolution) coast, the upper-stage engine restarts to perform the combined 3-D maneuver to change the plane and energy. The reader should note that a transfer to a GPS orbit requires a *total* inclination change of $55° − 28.5° = 26.5°$ for launch from Cape Canaveral (latitude $\phi = 28.5°$). In the example presented here, the launch vehicle performs a 12.5° plane change (note that $i_1 = 41°$ for LEO) and the upper-stage engine performs the remaining 14° plane change with the combined Δv maneuver. The allocation of the plane change between the launch vehicle and upper stage requires a numerical optimization process and is beyond the scope of this textbook.

7.7 Summary

In this chapter, we determined the velocity increment (Δv) required to change the shape and orientation of an orbit. This chapter has focused on *impulsive maneuvers* where firing a "large" rocket thrust produces an instantaneous change in velocity without any change in the satellite's orbital position. In general, the velocity increment $\Delta \mathbf{v}$ is the difference between the satellite's current velocity vector and a desired velocity vector. For many common coplanar orbital maneuvers (such as impulses that raise or lower periapsis or apoapsis), vector addition is not required because the satellite's current and targeted velocities are collinear. The *Hohmann transfer* is an elliptical transfer orbit where both apses are tangent to inner and outer circular orbits. Thus, the Hohmann transfer requires two impulses to perform a coplanar transfer from an inner circular orbit to an outer circular orbit (or vice versa). An impulse with a component normal to the orbit plane will change the inclination. The inclination-change Δv depends on the satellite's horizontal velocity component and the magnitude of the change in inclination angle.

We close this chapter by presenting a few important rules-of-thumb for impulsive orbital maneuvers:

1) For a fixed Δv capability, performing an impulsive burn at periapsis will produce the maximum change in energy.
2) An impulsive burn that only changes inclination *must* occur at either the ascending or descending node.
3) Performing an inclination-change maneuver at apoapsis (at the equatorial plane crossing) will minimize the impulsive Δv for a desired change in inclination angle.

References

1 Hohmann, W., *Die Erreichbarkeit der Himmelskörper*, R. Oldenbourg, Munich, 1925 (in German). (*The Attainability of Heavenly Bodies*, NASA Technical Translation TTF44, Washington, DC, 1960.)
2 Vallado, D.A., *Fundamentals of Astrodynamics and Applications*, 4th edn, Microcosm Press, Hawthorne, CA, 2013.

Problems

Conceptual Problems

7.1 A launch vehicle delivers a satellite to an orbit with an apogee altitude of 400 km and perigee altitude of 250 km. The satellite's initial mass in this elliptical orbit is 1,200 kg and it is equipped with a rocket engine that can deliver 800 N of thrust with a specific impulse of 325 s. Determine:
   a) The impulsive Δv required to establish a 400-km altitude circular orbit.
   b) The propellant mass required for the circularization burn.
   c) The circularization burn time (in min).

7.2   A satellite is in a 600-km altitude circular geocentric orbit. If the satellite has a rocket engine and propellant mass capable of providing a total velocity increment $\Delta v = 3$ km/s, determine the largest possible circular orbit that can be achieved.

7.3   A geocentric satellite is initially in a circular low-Earth orbit (LEO) with an altitude of 300 km and an inclination of 28.5°. The satellite's target orbit is elliptical with a perigee altitude of 300 km, an apogee altitude of 2,000 km, and an inclination of 40°. Compare the single-impulse and two-impulse strategies for shaping the satellite's orbit.

7.4   A launch vehicle delivers a spy satellite to a 325-km altitude circular parking orbit. The Air Force wants to place the satellite in an elliptical orbit with a 12-h period and perigee at 325 km altitude. Compute the Δv required to place the satellite in the desired orbit.

7.5   A spacecraft is in a 185-km altitude LEO. Compute the velocity increment Δv_{TLI} for translunar injection (TLI) if the sole purpose of the TLI burn is to raise apogee to a radial distance of 384,400 km (i.e., the approximate Earth–moon distance).

7.6   An Earth-orbiting satellite is in an elliptical orbit with perigee and apogee altitudes of 300 and 800 km, respectively. Compare the Δv required to achieve an escape (i.e., parabolic) trajectory if the impulse occurs at perigee or apogee.

7.7   An interplanetary satellite is in a 185-km altitude circular parking orbit about the Earth. Determine the velocity increment Δv required to establish a hyperbolic trajectory with launch energy $C_3 = 10$ km²/s².

7.8   We desire a coplanar geocentric transfer between inner and outer circular orbits with altitudes of 400 and 25,000 km, respectively. Determine if the following transfer ellipses are feasible or infeasible and justify your answers.
a) Semimajor axis $a = 19,078$ km, eccentricity $e = 0.651$.
b) Semimajor axis $a = 20,500$ km, eccentricity $e = 0.639$.
c) Semimajor axis $a = 19,100$ km, eccentricity $e = 0.645$.

7.9   Figure P7.9 shows two concentric, coplanar geocentric orbits. Determine the total velocity increment Δv for a transfer starting from the 600-km altitude circular orbit and ending at the 1,500-km altitude circular orbit. In addition, compute the transfer time.

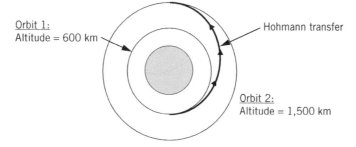

Orbit 1:
Altitude = 600 km

Hohmann transfer

Orbit 2:
Altitude = 1,500 km

Figure P7.9

7.10 Determine the total velocity increment Δv for a coplanar geocentric transfer starting from a 1,500-km altitude circular orbit and ending at a 600-km altitude circular orbit.

7.11 Figure P7.11 shows two satellites moving in concentric coplanar orbits about the Earth. Satellite A is in a 300-km altitude circular orbit and Satellite B is in a 700-km circular orbit. At epoch t_0, Satellite B is 60° ahead of Satellite A (as depicted in Figure P7.11). Determine the earliest epoch time t_1 such that a Hohmann transfer delivers Satellite A to the 700-km orbit so that Satellites A and B occupy the same position in orbit (i.e., rendezvous). Let initial epoch time t_0 be zero. In addition, determine the total Δv for the Hohmann-transfer rendezvous.

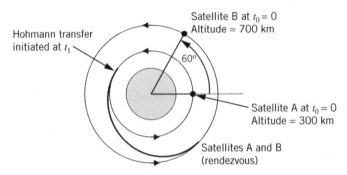

Figure P7.11

7.12 Determine the two velocity increments, Δv_1 and Δv_2, for the coplanar geocentric orbit transfer shown in Figure P7.12. The initial and final orbit radii are $r_1 = 6,878$ km and $r_2 = 34,000$ km, respectively. The first impulse is tangent to the initial circular orbit and it establishes a transfer ellipse with an apogee altitude of 35,000 km.

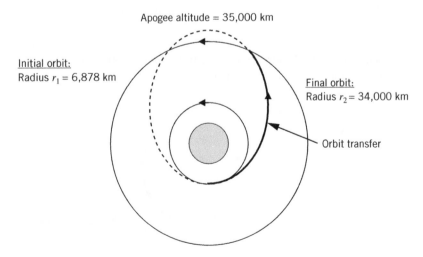

Figure P7.12

Problems 7.13 and 7.14 involve the two geocentric coplanar orbits shown in Figure P7.13. Both orbits share a common apse line

7.13 Determine the two-impulse orbit transfer from the initial orbit to the final orbit with the minimum total Δv. Compute the total Δv and the perigee and apogee of the orbit transfer between the initial and final orbits shown in Figure P7.13.

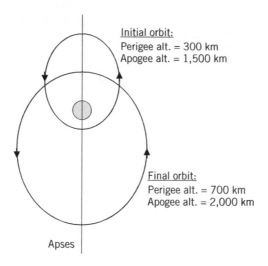

Initial orbit:
Perigee alt. = 300 km
Apogee alt. = 1,500 km

Final orbit:
Perigee alt. = 700 km
Apogee alt. = 2,000 km

Apses

Figure P7.13

7.14 Determine the single Δv (applied at the orbit-intersection point) required to transfer a satellite from the initial orbit to the final orbit. Compare the single-impulse Δv with the (minimum) total Δv for the two-impulse transfer found in Problem 7.13.

7.15 A communication satellite in a geostationary-equatorial orbit (GEO) has been perturbed by lunar and solar gravity forces and consequently its orbital elements are $a = 42,164$ km, $e = 0$, $i = 0.5°$, and $\Omega = 245°$. The GEO satellite is equipped with an onboard rocket engine with specific impulse $I_{sp} = 320$ s. If the satellite's current mass is 1,890 kg, determine the propellant mass required to re-establish GEO.

7.16 A launch vehicle delivers a satellite to a 800-km altitude circular LEO with an inclination of 88°. The goal is to achieve an 800-km circular sun-synchronous orbit (SSO). Determine the impulsive Δv required to establish the SSO. [Hint: see Example 5.3.]

7.17 An Earth-orbiting satellite is in a 300-km altitude circular orbit with an inclination of 28.5°. Its target orbit is a near-polar, 300-km altitude circular orbit with an inclination of 85°. Compare the total Δv for two orbit-transfer strategies: (a) a single impulse for a pure inclination change; and (b) an apogee-raising impulse that establishes an ellipse with an apogee altitude of 30,000 km, following by a pure inclination change to $i = 85°$ (applied at apogee), and finally an apogee-lowering impulse (applied at perigee) to re-establish a 300-km circular orbit.

7.18 Figure P7.18 shows an elliptical geocentric orbit with perigee and apogee altitudes of 1,200 and 1,900 km, respectively, an inclination of 10°, and an argument of perigee of 60°. Determine the minimum Δv that establishes an *equatorial* elliptical orbit without changing perigee or apogee altitudes. Be sure to describe the orbital location of the impulsive maneuver.

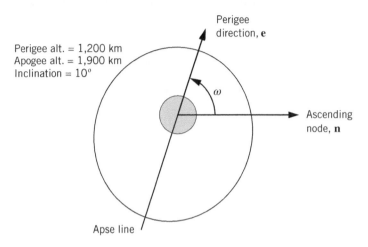

Perigee direction, **e**

Perigee alt. = 1,200 km
Apogee alt. = 1,900 km
Inclination = 10°

ω

Ascending node, **n**

Apse line

Figure P7.18

7.19 A geocentric satellite has been perturbed from its 800-km altitude circular, equatorial orbit. Figure P7.19 shows the perturbed orbit, which has the following orbital characteristics: perigee altitude = 700 km, apogee altitude = 780 km, inclination $i = 2°$, and argument of perigee $\omega = 135°$. Determine the *best* sequence of impulsive maneuvers that re-establishes the 800-km circular equatorial orbit and minimizes the total Δv. Carefully describe the sequence, location, and magnitude of each impulse.

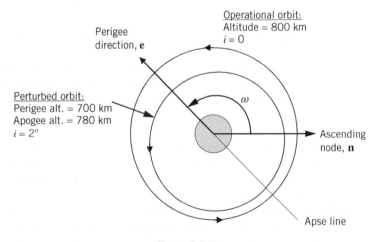

Operational orbit:
Altitude = 800 km
$i = 0$

Perigee direction, **e**

Perturbed orbit:
Perigee alt. = 700 km
Apogee alt. = 780 km
$i = 2°$

ω

Ascending node, **n**

Apse line

Figure P7.19

7.20 Figure P7.20 shows an orbit transfer between circular, inclined geocentric orbits. The first impulsive, $\Delta v_1 = 2$ km/s, is applied at the descending node and is collinear with the inner circular orbital velocity vector. The initial orbit has an inclination of $28.5°$ and the final orbit inclination is $10°$. Determine the second velocity increment Δv_2.

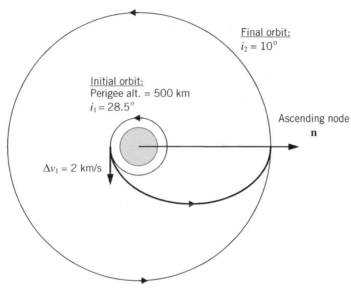

Final orbit:
$i_2 = 10°$

Initial orbit:
Perigee alt. = 500 km
$i_1 = 28.5°$

Ascending node
n

$\Delta v_1 = 2$ km/s

Figure P7.20

7.21 A launch vehicle delivers a 15,700-kg satellite to a circular 185-km altitude LEO with an inclination of $28.5°$. The satellite's upper stage consists of a liquid-propulsion rocket with $I_{sp} = 420$ s. If the structural coefficient of the upper stage is $\varepsilon = 0.09$, compute the payload mass m_{PL} delivered to geostationary-equatorial orbit (GEO) with an altitude of 35,786 km. Use a two-impulse geostationary transfer orbit (GTO).

MATLAB Problem

7.22 Write an M-file that will calculate the two velocity increments, Δv_1 and Δv_2, for an orbit transfer between arbitrary geocentric orbits. The initial and final orbits may be circular or elliptical with different inclinations. Assume that the two impulses occur at the appropriate apse locations and that the apse line is collinear with the line of nodes. The six inputs are the perigee and apogee altitudes (in km) and inclinations (in degrees) of the initial and final orbits. The desired outputs are the magnitudes of the two impulses (in km/s) and the transfer time between the orbits (in h). Test your M-file by solving Example 7.10.

Mission Applications

7.23 Figure P7.23 shows the US Space Shuttle's 320-km altitude circular orbit. Determine the Shuttle's de-orbit impulse Δv required to target a flight-path angle of $-1.3°$ at the so-called "entry interface" altitude of 122 km.

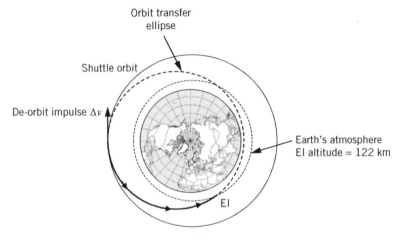

Figure P7.23

7.24 An Ariane 5 is launched from Europe's Spaceport in French Guiana and delivers a 6,640-kg satellite to a GTO with perigee and apogee altitudes of 560 and 35,786 km, respectively, and an inclination of 7.0°. Determine:
a) The minimum impulsive Δv required to establish an equatorial GTO.
b) The minimum impulsive Δv required to establish GEO.
c) The satellite's mass in GEO (assume an onboard rocket engine with $I_{sp} = 320$ s).

7.25 The Meridian 4 is a Russian communication satellite that was launched in May 2011 on a Soyuz-2 rocket. The launch vehicle initially delivered the satellite to a 203-km circular LEO. The first impulsive Δv burn of the Fregat upper stage raised apogee to its target altitude of 39,724 km. The second impulsive Δv at apogee of the transfer orbit raised its perigee altitude to 998 km. Figure P7.25 presents the two impulsive maneuvers, the transfer orbit, and the final target orbit for the Meridian 4 satellite (not to scale). Determine both impulsive Δv burns and the coasting time between the two impulses.

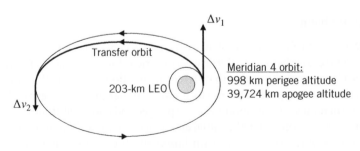

Figure P7.25

7.26 Lunar Orbiter 1 (1966) was the first US spacecraft to orbit the moon. Initially the Lunar Orbit 1 entered an elliptical lunar orbit with periapsis ("perilune") and apoapsis ("apolune") altitudes of 189 and 1,867 km, respectively. One week later, an onboard rocket burn lowered the perilune altitude to 58 km while maintaining apolune altitude at 1,867 km. Four days after the first orbital maneuver, a second burn lowered perilune altitude to 40.5 km (apolune did not change). Compute the two Δv impulses required for the two periapsis-lowering maneuvers.

Problems 7.27 and 7.28 involve the Chandra X-ray Observatory (CXO), which is an Earth-orbiting observation satellite.

7.27 In July 1999, the two-stage Inertial Upper Stage (IUS) booster rocket was used to transfer the CXO from a 300-km altitude circular LEO to a highly elliptical orbit as shown in Figure P7.27 (not to scale). Igniting the first IUS stage established an elliptical orbit (Transfer orbit 1) with an apogee altitude of 13,200 km as shown in Figure P7.27. After jettisoning the first IUS stage, the second stage was fired at perigee (300 km altitude) after one orbital revolution. The second impulse established Transfer orbit 2 with an apogee altitude of 72,000 km as shown in Figure P7.27. Both orbit transfers are coplanar. Determine Δv for each of the two IUS burns.

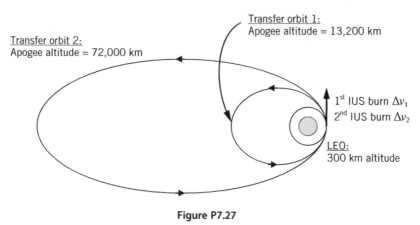

Figure P7.27

7.28 After a sequence of coplanar maneuvers, the CXO was placed in highly elliptical orbit with apogee and perigee radii of 140,906 and 20,686 km, respectively, and an inclination of 28.5°. Compute the Δv required to increase the inclination to 76.72°. Assume that the plane-change Δv occurs at apogee (at a nodal crossing) and only affects the orbital inclination.

Problems 7.29 and 7.30 involve the Lunar Atmosphere and Dust Environment Explorer (LADEE) spacecraft, which was launched in September 2013 and was eventually placed in an orbit about the moon.

7.29 The LADEE spacecraft was launched into a highly elliptical orbit by a Minotaur V booster. This elliptical orbit had a period of 6.4 days and perigee altitude of 200 km (see Figure P7.29). After coasting for one revolution, the LADEE spacecraft fired an onboard rocket at perigee to increase its orbital period to 8.2 days.

After coasting for another revolution, the LADEE spacecraft fired a second perigee burn to increase its period to 9.9 days. Determine the two Δv increments for the two impulsive perigee burns. In addition, compute the apogee radius after the second perigee burn and compare it to the mean Earth–moon distance.

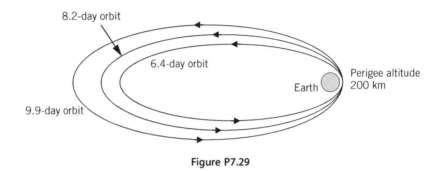

8.2-day orbit

6.4-day orbit

9.9-day orbit

Earth

Perigee altitude
200 km

Figure P7.29

7.30 After a coasting translunar trajectory, the LADEE spacecraft was inserted into a highly elliptical "capture orbit" about the moon by firing a rocket at an altitude of 400 km above the lunar surface. The resulting capture orbit had perilune and apolune altitudes of 400 and 15,588 km, respectively (see Figure P7.30). After coasting for three revolutions in this highly elliptical orbit, LADEE fired its onboard rocket at periapsis to create an elliptical orbit with a period of 4 h. Determine the Δv for this impulsive maneuver (note that the moon's radius is $R_m = 1{,}738$ km and its gravitational parameter is $\mu_m = 4{,}903$ km^3/s^2).

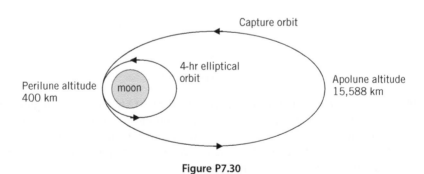

Capture orbit

4-hr elliptical
orbit

moon

Perilune altitude
400 km

Apolune altitude
15,588 km

Figure P7.30

7.31 The Apollo astronauts departed the moon's surface in the lunar module (LM) ascent stage. After over 7 min of powered flight, the LM ascent stage achieved the elliptical lunar orbit shown in Figure P7.31 (not to scale) with perilune and apolune altitudes of 15.8 and 89.4 km, respectively. The Apollo command and service module (CSM) orbited the moon in a circular orbit 116 km above the lunar surface. Determine the Δv impulse performed by the LM stage at apolune so that the transfer orbit's perilune and apolune are 89.4 and 116 km, respectively.

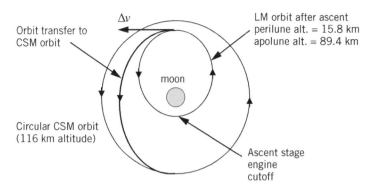

Figure P7.31

7.32 In March 2006, the Mars Reconnaissance Orbiter (MRO) spacecraft approached Mars on a hyperbolic trajectory with asymptotic approach speed $v_\infty^- = 2.9572$ km/s. Mars orbit insertion (MOI) occurred at periapsis by firing six rocket engines. The MOI burn produced an elliptical orbit with periapsis and apoapsis altitudes of 426 and 44,500 km, respectively. Determine the Δv for this impulsive maneuver [note that Mars' radius is $R_M = 3,396$ km and its gravitational parameter is $\mu_M = 4.2828(10^4)$ km^3/s^2].

Problems 7.33 and 7.34 involve the Juno spacecraft, which departed Earth in early August 2011 and arrived at Jupiter in early July 2016.

7.33 The Juno spacecraft approached Jupiter on a hyperbolic trajectory with eccentricity $e = 1.0172$ and semimajor axis $a = -4.384(10^6)$ km. Determine the Δv impulse applied at the periapsis ("perijove") passage that established an elliptical orbit about Jupiter with semimajor axis $a = 4,092,211$ km and eccentricity $e = 0.981574$ (the impulsive maneuver did not change the perijove radius). Jupiter's gravitational parameter is $\mu_J = 1.266865(10^8)$ km^3/s^2.

7.34 Mission operators had hoped to fire Juno's onboard rocket in October 2016 to reduce its orbital period from 53.486 to 14 days (a valve malfunction cancelled the burn). The single impulse would have occurred at perijove. Compute the (planned) impulsive Δv for the period reduction. Use the information in Problem 7.33 to determine the perijove radius, and assume that the burn does not change perijove radius. Jupiter's gravitational parameter is $\mu_J = 1.266865(10^8)$ km^3/s^2.

8

Relative Motion and Orbital Rendezvous

8.1 Introduction

Many space flight applications involve rendezvous, where a maneuvering spacecraft (e.g., the SpaceX Dragon) approaches and eventually docks with a target satellite (e.g., the International Space Station). For such cases, it is important to analyze the relative motion between the two satellites. Relative motion analysis also applies to the terminal orbit-insertion phase. In this scenario, a satellite is moving with respect to its intended "slot" in a target orbit. A third example of relative motion is orbital station-keeping, where a satellite must perform small thrusting maneuvers in order to return to its intended position in an orbit (station-keeping is periodically required because satellites are perturbed from their operational orbits by third-body forces, atmospheric drag, and solar radiation pressure).

Up to this point in the textbook, we have used conic sections to describe the two-body motion of a satellite in an inverse-square gravity field. In Chapter 7, we determined the impulsive Δv required to impart changes in a satellite's orbital elements. For relative motion analysis, this approach would be cumbersome if we demanded that differences in orbital elements represent deviations between a satellite and its target orbit. In this chapter, we take a different approach by developing linear ordinary differential equations (ODEs) for the relative position and velocity coordinates. These linear ODEs allow us to obtain analytical closed-form expressions for relative motion as a function of time. One caveat is that the linear ODEs and their solutions are only accurate when certain position and velocity components remain "small" with respect to the target orbit.

8.2 Linear Clohessy–Wiltshire Equations

Figure 8.1 shows the scenario where a satellite is moving relative to a target in a circular orbit with radius r^*. The polar coordinates of the moving target and satellite are (r^*, θ^*) and (r, θ), respectively. It should be clear to the reader that the target is moving in the circular orbit with angular velocity $n = \dot{\theta}^* = \sqrt{\mu/r^{*3}}$. Furthermore, the target may simply be a "moving slot" in a desired circular orbit or it may be another space vehicle, such as a space station. In either case, we wish to express the satellite's dynamical equations of motion relative to the target.

Space Flight Dynamics, First Edition. Craig A. Kluever.
© 2018 John Wiley & Sons Ltd. Published 2018 by John Wiley & Sons Ltd.
Companion website: www.wiley.com/go/Kluever/spaceflightmechanics

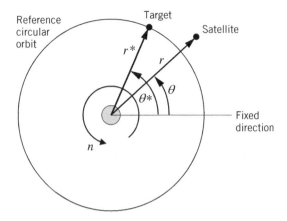

Figure 8.1 Satellite relative to a moving target position in a reference orbit.

Let us assume that two forces are acting on the satellite: inverse-square gravity and a rocket thrust force. The absolute acceleration of the satellite expressed in the moving polar coordinates [see Eq. (C.23) in Appendix C] is

$$\mathbf{a} = \left(\ddot{r} - r\dot{\theta}^2 \right)\mathbf{u}_r + \left(r\ddot{\theta} + 2\dot{r}\dot{\theta} \right)\mathbf{u}_\theta \tag{8.1}$$

where $\mathbf{u}_r$ and $\mathbf{u}_\theta$ are unit vectors along the radial and transverse directions, respectively. Next, let us consider the forces (or accelerations) that act on the satellite. The radial component of absolute acceleration is the sum of gravitational acceleration $(-\mu/r^2)$ and the radial component of thrust acceleration (a_r), whereas the transverse thrust acceleration (a_θ) is the sole acceleration term in the $\mathbf{u}_\theta$ direction. Therefore, we may rewrite Eq. (8.1) in terms of its radial and transverse components and employ Newton's second law:

$$\ddot{r} - r\dot{\theta}^2 = -\frac{\mu}{r^2} + a_r \tag{8.2}$$

$$r\ddot{\theta} + 2\dot{r}\dot{\theta} = a_\theta \tag{8.3}$$

The left-hand sides of Eqs. (8.2) and (8.3) are the absolute acceleration components and the right-hand sides are the external forces divided by the satellite mass. It is clear that Eqs. (8.2) and (8.3) constitute a fourth-order, nonlinear system. A good starting point for deriving a *linear* system is to express Eqs. (8.2) and (8.3) using state-variable equations. To show this, let us define a four-element state vector $\mathbf{x}$ as

$$\mathbf{x} = \begin{bmatrix} x_1 \\ x_2 \\ x_3 \\ x_4 \end{bmatrix} = \begin{bmatrix} r \\ \dot{r} \\ r^*\theta \\ r^*\dot{\theta} \end{bmatrix} \tag{8.4}$$

The first state variable x_1 is the radius of the satellite, the second state variable x_2 is radial velocity, the third state variable x_3 is the arc length projected along the reference circle (see Figure 8.1), and the fourth state variable x_4 is the satellite's transverse (or circumferential) velocity projected along the reference circle. The input (or control) vector $\mathbf{u}$ consists of the radial and transverse thrust acceleration components

$$\mathbf{u} = \begin{bmatrix} u_1 \\ u_2 \end{bmatrix} = \begin{bmatrix} a_r \\ a_\theta \end{bmatrix} \tag{8.5}$$

We can derive state-variable equations by taking the first time derivative of each state x_i and expressing the right-hand sides solely in terms of state variables x_i and input variables u_1 and u_2

$$\dot{x}_1 = \dot{r} = x_2 \tag{8.6}$$

$$\dot{x}_2 = \ddot{r} = -\frac{\mu}{x_1^2} + \frac{x_1 x_4^2}{r^{*2}} + u_1 \tag{8.7}$$

$$\dot{x}_3 = r^*\dot{\theta} = x_4 \tag{8.8}$$

$$\dot{x}_4 = r^*\ddot{\theta} = -\frac{2x_2 x_4}{x_1} + \frac{r^*}{x_1} u_2 \tag{8.9}$$

Equations (8.6)–(8.9) are four first-order state-variable equations. The reader should verify that Eqs. (8.6)–(8.9) are simply another representation of the two second-order differential equations (8.2) and (8.3); the dynamics have not been altered in any way. We can use a compact vector notation to express the state-variable equations as

$$\dot{\mathbf{x}} = \mathbf{f}(\mathbf{x}, \mathbf{u}) = \begin{bmatrix} x_2 \\ -\dfrac{\mu}{x_1^2} + \dfrac{x_1 x_4^2}{r^{*2}} + u_1 \\ x_4 \\ -\dfrac{2x_2 x_4}{x_1} + \dfrac{r^*}{x_1} u_2 \end{bmatrix} \tag{8.10}$$

Of course, Eq. (8.10) is still a nonlinear system. The first step in the linearization process is to define the reference state $\mathbf{x}^*$ and reference input $\mathbf{u}^*$ vectors:

$$\mathbf{x}^* = \begin{bmatrix} r^* \\ 0 \\ r^*\theta^* \\ r^*n \end{bmatrix} \quad \text{and} \quad \mathbf{u}^* = \begin{bmatrix} 0 \\ 0 \end{bmatrix} \tag{8.11}$$

The reference state vector $\mathbf{x}^*$ defines the position and velocity of the target in the reference circular orbit depicted in Figure 8.1. The reference input $\mathbf{u}^*$ is zero thrust acceleration so that a satellite with state vector $\mathbf{x}^*$ will remain in the reference circular orbit. It is easy to show that the reference states and inputs, Eq. (8.11), satisfy the state-variable equations (8.10).

The next step in the linearization process is to define the *perturbation vectors* or deviations between the satellite's states and inputs and their respective reference values:

$$\delta\mathbf{x} = \mathbf{x} - \mathbf{x}^* \quad \text{and} \quad \delta\mathbf{u} = \mathbf{u} - \mathbf{u}^* \tag{8.12}$$

The perturbation state vector, $\delta\mathbf{x}$, is a collection of the differences in position and velocity components between the satellite and target. Next, we rewrite the state-variable equations (8.10) in terms of the reference and perturbation variables using $\mathbf{x} = \delta\mathbf{x} + \mathbf{x}^*$ and $\mathbf{u} = \delta\mathbf{u} + \mathbf{u}^*$

$$\delta\dot{\mathbf{x}} + \dot{\mathbf{x}}^* = \mathbf{f}(\delta\mathbf{x} + \mathbf{x}^*, \delta\mathbf{u} + \mathbf{u}^*) \tag{8.13}$$

Expanding the right-hand side of Eq. (8.13) in a first-order Taylor series about the references $\mathbf{x}^*$ and $\mathbf{u}^*$ yields

$$\delta\dot{\mathbf{x}} + \dot{\mathbf{x}}^* = \mathbf{f}(\mathbf{x}^*, \mathbf{u}^*) + \left.\frac{\partial\mathbf{f}}{\partial\mathbf{x}}\right|_* \delta\mathbf{x} + \left.\frac{\partial\mathbf{f}}{\partial\mathbf{u}}\right|_* \delta\mathbf{u} \tag{8.14}$$

Because the reference is a solution to the nonlinear dynamic equation, the terms $\dot{\mathbf{x}}^*$ and $\mathbf{f}(\mathbf{x}^*, \mathbf{u}^*)$ cancel each other, and Eq. (8.14) becomes

$$\delta\dot{\mathbf{x}} = \left.\frac{\partial\mathbf{f}}{\partial\mathbf{x}}\right|_* \delta\mathbf{x} + \left.\frac{\partial\mathbf{f}}{\partial\mathbf{u}}\right|_* \delta\mathbf{u} \tag{8.15}$$

Equation (8.15) is a *linear* differential equation that approximates the motion of the satellite relative to the target in the reference circular orbit. It is very important to note that the solution of Eq. (8.15) provides the satellite's position and velocity *deviations* from the reference orbit.

Next, let us expand and show all terms in the second and fourth rows of the linearized equation (8.15)

$$\delta\dot{x}_2 = \left.\frac{\partial f_2}{\partial x_1}\right|_* \delta x_1 + \left.\frac{\partial f_2}{\partial x_2}\right|_* \delta x_2 + \left.\frac{\partial f_2}{\partial x_3}\right|_* \delta x_3 + \left.\frac{\partial f_2}{\partial x_4}\right|_* \delta x_4 + \left.\frac{\partial f_2}{\partial u_1}\right|_* \delta u_1 + \left.\frac{\partial f_2}{\partial u_2}\right|_* \delta u_2 \tag{8.16}$$

$$\delta\dot{x}_4 = \left.\frac{\partial f_4}{\partial x_1}\right|_* \delta x_1 + \left.\frac{\partial f_4}{\partial x_2}\right|_* \delta x_2 + \left.\frac{\partial f_4}{\partial x_3}\right|_* \delta x_3 + \left.\frac{\partial f_4}{\partial x_4}\right|_* \delta x_4 + \left.\frac{\partial f_4}{\partial u_1}\right|_* \delta u_1 + \left.\frac{\partial f_4}{\partial u_2}\right|_* \delta u_2 \tag{8.17}$$

We only consider the partial derivatives of the second and fourth rows of $\mathbf{f}(\mathbf{x},\mathbf{u})$ because these two rows represent the dynamical equations of motion. Let us complete the linearization process by computing the partial derivatives of $\mathbf{f}(\mathbf{x},\mathbf{u})$ in Eq. (8.10) with respect to states and inputs. The partial derivatives of $f_2(\mathbf{x},\mathbf{u})$ are

$$\frac{\partial f_2}{\partial\mathbf{x}} = \left[\frac{\partial f_2}{\partial x_1} \ \frac{\partial f_2}{\partial x_2} \ \frac{\partial f_2}{\partial x_3} \ \frac{\partial f_2}{\partial x_4}\right]$$

$$= \left[\frac{2\mu}{x_1^3} + \frac{x_4^2}{r^{*2}} \ \ 0 \ \ 0 \ \ \frac{2x_1 x_4}{r^{*2}}\right] \tag{8.18}$$

$$\frac{\partial f_2}{\partial\mathbf{u}} = \left[\frac{\partial f_2}{\partial u_1} \ \frac{\partial f_2}{\partial u_2}\right] = [1 \ 0] \tag{8.19}$$

Evaluating Eq. (8.18) at the reference states $x_1^* = r^*$ and $x_4^* = r^* n$ yields

$$\left.\frac{\partial f_2}{\partial x_1}\right|_* = \frac{2\mu}{r^{*3}} + \frac{(r^* n)^2}{r^{*2}} = 3n^2 \tag{8.20}$$

$$\left.\frac{\partial f_2}{\partial x_4}\right|_* = \frac{2r^*(r^* n)}{r^{*2}} = 2n \tag{8.21}$$

Using these partial derivatives in Eq. (8.16), we obtain the first linearized dynamical equation

$$\delta \dot{x}_2 = 3n^2 \delta x_1 + 2n \delta x_4 + \delta u_1 \tag{8.22}$$

The partial derivatives of the fourth row, $f_4(\mathbf{x}, \mathbf{u})$, are

$$\frac{\partial f_4}{\partial \mathbf{x}} = \begin{bmatrix} \dfrac{\partial f_4}{\partial x_1} & \dfrac{\partial f_4}{\partial x_2} & \dfrac{\partial f_4}{\partial x_3} & \dfrac{\partial f_4}{\partial x_4} \end{bmatrix} \tag{8.23}$$

$$= \begin{bmatrix} \dfrac{2x_2 x_4}{x_1^2} - \dfrac{r^*}{x_1^2} u_2 & -\dfrac{2x_4}{x_1} & 0 & -\dfrac{2x_2}{x_1} \end{bmatrix}$$

$$\frac{\partial f_4}{\partial \mathbf{u}} = \begin{bmatrix} \dfrac{\partial f_4}{\partial u_1} & \dfrac{\partial f_4}{\partial u_2} \end{bmatrix} = \begin{bmatrix} 0 & \dfrac{r^*}{x_1} \end{bmatrix} \tag{8.24}$$

Evaluating Eqs. (8.23) and (8.24) at the references $x_1{}^* = r^*$, $x_2{}^* = 0$, $x_4{}^* = r^* n$, and $u_2{}^* = 0$ yields

$$\left. \frac{\partial f_4}{\partial x_1} \right|_* = 0 \tag{8.25}$$

$$\left. \frac{\partial f_4}{\partial x_2} \right|_* = \frac{-2(r^* n)}{r^*} = -2n \tag{8.26}$$

$$\left. \frac{\partial f_4}{\partial x_4} \right|_* = 0 \tag{8.27}$$

$$\left. \frac{\partial f_4}{\partial u_2} \right|_* = 1 \tag{8.28}$$

Using these partials in Eq. (8.17), we obtain the second linearized equation

$$\delta \dot{x}_4 = -2n \delta x_2 + \delta u_2 \tag{8.29}$$

Finally, let us rewrite the two linearized dynamical equations in terms of relative coordinates. Figure 8.2 shows a rotating coordinate frame with its origin fixed at the target as it moves in the reference circular orbit. The $+x$ axis is along the target's outward radial direction, and the $+y$ axis is along the circular arc of the reference orbit in the direction of motion (note the slight abuse in notation here; x is the radial perturbation from the reference circle even though we used x_i for the state variables in our linearization process). It should be clear from Figure 8.2 that $x = \delta r$ is the satellite's radial displacement from the reference circle, and $y = r^* \delta \theta$ is the satellite's circumferential displacement from the moving target location (projected onto the reference circular orbit). The scenario depicted in Figures 8.1 and 8.2 shows $x > 0$ (the satellite is above the reference circle), and $y < 0$ (the satellite lags behind the target). Using these new coordinates, we can make the following substitutions for the perturbation state and input variables: $x = \delta x_1$, $\dot{x} = \delta x_2$, $y = \delta x_3$, $\dot{y} = \delta x_4$, $a_x = \delta u_1$, and $a_y = \delta u_2$. Hence, Eqs. (8.22) and (8.29) become

$$\ddot{x} = 3n^2 x + 2n\dot{y} + a_x \tag{8.30}$$

$$\ddot{y} = -2n\dot{x} + a_y \tag{8.31}$$

Equations (8.30) and (8.31) are commonly known as the Clohessy–Wiltshire (CW) equations. Hill [1] first derived these relative-motion equations in 1878; Clohessy and Wiltshire

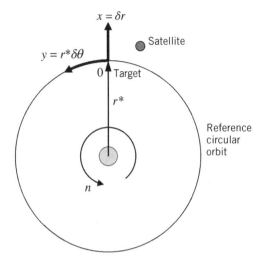

Figure 8.2 Coordinate frame for relative motion.

rediscovered them in a 1960 study on orbital rendezvous [2]. Equations (8.30) and (8.31) are a set of coupled, linear, time-invariant differential equations that describe the planar motion of a satellite relative to a reference circular orbit. Although our derivation has only considered planar motion, it is possible to derive an additional linearized dynamical equation for out-of-plane motion

$$\ddot{z} = -n^2 z + a_z \tag{8.32}$$

where z is the satellite's displacement from the circular orbit in a direction normal to the orbital plane. We can attach a $+z$ axis to the target (origin) in Figure 8.2 that points out of the page. Referring again to Figure 8.2, we may think of the x, y, and z coordinates as the radial, along-track, and cross-track displacements of the satellite relative to the target moving in a reference circular orbit. The thrust acceleration components in the radial, along-track, and cross-track directions are a_x ($= a_r$), a_y ($= a_\theta$), and a_z, respectively.

8.3 Homogeneous Solution of the Clohessy–Wiltshire Equations

The CW equations (8.30)–(8.32) are a set of linear, time-invariant differential equations (at first glance the reader may think that the existence of products including the n and n^2 terms results in a nonlinear system; recall that n and n^2 are simply constant coefficients in this case). We should emphasize that the CW equations are an *approximation* of the two-body dynamics, and, like all linearized systems, this approximation is only accurate for "small" deviations from the reference state. Therefore, we should identify the limitations of the CW differential equations. Referring again to Figure 8.2 (with the $+z$ axis out of the page), we see that (xyz) is a cylindrical coordinate frame. We also note that the only displacements appearing in the right-hand sides of the CW equations are x and z; along-track displacement (y) does *not* appear in the CW equations. Therefore, the accuracy of the linearized CW equations is confined to a toroidal or doughnut-shaped ring about the reference orbit where x and z displacements remain "small." The CW equations accurately describe the satellite's relative motion for scenarios with large along-track (y) displacements as long as x and z displacements are "small."

Because the CW equations are linear, we may obtain analytical closed-form solutions. Recall that the complete solution to a differential equation is composed of the homogeneous and particular solutions. The homogenous solution is obtained by solving the differential equation with zero input, while the particular solution depends on the nature of the input (or forcing) function. For our relative-motion equations, the thrust acceleration components (a_x, a_y, and a_z) are the input functions. We will develop and focus on the homogeneous solution of the CW equations with zero thrust acceleration. The homogeneous solution will describe the "natural" motion of the satellite relative to the target where gravity is the only force.

To begin, note that the out-of-plane motion, Eq. (8.32), is uncoupled from the in-plane Eqs. (8.30) and (8.31); that is, coordinate z only appears in Eq. (8.32). The homogenous cross-track CW equation is

$$\ddot{z} + n^2 z = 0 \tag{8.33}$$

This differential equation is an *undamped harmonic oscillator*, and its solution consists of sine and cosine terms

$$z(t) = C_1 \cos nt + C_2 \sin nt \tag{8.34}$$

where the two constants depend on the initial cross-track position and velocity components, that is, $C_1 = z_0$ and $C_2 = \dot{z}_0/n$. The solution for out-of-plane motion is

$$z(t) = z_0 \cos nt + \frac{\dot{z}_0}{n} \sin nt \tag{8.35}$$

The reader can verify this solution by substituting it (along with its second time derivative) into the governing differential equation (8.33). The cross-track velocity is the first time derivative of Eq. (8.35):

$$\dot{z}(t) = -z_0 n \sin nt + \dot{z}_0 \cos nt \tag{8.36}$$

Next, we turn our attention to the along-track (y) solution. Direct integration of the homogeneous along-track acceleration (8.31) with $a_y = 0$ yields the velocity

$$\dot{y}(t) = -2nx + c \tag{8.37}$$

The integration constant *must* be $c = 2nx_0 + \dot{y}_0$ so that Eq. (8.37) is satisfied at time $t = 0$, that is, $\dot{y}(0) = \dot{y}_0$. Using Eq. (8.37) in the homogeneous radial acceleration equation (8.30), we obtain

$$\ddot{x} = 3n^2 x + 2n(-2nx + c)$$

or

$$\ddot{x} = -n^2 x + 2nc$$

Placing the terms involving x on the left-hand side yields

$$\ddot{x} + n^2 x = 2nc \tag{8.38}$$

Equation (8.38) is an undamped second-order differential equation with a constant forcing (or input) function on the right-hand side. The complete solution $x(t)$ will be the sum of the homogeneous and particular solutions. We know that the homogenous solution will involve sine and cosine terms (a harmonic oscillator), while the particular solution will be a constant because the forcing function is a constant. Therefore, the *form* of the complete solution of Eq. (8.38) is

$$x(t) = A_1 \cos nt + A_2 \sin nt + A_3 \tag{8.39}$$

We need three equations to determine the three constants A_1, A_2, and A_3. The first and second time derivatives of Eq. (8.39) provide two additional equations:

$$\dot{x}(t) = -A_1 n \sin nt + A_2 n \cos nt \tag{8.40}$$

$$\ddot{x}(t) = -A_1 n^2 \cos nt - A_2 n^2 \sin nt \tag{8.41}$$

Evaluating Eqs. (8.39)–(8.41) at time $t = 0$, we obtain

$$x(0) = A_1 + A_3 = x_0 \tag{8.42}$$

$$\dot{x}(0) = A_2 n = \dot{x}_0 \tag{8.43}$$

$$\ddot{x}(0) = -A_1 n^2 = -n^2 x_0 + 2nc \tag{8.44}$$

where Eq. (8.38) is used to derive Eq. (8.44). The three constants are

$$A_1 = x_0 - \frac{2c}{n} \tag{8.45}$$

$$A_2 = \frac{\dot{x}_0}{n} \tag{8.46}$$

$$A_3 = \frac{2c}{n} \tag{8.47}$$

Finally, substituting the three A_i constants (with $c = 2nx_0 + \dot{y}_0$) into Eq. (8.39) yields the homogeneous solution for radial position:

$$x(t) = \left(-3x_0 - \frac{2\dot{y}_0}{n} \right) \cos nt + \frac{\dot{x}_0}{n} \sin nt + \left(4x_0 + \frac{2\dot{y}_0}{n} \right) \tag{8.48}$$

The relative radial velocity is the first time derivative of Eq. (8.48)

$$\dot{x}(t) = \dot{x}_0 \cos nt + (3nx_0 + 2\dot{y}_0) \sin nt \tag{8.49}$$

It is easy to see that Eqs. (8.48) and (8.49) are satisfied at time $t = 0$; that is, $x(0) = x_0$ and $\dot{x}(0) = \dot{x}_0$.

The along-track solution, $y(t)$, is obtained by integrating Eq. (8.37) with the substitutions $x = A_1 \cos nt + A_2 \sin nt + A_3$ and $c = 2nx_0 + \dot{y}_0$

$$y(t) = -2n \int (A_1 \cos nt + A_2 \sin nt + A_3) dt + \int (2nx_0 + \dot{y}_0) dt + B$$

Performing the integration yields

$$y(t) = -2n\left(\frac{A_1 \sin nt}{n} - \frac{A_2 \cos nt}{n} + A_3 t\right) + (2nx_0 + \dot{y}_0)t + B \tag{8.50}$$

where B is a constant of integration. Substituting the previous expressions for A_1, A_2, and A_3 and using $y(0) = y_0$ to solve for the constant B yields the solution:

$$y(t) = \left(6x_0 + \frac{4\dot{y}_0}{n}\right)\sin nt + \frac{2\dot{x}_0}{n}\cos nt - (6nx_0 + 3\dot{y}_0)t + \left(y_0 - \frac{2\dot{x}_0}{n}\right) \tag{8.51}$$

Equation (8.51) is the homogeneous solution for the along-track motion of the satellite relative to the target. Taking a time derivative of Eq. (8.51), we obtain

$$\dot{y}(t) = (6nx_0 + 4\dot{y}_0)\cos nt - 2\dot{x}_0 \sin nt - 6nx_0 - 3\dot{y}_0 \tag{8.52}$$

Equations (8.35), (8.36), (8.48), (8.49), (8.51), and (8.52) describe the satellite's position and velocity relative to a target moving on the reference circular orbit. These six equations are collected and comprise equation set (8.53):

$$x(t) = \left(-3x_0 - \frac{2\dot{y}_0}{n}\right)\cos nt + \frac{\dot{x}_0}{n}\sin nt + \left(4x_0 + \frac{2\dot{y}_0}{n}\right) \tag{8.53a}$$

$$y(t) = \left(6x_0 + \frac{4\dot{y}_0}{n}\right)\sin nt + \frac{2\dot{x}_0}{n}\cos nt - (6nx_0 + 3\dot{y}_0)t + \left(y_0 - \frac{2\dot{x}_0}{n}\right) \tag{8.53b}$$

$$z(t) = z_0 \cos nt + \frac{\dot{z}_0}{n}\sin nt \tag{8.53c}$$

$$\dot{x}(t) = \dot{x}_0 \cos nt + (3nx_0 + 2\dot{y}_0)\sin nt \tag{8.53d}$$

$$\dot{y}(t) = (6nx_0 + 4\dot{y}_0)\cos nt - 2\dot{x}_0 \sin nt - 6nx_0 - 3\dot{y}_0 \tag{8.53e}$$

$$\dot{z}(t) = -z_0 n \sin nt + \dot{z}_0 \cos nt \tag{8.53f}$$

Equations (8.53a–f) represent relative motion of the satellite without thrust forces. The reader should verify that setting $t = 0$ in the right-hand sides of Eqs. (8.53a–f) does indeed yield the expected initial conditions for each position and velocity component. Let us list the characteristics of relative satellite motion described by the linear CW equations:

1) A satellite that is initially at the origin ($x_0 = y_0 = z_0 = 0$) with zero relative velocity ($\dot{x}_0 = \dot{y}_0 = \dot{z}_0 = 0$) will remain at the origin or target orbit for all time $t > 0$.
2) A satellite that is *on* the reference circular orbit with zero radial and cross-track displacement (i.e., $x_0 = z_0 = 0$) and zero relative velocity (i.e., $\dot{x}_0 = \dot{y}_0 = \dot{z}_0 = 0$) but displaced in the along-track direction (i.e., $y_0 \neq 0$) will remain in this relative position for all time. In other words, a satellite in the same circular orbit as the target but at a different angular position will simply remain at a fixed along-track offset distance y_0 at all times. This is true because y_0 only appears as the initial condition for solution $y(t)$ and does not appear in any harmonic terms or terms that are linear in time t.
3) All solutions contain harmonic terms with a common angular frequency n that is equal to the mean motion of the reference circular orbit, or 2π divided by the period.
4) The cross-track (or out-of-plane) position and velocity solutions $z(t)$ and $\dot{z}(t)$ are uncoupled from the in-plane motion; that is, z_0 and $\dot{z}_0$ only influence Eqs. (8.53c) and (8.53f).

5) The along-track solution (8.53b) is the only relative motion that involves a secular "drift" term that is linear in time t.

The following examples utilize the homogeneous solution to the CW equations to illustrate relative motion.

Example 8.1   Figure 8.3 shows a satellite in a circular orbit (radius $r = 8{,}590$ km) that is "lagging behind" a target position (the origin) in the reference circular orbit (radius $r^* = 8{,}600$ km). The two neighboring circular orbits are coplanar. At time $t = 0$, the satellite has an angular separation $\delta\theta_0 = -2°$ (lag) behind the target. Compute the initial relative position and velocity components in the CW frame and determine the satellite's relative position and velocity components 30 min later.

Because the satellite's orbit is coplanar with the reference orbit, the cross-track position and velocity components (z and $\dot{z}$) are always zero. The initial radial displacement in the CW frame is

$$x_0 = \delta r_0 = r - r^* = \boxed{-10 \text{ km}}$$

The initial along-track displacement is

$$y_0 = r^* \delta\theta_0 = (8{,}600\,\text{km})(-0.0349\,\text{rad}) = \boxed{-300.2 \text{ km}}$$

These two calculations show that the satellite is below and lagging behind the target as seen in Figure 8.3.

Because the reference orbit and the neighboring satellite orbit are circular, the initial relative radial velocity is $\dot{x}_0 = 0$. Furthermore, the relative radial velocity $\dot{x}$ remains constant (at zero) due to the satellite's circular orbit, and therefore $\ddot{x} = 0$ holds at all times. Equation (8.30) shows that for this special case of neighboring circular orbits (and no thrust), we may write

$$\ddot{x} = 3n^2 x + 2n\dot{y} = 0$$

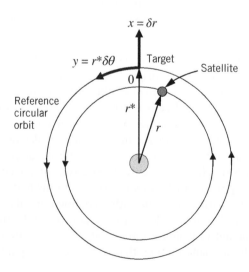

Figure 8.3 Satellite in a neighboring circular orbit (Example 8.1; not to scale).

Using this expression at time $t = 0$, the initial along-track relative velocity must be

$$\dot{y}_0 = -1.5nx_0 = \boxed{0.011874 \text{ km/s}}$$

where $n = \sqrt{\mu/r^{*3}} = 7.9163(10^{-4})$ rad/s is the angular velocity of the reference orbit. Equation (8.31) shows that along-track velocity is constant (i.e., $\ddot{y} = 0$) because radial velocity $\dot{x}$ is always equal to zero.

Even though we expect x, $\dot{x}$, and $\dot{y}$ to remain constant, let us evaluate the homogeneous solutions (8.53) at time $t = 1,800$ s ($=30$ min). The harmonic terms evaluated at angle $nt = 1.4249$ rad are $\cos nt = 0.145349$ and $\sin nt = 0.989380$. Using these values along with the appropriate initial conditions, we obtain

$$x(1,800) = \left(-3x_0 - \frac{2\dot{y}_0}{n}\right)\cos nt + \frac{\dot{x}_0}{n}\sin nt + \left(4x_0 + \frac{2\dot{y}_0}{n}\right) = \boxed{-10 \text{ km (no change)}}$$

$$y(1,800) = \left(6x_0 + \frac{4\dot{y}_0}{n}\right)\sin nt + \frac{2\dot{x}_0}{n}\cos nt - (6nx_0 + 3\dot{y}_0)t + \left(y_0 - \frac{2\dot{x}_0}{n}\right) = \boxed{-278.8 \text{ km}}$$

$$\dot{x}(1,800) = \dot{x}_0 \cos nt + (3nx_0 + 2\dot{y}_0)\sin nt = \boxed{0 \text{ (no change)}}$$

$$\dot{y}(1,800) = (6nx_0 + 4\dot{y}_0)\cos nt - 2\dot{x}_0 \sin nt - 6nx_0 - 3\dot{y}_0 = \boxed{0.011874 \text{ km/s (no change)}}$$

We see that in 30 min the satellite has moved 21.4 km closer to the target. Of course, for this simple case we could have obtained the along-track displacement using $y(t) = y_0 + \dot{y}_0 t$ with $t = 1,800$ s. We may also use the *inertial* displacements to compute the final relative displacement. Figure 8.4 presents the along-track *inertial* displacements of the target and satellite in their respective orbits. The total along-track distance of the target is easy to compute from the product of the constant angular rate (n), radius r^*, and time t; i.e., $r^* nt = 12,254.4$ km. Similarly, the total along-track travel distance of the satellite is $r^* \omega t = 12,275.8$ km where the constant angular rate of the inner circular orbit is $\omega = \sqrt{\mu/r^3} = 7.9301(10^{-4})$ rad/s. Remember that we must use the satellite's *projected* along-track distance $r^* \omega t$ (instead of the inner circle arc length $r \omega t$) due to the definition

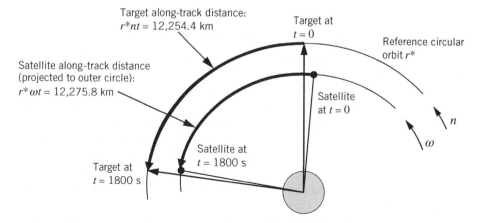

Figure 8.4 Target and satellite along-track distances (Example 8.1; not to scale).

of the along-track distance, $y = r^* \delta\theta$ (see Figures 8.1 and 8.2). The difference in the two inertial along-track distances, $r^*(\omega t - nt) = 21.4$ km, is the change in along-track displacement of the satellite relative to the target. Adding this relative displacement to y_0 yields −278.8 km, which matches the solution using the linearized CW equations.

This simple example has demonstrated that when the satellite is in a neighboring circular orbit with a radius lower than the target orbit, the satellite will maintain a constant *positive* along-track relative velocity. Thus, if the satellite is initially behind the target (as shown in Figures 8.3 and 8.4), it will continually reduce the along-track (y) separation. Given enough time, the satellite will reach $y = 0$ and thereafter it will "lead" the target. Similarly, when the satellite is in a higher neighboring circular orbit ($x_0 > 0$), the along-track relative velocity is negative (i.e., $\dot{y}_0 = -1.5nx_0$) and a satellite starting behind the target will continuously drift away at a constant rate.

Example 8.2 An astronaut in the open payload bay of the Space Shuttle throws a wrench with positive radial velocity of 0.04 km/s (nearly 90 mph – imagine Randy Johnson in space in the 1990s). If the Space Shuttle is in a 320-km circular orbit, determine the position and velocity components of the wrench after one orbital revolution of the Shuttle.

The circular Space-Shuttle orbit is the reference orbit with radius $r^* = 320$ km $+ R_E = 6{,}698$ km and the Shuttle is at the origin of the CW frame. Because the wrench is thrown from the Shuttle, its initial position coordinates are zero: $x_0 = y_0 = z_0 = 0$. The wrench's initial velocity relative to the Shuttle (target) is purely radial and therefore $\dot{x}_0 = 0.04$ km/s and $\dot{y}_0 = \dot{z}_0 = 0$. The cross-track components (z and $\dot{z}$) are always zero because the wrench is initially in the reference orbital plane with zero out-of-plane velocity. Angular velocity of the reference Shuttle orbit is $n = \sqrt{\mu/r^{*3}} = 1.1517(10^{-3})$ rad/s, and the corresponding period is $T_{period} = 2\pi/n = 5{,}455.43$ s ($= 90.9$ min). The relative position and velocity components after one orbit are determined by evaluating Eq. (8.53) with $t = T_{period}$

$$x(5{,}455.43) = \left(-3x_0 - \frac{2\dot{y}_0}{n}\right)\cos nt + \frac{\dot{x}_0}{n}\sin nt + \left(4x_0 + \frac{2\dot{y}_0}{n}\right) = 0 \text{ km}$$

$$y(5{,}455.43) = \left(6x_0 + \frac{4\dot{y}_0}{n}\right)\sin nt + \frac{2\dot{x}_0}{n}\cos nt - (6nx_0 + 3\dot{y}_0)t + \left(y_0 - \frac{2\dot{x}_0}{n}\right) = 0 \text{ km}$$

$$\dot{x}(5{,}455.43) = \dot{x}_0\cos nt + (3nx_0 + 2\dot{y}_0)\sin nt = 0.04 \text{ km/s}$$

$$\dot{y}(5{,}455.43) = (6nx_0 + 4\dot{y}_0)\cos nt - 2\dot{x}_0\sin nt - 6nx_0 - 3\dot{y}_0 = 0 \text{ km/s}$$

Clearly, the wrench has returned to the Shuttle (target) with the same velocity components that existed at time $t = 0$.

We may use Eqs. (8.53a) and (8.53b) to obtain the wrench's radial and along-track relative position coordinates at arbitrary time $0 \le t \le T_{period}$. Figure 8.5 shows the radial and along-track deviations of the wrench relative to the Shuttle over one orbit (the $+x$ and $+y$ axes have been added for clarity). Note that the wrench initially moves radial outward from the origin and drifts behind the Shuttle (solid path in Figure 8.5). At half an orbit, the wrench crosses the reference orbit (i.e., $x = 0$) at a relative along-track distance of −138.9 km. During the second half of the orbit, the wrench is below the Shuttle orbit ($x < 0$) and drifts toward the origin and eventually returns to the Shuttle in exactly one orbital period. If unobstructed, the wrench would continue to follow the "elliptical

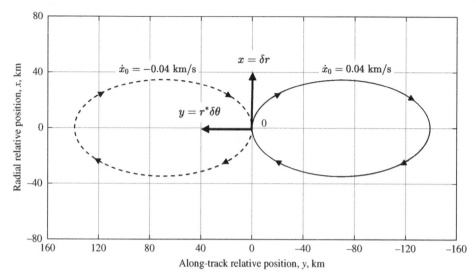

Figure 8.5 Relative position of an object with initial radial velocity during one orbital revolution (Example 8.2).

trace" in the CW coordinates shown in Figure 8.5. Figure 8.5 also shows the case where the wrench has negative initial radial velocity, $\dot{x}_0 = -0.04$ km/s. As expected, the relative motion (dashed path in Figure 8.5) is the mirror image of the previous case reflected across the x axis: the wrench is initially below the reference orbit and drifts ahead of the Shuttle; after half an orbit the wrench is above the Shuttle and drifts back to the origin.

Example 8.3   Consider again the scenario of Example 8.2 but in this case, an astronaut throws a wrench from the Shuttle with an along-track velocity magnitude of 0.04 km/s. Plot the relative position coordinates of the wrench during two orbital revolutions of the Shuttle for the initial conditions $\dot{y}_0 = 0.04$ km/s and $\dot{y}_0 = -0.04$ km/s.

As with Example 8.2, we use Eqs. (8.53a) and (8.53b) to obtain the wrench's radial and along-track relative coordinates for values of time between zero and two periods, or $0 \le t \le 2T_{\text{period}}$. Figure 8.6 shows the radial and along-track deviations of the wrench relative to the Shuttle over two orbits for $\dot{y}_0 = \pm 0.04$ km/s. When the initial along-track velocity is positive (solid line), the wrench's orbit is more energetic than the Shuttle's orbit, and therefore it initially drifts above the reference orbit because its apogee is larger. After one revolution, the wrench returns to its perigee which is at the same altitude as the Shuttle orbit and therefore $x = 0$. However, after one revolution the wrench is about −655 km behind the Shuttle because its orbital speed near apogee is slower than the Shuttle's circular velocity, and hence it drifts behind the target. When the wrench is near its perigee (the Shuttle's orbit), its speed is greater than circular speed and this effect accounts for the brief "loops" observed in Figure 8.6 near $x = 0$. The wrench will continue to drift behind the Shuttle losing 655 km in the along-track (y) direction every revolution. When the initial along-track velocity is negative (dashed path in Figure 8.6), we observe the reverse effect: the wrench's orbit is less energetic than the Shuttle's orbit and the

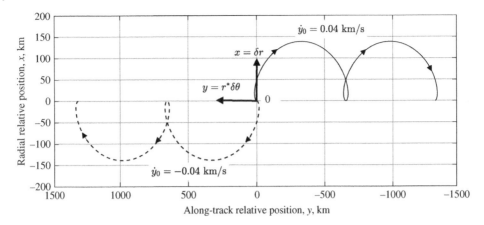

Figure 8.6 Relative position of an object with initial along-track velocity during two orbital revolutions (Example 8.3).

wrench's perigee is below the reference circle. In this case, the wrench is moving faster than the Shuttle (when near its perigee), and therefore it drifts ahead in the along-track direction. However, when the wrench is near its apogee ($x = 0$), it is moving slower than the target and a brief "loop" is observed in the relative coordinate frame.

8.4 Orbital Rendezvous Using the Clohessy–Wiltshire Equations

It is convenient to express the analytical solutions (8.53) in a matrix-vector format. We shall soon see that this format facilitates solutions for orbital rendezvous. Before we do so, let us simplify the discussion by considering *planar* relative motion where $z = \dot{z} = 0$ at all times. Note that separately analyzing the x–y motion is not restrictive because the out-of-plane coordinates z and $\dot{z}$ do not appear in the planar solutions (8.53a), (8.53b), (8.53d), and (8.53e).

To begin, let us define two 2×1 column vectors: the relative position vector $\delta\mathbf{r}$ and relative velocity vector $\delta\mathbf{v}$

$$\delta\mathbf{r} = \begin{bmatrix} x \\ y \end{bmatrix}, \quad \delta\mathbf{v} = \begin{bmatrix} \dot{x} \\ \dot{y} \end{bmatrix} \tag{8.54}$$

Next, we can express the planar *position* solutions, Eqs. (8.53a) and (8.53b), in the following matrix-vector format

$$\delta\mathbf{r}(t) = \begin{bmatrix} -3\cos nt + 4 & 0 \\ 6(\sin nt - nt) & 1 \end{bmatrix} \delta\mathbf{r}_0 + \begin{bmatrix} \dfrac{\sin nt}{n} & \dfrac{-2}{n}(\cos nt - 1) \\ \dfrac{2}{n}(\cos nt - 1) & \dfrac{4}{n}\sin nt - 3t \end{bmatrix} \delta\mathbf{v}_0 \tag{8.55}$$

where the initial relative position and velocity vectors are

$$\delta\mathbf{r}_0 = \begin{bmatrix} x_0 \\ y_0 \end{bmatrix}, \qquad \delta\mathbf{v}_0 = \begin{bmatrix} \dot{x}_0 \\ \dot{y}_0 \end{bmatrix} \tag{8.56}$$

The reader should verify that carrying out the matrix-vector multiplication of the top rows of Eq. (8.55) yields Eq. (8.53a), the solution for relative radial position $x(t)$. Multiplication of the bottom rows yields Eq. (8.53b), the solution for $y(t)$. Similarly, the planar relative velocity solutions are

$$\delta\mathbf{v}(t) = \begin{bmatrix} 3n\sin nt & 0 \\ 6n(\cos nt - 1) & 0 \end{bmatrix} \delta\mathbf{r}_0 + \begin{bmatrix} \cos nt & 2\sin nt \\ -2\sin nt & 4\cos nt - 3 \end{bmatrix} \delta\mathbf{v}_0 \tag{8.57}$$

Let us write Eqs. (8.55) and (8.57) as

$$\delta\mathbf{r}(t) = \mathbf{\Phi}_{11}\delta\mathbf{r}_0 + \mathbf{\Phi}_{12}\delta\mathbf{v}_0 \tag{8.58}$$

$$\delta\mathbf{v}(t) = \mathbf{\Phi}_{21}\delta\mathbf{r}_0 + \mathbf{\Phi}_{22}\delta\mathbf{v}_0 \tag{8.59}$$

where the four 2×2 matrices are

$$\mathbf{\Phi}_{11} = \begin{bmatrix} -3\cos nt + 4 & 0 \\ 6(\sin nt - nt) & 1 \end{bmatrix} \tag{8.60a}$$

$$\mathbf{\Phi}_{12} = \begin{bmatrix} \dfrac{\sin nt}{n} & \dfrac{-2}{n}(\cos nt - 1) \\ \dfrac{2}{n}(\cos nt - 1) & \dfrac{4}{n}\sin nt - 3t \end{bmatrix} \tag{8.60b}$$

$$\mathbf{\Phi}_{21} = \begin{bmatrix} 3n\sin nt & 0 \\ 6n(\cos nt - 1) & 0 \end{bmatrix} \tag{8.60c}$$

$$\mathbf{\Phi}_{22} = \begin{bmatrix} \cos nt & 2\sin nt \\ -2\sin nt & 4\cos nt - 3 \end{bmatrix} \tag{8.60d}$$

Up to this point, we have done nothing more than to express the planar homogeneous solutions (8.53) in terms of the two sets of matrix-vector equations (8.58) and (8.59).

Suppose we want to determine an orbit transfer from an arbitrary initial state, $\delta\mathbf{r}_0$ and $\delta\mathbf{v}_0$, to the origin of the CW frame with a prescribed transfer time t. Such a maneuver might represent an orbital insertion into a "slot" in a target orbit or orbital rendezvous with another satellite such as a space station. Consider first the relative position solution Eq. (8.58). For arbitrary initial position and velocity $\delta\mathbf{r}_0$ and $\delta\mathbf{v}_0$, there is no guarantee of rendezvous [i.e., $\delta\mathbf{r}(t) = \mathbf{0}$] at time t. However, it is possible to determine the *required* initial relative velocity for rendezvous by solving Eq. (8.58) with $\delta\mathbf{r}(t) = \mathbf{0}$

$$\delta\mathbf{v}_0^{\text{req}} = -\mathbf{\Phi}_{12}^{-1}\mathbf{\Phi}_{11}\delta\mathbf{r}_0 \tag{8.61}$$

where $\mathbf{\Phi}_{12}^{-1}$ is the inverse of the matrix shown in Eq. (8.60b). A satellite with initial relative velocity defined by Eq. (8.61) will reach the origin at the prescribed time t. However, the satellite's *actual* initial relative velocity is $\delta\mathbf{v}_0$. Therefore, the following impulsive maneuver is required at time $t = 0$:

$$\Delta\mathbf{v}_0 = \delta\mathbf{v}_0^{\text{req}} - \delta\mathbf{v}_0 \tag{8.62}$$

Equation (8.62) determines the radial and along-track components of the initial impulsive burn for rendezvous. It is important to reiterate that the *required* initial velocity is computed using Eq. (8.61) for a specified transfer time t, and that matrices Φ_{11} and Φ_{12} depend solely on time t and angular velocity of the reference orbit n; see Eqs. (8.60a) and (8.60b).

After the first impulse has established initial relative velocity δv_0^{req}, the satellite coasts and reaches the origin $\delta r(t) = 0$ at time t. Equation (8.59) determines the satellite's terminal relative velocity when it arrives at the origin:

$$\delta v(t) = \Phi_{21}\delta r_0 + \Phi_{22}\delta v_0^{req}$$

Because we desire zero relative velocity (i.e., rendezvous) at time t, a second impulsive maneuver is required:

$$\Delta v_f = -\delta v(t) \tag{8.63}$$

This second impulse simply cancels out the satellite's residual relative velocity when it arrives at the origin or target. The magnitude of the total velocity increment for the two-impulse maneuver is

$$\Delta v = \|\Delta v_0\| + \|\Delta v_f\| \tag{8.64}$$

Using the total Δv in the rocket equation (6.13) determines the total propellant mass required for rendezvous.

A final note is in order. Computing the initial relative velocity for rendezvous via Eq. (8.61) requires the inverse of matrix Φ_{12}. Referring to Eq. (8.60b), we see that when the transfer time is an *exact* multiple of the period of the reference orbit (i.e., $t = kT_{period}$), the harmonic terms are $\sin nt = 0$ and $\cos nt = 1$ and thus matrix Φ_{12} becomes

$$\Phi_{12}\left(kT_{period}\right) = \begin{bmatrix} 0 & 0 \\ 0 & -3kT_{period} \end{bmatrix}$$

Under these conditions, the matrix Φ_{12} is singular (its determinant is zero) and its inverse does not exist. However, if one avoids transfer times that are exact multiples of T_{period}, then it is possible to invert matrix Φ_{12}. For example, suppose we have a reference orbit with period $T_{period} = 95$ min ($= 5,700$ s) and $n = 2\pi/T_{period} = 0.001102$ rad/s. For transfer time $t = 2.99T_{period}$, the matrix Φ_{12} is

$$\Phi_{12}\left(2.99T_{period}\right) = \begin{bmatrix} -56.9625 & 3.5802 \\ -3.5802 & -51,356.85 \end{bmatrix}$$

In this case, Φ_{12} is nonsingular and therefore its inverse exists. The following examples illustrate orbital rendezvous scenarios.

Example 8.4 Figure 8.7 shows a scenario where a "chaser" satellite is 20 km above and 40 km ahead of a target in a 300-km altitude circular orbit. Determine the satellite's necessary velocity components at this instant so that it performs a rendezvous with the target in one-quarter of the period of the reference orbit.

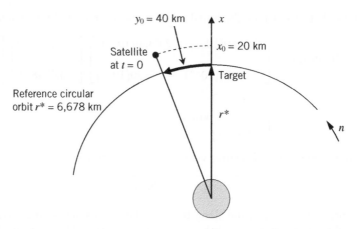

Figure 8.7 Satellite initial position for orbital rendezvous (Example 8.4; not to scale).

Equation (8.61) determines the initial relative velocity *required* for rendezvous

$$\delta \mathbf{v}_0^{\text{req}} = -\Phi_{12}^{-1} \Phi_{11} \delta \mathbf{r}_0$$

where the initial relative position vector is

$$\delta \mathbf{r}_0 = \begin{bmatrix} x_0 \\ y_0 \end{bmatrix} = \begin{bmatrix} 20 \\ 40 \end{bmatrix} \text{ km}$$

and the matrices Φ_{11} and Φ_{12} are defined by Eqs. (8.60a) and (8.60b), respectively

$$\Phi_{11} = \begin{bmatrix} -3\cos nt + 4 & 0 \\ 6(\sin nt - nt) & 1 \end{bmatrix}$$

$$\Phi_{12} = \begin{bmatrix} \dfrac{\sin nt}{n} & \dfrac{-2}{n}(\cos nt - 1) \\ \dfrac{2}{n}(\cos nt - 1) & \dfrac{4}{n}\sin nt - 3t \end{bmatrix}$$

The angular velocity of the reference orbit ($r^* = 6{,}678$ km) is $n = \sqrt{\mu/r^{*3}} = 1.1569(10^{-3})$ rad/s. The orbital period is $T_{\text{period}} = 2\pi/n = 5{,}431.01$ s ($= 90.52$ min), so the desired rendezvous time is $t = T_{\text{period}}/4 = 1{,}357.75$ s ($= 22.63$ min). Evaluating the matrices using these values for n and t, we obtain

$$\Phi_{11} = \begin{bmatrix} 4 & 0 \\ -3.4248 & 1 \end{bmatrix}$$

$$\Phi_{12} = \begin{bmatrix} 864.3726 & 1{,}728.7451 \\ -1{,}728.7451 & -615.7695 \end{bmatrix}$$

The inverse of Φ_{12} is

$$\Phi_{12}^{-1} = \begin{bmatrix} -2.5069(10^{-4}) & -7.0380(10^{-4}) \\ 7.0380(10^{-4}) & 3.5190(10^{-4}) \end{bmatrix}$$

We use Eq. (8.61) to determine the initial relative velocity required for rendezvous

$$\delta v_0^{req} = -\Phi_{12}^{-1}\Phi_{11}\delta r_0 = \begin{bmatrix} 2.5069(10^{-4}) & 7.0380(10^{-4}) \\ -7.0380(10^{-4}) & -3.5190(10^{-4}) \end{bmatrix} \begin{bmatrix} 4 & 0 \\ -3.4248 & 1 \end{bmatrix} \begin{bmatrix} 20 \\ 40 \end{bmatrix}$$

$$= \begin{bmatrix} 0 \\ -0.0463 \end{bmatrix} \text{km/s}$$

Therefore, the required initial relative velocity components are

$$\boxed{\dot{x}_0 = 0, \quad \dot{y}_0 = -0.0463 \text{ km/s}}$$

Figure 8.8 presents the radial and along-track displacements of the orbital rendezvous using the satellite's initial states $x_0 = 20$ km, $y_0 = 40$ km, $\dot{x}_0 = 0$, and $\dot{y}_0 = -0.0463$ km/s and rendezvous time $t = 1{,}357.75$ s ($= 22.63$ min). Note that we have added "time-to-go" (t_{go}) tags to the relative position coordinates (t_{go} is the time remaining until rendezvous). Figure 8.8 shows that the satellite is initially at apogee with a maximum radial relative displacement of 20 km, zero radial velocity, and negative along-track velocity. Figure 8.8 also implies that the chaser satellite approaches the target "from above" along the positive radial direction during the terminal phase of its rendezvous. The final velocity components of the orbital rendezvous are $\dot{x}_f = -0.0231$ km/s (-23.1 m/s) and $\dot{y}_f = 0$. Hence, an onboard rocket must provide a purely radial $\Delta v = 23.1$ m/s when the satellite reaches the origin at 22.63 min to complete the rendezvous maneuver. The reader should compare Figure 8.8 with the dashed-line relative-motion trajectory shown in Figure 8.5 (Example 8.2). Figure 8.8 is essentially the final one-quarter of the "elliptical trace" (Figure 8.5) as the satellite moves from its apogee to the origin.

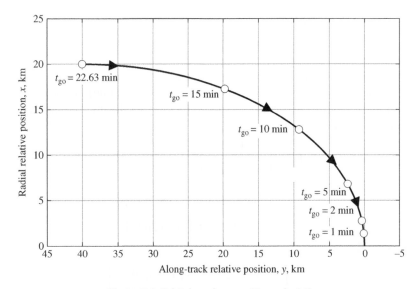

Figure 8.8 Orbital rendezvous (Example 8.4).

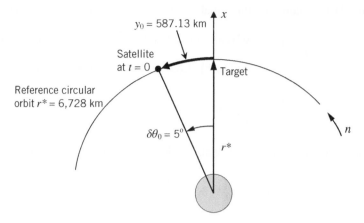

Figure 8.9 Satellite initial position for orbital rendezvous (Example 8.5; not to scale).

Example 8.5 An 872-kg satellite (the "chaser") is in a 350-km circular orbit. A target satellite is in the same circular orbit. The chaser satellite "leads" the target by 5° in the along-track (y) direction (see Figure 8.9). Determine the chaser satellite's total Δv and propellant mass required to perform a rendezvous maneuver with the target satellite. The transfer time is 250 min and the specific impulse of the chaser's hydrazine thruster is 225 s.

Angular velocity of the reference orbit ($r^* = 6{,}728$ km) is $n = \sqrt{\mu/r^{*3}} = 1.1440(10^{-3})$ rad/s. Because the chaser satellite is initially in the reference circular orbit, its relative radial position is zero ($x_0 = 0$) and its velocity components are zero ($\dot{x}_0 = \dot{y}_0 = 0$). The chaser satellite is 5° (0.0873 rad) *ahead* of the target in the direction of orbital motion, and therefore the initial along-track position is $y_0 = r^* \delta\theta_0 = 587.13$ km. We can use Eq. (8.61) to compute the initial relative velocity *required* for rendezvous

$$\delta \mathbf{v}_0^{\text{req}} = -\Phi_{12}^{-1}\Phi_{11}\delta\mathbf{r}_0$$

where the initial relation position vector is

$$\delta\mathbf{r}_0 = \begin{bmatrix} x_0 \\ y_0 \end{bmatrix} = \begin{bmatrix} 0 \\ 587.13 \end{bmatrix} \text{ km}$$

Using transfer time $t = 250$ min $= 15{,}000$ s in Eqs. (8.60a) and (8.60b), the matrices are

$$\Phi_{11} = \begin{bmatrix} 4.3538 & 0 \\ -108.9214 & 1 \end{bmatrix}$$

$$\Phi_{12} = \begin{bmatrix} -867.9973 & 1{,}954.3881 \\ -1{,}954.3881 & -48{,}471.9892 \end{bmatrix}$$

and the inverse of Φ_{12} is

$$\Phi_{12}^{-1} = \begin{bmatrix} -0.00106 & -4.2586(10^{-5}) \\ 4.2586(10^{-5}) & -1.8913(10^{-5}) \end{bmatrix}$$

We use Eq. (8.61) to compute the initial relative velocity required for rendezvous:

$$\delta v_0^{req} = -\Phi_{12}^{-1}\Phi_{11}\delta r_0 = \begin{bmatrix} 0.00106 & 4.2586(10^{-5}) \\ -4.2586(10^{-5}) & 1.8913(10^{-5}) \end{bmatrix} \begin{bmatrix} 4.3538 & 0 \\ -108.9214 & 1 \end{bmatrix} \begin{bmatrix} 0 \\ 587.13 \end{bmatrix}$$

$$= \begin{bmatrix} 0.0250 \\ 0.0111 \end{bmatrix} \text{km/s}$$

Because the chaser satellite's initial relative velocity is zero, the first impulse is

$$\Delta v_0 = \delta v_0^{req} - \delta v_0 = \begin{bmatrix} 0.0250 \\ 0.0111 \end{bmatrix} \text{km/s}$$

The chaser satellite's terminal relative velocity is

$$\delta v(t) = \Phi_{21}\delta r_0 + \Phi_{22}\delta v_0^{req}$$

where the two matrices are defined by Eqs. (8.60c) and (8.60d). Evaluating these matrices at time $t = 15{,}000$ s, we obtain

$$\Phi_{21} = \begin{bmatrix} -0.003408 & 0 \\ -0.007674 & 0 \end{bmatrix}$$

and

$$\Phi_{22} = \begin{bmatrix} -0.1179 & -1.9860 \\ 1.9860 & -3.4718 \end{bmatrix}$$

Therefore, the terminal velocity at time $t = 15{,}000$ s is

$$\delta v(15{,}000) = \Phi_{21}\delta r_0 + \Phi_{22}\delta v_0^{req}$$

$$= \begin{bmatrix} -0.003408 & 0 \\ -0.007674 & 0 \end{bmatrix} \begin{bmatrix} 0 \\ 587.13 \end{bmatrix} + \begin{bmatrix} -0.1179 & -1.9860 \\ 1.9860 & -3.4718 \end{bmatrix} \begin{bmatrix} 0.0250 \\ 0.0111 \end{bmatrix}$$

$$= \begin{bmatrix} -0.0250 \\ 0.0111 \end{bmatrix} \text{km/s}$$

The second impulse must cancel out the terminal relative velocity

$$\Delta v_f = -\delta v(15{,}000) = \begin{bmatrix} 0.0250 \\ -0.0111 \end{bmatrix} \text{km/s}$$

The vector norms of the two impulses are

$$\|\Delta v_0\| = 0.02736 \text{ km/s} \quad (= 27.36 \text{ m/s})$$
$$\|\Delta v_f\| = 0.02736 \text{ km/s} \quad (= 27.36 \text{ m/s})$$

Therefore, the total velocity increment for rendezvous is $\Delta v = \|\Delta\mathbf{v}_0\| + \|\Delta\mathbf{v}_f\|$ = $\boxed{54.72\,\text{m/s}}$.

Finally, we can use a variation of the rocket equation, Eq. (6.13), to determine the propellant mass required for the rendezvous maneuver. Using exhaust speed $g_0 I_{sp} = 2,206.5$ m/s, we obtain

$$m_p = m_0\left[1 - \exp\left(\frac{-\Delta v}{g_0 I_{sp}}\right)\right] = 872\,\text{kg}[1 - \exp(-54.72/2,206.5)] = \boxed{21.36\,\text{kg}}$$

Example 8.6 Consider again the rendezvous scenario in Example 8.5. Plot the relative position coordinates of the chaser satellite during rendezvous for transfer times of 250 and 274 min.

As with the previous examples, we can use Eqs. (8.53a) and (8.53b) to obtain the position coordinates (x,y) of the chaser relative to the target for any arbitrary time t

$$x(t) = \left(-3x_0 - \frac{2\dot{y}_0}{n}\right)\cos nt + \frac{\dot{x}_0}{n}\sin nt + \left(4x_0 + \frac{2\dot{y}_0}{n}\right)$$

$$y(t) = \left(6x_0 + \frac{4\dot{y}_0}{n}\right)\sin nt + \frac{2\dot{x}_0}{n}\cos nt - (6nx_0 + 3\dot{y}_0)t + \left(y_0 - \frac{2\dot{x}_0}{n}\right)$$

The reader should note that the required initial relative velocity at the beginning of the rendezvous maneuver is

$$\begin{bmatrix}\dot{x}_0 \\ \dot{y}_0\end{bmatrix}^{\text{req}} = \delta\mathbf{v}_0^{\text{req}} = -\Phi_{12}^{-1}\Phi_{11}\delta\mathbf{r}_0$$

This initial velocity is the result of the first impulsive burn. For a 250-min transfer, the *required* initial velocity is $\delta\mathbf{v}_0^{\text{req}} = [0.0250 \quad 0.0111]^T$ km/s, and therefore the magnitude of the first impulse is 27.36 m/s (see Example 8.5 for details).

Figure 8.10 shows the radial and along-track coordinates of the chaser satellite during the 250-min rendezvous maneuver. The first impulse $\Delta\mathbf{v}_0$ shown in Figure 8.10 has positive x (radial) and y (along-track) components. Hence, the chaser satellite would need to be rotated prior to the first impulse so that its thrust vector is at an elevation angle $\tan^{-1}(0.025/0.011) = 66°$ above the local horizon in the direction of motion. Because the first impulse produces positive radial and along-track relative velocity components, the chaser's orbit has an apogee greater than r^* and the satellite initially drifts above the reference orbit. When the chaser is above the reference orbit (i.e., $x > 0$), it drifts "back" towards the target (the chaser has a *very* small positive $\dot{y}$ component as it crosses the reference orbit). After completing nearly three "periodic paths" in the relative (x,y) frame, the chaser reaches the target (origin) and the symmetric second impulse $\Delta\mathbf{v}_f$ cancels out the relative velocity components to complete the rendezvous. Because the period is

$$T_{\text{period}} = \frac{2\pi}{n} = 5,492.1\,\text{s}\,(=91.5\,\text{min})$$

the number of orbital revolutions is $t/T_{\text{period}} = 250\,\text{min}/91.5\,\text{min} = 2.73$. Figure 8.10 shows that the chaser satellite completes two "periodic paths" in the CW frame but does not quite complete the third path before reaching the target.

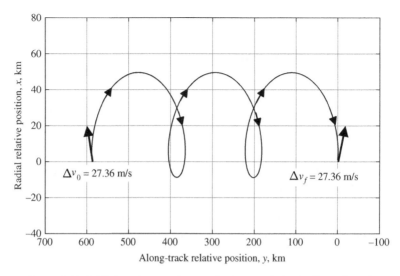

Figure 8.10 Orbital rendezvous for transfer time t = 250 min (Example 8.6).

We repeat the rendezvous analysis for a 274-min transfer. Using Eq. (8.61) with t = 274 min = 16,440 s, the initial relative velocity required for rendezvous is

$$\delta v_0^{req} = -\Phi_{12}^{-1}\Phi_{11}\delta r_0 = \begin{bmatrix} 0.0275 & 8.4121(10^{-7}) \\ -8.4121(10^{-7}) & 2.0216(10^{-5}) \end{bmatrix}\begin{bmatrix} 1.0026 & 0 \\ -113.0973 & 1 \end{bmatrix}\begin{bmatrix} 0 \\ 587.13 \end{bmatrix}$$

$$= \begin{bmatrix} 0.00049 \\ 0.01187 \end{bmatrix} km/s$$

The satellite's terminal velocity at time t = 16,440 s is

$$\delta v(16,440) = \Phi_{21}\delta r_0 + \Phi_{22}\delta v_0^{req}$$

$$= \begin{bmatrix} -0.000143 & 0 \\ -5.94(10^{-6}) & 0 \end{bmatrix}\begin{bmatrix} 0 \\ 587.13 \end{bmatrix} + \begin{bmatrix} 0.9991 & -0.0832 \\ 0.0832 & 0.9965 \end{bmatrix}\begin{bmatrix} 0.00049 \\ 0.01187 \end{bmatrix}$$

$$= \begin{bmatrix} -0.00049 \\ 0.01187 \end{bmatrix} km/s$$

Figure 8.11 shows the radial and along-track coordinates of the chaser satellite during the 274-min rendezvous maneuver. Here the first impulse is essentially along the circumferential (y) direction in order to raise the chaser's apogee. Figure 8.11 shows that the chaser always remains above the reference orbit as it drifts back and completes three "periodic paths" before it reaches the target (origin). The second impulse is essentially "anti-circumferential" and cancels out the terminal relative velocity. Because the period of the reference orbit is 91.5 min, the number of orbital revolutions is 274/91.5 = 2.99 (essentially three orbits).

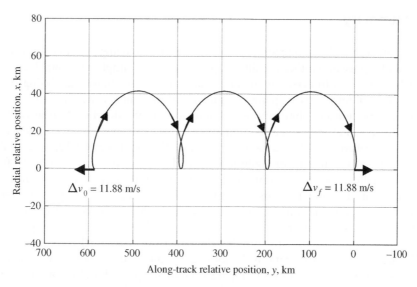

Figure 8.11 Orbital rendezvous for transfer time $t = 274$ min (Example 8.6).

Finally, it is instructive to characterize the total Δv as a function of transfer time for this particular rendezvous scenario. It is possible to repeat the $\Delta \mathbf{v}_0$ and $\Delta \mathbf{v}_f$ calculations for a wide range of transfer times by using MATLAB and a simple "looping" script. Figure 8.12 shows the total Δv (the sum of $\|\Delta \mathbf{v}_0\|$ and $\|\Delta \mathbf{v}_f\|$) plotted against the number of revolutions of the reference orbit (note that the longest transfer time shown in Figure 8.12 is $8T_{period} = 732$ min or 12.2 h). Figure 8.12 clearly shows that total Δv reaches a (local) minimum value when the transfer time is a multiple of the target orbit's period. Furthermore, increasing the number of target-orbit revolutions decreases the total Δv. The worst possible scenario is to use a transfer time that is approximately half-way between a multiple of the reference period (e.g., $t = 6.5T_{period}$). Figure 8.12 also shows the 250 and 274-min rendezvous maneuvers.

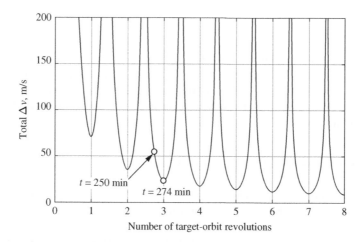

Figure 8.12 Total Δv for rendezvous vs. number of revolutions of the target orbit (Example 8.6).

8.5 Summary

This chapter dealt with satellite motion relative to a target moving in a circular orbit. The target could represent another satellite, a space station, or simply an orbital location or "slot" where we wish to insert the maneuvering satellite. Accurately analyzing relative motion is critical for in-space operations such as orbital rendezvous between two satellites or repositioning a satellite in a new orbital location. We developed linear ODEs for the satellite's position and velocity coordinates relative to the moving target. These linear equations are called the CW equations and because they are linear ODEs, we are able to obtain analytical or closed-form solutions. We chose a cylindrical coordinate frame where the satellite's position deviations from the target are in radial, along-track, and cross-track (out-of-plane) directions. As with any linearized system of equations, retaining adequate accuracy requires that deviations from the reference trajectory remain "small" at all times. However, due to the nature of an inverse-square gravity field, the linearized CW equations only require that the radial and cross-track deviations remain "small"; large along-track deviations do not degrade the accuracy of the linear solution. Finally, we ended this chapter by developing a systematic procedure for obtaining a two-impulse rendezvous maneuver. In general, the total Δv of the two-impulse rendezvous maneuver will depend on the satellite's initial position and velocity relative to the target and the allocated transfer time.

References

1 Hill, G.W., "Researches in Lunar Theory," *American Journal of Mathematics*, Vol. **1**, 1878, pp. 5–26.
2 Clohessy, W.H. and Wiltshire, R.S., "Terminal Guidance System for Satellite Rendezvous," *Journal of the Aerospace Sciences*, Vol. **27**, No. 9, 1960, pp. 653–658.

Problems

Conceptual Problems

Problems 8.1–8.8 involve relative satellite motion in the Clohessy-Wiltshire (CW) frame. Each problem specifies the reference (target) circular orbital radius r^*, the initial CW-frame position coordinates (x_0, y_0, and z_0), and the initial CW-frame velocity components ($\dot{x}_0$, $\dot{y}_0$, and $\dot{z}_0$). Determine the satellite's position and velocity coordinates relative to the CW frame at end-time t_f.

8.1 Reference radius r^* = 6,678 km, $x_0 = -2$ km, $y_0 = -500$ km, $z_0 = 0$, $\dot{x}_0 = 0$, $\dot{y}_0 = 0.008$ km/s, $\dot{z}_0 = 0$. End-time $t_f = 140$ min.

8.2 Reference radius r^* = 6,678 km, $x_0 = 5$ km, $y_0 = 200$ km, $z_0 = 0$, $\dot{x}_0 = 0$, $\dot{y}_0 = -0.01$ km/s, $\dot{z}_0 = 0$. End-time $t_f = 140$ min.

8.3 Reference radius $r^* = 6{,}678$ km, $x_0 = 20$ km, $y_0 = -100$ km, $z_0 = 0$, $\dot{x}_0 = -0.002$ km/s, $\dot{y}_0 = -0.03$ km/s, $\dot{z}_0 = 0$. End-time $t_f = 3T_{\text{period}}$

8.4   Reference radius $r^* = 6{,}678$ km, $x_0 = 20$ km, $y_0 = -100$ km, $z_0 = 5$ km, $\dot{x}_0 = -0.002$ km/s, $\dot{y}_0 = -0.03$ km/s, $\dot{z}_0 = 0.004$ km/s. End-time $t_f = 3T_{\text{period}}$

8.5   Reference radius $r^* = 20{,}500$ km, $x_0 = 80$ km, $y_0 = 600$ km, $z_0 = 0$, $\dot{x}_0 = -0.001$ km/s, $\dot{y}_0 = -0.026$ km/s, $\dot{z}_0 = 0$. End-time $t_f = 1.5T_{\text{period}}$

8.6   Reference radius $r^* = 20{,}500$ km, $x_0 = -80$ km, $y_0 = 0$, $z_0 = 20$ km, $\dot{x}_0 = 0$, $\dot{y}_0 = 0.027$ km/s, $\dot{z}_0 = -0.005$ km/s. End-time $t_f = 1.5T_{\text{period}}$

8.7   Reference radius $r^* = 20{,}500$ km, $x_0 = -4$ km, $y_0 = 0$, $z_0 = 2$ km, $\dot{x}_0 = -0.002$ km/s, $\dot{y}_0 = 0.0013$ km/s, $\dot{z}_0 = 0.001$ km/s. End-time $t_f = 30$ min.

8.8   Reference radius $r^* = 20{,}500$ km, $x_0 = 0$, $y_0 = 20$ km, $z_0 = 0$, $\dot{x}_0 = 0.002$ km/s, $\dot{y}_0 = 0.001$ km/s, $\dot{z}_0 = 0.001$ km/s. End-time $t_f = 30$ min.

   Problems 8.9–8.14 involve the scenario depicted in Figure P8.9. Two satellites are in coplanar orbits about the Earth. At the instant shown, Satellite A is 10 km directly above Satellite B. The *inertial* velocity of Satellite A is 7.71 km/s and its flight-path angle is zero at this instant. Satellite B is in a circular orbit with a radius of 6,678 km.

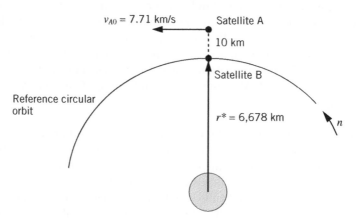

Figure P8.9

8.9   Determine the initial conditions of Satellite A relative to a CW frame that moves with Satellite B.

8.10   Determine the position and velocity coordinates of Satellite A relative to the CW frame at time $t = T_{\text{period}}$.

8.11   Analytically determine the perigee and apogee radii of Satellite A.

8.12   Verify the solutions to Problems 8.10 and 8.11 by plotting the time histories of the relative position and velocity coordinates of Satellite A for one orbital period. In addition, present the relative motion of Satellite A by plotting radial displacement (x) vs. along-track displacement (y).

8.13 Compute the two impulsive Δv vectors such that Satellite A performs an orbital rendezvous with Satellite B in one-quarter period of the reference orbit. Present the orbital rendezvous by plotting x vs. y.

8.14 Repeat Problem 8.13 for a rendezvous time of 60 min (about two-thirds of a period). Present the orbital rendezvous by plotting x vs. y.

8.15 A satellite's initial states relative to the CW frame are $x_0 = 3$ km, $y_0 = 0$, $z_0 = 0$, $\dot{x}_0 = 0.002$ km/s, $\dot{y}_0 = -0.005$ km/s, and $\dot{z}_0 = 0$. The reference circular orbit has a period of 91.3 min. Determine the epoch times when the satellite next reaches perigee and apogee. Which apse passage is first?

MATLAB Problems

8.16 Write an M-file that will determine the propagated relative position and velocity coordinates of a satellite with respect to the origin of the CW frame. The inputs to the M-file should be propagation time t (in min), radius of the reference circular orbit r^* (in km), the initial radial, along-track, and cross-track positions in the CW frame (x_0, y_0, and z_0, in km), and the initial CW-frame velocity components ($\dot{x}_0$, $\dot{y}_0$, and $\dot{z}_0$, in km/s). The output of the M-file should be the three position and velocity coordinates at the end time t as expressed in the CW frame (in km and km/s). Test your M-file by solving Examples 8.1 and 8.2.

8.17 Write an M-file that will plot the satellite's relative motion in the CW frame. The M-file should create two plots: (a) along-track displacement (y) on the horizontal axis and radial displacement (x) on the vertical axis; and (b) along-track displacement (y) on the horizontal axis and cross-track displacement (z) on the vertical axis. The inputs to the M-file should be propagation time t (in min), radius of the reference circular orbit r^* (in km), the initial radial, along-track, and cross-track positions in the CW frame (x_0, y_0, and z_0, in km), and the initial velocity components ($\dot{x}_0$, $\dot{y}_0$, and $\dot{z}_0$, in km/s). This M-file should make use of the relative-motion solver M-file described in Problem 8.16. Test your M-file by recreating Figures 8.5 and 8.6.

8.18 Write an M-file that will compute the two impulsive Δv vectors for a planar orbital rendezvous. The inputs to the M-file should be rendezvous time t_f (in min), radius of the reference circular orbit r^* (in km), and the initial position and velocity components relative to the CW frame (x_0, y_0, $\dot{x}_0$, and $\dot{y}_0$) in kilometers and kilometers per second, respectively. The output of the M-file should be the two impulsive Δv vectors (radial and along-track components) that are applied at times $t = 0$ and $t = t_f$.

Mission Applications

8.19 Satellite operators want to design a rendezvous strategy for a satellite that is displaced from a targeted "slot" in a 500-km altitude low-Earth orbit (LEO). The satellite's initial relative position and velocity components in the CW frame are $x_0 = -2$ km, $y_0 = 1.4$ km, $z_0 = 0.8$ km, $\dot{x}_0 = 0.001$ km/s, $\dot{y}_0 = 0.003$ km/s, and $\dot{z}_0 = 0.002$ km/s. The satellite operators decide on a two-phase maneuver strategy: (1) command an out-of-plane impulse (Δv) that will correct the orbital plane; and (2) command two impulses for a planar rendezvous with the target orbit slot.
  a) Determine the timing and magnitude of the out-of-plane (z axis) impulse; i.e., find the time t_1 when the satellite crosses the orbital plane of the reference LEO.
  b) Determine the two impulsive vectors required for a rendezvous maneuver so that the satellite reaches the origin at time $t_2 = T_{period}$, i.e., the orbital period of the reference LEO (hence the transit time for the planar rendezvous phase is $t_2 - t_1$).

8.20 Satellite operators want to change the longitude of a communication satellite in geostationary-equatorial orbit by 30° E. Determine the total magnitude of the two Δv impulses required for orbital-rendezvous maneuvers that take 1, 7, 14, 30, and 60 days. Is there a dramatic difference in total Δv for this range of maneuver times? Which maneuver time appears to provide a good tradeoff between total Δv and transfer time?

8.21 A geocentric satellite is in a 300-km altitude circular orbit. Satellite operators want to reposition the satellite so that it occupies a location in the same circular orbit that is 60° "behind" its current position (see Figure P8.21). The satellite is equipped with thrusters that can only provide impulses along the horizontal (y) direction, and its propellant budget limits the total Δv to 80 m/s for the rendezvous maneuver. Determine the shortest possible rendezvous time and the corresponding total Δv for the 60° repositioning maneuver.

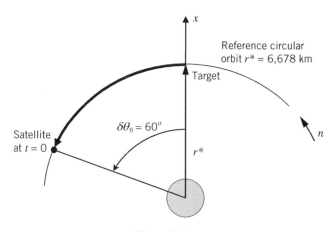

Figure P8.21

8.22 Consider again the orbital scenario depicted in Figure P8.9 and Problems 8.9–8.14. Perform a trade study by computing the total magnitude of the two impulsive $\Delta \mathbf{v}$ vectors for orbital rendezvous for a range of rendezvous times (the rendezvous M-file from Problem 8.18 will be very useful here). Let rendezvous time range from 5 to 85 min (nearly one orbit). Plot total Δv vs. rendezvous time to determine the optimal time. In addition, plot the optimal orbital rendezvous (radial distance x vs. along-track distance y).

9

Low-Thrust Transfers

9.1 Introduction

All orbital maneuvers presented in Chapters 7 and 8 utilized impulsive velocity changes. Our fundamental assumption is that firing a high-thrust chemical rocket will change velocity (and hence the orbital elements) instantly with no change in position (the reader may wish to review Section 6.7). A very different propulsion mode is *low-thrust propulsion*, where charged particles or plasmas are accelerated using electrostatic or electromagnetic forces and ejected at very high exhaust velocities. Ion and Hall-effect thrusters are two examples of electric propulsion (EP) devices currently being used for space missions. The high EP exhaust speeds translate to specific impulses (I_{sp}) that are nearly 10 times greater than conventional chemical rockets. The rocket equation, Eq. (6.13), shows that increasing I_{sp} significantly reduces the propellant mass. However, because the EP mass-flow rate is extremely small, the thrust magnitude is very low. Consequently, an EP device must operate continuously so that the very low thrust acceleration will produce a sizeable velocity (or orbital) change when integrated over a long time (often days or months). We will present the fundamental equation that relates EP thrust to electric power and I_{sp}. The interested reader may consult Sutton and Biblarz [1; pp. 622–656] for additional details regarding electric propulsion devices.

Analyzing low-thrust trajectories is challenging because Keplerian two-body motion is no longer valid and hence the orbital elements are continuously changing over time. Recall that in Section 5.4 we developed Gauss' form of the variation of parameters method that describes how the orbital elements change due to perturbing accelerations. We will develop analytical solutions for special low-thrust transfers by using Gauss' variation of parameters and treating the EP thrust as a perturbation.

Figure 9.1 shows a continuous-thrust orbit transfer between circular orbits. Here a small propulsive force continually acts on the satellite to slowly increase the altitude (and energy) of the inner orbit until it reaches the target orbit. Hence, the orbit transfer is an unwinding spiral trajectory where each orbital revolution is a nearly circular orbit. An onboard electric propulsion system (such as an ion or Hall-effect thruster) provides the small, continuous thrust magnitude for the low-thrust transfer. Characterizing a low-thrust transfer is more difficult than analyzing an impulsive orbit transfer because the orbital elements are continuously changing and therefore we cannot use the constants of motion (such as specific energy and angular momentum) to define the

Space Flight Dynamics, First Edition. Craig A. Kluever.
© 2018 John Wiley & Sons Ltd. Published 2018 by John Wiley & Sons Ltd.
Companion website: www.wiley.com/go/Kluever/spaceflightmechanics

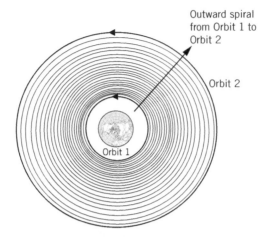

Figure 9.1 Continuous-thrust, coplanar orbit transfer.

transfer orbit between the two circles. We will develop analytical expressions for low-thrust transfers, such as coplanar circle-to-circle transfers and pure inclination-change maneuvers.

9.2 Electric Propulsion Fundamentals

Chapter 6 presented the fundamental relationships for rocket propulsion, and these basic relationships still hold for low-thrust EP devices. Let us repeat the fundamental equation that expresses the thrust force T in terms of effective exhaust velocity, v_{eff}, and mass-flow rate, $\dot{m}$:

$$T = \dot{m} v_{\text{eff}} = \dot{m} c \tag{9.1}$$

In this chapter, we will use symbol c to denote the EP thruster's effective exhaust velocity. Recall that Eq. (6.4) defines effective exhaust velocity:

$$c = g_0 I_{\text{sp}} \tag{9.2}$$

EP devices convert the electric power input (typically from solar cells) to the output power of the jet exhaust (i.e., accelerated ions or plasma). For constant effective exhaust velocity, the output jet power is the time-rate of the exhaust jet's kinetic energy:

$$P_{\text{out}} = \frac{1}{2} \dot{m} c^2 \tag{9.3}$$

EP thruster efficiency η is the ratio of the output power P_{out} (in the exhaust jet) to the input power P_{in}:

$$\eta = \frac{P_{\text{out}}}{P_{\text{in}}} = \frac{\frac{1}{2} \dot{m} c^2}{P_{\text{in}}} \tag{9.4}$$

Substituting Eq. (9.1) in the numerator of Eq. (9.4), we obtain

$$\eta = \frac{\frac{1}{2} T c}{P_{\text{in}}} \tag{9.5}$$

Substituting $c = g_0 I_{sp}$ in Eq. (9.5) and solving for thrust T yields

$$T = \frac{2\eta P_{in}}{g_0 I_{sp}} \tag{9.6}$$

Equation (9.6) is the fundamental thrust equation for an electric propulsion device. Thruster efficiency accounts for the energy losses associated with converting electric input power to jet power; typically, experimental trials quantify η. One odd feature of Eq. (9.6) is that increasing I_{sp} (or, increasing the exhaust speed of the accelerated particles) decreases the magnitude of the thrust for constant input power and constant thruster efficiency. Substituting Eq. (9.1) into (9.6) and solving for mass-flow rate yields

$$\dot{m} = \frac{2\eta P_{in}}{c^2} \tag{9.7}$$

Therefore, increasing I_{sp} (i.e., increasing c) will also reduce the mass-flow rate when efficiency and input power are constant.

Three basic categories of EP devices exist: (1) electrothermal, (2) electrostatic, and (3) electromagnetic thrusters. In this textbook, we will focus on two EP devices with proven flight performance as primary propulsion for orbit transfers. Ion thrusters are electrostatic devices that accelerate charged particles and Hall-effect thrusters are electromagnetic devices that accelerate a gas heated to a plasma state. Both thrusters typically use xenon as the propellant because it is the stable inert gas with the highest molecular mass. Ion and Hall-effect thrusters create a thrust force through the momentum flux of the exhausted particles. Because the exiting particles have very low mass (with very high velocities), the resulting thrust magnitude is extremely small. Table 9.1 summarizes the performance of two representative EP thrusters, NASA's NSTAR ion thruster and the Aerojet BPT-4000 Hall-effect thruster [2,3]. Note that the specific impulses of EP thrusters are four to ten times greater than the I_{sp} for chemical-propulsion rockets. The reader should also note that the thrust magnitude of a single thruster is less than 0.1 N for low input power and between 0.1 N and 0.25 N at the highest input power setting. Clearly, these EP devices produce a very low thrust force and consequently a thruster must operate continuously for months in order to produce a significant change in a spacecraft's orbit.

The basic "rocket equation" developed in Chapter 6 is still valid for low-thrust devices. Recall that Eq. (6.13) is a form of the rocket equation that allows the calculation of the propellant mass m_p required to provide the velocity change Δv

$$m_p = m_0 \left[1 - \exp\left(\frac{-\Delta v}{g_0 I_{sp}}\right) \right] \tag{9.8}$$

Table 9.1 Performance specifications for two electric propulsion thrusters.

Thruster	Specific impulse (s)	Input power (kW)	Thruster efficiency	Thrust (mN)
NSTAR ion	1,972–3,120	0.47–2.29	0.424–0.617	20.6–92.4
BPT-4000 Hall	1,220–2,150	1.0–4.5	0.473–0.595	79–254

Source: Adapted from Refs [2] and [3].

where m_0 is the satellite's initial mass. For in-space impulsive maneuvers, Δv is simply the instantaneous change in the velocity vector determined from the orbital elements of the initial orbit and transfer orbit. For low-thrust transfers, the calculation of Δv is not so straightforward. Equation (9.8) shows that increasing I_{sp} (i.e., increasing the jet exhaust speed) reduces the propellant mass. Let us demonstrate the potential mass savings of using low-thrust EP by considering a satellite with initial mass $m_0 = 2,000$ kg and a maneuver that calls for $\Delta v = 3$ km/s (3,000 m/s). For a chemical-propulsion rocket with $I_{sp} = 310$ s, the propellant mass required for $\Delta v = 3,000$ m/s is 1,254 kg, which is 63% of the initial mass of the satellite. Therefore, relatively little payload mass remains after the impulsive burn. For an ion thruster with $I_{sp} = 3,100$ s, the required xenon propellant mass for $\Delta v = 3,000$ m/s is 188 kg! Hence, the xenon propellant mass is only 9% of the initial mass, and a much larger payload mass is delivered after completing the orbit transfer. We should note that this mass comparison is not as straightforward as presented by this simple example because the low-thrust Δv for a given orbit transfer does not match the impulsive Δv. For example, the low-thrust Δv for a circle-to-circle orbit transfer is greater than the impulsive Δv for the corresponding Hohmann transfer. The following sections will characterize the low-thrust Δv so that we may use variants of the rocket equation to determine the xenon propellant mass and transfer time required to complete a given orbit transfer.

9.3 Coplanar Circle-to-Circle Transfer

Because the electric-propulsion thrust force is so low, a continuous-thrust maneuver follows an unwinding spiral trajectory when starting from a circular orbit. The "quasi-circular" nature of the low-thrust spiral transfer allows us to develop analytical solutions. To show the quasi-circular spiral transfer, let us begin with Gauss' variational equation for eccentricity, Eq. (5.95)

$$\frac{de}{dt} = \frac{1}{v}\left[2(e + \cos\theta)a_t + \frac{r\sin\theta}{a}a_n\right] \tag{9.9}$$

where a_t and a_n are the perturbing acceleration components from the low-thrust propulsion system. Perturbing acceleration a_t is the thrust acceleration component along the velocity vector, while thrust acceleration component a_n is normal to the velocity vector and in the orbital plane (where the radial outward direction is considered to be positive). Figure 9.2 shows a schematic diagram of an EP spacecraft with its thrust vector orientation relative, to the rotating normal-tangent **NTW** frame. The **T** axis (not to be confused with the thrust vector) is always tangent to the orbit; the **N** axis is in the orbital plane normal to the **T** axis and points away from the central body. The **W** axis is normal to the orbital plane (i.e., $\mathbf{W} = \mathbf{N} \times \mathbf{T}$). The EP spacecraft in Figure 9.2 is oriented so that it has positive thrust components in the **T** and **N** directions (i.e., $a_t > 0$ and $a_n > 0$) and $a_w = 0$ because the thrust vector is in the orbital plane. Because the EP thruster is fixed to the spacecraft, the thrust vector may be "steered" in any direction by rotating the entire satellite about its center of mass.

Let us investigate a low-thrust transfer that departs from a circular orbit. We will assume that thrust is always aligned with the velocity vector (the **T** axis) so that $a_n = 0$ and $a_t = T/m$ where T is the propulsive thrust force. Recall from basic mechanics that

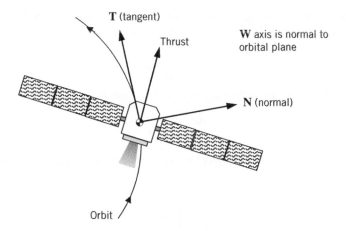

Figure 9.2 Electric-propulsion thrust vector in the orbital plane.

the time-rate of energy (i.e., power) is the dot product of the force and velocity vectors. Hence, thrusting along the velocity vector will maximize the time-rate of orbital energy; this strategy is beneficial for transferring from a low circular orbit to a higher orbit. Assuming $a_n = 0$ and $e = 0$ (circular orbit), Eq. (9.9) becomes

$$\frac{de}{dt} = \frac{2\cos\theta}{v}a_t \tag{9.10}$$

Instead of integrating Eq. (9.10) with respect to time, we choose to compute the change in eccentricity over a single orbital revolution. Therefore, we can eliminate dt by dividing Eq. (9.10) by the time-rate of angular position (true anomaly) for a circular orbit, $d\theta/dt = v/r$; the result is

$$\frac{de}{d\theta} = \frac{2r\cos\theta}{v^2}a_t \tag{9.11}$$

After separating variables, Eq. (9.11) becomes

$$de = \frac{2r\cos\theta}{v^2}a_t d\theta \tag{9.12}$$

Equation (9.12) is easy to integrate if we assume that radius r, velocity v, and thrust acceleration a_t remain constant over one orbital revolution. This assumption is acceptable for quasi-circular low-thrust transfers because the orbital elements change very little during a single powered orbit. Using this assumption, the integral of Eq. (9.12) is

$$de = \frac{2r}{v^2}a_t \int_0^{2\pi} \cos\theta d\theta = 0 \tag{9.13}$$

Thus, we can conclude that the net change in eccentricity is zero over a single powered orbital revolution and therefore a low-thrust transfer starting from a circular orbit follows a quasi-circular unwinding spiral trajectory. The reader should keep in mind that we have assumed that the thrust vector remains aligned with the velocity vector in order to maximize the rate of energy change.

We can investigate the change in semimajor axis over a single powered orbit by beginning with Gauss' variational equation for semimajor axis, Eq. (5.82):

$$\frac{da}{dt} = \frac{2a^2 v}{\mu} a_t \tag{9.14}$$

As before, we eliminate dt by dividing Eq. (9.14) by $d\theta/dt = v/a$

$$\frac{da}{d\theta} = \frac{2a^3}{\mu} a_t \tag{9.15}$$

Note that $a = r$ because the orbit is circular. Separating variables and integrating over one orbital revolution yields

$$da = \frac{2a^3}{\mu} a_t \int_0^{2\pi} d\theta = \frac{4\pi a^3}{\mu} a_t \tag{9.16}$$

Again, we have assumed that orbital element a is essentially constant over a single revolution. Equation (9.16) shows that semimajor axis (and hence energy) does indeed change after each powered revolution. As a quick example, consider a low-thrust engine with thrust $T = 0.3$ N propelling a spacecraft with initial mass $m = 3,000$ kg; hence the thrust acceleration is $a_t = T/m = 10^{-4}$ m/s^2 $= 10^{-7}$ km/s^2. If the starting orbit is a 400-km altitude circular orbit ($a = 6,778$ km), then Eq. (9.16) shows that $da = 0.98$ km after one powered orbital revolution, which is a 0.01% change in semimajor axis. Hence, the constant-semimajor axis assumption used to achieve Eq. (9.16) is valid. The low-thrust transfer will require many, many revolutions in order to impart a significant change in orbital energy.

We can extend the previous analyses to determine the time required to perform a low-thrust coplanar transfer between two circular orbits. Of course, knowing transfer time allows the calculation of propellant mass if we assume that the low-thrust engine is continuously operating during the transfer with a constant mass-flow rate. To begin, rewrite Eq. (9.14) with circular velocity $v = \sqrt{\mu/a}$

$$\frac{da}{dt} = \frac{2a^{3/2}}{\sqrt{\mu}} a_t \tag{9.17}$$

Separate variables and rewrite thrust acceleration as $a_t = T/m$

$$\frac{\sqrt{\mu}}{2} a^{-3/2} da = \frac{T}{m_0 - \dot{m}t} dt \tag{9.18}$$

Note that the diminishing spacecraft mass is $m = m_0 - \dot{m}t$, where $\dot{m}$ is the mass-flow rate (magnitude) of the low-thrust engine. Substituting Eq. (9.1) for thrust in Eq. (9.18) yields

$$\frac{\sqrt{\mu}}{2} a^{-3/2} da = \frac{\dot{m}c}{m_0 - \dot{m}t} dt \tag{9.19}$$

It is convenient to define the "time constant" $\tau = m_0/\dot{m}$ which allows us to express the right-hand side of Eq. (9.19) as

$$\frac{\sqrt{\mu}}{2} a^{-3/2} da = \frac{c}{\tau - t} dt \tag{9.20}$$

Integrating Eq. (9.20) from semimajor axis a_0 to a_f, and from time $t_0 = 0$ to final time $t = t_f$ yields

$$\sqrt{\frac{\mu}{a_0}} - \sqrt{\frac{\mu}{a_f}} = c\ln\left(\frac{\tau}{\tau - t_f}\right) \tag{9.21}$$

Note that the left-hand side is the initial circular orbital velocity, $v_0 = \sqrt{\mu/a_0}$, minus the final (target) circular velocity, $v_f = \sqrt{\mu/a_f}$. Solving Eq. (9.21) for the low-thrust transfer time t_f, we obtain

$$t_f = \tau\left[1 - \exp\left(\frac{-\Delta v}{c}\right)\right] \tag{9.22}$$

where $\Delta v = v_0 - v_f$ is the difference in circular velocities for the low-thrust coplanar transfer. Equation (9.22) is actually a variation of the rocket equation: dividing Eq. (6.13) by mass-flow rate $\dot{m}$ yields Eq. (9.22) because propellant mass is $m_p = \dot{m}t_f$. Therefore, Eq. (9.22) shows that the low-thrust Δv for a coplanar circle-to-circle transfer is simply the difference in circular orbital speeds:

$$\Delta v = v_0 - v_f \tag{9.23}$$

Manipulating Eq. (9.21) will produce an analytical expression for semimajor axis as a function of transfer time. Substituting $v_0 = \sqrt{\mu/a_0}$, $a(t) = a_f$, and $t = t_f$ into Eq. (9.21) yields

$$\sqrt{\frac{\mu}{a(t)}} = v_0 - c\ln\left(\frac{\tau}{\tau - t}\right) \tag{9.24}$$

Solving Eq. (9.24) for semimajor axis results in

$$\text{Outward spiral:} \quad a(t) = \frac{\mu}{\left[v_0 - c\ln\left(\dfrac{\tau}{\tau - t}\right)\right]^2} \tag{9.25}$$

Equation (9.25) shows how semimajor axis changes with time during an *outward* low-thrust quasi-circular transfer (i.e., energy increases during the transfer). Recall that the time constant $\tau = m_0/\dot{m}$ is the initial spacecraft mass divided by the engine mass-flow rate, and that $c = T/\dot{m}$ is the effective jet exhaust velocity. Therefore, the initial thrust acceleration is $T/m_0 = c/\tau$.

An inward spiral transfer to a lower circular orbit is essentially an outward spiral in reverse. Here the thrust acceleration a_t is directed 180° opposite of the velocity vector so that energy decreases at a maximum rate. Therefore, we simply insert a minus sign on one side of Eq. (9.17) because $da/dt < 0$. The remaining derivation is the same as the outward spiral except that now the (positive) velocity change is $\Delta v = v_f - v_0$. Using Eq. (9.22) to compute the inward transfer time, the inward (decreasing) semimajor axis is

$$\text{Inward spiral:} \quad a(t) = \frac{\mu}{\left[v_0 + c\ln\left(\dfrac{\tau}{\tau - t}\right)\right]^2} \tag{9.26}$$

Example 9.1   A cluster of three Hall-effect thrusters produces 0.6 N of total thrust. The Hall thruster has specific impulse $I_{sp} = 1,700$ s. Compute the low-thrust transfer time and propellant mass for a coplanar transfer from low-Earth orbit (LEO) to geostationary-equatorial orbit (GEO). The LEO altitude is 300 km and the initial mass of the spacecraft in LEO is 2,500 kg.

We use Eq. (9.22) to compute the time required to complete the continuous-thrust LEO–GEO transfer:

$$t_f = \tau \left[1 - \exp\left(\frac{-\Delta v}{c} \right) \right]$$

where Δv is the difference between the initial and target circular velocities, and $c = g_0 I_{sp} = 16,671$ m/s (recall that $g_0 = 9.80665$ m/s^2). The initial LEO circular speed is $v_{LEO} = \sqrt{\mu/a_{LEO}} = 7,726$ m/s, the final GEO circular speed is $v_{GEO} = \sqrt{\mu/a_{GEO}} = 3,075$ m/s, and therefore the low-thrust velocity increment is $\Delta v = v_{LEO} - v_{GEO} = 4,651$ m/s. In order to compute the time constant $\tau = m_0/\dot{m}$, we need the (total) mass-flow rate of the combined operation of three Hall thrusters:

$$\dot{m} = \frac{T}{c} = 0.6\,\text{N}/16,671\,\text{m/s} = 3.5990\left(10^{-5}\right) \text{kg/s} = 3.1095\,\text{kg/day}$$

Therefore, $\tau = 6.9464(10^7)$ s (=803.98 days). Using τ, Δv, and c, we can compute the continuous-thrust transfer time:

$$t_f = 803.98 \left[1 - \exp\left(\frac{-4,651}{16,671} \right) \right] = \boxed{195.73 \text{ days}}$$

The propellant mass is $m_p = \dot{m} t_f = (3.1095\,\text{kg/day})(195.73\,\text{days}) = \boxed{608.6\,\text{kg}}$

As an aside, let us compare a numerically integrated low-thrust transfer with the analytical results developed in this section. Gauss' variational equations for a, e, and true anomaly θ [Eqs. (5.82), (5.95), and (5.108)] can be numerically integrated to determine the coplanar transfer. These differential equations with tangential thrust (i.e., $a_n = 0$) are

$$\frac{da}{dt} = \frac{2a^2 v}{\mu} a_t \tag{9.27}$$

$$\frac{de}{dt} = \frac{2(e + \cos\theta)}{v} a_t \tag{9.28}$$

$$\frac{d\theta}{dt} = \frac{h}{r^2} - \frac{2\sin\theta}{ev} a_t \tag{9.29}$$

where angular momentum is $h = \sqrt{\mu p}$ and parameter p can be determined from a and e. Radius r is determined from the trajectory equation (2.45), and velocity v is determined from the energy equation (2.29). Tangential thrust acceleration is $a_t = T/m$, and the diminishing spacecraft mass is determined by integrating

$$\frac{dm}{dt} = -\frac{T}{c} \tag{9.30}$$

Note the minus sign in Eq. (9.30) because mass is decreasing. We use MATLAB's M-file ode45.m to numerically integrate Eqs. (9.27)–(9.30) starting from the initial

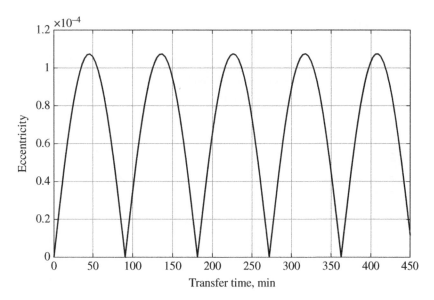

Figure 9.3 Eccentricity history during the initial low-thrust transfer from LEO (Example 9.1).

conditions given in this example problem (we are essentially applying the special perturbation method discussed in Chapter 5 to our low-thrust transfer problem). Figure 9.3 shows the orbital eccentricity resulting from numerical integration of Eq. (9.28) for a 450 min transfer time. Note that eccentricity exhibits a periodic profile where its peak value is about 10^{-4} and its minimum value returns to zero (i.e., a circular orbit) roughly every 91 min. The reader should note that the period of the 300-km altitude LEO is 91 min, and therefore Figure 9.3 verifies Eq. (9.13): the net change in eccentricity due to tangential thrust is zero after each orbital revolution.

Figure 9.4 shows the time histories of the semimajor axis determined by numerical integration and the analytical semimajor axis computed using Eq. (9.25). The transfer time is 195.73 days, that is, the time required to reach GEO from LEO. The numerical and analytical solutions for $a(t)$ are indistinguishable in Figure 9.4 and differ by less than 0.2 km throughout the transfer. These comparisons validate the quasi-circular approximation used to develop Eqs. (9.22) and (9.25).

Example 9.2   Two advanced ion thrusters produce a total 0.25 N of thrust, and each engine has a specific impulse $I_{sp} = 3{,}200$ s. Plot semimajor axis vs. transfer time for two coplanar circle-to-circle transfers where the initial spacecraft mass is 1,200 kg and 2,400 kg. The initial and final circular orbit radii are $2R_E$ and $4R_E$, respectively (recall that Earth's radius is $R_E = 6{,}378$ km).

First, we use Eq. (9.22) to compute the continuous-thrust transfer time

$$t_f = \tau\left[1 - \exp\left(\frac{-\Delta v}{c}\right)\right]$$

where $\Delta v = v_0 - v_f$ is the difference between the initial and target circular velocities, and exhaust speed is $c = g_0 I_{sp} = 31{,}381$ m/s. The initial and final circular speeds are easily computed using the initial and final radii ($a_0 = 2R_E = 12{,}756$ km, $a_f = 4R_E = 25{,}512$ km):

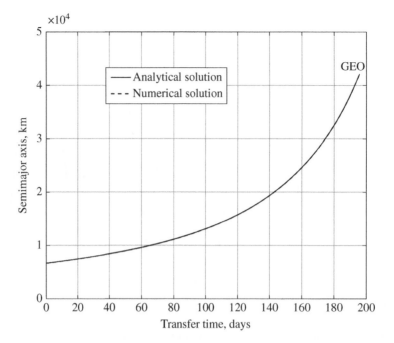

Figure 9.4 Semimajor axis history for a low-thrust LEO-GEO transfer (Example 9.1).

$$v_0 = \sqrt{\mu/a_0} = 5{,}590 \text{ m/s} \quad \text{and} \quad v_f = \sqrt{\mu/a_f} = 3{,}953 \text{ m/s}$$

hence $\Delta v = 1{,}637$ m/s. The time constant is $\tau = m_0/\dot{m}$, where the mass-flow rate is computed from the total thrust magnitude and exhaust speed c:

$$\dot{m} = \frac{T}{c} = 0.25 \text{ N}/31{,}381 \text{ m/s} = 7.9666\left(10^{-6}\right) \text{ kg/s} = 0.6883 \text{ kg/day}$$

Therefore, the time constant is $\tau = 1.5063(10^8)$ s (=1,743.4 days) for $m_0 = 1{,}200$ kg, and $\tau = 3.0126(10^8)$ s (=3,486.8 days) for $m_0 = 2{,}400$ kg. Using the two time constants yields transfer times of $t_f = 88.63$ and 177.25 days, respectively. Clearly, the second transfer time is twice as long as the first transfer time because the factor of two in the initial mass (which leads to a factor of two in the time constant τ).

Equation (9.25) determines semimajor axis as a function of transfer time t

$$a(t) = \frac{\mu}{\left[v_0 - c\ln\left(\dfrac{\tau}{\tau - t}\right)\right]^2}$$

Figure 9.5 shows semimajor axis (in units of R_E) for the "short transfer" ($m_0 = 1{,}200$ kg and $t_f = 88.63$ days) and the "long transfer" ($m_0 = 2{,}400$ kg and $t_f = 177.25$ days).

We can compute the mass ratio m_0/m_f using the rocket equation (6.11):

$$\frac{m_0}{m_f} = \exp\left(\frac{\Delta v}{g_0 I_{sp}}\right) = 1.054$$

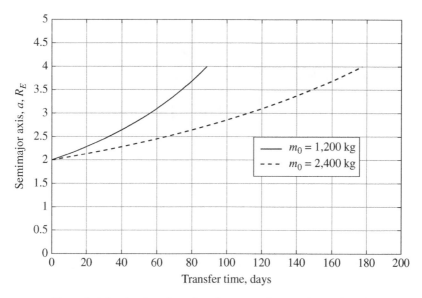

Figure 9.5 Semimajor axis vs. low-thrust transfer time (Example 9.2).

Because both transfers involve the same initial and target orbits, and employ the same low-thrust engine, the mass ratio m_0/m_f (and propellant mass ratio m_p/m_0) is constant. Therefore, the propellant mass required for the spacecraft with $m_0 = 1,200$ kg is half of the propellant mass required for the 2,400 kg spacecraft (of course this result can be gleaned from the product of the constant mass-flow rate and the two transfer times).

9.3.1 Comparing Impulsive and Low-Thrust Transfers

Thus far, we have developed the basic performance equations for low-thrust transfer time [Eq. (9.22)] and low-thrust Δv [Eq. (9.23)] for a quasi-circular coplanar transfer with continuous thrust. The xenon propellant mass m_p required for a continuous transfer may be computed using the rocket equation (9.8) or by the product of mass-flow rate $\dot{m}$ and transfer time t_f.

Let us compare a coplanar LEO–GEO transfer using a conventional chemical rocket and a low-thrust EP device. For a 300-km altitude LEO, the two-impulse Hohmann transfer Δv is 3,893 m/s (see Section 7.3 for the Hohmann transfer formulas). Assuming a chemical stage with $I_{sp} = 320$ s, we can compute the final/initial mass ratio using the rocket equation (6.11); the Hohmann-transfer result is $m_f/m_0 = 0.289$. For the low-thrust transfer, the velocity increment is the difference in circular speeds, that is, $\Delta v = 4,651$ m/s (see Example 9.1). Hence, the low-thrust Δv is 19.5% greater than the total impulsive Δv from the Hohmann transfer. Using the Hall-effect thruster from Example 9.1 ($I_{sp} = 1,700$ s), we find that the final/initial mass ratio is $m_f/m_0 = 0.757$. Therefore, a spacecraft equipped with Hall thrusters can deliver much more mass to GEO compared with a chemical rocket (despite the larger Δv) because its specific impulse is greater by a factor of five. A spacecraft equipped with ion thrusters ($I_{sp} = 3,200$ s) will deliver even

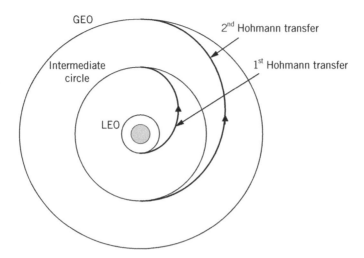

Figure 9.6 Low-Earth orbit (LEO) to geostationary-equatorial orbit (GEO) using two Hohmann transfers (not to scale).

more payload mass to GEO because $m_f/m_0 = 0.862$. The tradeoff is the LEO–GEO transfer time: the low-thrust transfers will take hundreds of days to reach GEO.

Why is the low-thrust Δv greater than the two-impulse Hohmann-transfer Δv? The low-thrust transfer suffers *trajectory losses* due to the quasi-circular nature of the transfer. To illustrate these trajectory losses, let us consider performing multiple Hohmann transfers from LEO to GEO. Figure 9.6 shows a LEO–GEO scenario with *two* Hohmann transfers: the first Hohmann transfer reaches an intermediate circular orbit halfway between LEO and GEO; the second Hohmann transfer delivers the satellite to GEO. Two Hohmann transfers (as shown in Figure 9.6) will require four impulses (two pairs of impulses at periapsis and apoapsis). Using the formulas in Section 7.3, we find that the total impulsive Δv is 4,297 m/s for two Hohmann transfers which is more than 10% greater than the single Hohmann-transfer case. For three Hohmann transfers, we obtain $\Delta v = 4{,}449$ m/s. Computing the total impulsive Δv for multiple transfers is relatively easy using a MATLAB M-file that repeatedly performs the calculations for a single Hohmann transfer. Figure 9.7 shows the total impulsive Δv for multiple Hohmann transfers from LEO to GEO. It is clear that the total impulsive Δv steadily increases as the number of intermediate Hohmann transfers increases (the largest increase occurs between the single and double Hohmann-transfer scenarios). Figure 9.7 also shows that when the LEO–GEO transfer is divided into 15 or more Hohmann transfers, the total impulsive Δv asymptotically approaches the low-thrust Δv (=4,651 m/s). This analysis illustrates the trajectory losses associated with a quasi-circular transfer: the total impulsive Δv increases if we force a circularization burn at the intermediate orbits. As the number of intermediate orbits (or, Hohmann transfers) increases, the overall transfer to GEO resembles an unwinding spiral trajectory.

One way to circumvent low-thrust trajectory losses would be to concentrate powered low-thrust maneuvers at the apses in the same manner as the impulsive Hohmann transfer. However, because the EP thrust acceleration is so low, this strategy would require many powered arcs. To show this, consider operating the EP thruster for short segments near each LEO perigee passage and subsequently coasting through apogee until the

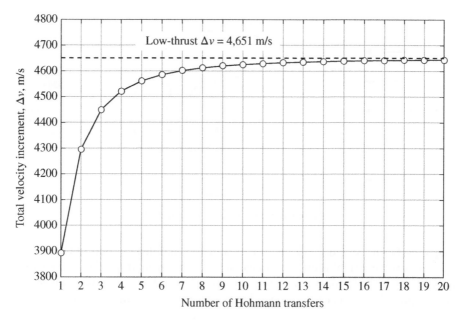

Figure 9.7 Total impulsive Δv for multiple Hohmann transfers (low-Earth orbit to geostationary-equatorial orbit).

spacecraft returns to perigee for another powered arc. This perigee-thrusting strategy would persist until the orbit's apogee reached GEO altitude, and then a series of apogee-thrusting arcs would follow to raise perigee to GEO. In order to quantify this approach, let us consider the Hall-thruster system presented in Example 9.1. The initial thrust acceleration is $T/m_0 = 2.4(10^{-4})$ m/s^2 which is approximately equal to $\Delta v/\Delta t$ where Δt is the thrusting time near each perigee pass. If we assume that the EP device thrusts for a 10-min arc centered on each perigee pass (the initial LEO period is 91 min), then the "impulsive" velocity increment for *each* perigee pass is merely $\Delta v = (T/m_0)\Delta t = 0.144$ m/s. Recall that the velocity increment required at perigee for an impulsive Hohmann transfer is 2,426 m/s. Therefore, this low-thrust perigee-thrusting strategy will require nearly 17,000 perigee-thrusting arcs! The total transfer time using the perigee-thrusting strategy would be greater than 5 years. Of course, this 5-year maneuver does not result in GEO; it has only raised apogee to GEO altitude. Next, the spacecraft performs a series of concentrated apogee-thrusting maneuvers to raise perigee to GEO altitude; this phase would also take more than 5 years. This simple example shows why low-thrust trajectory losses are inevitable – GEO satellite owners cannot wait 10 years after launch to begin operating their satellite! If low-thrust devices are utilized for large-scale orbit transfers (like LEO–GEO), then it is likely they will be operated continuously in order to keep the total transfer time to a minimum value.

9.4 Coplanar Transfer with Earth-Shadow Effects

Electric-propulsion engines receive their power from solar arrays. Therefore, a realistic low-thrust spiral transfer has unpowered coasting arcs when the spacecraft passes

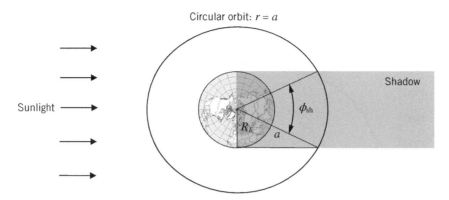

Figure 9.8 Simple Earth-shadow model.

through the Earth's shadow. For low-thrust transfers that depart from LEO, the Earth-eclipse arc can be a significant percentage of an orbital revolution. Figure 9.8 shows a simple Earth-shadow model where the circular orbit (radius $r = a$) is assumed to be in the ecliptic plane. It is easy to show that the Earth-shadow angle ϕ_{sh} is

$$\phi_{sh} = 2\sin^{-1}\left(\frac{R_E}{a}\right) \tag{9.31}$$

Recall that R_E is the Earth's radius. We can define a "sunlight weighting function" w that denotes the percentage of time the spacecraft spends in sunlight over one orbital revolution:

$$w = 1 - \frac{\phi_{sh}}{2\pi} = 1 - \frac{1}{\pi}\sin^{-1}\left(\frac{R_E}{a}\right) \tag{9.32}$$

Clearly, $w = 0.5$ when $a = R_E$ (half of the orbit is in sunlight), and $w \approx 1$ (no shadow) for a circular orbit with a very large radius. The reader should note that this simple shadow model is restrictive because it assumes that the circular orbit remains in the ecliptic plane. Actual shadow conditions for a circular orbit will depend on the calendar date, inclination i, and longitude of the ascending node Ω. For example, a satellite in GEO will experience a short Earth-shadow arc on the spring and fall equinox dates but no shadow will exist for the summer and winter solstice dates.

We can estimate the low-thrust transfer time with Earth-shadow conditions by factoring the weighting function w into the time-rate of semimajor axis:

$$\frac{da}{dt} = \frac{2a^{3/2}}{\sqrt{\mu}}wa_t \tag{9.33}$$

Clearly, if $w = 1$ (continuous thrust), then Eq. (9.33) becomes Eq. (9.17). Our previous approach of separating variables and integrating each side of Eq. (9.33) does not work if we use Eq. (9.32) to define w because analytical integrals do not exist. One remedy is to use a small-angle approximation for w

$$w = 1 - \frac{R_E}{\pi a} \tag{9.34}$$

Substituting Eq. (9.34) into Eq. (9.33) and separating variables yields

$$\frac{\sqrt{\mu}}{2a^{3/2}}\left(1-\frac{R_E}{\pi a}\right)^{-1}da = \frac{T}{m_0 - \dot{m}t}dt \tag{9.35}$$

Integrating Eq. (9.35), we obtain

$$\sqrt{\frac{\pi\mu}{R_E}}\left[\tanh^{-1}\left(\sqrt{\frac{\pi a_0}{R_E}}\right) - \tanh^{-1}\left(\sqrt{\frac{\pi a_f}{R_E}}\right)\right] = c\ln\left(\frac{\tau}{\tau - \tilde{t}_f}\right) \tag{9.36}$$

Note that both sides of Eq. (9.36) have units of velocity because of the term $\sqrt{\mu/R_E}$ on the left-hand side, and the effective exhaust velocity c on the right-hand side. Finally, we can manipulate Eq. (9.36) and solve for transfer time in the presence of shadow eclipses

$$\tilde{t}_f = \tau\left[1 - \exp\left(\frac{-\tilde{c}}{c}\right)\right] \tag{9.37}$$

where $\tilde{c}$ is the left-hand side of Eq. (9.36)

$$\tilde{c} = \sqrt{\frac{\pi\mu}{R_E}}\left[\tanh^{-1}\left(\sqrt{\frac{\pi a_0}{R_E}}\right) - \tanh^{-1}\left(\sqrt{\frac{\pi a_f}{R_E}}\right)\right] \tag{9.38}$$

Equation (9.37) approximates the total low-thrust transfer time with interrupted thrust caused by Earth-shadow eclipses. Total transfer time $\tilde{t}_f$ is the sum of the powered time (engine on) and coasting time (engine off during eclipses). The total *powered* time is essentially the same for continuous- and interrupted-thrust transfers because both involve the same change in orbital energy. Hence, the propellant mass required for either case is the product of the mass-flow rate $\dot{m}$ and the continuous-thrust transfer time determined by Eq. (9.22). The following examples illustrate low-thrust transfers with and without Earth-shadow eclipses.

Example 9.3　Consider again the low-thrust coplanar LEO–GEO transfer outlined in Example 9.1. Compute the total transfer time for the case where Earth's shadow interrupts thrust. Compare this result with the continuous-thrust scenario.

To determine the transfer time with shadow eclipses, we start by using Eq. (9.38) to compute $\tilde{c}$ for a_0 = 6,678 km (LEO) and a_f = 42,164 km (GEO):

$$\tilde{c} = \sqrt{\frac{\pi\mu}{R_E}}\left[\tanh^{-1}\left(\sqrt{\frac{\pi a_0}{R_E}}\right) - \tanh^{-1}\left(\sqrt{\frac{\pi a_f}{R_E}}\right)\right] = 5{,}567 \text{ m/s}$$

Next, use Eq. (9.37) to compute the transfer time with interrupted thrust

$$\tilde{t}_f = \tau\left[1 - \exp\left(\frac{-\tilde{c}}{c}\right)\right]$$

Recall that $c = g_0 I_{sp} = 16{,}671$ m/s for the Hall-effect thruster in Example 9.1 with a specific impulse of 1,700 s. We computed the time constant $\tau = m_0/\dot{m} = 803.98$ days in Example 9.1. Therefore, the transfer time with Earth-shadow eclipses is

$$\tilde{t}_f = 803.98\left[1 - \exp\left(\frac{-5{,}567}{16{,}671}\right)\right] = \boxed{228.2 \text{ days}}$$

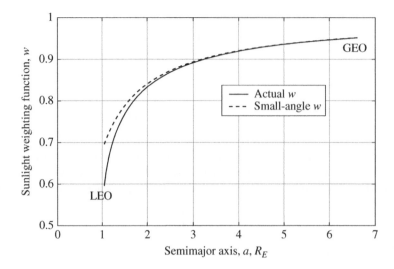

Figure 9.9 Sunlight weighting function *w* for low-Earth orbit (LEO) to geostationary-equatorial orbit (GEO) transfer (Example 9.3).

Hence, including the Earth-shadow eclipses has increased the transfer time by more than 32 days compared with the continuous-thrust transfer determined in Example 9.1.

As an aside, we can plot the "sunlight weighting function" *w* for this LEO–GEO transfer. Equation (9.32) presents the accurate calculation of *w* as a function of semimajor axis (using the simple ecliptic shadow model), while Eq. (9.34) uses the small-angle approximation (recall that the small-angle approximation is required for the analytical integral of the transfer time with interrupted thrust). Figure 9.9 plots the actual and approximate weighting functions vs. semimajor axis. Note that at LEO ($a_{LEO} = 1.047R_E$), the Earth-shadow angle is $\phi_{sh} = 146°$ (thus $w = 0.59$) and the small-angle method overestimates the actual sunlight function by about 17%. However, when the spacecraft reaches a radius of about $2R_E$, the Earth-shadow arc has diminished to $\phi_{sh} = 60°$ and the small-angle w shows a good match with the actual weighting function w.

9.5 Inclination-Change Maneuver

We can develop analytical solutions for a low-thrust maneuver that only changes orbital inclination. To begin, consider Gauss' variational equation for inclination, Eq. (5.102):

$$\frac{di}{dt} = \frac{r}{h}a_w \cos(\omega + \theta) \tag{9.39}$$

where a_w is the perturbing thrust acceleration component in the direction of the angular momentum vector **h**, or normal to the orbital plane. For a pure plane change that maximizes the inclination rate di/dt, the thrust vector is always aligned normal to the orbital plane. In such a case, the thrust components along the velocity and in-plane normal directions, a_t and a_n, are zero and hence semimajor axis and eccentricity remain constant [see Eqs. (9.9) and (9.14)]. Note that the cosine term $\cos(\omega + \theta)$ in Eq. (9.39) switches signs as the spacecraft passes through the argument of latitude $u = \omega + \theta = \pm 90°$ (i.e., $\pm 90°$ from the ascending node). Therefore, in order to maintain a positive inclination rate $di/dt > 0$, the thrust acceleration a_w is aligned with the angular momentum vector **h** when $\cos(\omega + \theta) > 0$, and aligned opposite of **h** when $\cos(\omega + \theta) < 0$ (of course, this strategy is reversed if we desire $di/dt < 0$). Using this thrust-steering strategy, the magnitude of the inclination rate is

$$\frac{di}{dt} = \frac{r}{h}a_w|\cos u| \tag{9.40}$$

where $u = \omega + \theta$ and $a_w = T/m$ is the magnitude of the thrust acceleration. For a circular orbit, we can substitute $h = rv$ in Eq. (9.40). We can change the independent variable from time t to argument of latitude u by dividing Eq. (9.40) by $du/dt = d\theta/dt = v/r$, and the result is

$$\frac{di}{du} = \frac{r^2}{\mu}a_w|\cos u| \tag{9.41}$$

We have substituted circular orbital speed $v = \sqrt{\mu/r}$ to develop Eq. (9.41). Finally, we can separate variables and integrate both sides of Eq. (9.41) over one orbital revolution:

$$\int di = \frac{r^2}{\mu}a_w \int_0^{2\pi} |\cos u| du = \frac{4r^2}{\mu}a_w \tag{9.42}$$

Equation (9.42) is the inclination change after one circular orbital revolution of continuous thrust. We can divide Eq. (9.42) by the period of a circular orbit, $2\pi r/v$, to obtain the *mean* time rate for inclination change:

$$\frac{d\bar{i}}{dt} = \frac{2}{\pi v}a_w \tag{9.43}$$

The overbar in Eq. (9.43) denotes the *mean* time rate of inclination change, and therefore we may use Eq. (9.43) to estimate large plane changes after many powered orbital revolutions. In contrast, Eqs. (9.39) and (9.40) present the *instantaneous* time-rate of inclination, which depends on the satellite's position in the orbit. Finally, we can separate variables in Eq. (9.43)

$$\frac{\pi v}{2}d\bar{i} = \frac{\dot{m}c}{m_0 - \dot{m}t}dt \tag{9.44}$$

Note that we have substituted $a_w = T/m$, similar to the steps surrounding Eqs. (9.18) and (9.19). Integrating both sides and some algebraic manipulation leads to the transfer time

$$t_f = \tau\left[1 - \exp\left(\frac{-\Delta i \pi v}{2c}\right)\right] \tag{9.45}$$

where Δi is the magnitude of the desired plane change (in rad) and $\tau = m_0/\dot{m}$ as before. Recall that v is the (constant) circular orbital speed because the low-thrust maneuver only changes the inclination. Equation (9.45) is the low-thrust transfer time required for inclination change Δi for a circular orbit with velocity v. This equation assumes continuous thrust (i.e., thrust is not interrupted by Earth-shadow eclipses). If we compare the inclination-change transfer time equation (9.45) with the coplanar circle-to-circle equation (9.22), we see that the low-thrust Δv for a pure inclination change is

$$\Delta v_i = \frac{\Delta i \pi v}{2} \tag{9.46}$$

Here we have included subscript i to indicate the low-thrust Δv for a pure inclination change (keep in mind that Δi is the *magnitude* of the change in orbital inclination expressed in radians).

Example 9.4 Consider again a spacecraft equipped with three Hall-effect thrusters with 0.6 N of total thrust and specific impulse $I_{sp} = 1{,}700$ s. Compute the low-thrust transfer time and propellant mass required for a 20° plane-change maneuver for two circular orbits: (a) radius $r = 10{,}000$ km; and (b) radius $r = 30{,}000$ km. The initial mass of the spacecraft is 2,500 kg in both cases.

The exhaust speed and mass-flow rate for the Hall-effect thruster are

$$c = g_0 I_{sp} = 16{,}671 \text{ m/s} \quad \text{and} \quad \dot{m} = \frac{T}{c} = 3.5990\left(10^{-5}\right) \text{ kg/s} = 3.1095 \text{ kg/day}$$

Therefore, the time constant is $\tau = m_0/\dot{m} = 6.9464(10^7)$ s ($= 803.98$ days).

a) For the lower orbital radius $r = 10{,}000$ km, the circular speed is $v = \sqrt{\mu/r} = 6{,}314$ m/s. Equation (9.45) provides the low-thrust transfer time for $\Delta i = 0.3491$ rad:

$$t_f = \tau \left[1 - \exp\left(\frac{-\Delta i \pi v}{2c}\right)\right] = \boxed{150.75 \text{ days}}$$

The corresponding propellant mass is $m_p = \dot{m} t_f = \boxed{468.8 \text{ kg}}$

b) For the higher orbital radius $r = 30{,}000$ km, the circular speed is $v = \sqrt{\mu/r} = 3{,}645$ m/s. Hence, the low-thrust transfer time is $t_f = 90.83$ days, and the associated propellant mass is $m_p = 282.4$ kg. Like impulsive maneuvers, performing the plane change at a higher altitude (and lower orbital speed) significantly reduces the propellant mass.

9.6 Transfer Between Inclined Circular Orbits

In the previous sections, we developed simple analytical expressions for the low-thrust Δv, transfer time, and propellant mass for two scenarios: a coplanar circle-to-circle transfer, and an inclination-change maneuver. Uncoupling the changes in orbital elements allowed us to solve Gauss' variational equations using separation of variables and analytical integration. In 1961, T. N. Edelbaum obtained a closed-form solution for the *general* three-dimensional low-thrust transfer between inclined circular orbits [4]. Edelbaum utilized optimization theory and the calculus of variations to develop the minimum-propellant transfer between circular orbits with a plane change. His approach assumed that the out-of-plane thrust acceleration component a_w remained constant over an orbital revolution (switching signs at $\pm 90°$ from the nodal crossings) while the in-plane thrust component remained aligned with the velocity vector (i.e., $a_n = 0$). However, the magnitude of the out-of-plane component a_w varies in an optimal fashion so that the majority of the inclination change occurs at higher orbital altitudes. Using in-plane tangential thrust assures that the orbit transfer retains its quasi-circular "spiral" nature that we observed in Section 9.3. This thrust-steering strategy will cause simultaneous changes in semimajor axis and inclination during the orbit transfer.

Because the optimization theory employed by Edelbaum is beyond the scope of this textbook, we will not discuss or present the details here (the interested reader may consult Edelbaum's original work [4] or Kechichian's excellent reexamination of this classic solution using optimal control theory [5]). Instead, we will simply present Edelbaum's main result for the low-thrust velocity increment for a circle-to-circle low-thrust transfer with an inclination change:

$$\Delta v = \sqrt{v_0^2 + v_f^2 - 2v_0 v_f \cos\left(\frac{\Delta i \pi}{2}\right)} \tag{9.47}$$

Here v_0 and v_f are the initial and final velocities of the respective circular orbits and Δi is the *magnitude* of the inclination change between the two circular orbits. Because the factor π exists in Eq. (9.47), the reader should take care to perform the cosine operation with its argument in radians. Note that for a coplanar transfer ($\Delta i = 0$), the low-thrust velocity increment determined by Eq. (9.47) is simply the difference in circular velocities (i.e., $\Delta v = v_0 - v_f$). Therefore, Edelbaum's equation (9.47) with $\Delta i = 0$ matches the coplanar circle-to-circle Δv that we developed in Section 9.3. However, Eq. (9.47) is *not* correct for a pure inclination-change maneuver where $v_0 = v_f$. For a low-thrust maneuver that only changes inclination, the low-thrust velocity increment is

$$\text{Inclination change: } \quad \Delta v = \frac{\Delta i \pi v_0}{2} \tag{9.48}$$

Equation (9.48) matches Eq. (9.46), our result from Section 9.5. Table 9.2 summarizes the appropriate low-thrust Δv equation for a variety of orbit transfers.

Transfer time for the low-thrust maneuver between inclined circular orbits is

$$t_f = \tau \left[1 - \exp\left(\frac{-\Delta v}{c}\right) \right] \tag{9.49}$$

Table 9.2 Low-thrust Δv equations for orbital maneuvers with continuous thrust.[a]

Orbit transfer	Low-thrust Δv
Coplanar circle-to-circle	$\Delta v = v_0 - v_f$
Inclination-change only	$\Delta v = \dfrac{\Delta i \pi v_0}{2}$
Circle-to-circle with inclination change	$\Delta v = \sqrt{v_0^2 + v_f^2 - 2v_0 v_f \cos\left(\dfrac{\Delta i \pi}{2}\right)}$

[a] Note that v_0 and v_f are the circular orbital speeds of the initial and final orbits, respectively.

where the time constant τ and exhaust speed c are defined in the same manner as in Sections 9.2 and 9.3. We may compute the low-thrust Δv using the appropriate expression in Table 9.2 that corresponds to the desired orbital transfer. Propellant mass is the product of transfer time t_f and the constant mass-flow rate of the EP device. It is important to remember that Edelbaum's results (and the results developed in Sections 9.3 and 9.5) are valid for low-thrust transfers with continuous thrust (i.e., no Earth eclipses). The following example demonstrates a low-thrust transfer between inclined circular orbits.

Example 9.5   Using the electric-propulsion spacecraft characteristics from Example 9.1, determine the low-thrust Δv, transfer time, and xenon propellant mass for a LEO–GEO transfer with a plane change. Assume that the LEO inclination is 28.5° (i.e., the latitude of the Cape Canaveral launch facility).

The LEO and GEO circular velocities are $v_{LEO} = 7{,}726$ m/s and $v_{GEO} = 3{,}075$ m/s, respectively (for calculations see Example 9.1). We know that the inclination change is $\Delta i = 28.5°$ because GEO is an equatorial orbit. Using Edelbaum's equation (9.47), the low-thrust velocity increment is

$$\Delta v = \sqrt{v_{LEO}^2 + v_{GEO}^2 - 2v_{LEO}v_{GEO}\cos\left(\frac{\Delta i\pi}{2}\right)} = \boxed{5{,}950.8 \,\text{m/s}}$$

where $\Delta i = 0.4974$ rad. We compute the transfer time using Eq. (9.49) with $\tau = 803.98$ days and $c = g_0 I_{sp} = 16{,}671.3$ m/s (see Example 9.1):

$$t_f = \tau\left[1 - \exp\left(\frac{-\Delta v}{c}\right)\right] = 803.98\left[1 - \exp\left(\frac{-5{,}950.8}{16{,}671.3}\right)\right] = \boxed{241.3 \,\text{days}}$$

Recall that the *coplanar* LEO–GEO transfer time is 195.7 days for this EP spacecraft. Thus, the 28.5° plane change added over 45 days to the transfer time.

Because we are assuming a continuous-thrust transfer, the propellant mass is the product of transfer time and mass-flow rate $\dot{m} = 3.1095$ kg/day (see Example 9.1 for the mass-flow rate calculation):

$$m_p = \dot{m}t_f = \boxed{750.5 \text{ kg}}$$

9.7 Combined Chemical-Electric Propulsion Transfer

The high jet exhaust speed associated with electric propulsion allows a spacecraft propelled by EP devices to deliver more payload mass to a target orbit when compared with a transfer using conventional chemical-propulsion engines. The tradeoff, however, is that a low-thrust transfer requires a great deal of time which ultimately delays the satellite's intended operation in orbit. Table 9.3 presents the velocity increment Δv, transfer time, and mass ratio for a LEO–GEO transfer with a 28.5° inclination change for a two-impulse Hohmann transfer and low-thrust transfers. The total impulsive Δv is determined using the three-dimensional Hohmann-transfer method illustrated by Example 7.10c in Chapter 7 (i.e., the second impulse at apogee raises perigee and performs the

Table 9.3 Low-Earth orbit (LEO) to geostationary-equatorial orbit (GEO) transfer with 28.5° inclination change.[a]

Propulsion mode	Total Δv (m/s)	Transfer time, t_f (days)	Final-to-initial mass ratio, m_{GEO}/m_{LEO}
Chemical ($I_{sp} = 320$ s)	4,256.0	0.2	0.2576
Ion ($T = 0.6$ N, $I_{sp} = 3,200$ s)[b]	5,950.8	261.4	0.8273
Hall ($T = 0.6$ N, $I_{sp} = 1,700$ s)[b]	5,950.8	241.3	0.6998

[a] Initial LEO is 300-km altitude.
[b] Initial mass in LEO is $m_{LEO} = 2,500$ kg.

28.5° plane change). Edelbaum's equation (9.47) determines the total low-thrust Δv. Of course, the impulsive and low-thrust Δv depend only on the characteristics of the LEO and GEO orbits. A Hohmann transfer reaches GEO in a matter of hours (about 5 h), whereas low-thrust transfers require over 8 months of continuous thrust. Table 9.3 shows the potential payload advantages of using low-thrust propulsion: the mass ratios m_{GEO}/m_{LEO} for LEO–GEO transfers using EP devices are much greater than the mass ratio for an impulsive transfer.

Table 9.3 highlights the advantages and disadvantages of using chemical-propulsion engines and EP devices for a LEO–GEO transfer. A propulsion option worth considering is the *combined* use of chemical and EP stages for the LEO–GEO transfer. The chemical stage would transfer the spacecraft from LEO to an intermediate (higher) orbit at which point the EP stage would take over and complete the remaining transfer to GEO. This combined-propulsion strategy would increase the payload mass of an all-chemical (Hohmann) transfer and reduce the transfer time of an all-EP low-thrust transfer. Figure 9.10 shows a schematic diagram of this proposed combined-propulsion strategy. The Hohmann transfer takes the spacecraft from LEO to an intermediate circular orbit with radius r_1 and inclination i_1. After discarding the spent chemical stage, the EP thruster completes the spiral transfer from circle r_1 to GEO with an inclination change equal to i_1. It should be clear that the radius and inclination of the intermediate circular orbit, r_1 and i_1, greatly affect the payload mass delivered to GEO and the EP transfer time. For example, if we choose r_1 so that it is only slightly greater than LEO and select $i_1 = 28.5°$ ($= i_{LEO}$), then the EP stage will essentially perform the entire LEO–GEO transfer. At the other extreme, selecting a very large r_1 ($\approx r_{GEO}$) and small i_1 (≈ 0) essentially eliminates the payload advantages associated with using the EP thruster.

Before continuing, we should note that this maneuver strategy is a bit artificial and somewhat constrained by forcing the second impulsive burn to establish an intermediate circular orbit. The intermediate orbit (i.e., the starting orbit for the EP stage) could be an inclined *elliptical* orbit. However, the only analytical formulas available for low-thrust transfers involve circular orbits as the initial and terminal conditions. Computing the minimum-propellant low-thrust transfer between an arbitrary inclined elliptical orbit and a circular target orbit requires numerical integration of the equations of motion coupled with a numerical optimization algorithm. This topic is well beyond the scope of this textbook; the interested reader may wish to consult Conway [6] for examples of various methods for low-thrust trajectory optimization.

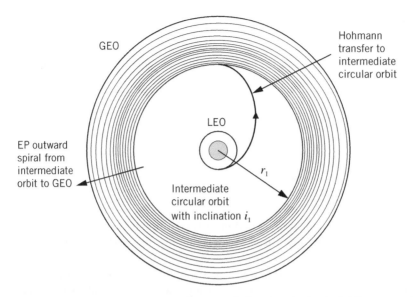

Figure 9.10 Combined chemical-electric propulsion transfer from low-Earth orbit (LEO) to geostationary-equatorial orbit (GEO).

Let us demonstrate the tradeoffs in payload mass and EP transfer time for a chemical-electric propulsion option for a LEO–GEO transfer. We can determine the mass ratio and transfer time using the following algorithm:

1) Given the intermediate circular radius r_1 and inclination i_1, compute the two-impulse, three-dimensional Hohmann transfer from LEO to the inclined circular orbit r_1 using the formulas developed in Chapter 7 (e.g., see Example 7.10c). Use the total impulsive Δv_c and the chemical-stage $I_{sp,c}$ to determine the chemical-stage propellant mass $m_{p,c}$

$$m_{p,c} = m_{\text{LEO}} \left[1 - \exp\left(\frac{-\Delta v_c}{g_0 I_{sp,c}} \right) \right] \tag{9.50}$$

2) Compute the mass of the spacecraft in the intermediate circular orbit using

$$m_1 = m_{\text{LEO}} - m_{p,c} - K_t m_{p,c} \tag{9.51}$$

where K_t is the so-called *tankage fraction*. The product $K_t m_{p,c}$ represents the "dry mass" of the chemical stage (engine, tank, and structure) that is jettisoned after the transfer to intermediate orbit r_1.

3) Use Edelbaum's equation (9.47) to compute the low-thrust Δv_{EP} for a transfer from the intermediate orbit (r_1 and i_1) to GEO. The final spacecraft mass in GEO is

$$m_{\text{GEO}} = m_1 - m_{p,\text{EP}}$$

where the xenon propellant mass, $m_{p,\text{EP}}$, is determined by Eq. (9.8). The low-thrust transfer time is

$$t_{\text{EP}} = \tau \left[1 - \exp\left(\frac{-\Delta v_{\text{EP}}}{g_0 I_{sp,\text{EP}}} \right) \right]$$

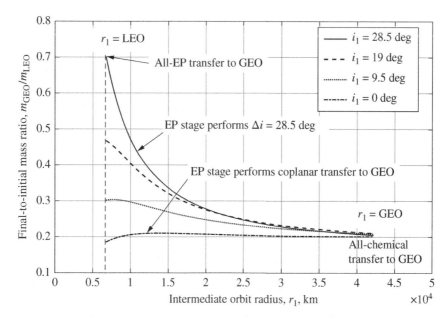

Figure 9.11 Final mass ratio for combined chemical-electric propulsion low-Earth orbit (LEO) to geostationary-equatorial orbit (GEO) transfer.

where $I_{sp,EP}$ is the EP stage specific impulse and the time constant is $\tau = m_1/\dot{m}$. The mass-flow rate of the EP device is computed using EP thrust T and $I_{sp,EP}$.

To demonstrate combined chemical-electric transfers, we assume that a Delta IV (Medium) launch vehicle delivers 13,300 kg to a 400-km circular LEO with inclination $i_{LEO} = 28.5°$. Suppose the spacecraft is equipped with Hall-effect thrusters with a total input power of 15 kW, thruster efficiency of 53%, and specific impulse $I_{sp,EP} = 1,700$ s. Using Eq. (9.6), we find that the total thrust is $T = 0.9537$ N. The chemical stage has specific impulse $I_{sp,c} = 320$ s and tankage fraction $K_t = 0.08$. Figure 9.11 presents the final-to-initial mass ratio m_{GEO}/m_{LEO} for different values of the intermediate orbital radius r_1 and inclination i_1. The extreme left end of the four curves presents EP transfers that begin at $r_1 = 6,778$ (LEO), and thus the chemical stage has not increased the orbital energy. The top (solid) curve in Figure 9.11 represents EP transfers that perform the entire 28.5° inclination change, and hence the top left point ($m_{GEO}/m_{LEO} = 0.7$) is an all-EP transfer to GEO (the chemical stage does not exist). The bottom curve in Fig. 9.11 represents coplanar EP transfers (the chemical stage has performed the 28.5° inclination change). It should be clear to the reader that the lower right point is an all-chemical-propulsion (Hohmann) transfer to GEO (there is no EP stage). Although the all-chemical impulsive transfer exhibits a very poor mass ratio ($m_{GEO}/m_{LEO} = 0.2$), its performance is slightly better than the low-thrust LEO–GEO coplanar transfer that begins after the chemical stage performs the 28.5° inclination change. The reader should note that the mass ratio for the Hohmann transfer in Table 9.3 ($m_{GEO}/m_{LEO} = 0.26$) is larger than the all-chemical value in Figure 9.11 because it includes the spent dry mass of the chemical stage. Figure 9.11 also shows that when the intermediate radius r_1 is greater than 30,000 km, the mass ratio ranges from 0.20 to 0.24 and therefore using the EP stage does not show any significant benefit.

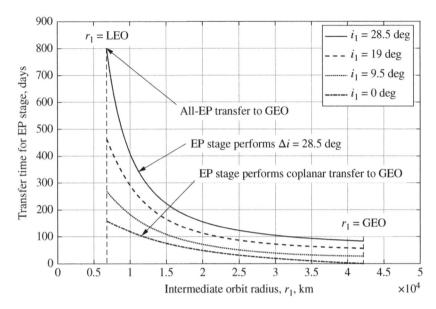

Figure 9.12 Electric propulsion transfer time for combined chemical-electric propulsion low-Earth orbit (LEO) to geostationary-equatorial orbit (GEO) transfer.

Figure 9.12 shows the low-thrust transfer time for the EP stage to reach GEO. The top curve in Figure 9.12 represents low-thrust transfers that perform the entire 28.5° change in inclination. The upper left point in Figure 9.12 is the all-EP transfer, and its 800-day transfer time makes it an unattractive option for satellite operators. The reader should note that the all-EP LEO-GEO transfer presented in Figure 9.12 exhibits a much higher transfer time (800 days) compared with the Hall-effect LEO–GEO transfer time in Table 9.3 (241 days). This dramatic time increase is due to the difference in time constants: $\tau = m_0/\dot{m} = 2{,}691$ days for the EP system used for Figure 9.12, and $\tau = m_0/\dot{m} = 804$ days for the EP system used for Table 9.3. The time constant τ is very large for the all-EP transfers presented here because the initial mass in LEO is very large (i.e., $m_0 = 13{,}300$ kg). Note that the initial mass in LEO is $m_0 = 2{,}500$ kg for the low-thrust LEO–GEO transfers presented in Table 9.3.

Mission operators can use Figures 9.11 and 9.12 to make a trade between the GEO payload mass and the EP transfer time. For example, suppose that mission designers require a mass ratio of at least 0.4 in order to justify the use of the additional EP stage. Figure 9.11 shows that the intermediate orbit $r_1 = 12{,}040$ km, $i_1 = 28.5°$ will provide this mass ratio, and Figure 9.12 shows that the corresponding EP transfer time is 306 days. However, we could also select the intermediate orbit $r_1 = 10{,}210$ km, $i_1 = 19°$ and achieve a mass ratio of 0.4 with an EP transfer time of 283 days. Table 9.4 summarizes combinations of r_1 and i_1 that result in a final mass ratio of 0.4. The smallest EP transfer time is about 276 days for an intermediate orbit where $r_1 = 11{,}072$ km and $i_1 = 21°$. We could perform a similar analysis where we vary r_1 and i_1 to achieve a desired EP transfer time (say, 150 days) and select the combination that maximizes the final mass ratio. Table 9.5 presents this tradeoff where $r_1 = 15{,}164$ km and $i_1 = 17°$ provides the maximum mass ratio ($m_{GEO}/m_{LEO} = 0.3106$) for an EP transfer time of 150 days.

Table 9.4 Combined chemical-electric propulsion low-Earth orbit (LEO) to geostationary-equatorial orbit (GEO) transfers with $m_{GEO}/m_{LEO} = 0.4$.

Intermediate orbit radius, r_1 (km)	Intermediate inclination, i_1	Electric propulsion transfer time (days)
12,042	28.5°	306.4
12,089	27.0°	295.5
11,973	25.0°	284.4
11,641	23.0°	277.6
11,072	21.0°	276.2
10,210	19.0°	283.2
8,882	17.0°	307.3

Table 9.5 Combined chemical-electric propulsion low-Earth orbit (LEO) to geostationary-equatorial orbit (GEO) transfers with electric propulsion transfer time = 150 days.

Intermediate orbit radius, r_1 (km)	Intermediate inclination, i_1	Final mass ratio, m_{GEO}/m_{LEO}
20,652	28.5°	0.2733
19,961	27.0°	0.2805
19,016	25.0°	0.2895
18,055	23.0°	0.2976
17,087	21.0°	0.3042
16,120	19.0°	0.3087
15,164	17.0°	0.3106
14,228	15.0°	0.3093
13,317	13.0°	0.3045
12,432	11.0°	0.2962

The ARTEMIS spacecraft demonstrated the combined use of chemical and electric propulsion modes to reach its target orbit. In July 2001, the European Space Agency (ESA) attempted to send the ARTEMIS telecommunications satellite to GEO using the Ariane 5 launch vehicle [7]. Due to a partial failure of the Ariane booster, the ARTEMIS spacecraft reached a sub-nominal elliptical orbit with perigee and apogee altitudes of 580 and 17,350 km, respectively. The Ariane's target was a geostationary transfer orbit (GTO) with an apogee altitude of 35,786 km (i.e., GEO altitude). Although the spacecraft was equipped with a chemical upper stage (intended for the GTO apogee burn to complete the transfer to GEO), this stage did not have enough liquid propellant to complete the transfer from its low-energy elliptical orbit. Mission operators at ESA redesigned the transfer strategy to GEO by utilizing the chemical-propulsion upper stage and the onboard EP ion thrusters (the ion thrusters were intended for on-orbit station-keeping). First, ARTEMIS fired its chemical stage at several perigee passes to raise apogee to an altitude of 31,000 km. Next, the chemical stage was fired at successive apogee passes to raise perigee and establish a circular orbit with an altitude of 31,000 km (Example 7.5 analyzed the impulsive Δv requirements of the chemical-propulsion stage). Finally, the ion thrusters ($T = 0.015$ N) performed the remaining orbit transfer from the 31,000-km altitude circle to GEO. ARTEMIS reached its GEO target orbit in late January 2003 more than 18 months after its launch (the total EP thrusting time was about 9 months). Despite the long transfer time and delay in satellite operations, the ion propulsion system essentially saved the ARTEMIS spacecraft from total failure. ARTEMIS successfully served as a communication satellite for Europe and the Middle East and exceeded its expected mission lifetime of 10 years. Problem 9.19 at the end of this chapter illustrates the low-thrust transfer phase of the ARTEMIS spacecraft.

9.8 Low-Thrust Transfer Issues

This chapter ends with a brief discussion of issues associated with low-thrust transfers. We have hinted at some of these issues in previous sections.

Obtaining closed-form (analytical) expressions for the low-thrust Δv and transfer time is only feasible for circle-to-circle transfers, pure inclination changes between circular orbits, and circle-to-circle transfers with an inclination change. Furthermore, these analytical solutions (particularly Edelbaum's method) assume that the low-thrust transfer involves continuous thrust without interruption. We were able to develop an approximate Earth-shadow model for coplanar circle-to-circle transfers in Section 9.4. However, it is difficult to extend this technique to maneuvers with plane changes. Determining the accurate Earth-shadow conditions is a complex calculation due to the continually changing three-dimensional orientation of the orbit relative to a geocentric frame and the continually changing sun–Earth vector with its seasonal variations. For example, it is not too difficult to visualize that a near-GEO orbit will *always* experience an Earth-shadow eclipse at the March and September equinox dates but will *not* enter the Earth's shadow at the June and December solstice dates (see Figure 3.1). Because solar arrays provide the power for EP systems, the Earth-eclipse periods of zero thrust will extend the total transfer time.

Low-thrust transfers between arbitrary orbits (such as elliptical orbits) do not have analytical solutions, and therefore numerical integration of the powered equations of motion is necessary. For example, there is no analytical solution for the low-thrust transfer from GTO to GEO. Furthermore, merely finding a *feasible* low-thrust transfer between the given initial and target orbits is usually not acceptable. Trajectory optimization techniques determine the optimal thrust-control steering program that minimizes or maximizes some desired performance index. It turns out that Edelbaum's solution (9.47) is the minimum Δv transfer between inclined circular orbits with continuous thrust. Conway [6] presents a variety of numerical trajectory optimization methods for low-thrust transfers.

Transit through the Van Allen radiation belts causes solar array power to degrade. High-energy particles (electrons and protons) exist in two toroidal "belt" regions in the Earth's magnetosphere. The more severe inner belt has peak-radiation dose rates at altitudes ranging from 2,000 to 6,000 km. Extended exposure of solar cells in this high-flux environment causes significant power degradation over time. It is possible, however, to limit the power degradation by shielding the solar cells but this approach adds mass to the power system. Modeling the radiation environment and its interaction with solar cells is extremely complicated and uncertain due to the variability in radiation flux caused by solar flares. Low-thrust transfers that pass through the radiation belts must account for power degradation effects because diminishing power will cause diminishing thrust acceleration, which in turn slows down the transfer rate through the belts! A combined chemical-electric propulsion strategy is useful in this scenario: the chemical-propulsion stage raises the orbit above the most severe region of the belts, and the EP stage completes the transfer with reduced power degradation.

9.9 Summary

In this chapter, we discussed the fundamentals of low-thrust orbit transfers. We presented a brief introduction to electric propulsion systems and the associated interactions between input power, thrust, specific impulse, and mass-flow rate. Next, we developed analytical solutions for low-thrust transfers by integrating Gauss' variational equations where the small propulsive thrust is a perturbation. These analytical solutions only exist for special orbit-transfer scenarios, such as quasi-circular spiral transfers between coplanar circular orbits and pure inclination changes for circular orbits. The analytical solutions allow us to calculate the low-thrust velocity increment (Δv), which through the rocket equation determines propellant mass and transfer time. We also developed a method that estimates the transfer time for quasi-circular, low-thrust transfers in the presence of Earth-shadow eclipses. Finally, we ended the chapter by presenting Edelbaum's analytical method for determining low-thrust transfers between inclined circular orbits, and combined chemical-electric propulsion strategies for reducing the transfer time.

References

1 Sutton, G.P. and Biblarz, O., *Rocket Propulsion Elements*, 8th edn, John Wiley & Sons, Inc., Hoboken, NJ, 2010.
2 Polk, J.E., Brinza, D., Kakuda, R.Y., Brophy, J.R., Katz, I, Anderson, J.R., Rawlin, V. K., Patterson, M. J, Sovey, J., and Hamley, J., "Demonstration of the NSTAR Ion Propulsion System on the Deep Space One Mission," Paper IEPC-01-075, International Electric Propulsion Conference, Pasadena, CA, October, 2001.
3 Welander, B., Carpenter, C., de Grys, K., Hofer, R. R., Randolph, T.M., "Life and Operating Range Extension of the BPT-4000 Qualification Model Hall Thruster," Paper AIAA-2006-5263, Joint Propulsion Conference and Exhibit, Sacramento, CA, July, 2006.
4 Edelbaum, T.N., "Propulsion Requirements for Controllable Satellites," *ARS Journal*, Vol. **31**,1961, pp. 1079–1089.
5 Kechichian, J.A., "Reformulation of Edelbaum's Low-Thrust Transfer Problem Using Optimal Control Theory," *Journal of Guidance, Control, and Dynamics*, Vol. **20**, No. 5, 1997, pp. 988–994.

6 Conway, B.A. (ed.), *Spacecraft Trajectory Optimization*, Cambridge University Press, New York, 2010.

7 Killinger, R., Gray, H., Kukies, R., Surauer, M., Saccoccia, G., Tomasetto, A., and Dunster, R., "ARTEMIS Orbit Raising In-Flight Experience with Ion Propulsion," Paper IEPC-2003-096, International Electric Propulsion Conference, Toulouse, France, March, 2003.

Problems

Conceptual Problems

9.1 The Aerojet BPT-4000 Hall thruster is operating with an input power of 3.5 kW and a specific impulse of 1,840 s. If the Hall thruster is producing 216 mN of thrust, determine its thruster efficiency.

9.2 An ion thruster is operating with an input power of 2 kW and is producing 79 mN of thrust. The thruster efficiency is 61%. Determine the specific impulse of the ion thruster.

9.3 In our derivation of the analytical low-thrust solutions we assumed that semimajor axis changed very little over a single powered orbital revolution (hence the quasi-circular nature of the unwinding spiral trajectory). However, as the powered spacecraft "spirals out" to greater orbital radii (and greater orbital energy), it will eventually violate the quasi-circular assumption. If the thrust acceleration is 10^{-7} km/s^2, determine the radius where the change in semimajor axis exceeds 1% over a single powered orbital revolution.

Problems 9.4–9.7 involve an electric-propulsion device with specific impulse I_{sp} = 2,900 s, thruster efficiency η = 60%, and input power P_{in} = 2 kW.

9.4 A 1,000-kg geocentric satellite is in a circular orbit with radius r = 8,500 km. Determine the orbital radius and spacecraft mass after 25 days of continuous thrusting along the velocity vector (i.e., thrust is uninterrupted).

9.5 Repeat Problem 9.4 for the case where thrust is aligned in the opposite direction as the satellite's velocity vector.

9.6 A 1,000-kg geocentric satellite is in a circular orbit with radius r = 8,500 km. Determine the total transfer time to reach an orbital radius of 9,500 km for two cases: (a) continuous thrust; and (b) thrust interrupted by Earth-shadow eclipses.

9.7 A 1,000-kg geocentric satellite is in a circular orbit with radius r = 35,000 km and inclination i = 28.5°. Determine the propellant mass required to reduce the orbital inclination to 15°.

9.8 An Earth-observation satellite is designed to operate in an 18,500-km altitude circular orbit with inclination i = 45°. However, the launch vehicle fails to achieve the target orbit and instead delivers the 2,000-kg satellite to a circular orbit with an altitude of 17,300 km and inclination of 45°. Fortunately, the satellite is equipped with ion thrusters (intended for station-keeping) that have a mass-flow rate of 0.13 kg/day and total thrust of 50 mN.

a) Determine the xenon propellant mass required to complete the orbit transfer using the ion thrusters.

b) Compute the maximum possible Earth-shadow angle that could exist for the satellite in its initial orbit.

c) Compute the low-thrust transfer time with and without Earth-shadow eclipses.

9.9 A spacecraft is using low-thrust electric propulsion to perform a coplanar orbit-raising maneuver. The spacecraft begins in a circular orbit with radius $r_0 = 12,500$ km and with mass $m_0 = 3,000$ kg. The ion propulsion system has a specific impulse $I_{sp} = 3,100$ s and its mass-flow rate is 1.12 kg/day. Compute the quasi-circular orbital radius after 100 days of continuous thrusting (i.e., there are no Earth-shadow effects).

9.10 A spacecraft has an initial mass of 1,400 kg in a 400-km altitude low-Earth orbit (LEO) with an inclination of 28.5°. The spacecraft's low-thrust propulsion system has a specific impulse of 3,100 s and a mass-flow rate of 0.61 kg/day. Determine the final spacecraft mass delivered to geostationary-equatorial orbit (GEO) and the total low-thrust transfer time for two transfer strategies: (a) a coplanar spiral to GEO altitude followed by a pure inclination change to achieve an equatorial orbit; and (b) a three-dimensional low-thrust transfer that simultaneously changes orbital energy and inclination. Assume continuous-thrust transfers (i.e., no shadow eclipses).

9.11 A 3,500-kg satellite is in an 8,000-km altitude circular Earth orbit with an inclination of 85°. The target orbit is an 8,000-km circular polar Earth orbit ($i = 90°$). Compare the propellant mass and powered-time requirements for two propulsive options: (a) a chemical rocket engine with thrust $T = 1,850$ N and $I_{sp} = 320$ s; and (b) a low-thrust Hall-effect thruster with $T = 0.7$ N and $I_{sp} = 1,600$ s (assume continuous thrusting).

9.12 A European Space Agency launch vehicle delivers a 10,700-kg communication satellite to an equatorial 200-km altitude circular orbit. The satellite consists of a chemical-propulsion stage ($I_{sp,c} = 320$ s, tankage fraction $K_t = 0.09$) mated to an electric-propulsion (EP) stage with a single Hall thruster ($I_{sp,EP} = 1,850$ s, $\eta = 0.55$, $P_{in} = 4$ kW). The satellite's target orbit is GEO and a mission constraint sets the low-thrust transfer time to 60 days (neglect Earth-shadow eclipses). Determine:

a) The radius of the starting orbit for the low-thrust transfer.

b) The impulsive velocity increment Δv_c and chemical-stage propellant mass.

c) The low-thrust velocity increment Δv_{EP} and xenon propellant mass.

d) The final satellite mass in GEO.

MATLAB Problem

9.13 Write an M-file that will compute the transfer time, propellant mass, and low-thrust Δv for a low-thrust transfer between inclined circular orbits. The inputs should be the initial and final circular-orbit altitudes (km), the initial and final orbital inclinations (degrees), the spacecraft's initial mass (kg), the electric-propulsion thrust magnitude (N), and the specific impulse (s). Assume continuous-thrust transfers with no thrust interruption.

Mission Applications

9.14 The initial mass of an electric-propulsion spacecraft is

$$m_0 = m_p + m_{pp} + m_{PL}$$

where m_p is the propellant mass, m_{pp} is the total power-plant mass (solar array, power processing units, thrusters, and xenon propellant tank), and m_{PL} is the payload mass. The ratio of the input array power P_{in} to the power-plant mass is called the "specific power," $\alpha = P_{in}/m_{pp}$. Suppose an electric-propulsion spacecraft has an input power of 8.5 kW, specific impulse $I_{sp} = 3000$ s, thruster efficiency $\eta = 64\%$, total xenon propellant mass $m_p = 90$ kg, specific power $\alpha = 40$ W/kg, and payload mass $m_{PL} = 1,000$ kg. Determine the low-thrust velocity increment Δv, the total low-thrust transfer time (in days), and the final orbit for a coplanar transfer starting at a circular 12,300 km radius geocentric orbit. Assume continuous (uninterrupted) thrust and complete usage of all xenon propellant.

9.15 Mission designers want to use Hall thrusters for a low-thrust transfer from a circular 8,000-km altitude mid-Earth orbit (MEO) to GEO. The spacecraft mass in MEO is 1,000 kg and the initial MEO has an inclination of 20°. The Hall thruster system has $I_{sp} = 1,650$ s and thruster efficiency $\eta = 0.56$. Determine the minimum input power so that the MEO–GEO transfer time is 120 days. Thrust is continuous during the low-thrust transfer.

9.16 An engineer is considering two electric propulsion options for transferring a spacecraft to GEO: (a) an ion thruster with specific impulse $I_{sp} = 3,100$ s and thruster efficiency $\eta = 0.65$; and (b) a Hall thruster with specific impulse $I_{sp} = 1,700$ s and thruster efficiency $\eta = 0.53$. Both options will use an 8-kW solar array for input power, and both options begin with a total spacecraft mass of 900 kg. The initial orbit is circular with an altitude of 10,000 km and inclination of 15°. The best option will provide the highest "transportation rate" which is defined as the spacecraft mass delivered to GEO divided by the low-thrust transfer time (in days). Which thruster option provides the greatest "transportation rate"? Assume that the thrusters operate continuously during the orbit transfer to GEO (i.e., ignore Earth-shadow effects).

9.17 Satellite operators want to reposition a satellite that is in a GEO. Because they want to initiate an eastward drift, they use the ion engine to spiral inward to a lower circular orbit with a radius of 41,500 km. After the proper coasting time in the lower orbit, the ion thruster raises the orbit back to GEO. If the satellite's initial mass in GEO is 2,500 kg and the ion thruster has a specific impulse of 3,100 s, determine the total xenon propellant mass for the repositioning maneuver.

9.18 Mission analysts want to design an orbit-transfer strategy that delivers a total satellite mass of 800 kg to a 20,000-km altitude circular orbit. They plan to use a Hall thruster for the final transfer phase to the target orbit. The Hall-effect device has a thrust magnitude of 160 mN and a specific impulse of 1,800 s. The satellite user (the customer) demands that the low-thrust orbit transfer take no more than 90 days, and therefore the mission planners decide to use a chemical-propulsion stage (I_{sp} = 325 s) to perform the initial orbit transfer from a 200-km circular LEO. Determine the total satellite mass in LEO, the chemical-stage propellant mass, the xenon propellant mass, and the radius of the starting orbit for the low-thrust transfer. All orbit transfers are coplanar. Determine the low-thrust transfer time with and without Earth-shadow eclipses.

9.19 In July 2001, the European Space Agency launched the ARTEMIS spacecraft on an Ariane 5 booster rocket. As discussed in Section 9.7, the Ariane launch vehicle encountered a partial failure and did not achieve the proper intermediate orbit required for a transfer to GEO. Consequently, the onboard chemical-propulsion upper stage raised the ARTEMIS spacecraft to a circular orbit with an altitude of 31,000 km. The ARTEMIS spacecraft then used its electric propulsion ion thrusters (originally intended for on-orbit inclination control) to complete the orbit transfer from the 31,000-km altitude circular orbit to GEO. Determine the total xenon propellant mass for this coplanar transfer and the total transfer time (include Earth-shadow effects). The ARTEMIS spacecraft's mass in the 31,000 km altitude circular orbit is 1,800 kg, its total thrust from the ion engines is 0.015 N, and the specific impulse is 3,200 s.

9.20 Design a feasible orbit-transfer strategy from LEO to GEO using combined chemical- and electric-propulsion stages. Assume that the Atlas V 521 is the launch vehicle of choice and therefore the total spacecraft mass in LEO is 12,725 kg (LEO is a 500-km altitude circular orbit with an inclination of 28.5°). Search the engineering literature and select an existing EP thruster. You will need to determine the EP input power, specific impulse, and total thrust magnitude (remember that you may use multiple thrusters simultaneously). In addition, select the appropriate parameters for the chemical stage. Design a feasible orbit-transfer strategy that delivers a payload mass of at least 2,000 kg to GEO with a low-thrust transfer time of less than 100 days. Compute payload mass, m_{PL}, by using the total mass of the spacecraft at the start of the EP-powered phase:

$$m_1 = m_{p,EP} + m_{pp} + m_{PL}$$

where $m_{p,EP}$ is the xenon propellant mass and m_{pp} is the total EP power-plant mass (solar array, power processing units, thrusters, and xenon propellant tank). Power-plant mass can be computed using the "specific power," $\alpha = P_{in}/m_{pp} = 40$ W/kg. Clearly document your orbit-transfer design by presenting all of the important characteristics of the chemical and EP stages, and the two orbit-transfer phases.

10

Interplanetary Trajectories

10.1 Introduction

All prior chapters have primarily focused on satellite motion influenced by a single central gravity field (the exception is the restricted three-body problem studied in Section 5.6). In fact, we can obtain closed-form analytical solutions only for satellite motion governed by the two-body problem (it may be useful for the reader to review the assumptions and limitations of the two-body problem presented in Chapter 2). Much of our discussion so far has dealt with determining and predicting the motion of Earth-orbiting satellites. In doing so, we have laid the foundation for computing orbital position and velocity, time of flight (TOF), and the velocity increment (Δv) associated with maneuvers between orbits.

As the title indicates, this chapter involves computing space trajectories from an initial planetary body to a target planet (of course, all interplanetary missions begin with departure from Earth). At first glance, the reader may wonder if we have the analytical tools to tackle such a daunting problem. Remember that our "analytical tool-kit" consists of equations that describe two-body motion: conservation of energy and angular momentum, and Kepler's laws. Therefore, a reasonable approach to obtaining an interplanetary trajectory is to break it into a sequence of two-body segments with respect to the appropriate central gravitational body. Using our "two-body tool-kit," we can obtain closed-form solutions of satellite motion relative to the dominant gravity field. Finally, we ensure the continuity of the interplanetary trajectory by piecing together (or "patching") the various two-body orbits (or conic sections). This strategy is the basis of the *patched-conic method*, which serves as a very useful technique for preliminary design of interplanetary missions. As with orbital maneuver analysis studied in Chapters 7–9, the ultimate goal of interplanetary mission analysis is to evaluate the total velocity increments required to carry out a space mission. It is extremely important that the reader keeps this objective in mind: minimizing the total Δv will minimize the propellant mass (through the rocket equation) and ultimately determine how much payload mass a space vehicle delivers to the planetary target.

Before beginning our discussion of the patched-conic method, it is instructive to survey a few actual interplanetary missions. Figure 10.1 shows the trajectory of the Mariner 2 mission to Venus in 1962. Mariner 2 was the first spacecraft to encounter a planetary target. Figure 10.1 shows the positions of Earth and Venus in their respective orbits about the sun at launch on August 27, 1962 (recall that the inertial $+X_H$ axis points along the

Space Flight Dynamics, First Edition. Craig A. Kluever.
© 2018 John Wiley & Sons Ltd. Published 2018 by John Wiley & Sons Ltd.
Companion website: www.wiley.com/go/Kluever/spaceflightmechanics

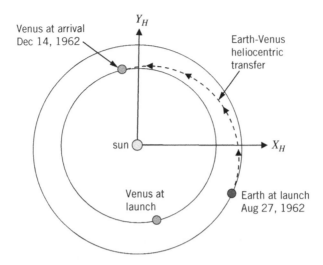

Figure 10.1 Mariner 2 heliocentric trajectory.

vernal equinox direction, and that Earth crosses the $+X_H$ axis on the first day of the autumnal equinox, roughly September 22 or 23). Figure 10.1 also shows the spacecraft's arrival at Venus 3.5 months later in mid-December. Clearly, the timing of the launch must be planned precisely so that the Mariner 2 spacecraft encounters Venus when its trajectory crosses Venus' orbit. This illustration also shows that the Mariner 2 spacecraft followed an inward elliptical transfer from Earth's orbit to Venus' orbit. However, unlike the orbit transfers discussed in Chapter 7, Figure 10.1 shows a *heliocentric* (or sun-centered) transfer where the sun is the primary gravitational body. Because Mariner 2 spent the great majority of its orbital transfer in heliocentric space, we may treat this trajectory as a two-body problem with respect to the sun. Of course, there are relatively brief periods (at the beginning and end of the mission) where the spacecraft is primarily under the gravitational influence of the departure and arrival planetary bodies. Figure 10.2 shows the heliocentric trajectory of the Mars Reconnaissance Orbiter (MRO), which was launched on August 12, 2005 and arrived at Mars 7 months later on March 10, 2006. In this case, the interplanetary trajectory is an outward heliocentric ellipse departing the inner orbit (Earth) and arriving at an outer orbit (Mars). Figure 10.2 also shows that Earth's orbit is nearly circular while Mars' orbit is noticeably eccentric; in fact, the MRO arrival appears to occur as Mars is approaching its aphelion. Figure 10.3 shows the Cassini mission to Saturn which departed Earth on October 15, 1997 and arrived at Saturn on July 1, 2004 [1]. The Cassini interplanetary trajectory is much more complicated when compared with the "direct" Earth–Venus and Earth–Mars missions illustrated in Figures 10.1 and 10.2. Figure 10.3 shows that the Cassini spacecraft completed two orbits about the sun that included three so-called "gravity assists" (labeled in Figure 10.3 as "flybys") with Venus and Earth. After the August 1999 gravity assist with Earth, the Cassini spacecraft encountered Jupiter in late December 2000 for its final gravity assist before reaching Saturn 3.5 years later. A gravity assist is a close encounter with a planetary body in order to change the spacecraft's heliocentric orbit; we will analyze gravity assists at the end of this chapter.

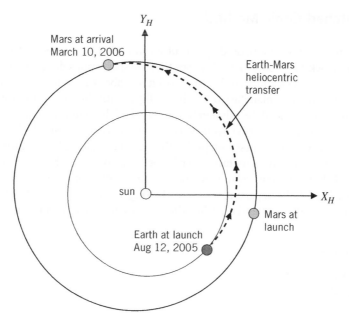

Figure 10.2 Mars Reconnaissance Orbiter heliocentric trajectory.

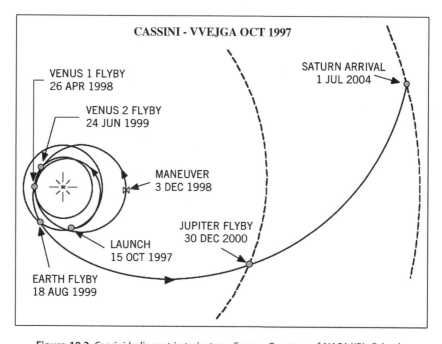

Figure 10.3 Cassini heliocentric trajectory. *Source*: Courtesy of NASA/JPL-Caltech.

10.2 Patched-Conic Method

Figure 10.4 presents a schematic diagram of an interplanetary mission from Earth to Mars. Using "clock-face coordinates," we see that the spacecraft's departure from Earth's orbit is at about 5 o'clock and arrival at Mars' orbit is at about 11 o'clock. The heliocentric transfer between the two planetary orbits is an ellipse with the sun as the primary gravitational body. This phase of the mission (typically called the "interplanetary cruise") lasts 7–9 months for an Earth–Mars trajectory. However, in order to begin the interplanetary cruise, the spacecraft must first escape Earth's gravitational pull and transition from "near-Earth space" to "interplanetary space." The insert figure in the lower right corner of Figure 10.4 shows a "zoomed-in view" of the spacecraft's departure from Earth orbit. Here we see a hyperbolic departure trajectory leaving a low-Earth orbit (LEO) and eventually crossing a fictitious boundary called the "Earth sphere of influence" (Earth SOI). In

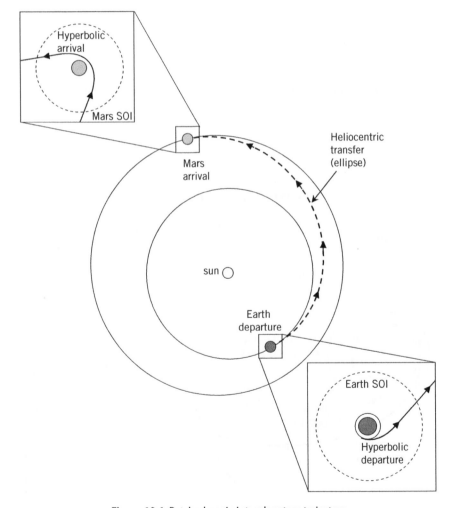

Figure 10.4 Patched-conic interplanetary trajectory.

a similar fashion, the upper left corner of Figure 10.4 shows a zoomed-in view of the hyperbolic arrival trajectory at Mars. In this case, the spacecraft crosses the fictitious "Mars sphere of influence" (Mars SOI) as it enters Mars' gravity field.

10.2.1   Sphere of Influence

The SOI is a fictitious boundary for two-body motion about a planet. If a satellite is within the Earth's SOI, then its orbit can be approximated as a two-body problem with the Earth as the central body. However, as a spacecraft leaves a planet and crosses its SOI, we can approximate its orbit as a two-body problem with the sun as the central body. Clearly, the SOI concept does not represent the true physics of motion in an N-body gravity field. However, the SOI is a convenient fabrication that allows us to analyze a complicated interplanetary trajectory as a sequence of two-body problems (after all, we can only obtain analytical solutions of two-body motion).

Laplace defined the SOI boundary using the ratio of the perturbing acceleration and the central-body acceleration. To show this concept, consider a spacecraft that is at a great distance from the Earth where we assume that the Earth and sun are the only gravitational bodies influencing the satellite's motion. The total gravitational acceleration of the satellite as expressed in an Earth-centered frame is

$$\mathbf{a}_1 = \mathbf{a}_E + \mathbf{a}_{p,s} \tag{10.1}$$

where $\mathbf{a}_E$ is the central-body acceleration due to the Earth's gravity and $\mathbf{a}_{p,s}$ is the *perturbing* acceleration due to the sun's gravity. In a similar fashion, the total gravitational acceleration of the satellite as expressed in a *sun*-centered frame is

$$\mathbf{a}_2 = \mathbf{a}_s + \mathbf{a}_{p,E} \tag{10.2}$$

where $\mathbf{a}_s$ is the central-body acceleration due to the sun's gravity and $\mathbf{a}_{p,E}$ is the perturbing acceleration due to Earth. Laplace equated the ratios of perturbing and central-body accelerations:

$$\frac{a_{p,s}}{a_E} = \frac{a_{p,E}}{a_s} \tag{10.3}$$

The SOI is the boundary where these two ratios are equal. We will not present the derivation defining the SOI radius; the interested reader may consult Battin [2; pp. 395–397], Vallado [3; pp. 945–948], or Prussing and Conway [4; pp. 155–158] for the details. The boundary where Eq. (10.3) holds is approximately spherical, and the radius of the Earth's SOI is

$$r_{\text{SOI}} = r_E \left(\frac{M_E}{M_s} \right)^{2/5} \tag{10.4}$$

where r_E is the mean Earth–sun distance, M_E is the mass of the Earth, and M_s is the mass of the sun. We may compute the SOI radius of any planet in our solar system by using Eq. (10.4) with the appropriate values for the mean planet–sun distance (i.e., semimajor axis) and planetary mass. Table 10.1 presents each planet's SOI radius as well as its semimajor axis in order to put the SOI into proper perspective. Note that Table 10.1 presents the planet–sun distance in terms of the *astronomical unit* (AU) which is the mean

Table 10.1 Planetary sphere of influence radii.

Planet	Semimajor axis, a (AU)	SOI radius, r_{SOI} (km)	SOI radius, r_{SOI} (AU)	r_{SOI}/a
Mercury	0.387	112,400	0.0008	0.0019
Venus	0.723	616,300	0.0041	0.0057
Earth	1.000	924,800	0.0062	0.0062
Mars	1.524	577,300	0.0039	0.0025
Jupiter	5.203	48,208,900	0.3223	0.0619
Saturn	9.555	54,655,500	0.3653	0.0382
Uranus	19.218	51,795,300	0.3462	0.0180
Neptune	30.110	86,895,000	0.5809	0.0193

distance from the Earth to the sun (essentially the semimajor axis of the Earth's heliocentric orbit, i.e., 1 AU = 149,597,871 km). While the Earth's SOI radius is nearly 1 million km, it is only 0.6% of the mean Earth–sun distance and hence it would be difficult to see the SOI on a heliocentric trajectory plot such as Figure 10.4. Table 10.1 shows that Jupiter has the largest SOI as a percentage of its semimajor axis.

The *patched-conic method* breaks the entire interplanetary mission into a sequence of distinct phases where each phase is a separate two-body problem. Figure 10.4 illustrates the three distinct phases summarized below:

1) Earth-departure phase: The spacecraft leaves the Earth's gravity field along a hyperbolic escape trajectory, which is analyzed using two-body motion in an Earth-centered inertial frame. When the spacecraft crosses the Earth's SOI, it has reached the first "patch point" where we must "turn off" Earth's gravity and "turn on" the sun's gravity.

2) Heliocentric phase: The spacecraft follows an elliptical transfer orbit where the sun is the primary gravitational body. This "interplanetary cruise" is analyzed using two-body motion in the heliocentric-ecliptic coordinate frame. Because the radius of each planet's SOI is small relative to the scale of the solar system, we can use the heliocentric position vectors of the departure and arrival planets as the boundary conditions of the heliocentric phase.

3) Planetary-arrival phase: The spacecraft crosses the target planet's SOI (the second "patch point") where we "turn off" the sun's gravity and "turn on" the planet's gravity. The arrival trajectory inside the planet's SOI is a hyperbola analyzed in an inertial planet-centered frame.

As previously mentioned, the heliocentric phase encompasses the great majority of the entire mission as the spacecraft cruises from Earth's orbit to the target planet's orbit. The duration of the interplanetary cruise depends on the target planet and may take several months to several years. In contrast, the planet-centered departure and arrival phases take only a few days. Therefore, the heliocentric phase is typically determined first in order to establish the launch and arrival dates and the mission duration.

Another consequence of the using the patched-conic method is that we must perform coordinate transformations between the various body-centered frames at the "SOI patch points." It is critical that we differentiate between the spacecraft's velocity relative to a planet and its velocity relative to the sun. The reader should remember that the two-body problem assumes that we have an inertial frame fixed at the center of the sole gravitational body and hence velocity is relative to the central body. Therefore, when we analyze the heliocentric trajectory, the spacecraft's velocity must be relative to the sun. When a spacecraft departs (or arrives at) a planet, we must be careful to use the proper vector addition to determine the appropriate velocity for the subsequent two-body problem.

Before we present the patched-conic method, it is useful to define a new set of reference units for heliocentric transfers. Unlike Earth-centered orbits, the radial distances of heliocentric orbits are hundreds of millions of kilometers. A convenient way to work with heliocentric orbits is to use a set of reference units (or "canonical units") to normalize the distances, velocities, and time. For heliocentric orbits, the reference distance unit is 1 AU, or $r_{ref} = 149{,}597{,}871$ km. The reference velocity is defined as the circular orbital speed in a heliocentric orbit at 1 AU:

$$v_{ref} = \sqrt{\frac{\mu_s}{r_{ref}}} = 29.7847 \text{ km/s} \tag{10.5}$$

where the sun's gravitational parameter is $\mu_s = 1.327124(10^{11}) \text{ km}^3/\text{s}^2$. This reference velocity is the *mean* velocity of Earth in its orbit about the sun. The reference time unit (TU) is defined such that $v_{ref} = 1 \text{ AU/TU} = 29.7847$ km/s and hence 1 TU = 5,022,643 s (=58.132 days). Note that if we normalize the sun's gravitational parameter, we obtain $\mu_s = 1 \text{ AU}^3/\text{TU}^2$ [e.g., Eq. (10.5) shows that μ_s must be 1 AU^3/TU^2 so that $v_{ref} = 1$ AU/TU for radius $r_{ref} = 1$ AU]. Table 10.2 summarizes the reference canonical units for heliocentric orbits. We will express heliocentric orbits in terms of the canonical units AU, TU, and AU/TU and use Table 10.2 to convert TU to days and AU/TU to units of kilometers per second.

10.2.2 Coplanar Heliocentric Transfers between Circular Orbits

As a first demonstration of the patched-conic method, let us use the simplest possible model of our solar system: circular, coplanar planetary orbits. Before we model the planets' orbits as concentric circles, we will present their orbital elements. Table 10.3 presents the (approximate) orbital elements for the eight planets for the epoch (i.e., reference date) January 1, 2000. Table 10.3 comprises an *ephemeris* that determines the locations of the planets in the heliocentric-ecliptic frame at any desired date. Recall the definition of the longitude of perihelion (Chapter 3):

Table 10.2 Canonical units for a heliocentric system.

Dimension	Reference unit
Distance	1 AU = 149,597,871 km
Time	1 TU = 5,022,643 s = 58.132444 days
Velocity	1 AU/TU = 29.784690 km/s

Table 10.3 Planetary ephemeris for the epoch January 1, 2000.

Planet	Semimajor axis, a (AU)	Eccentricity, e	Inclination, i (°)	Longitude of the ascending node, Ω (°)	Longitude of perihelion, ϖ (°)	True longitude at epoch, l_0 (°)
Mercury	0.38710	0.20563	7.005	48.331	77.456	252.251
Venus	0.72333	0.00677	3.394	76.680	131.564	181.980
Earth	1.00000	0.01671	0.000	Undefined	102.937	100.466
Mars	1.52368	0.09340	1.850	49.558	336.060	355.433
Jupiter	5.20260	0.04849	1.303	100.464	14.331	34.351
Saturn	9.55491	0.05551	2.489	113.666	93.057	50.077
Uranus	19.21845	0.04630	0.773	74.006	173.005	314.055
Neptune	30.11039	0.00899	1.770	131.784	48.124	304.349

Source: Adapted from Ref. [3].

$$\varpi \equiv \Omega + \omega \tag{10.6}$$

The longitude of perihelion is measured from the $+X_H$ axis to the ascending node in the ecliptic plane, and then from the node to perihelion in the planet's orbital plane. Because Earth's orbital plane is the ecliptic plane, Ω is undefined and therefore ϖ is the angle from $+X_H$ to Earth's perihelion. The true longitude at epoch is

$$l_0 \equiv \Omega + \omega + \theta_0 \tag{10.7}$$

True longitude defines the angular position of a planet at a given date (January 1, 2000 for Table 10.3). For Earth, true longitude is measured in the ecliptic plane from the $+X_H$ axis to its orbital position. We will treat the five planetary elements a, e, i, Ω, and ϖ as constants and use Kepler's equation to determine the true longitude l at a desired date.

Table 10.3 shows that Venus' orbit is very nearly circular, whereas the orbits of Mars and Mercury are eccentric. Venus and Mercury exhibit the largest inclinations with respect to the ecliptic plane (3.4° and 7°, respectively). Table 10.4 presents a simplified "two-dimensional (2-D) ephemeris" for Venus, Earth, and Mars, where the planets' orbits are coplanar and circular. We will use the concentric-coplanar planetary system summarized in Table 10.4 to demonstrate the steps required for the patched-conic method. In Section 10.5, we will apply the patched-conic method to the full three-dimensional (3-D) planetary ephemeris in Table 10.3.

We will first consider an Earth–Mars interplanetary mission. In Chapter 7, we identified the Hohmann transfer as the minimum-energy transfer between circular orbits in terms of total Δv. Figure 10.5 shows the Hohmann transfer ellipse from Earth's circular orbit to Mars' circular orbit. Following the methods developed in Section 7.3, the semimajor axis of the Hohmann transfer is half of the sum of the circular radii of Earth and Mars:

$$a_t = \frac{1}{2}(r_E + r_M) = 1.26185 \text{ AU} \tag{10.8}$$

Table 10.4 Concentric-coplanar planetary models for the epoch January 1, 2000.

Planet	Radius (AU)	Angular velocity (deg/day)	True longitude at epoch, l_0 (°)
Venus	0.7233	1.6021	181.980
Earth	1.0000	0.9856	100.466
Mars	1.5237	0.5240	355.433

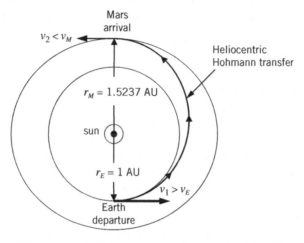

Figure 10.5 Heliocentric Hohmann transfer from Earth orbit to Mars orbit.

where $r_E = 1$ AU is the radius of Earth's orbit about the sun and $r_M = 1.5237$ AU is Mars' orbital radius (we will use subscript t to denote characteristics of the transfer). Energy of the Hohmann transfer ellipse is

$$\xi_t = \frac{-\mu_s}{2a_t} = -0.39624 \text{ AU}^2/\text{TU}^2 \tag{10.9}$$

where we have used heliocentric canonical units with the sun's gravitational parameter expressed as $\mu_s = 1 \text{ AU}^3/\text{TU}^2$. Finally, we compute the required velocities at each apse of the Hohmann transfer ellipse using the energy equation

$$\text{Perihelion:} \quad v_1 = \sqrt{2\left(\xi_t + \frac{\mu_s}{r_E}\right)} = \sqrt{\frac{-\mu_s}{a_t} + \frac{2\mu_s}{r_E}} = 1.09887 \text{ AU/TU} \tag{10.10}$$

$$\text{Aphelion:} \quad v_2 = \sqrt{2\left(\xi_t + \frac{\mu_s}{r_M}\right)} = \sqrt{\frac{-\mu_s}{a_t} + \frac{2\mu_s}{r_M}} = 0.72118 \text{ AU/TU} \tag{10.11}$$

Let us now carefully consider these velocity requirements for the heliocentric Hohmann transfer from Earth orbit to Mars orbit. The perihelion velocity $v_1 = 1.09887$ AU/TU is the spacecraft's velocity *relative to the sun* when it is at 1 AU (i.e., at Earth's orbit about the sun). Because Earth's orbital speed (relative to the sun) is $v_E = \sqrt{\mu_s/r_E} = 1$ AU/TU, the spacecraft must (somehow) achieve a velocity increment $\Delta v = v_1 - v_E = 0.09887$ AU/TU (=2.9448 km/s) in order to begin the heliocentric Hohmann transfer. The reader should remember that the heliocentric transfer *begins* when the spacecraft crosses the Earth's SOI as seen in Figure 10.4 (it is also important to remember that because the Earth's SOI is "small" on a solar-system scale, it is not shown in Figure 10.5).

Figure 10.6 shows a "zoomed-in" view of the Earth-departure hyperbola that begins the Hohmann transfer to Mars. We see that the spacecraft escapes Earth on a hyperbolic trajectory where the Earth is the central gravitational body. The spacecraft's *heliocentric* velocity at the first "patch point" (the Earth's SOI) can be stated in sentence form as

| Spacecraft's velocity | = | Earth's velocity | + | Spacecraft's velocity |
| relative to the sun | | relative to the sun | | relative to the Earth |

Or, expressed as a mathematical equation

$$v_1 = v_E + v_\infty^+ = 1.09887 \, \text{AU/TU} \tag{10.12}$$

where v_E is Earth's velocity relative to the sun (=1 AU/TU) and v_∞^+ is the hyperbolic excess velocity of the spacecraft as it crosses the SOI (remember that the superscript + indicates the hyperbolic *departure* asymptote). It is important to note that v_∞^+ is the spacecraft's velocity *relative to an Earth-centered frame*. Therefore, Eq. (10.12) is the first velocity transformation at the Earth's SOI: the spacecraft's velocity relative to the sun (v_1) is equal to the velocity of the moving coordinate frame (Earth's velocity relative to the sun, v_E) plus the spacecraft's velocity relative to the moving frame (v_∞^+). In general, Eq. (10.12) is a vector equation; however, for a Hohmann transfer, the departure asymptote is parallel with the Earth's velocity vector $\mathbf{v}_E$ as shown in Figure 10.6. Using Eq. (10.12), we see that the hyperbolic excess velocity required for the Hohmann transfer to Mars is

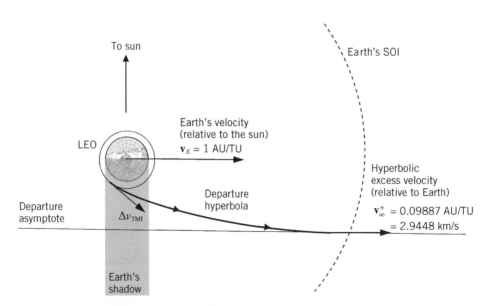

Figure 10.6 Earth-departure hyperbola for a Hohmann transfer to Mars.

$$v_\infty^+ = v_1 - v_E = 0.09887 \text{ AU/TU} = 2.9448 \text{ km/s} \tag{10.13}$$

This hyperbolic excess speed establishes the energy required for the Earth-departure hyperbola:

$$\xi = \frac{v_\infty^{+2}}{2} = \frac{v_p^2}{2} - \frac{\mu_E}{r_p} = 4.3359 \text{ km}^2/\text{s}^2 \tag{10.14}$$

where r_p and v_p are the radius and velocity of the Earth-departure hyperbola at perigee, respectively. Typically, a launch vehicle inserts the interplanetary probe into a circular low-Earth "parking orbit"; afterwards an upper stage is fired to establish the hyperbolic escape trajectory. Assuming a 185-km altitude LEO ($r_{\text{LEO}} = r_p = 6{,}563$ km), we see that the circular parking orbit velocity is $v_{\text{LEO}} = \sqrt{\mu_E/r_{\text{LEO}}} = 7.7932$ km/s. Using Eq. (10.14), the perigee velocity of the Earth-departure hyperbola is $v_p = 11.4079$ km/s. Finally, the velocity impulse in LEO for "trans-Mars injection" (TMI) is

$$\Delta v_{\text{TMI}} = v_p - v_{\text{LEO}} = 3.6147 \text{ km/s} \tag{10.15}$$

Figure 10.6 shows the TMI velocity impulse applied in LEO. It is important to emphasize that Δv_{TMI} is the only rocket burn required to send the spacecraft on its coasting trajectory to Mars. No orbital maneuvers occur at the Earth's SOI. The previous discussion of the Hohmann transfer shown in Figure 10.5 identified a velocity increment (0.09887 AU/TU = 2.9448 km/s) that must be added to Earth's sun-relative velocity v_E in order to initiate the *heliocentric* Hohmann transfer. The hyperbolic excess velocity v_∞^+ provides this velocity increment at the Earth's SOI but it is *not* the result of a rocket burn *performed at* the SOI boundary. Hyperbolic excess velocity v_∞^+ is a direct result of the TMI burn Δv_{TMI} applied in LEO as shown in Figure 10.6. In practice, an upper rocket stage provides the TMI impulse at the appropriate location in the low-Earth parking orbit. It is extremely important for the reader to understand the Earth-departure phase and the associated velocity transformations as the spacecraft moves from an Earth-centered frame (before the SOI) to a sun-centered frame (after the SOI).

Knowledge of Δv_{TMI}, initial mass in LEO, and the specific impulse of the upper stage determine the mass of the spacecraft sent to Mars on a Hohmann transfer. It is also of interest to note that the rocket burn to establish the escape hyperbola occurs when the spacecraft is in the Earth's shadow as seen in Figure 10.6.

Perhaps a more standardized method of characterizing the payload mass for an interplanetary transfer is to compute the launch energy C_3. Recall that we defined launch energy in Chapter 6 as the square of hyperbolic excess speed. For a coplanar Earth–Mars Hohmann transfer, the launch energy is

$$C_3 = \left(v_\infty^+\right)^2 = 8.6718 \text{ km}^2/\text{s}^2 \tag{10.16}$$

Now we may use the launch-energy performance curve for a specific launch vehicle to determine the payload mass. For example, Figure 6.9 shows the payload mass capabilities for the Atlas V 551 and Delta IV Heavy boosters. Reference [5] is an on-line launch-performance calculator provided by NASA. Table 10.5 presents the payload mass to Mars (via a coplanar Hohmann transfer) for four different launch vehicles.

Table 10.5 Payload mass for a Hohmann transfer to Mars with $C_3 = 8.67$ km²/s².

Launch vehicle	Payload mass to Mars (kg)
Atlas V 501	1,640
Falcon 9	2,750
Atlas V 551	5,195
Delta IV Heavy	9,090

Finally, we use Eq. (4.42) to compute the spacecraft's flight time on the Earth-departure hyperbola from perigee to the Earth's SOI

$$t - t_p = \sqrt{\frac{-a^3}{\mu_E}}(e \sinh F - F) \tag{10.17}$$

To use this equation, we must determine the semimajor axis and eccentricity of the hyperbolic trajectory and the hyperbolic anomaly (F) of the point in the escape trajectory when the spacecraft reaches the Earth's SOI. Semimajor axis is determined from the escape energy, Eq. (10.14):

$$\xi = \frac{v_\infty^{+2}}{2} = -\frac{\mu_E}{2a} = 4.3359 \text{ km}^2/\text{s}^2$$

Therefore, $a = -45{,}964.8$ km. Perigee position and velocity on the hyperbola determine the angular momentum, parameter, and eccentricity as follows:

Angular momentum: $h = r_p v_p = 74{,}870$ km²/s $= \sqrt{p\mu}$

Parameter: $p = \dfrac{h^2}{\mu} = 14{,}063$ km $= a(1 - e^2)$

Eccentricity: $e = \sqrt{1 - \dfrac{p}{a}} = 1.1428$

The trajectory equation (2.45) determines the spacecraft's true anomaly at the Earth's SOI:

$$r = \frac{p}{1 + e\cos\theta} = 924{,}800 \text{ km (Earth's SOI)}$$

which yields $\theta = 149.514°$. Using Eq. (4.43) to compute $\sinh F$ and F yields

$$\sinh F = \frac{\sqrt{e^2 - 1}\,\sin\theta}{1 + e\cos\theta} = 18.4539 \quad \text{and} \quad F = 3.6092$$

Using these values in Eq. (10.17), we find that the hyperbolic flight time from perigee (at LEO) to the Earth's SOI is 272,836 s = 75.8 h = 3.16 days. This calculation confirms our previous assertion that the Earth-departure phase lasts only a few days.

The previous discussion addresses the Earth-departure hyperbola; that is, the first phase of the patched-conic interplanetary trajectory. Now let us use Eq. (7.7) to determine the flight time on the *heliocentric* Hohmann transfer from Earth to Mars orbit:

$$t_H = \frac{\pi}{\sqrt{\mu_s}} a_t^{3/2}$$

Using $a_t = 1.2619$ AU and $\mu_s = 1$ AU^3/TU^2, we obtain $t_H = 4.4531$ TU. Using Table 10.2 to convert heliocentric time units (TU) to days, we see that the Hohmann-transfer flight time is 258.9 days or nearly 9 months.

Next, we turn our attention to the Mars-arrival phase. The spacecraft will follow a hyperbolic trajectory after it crosses Mars' SOI. To show this, we can calculate the Mars-relative velocity at Mars' SOI; that is, the second "patch point"

$$\mathbf{v}_\infty^- = \mathbf{v}_2 - \mathbf{v}_M = -0.08894 \text{ AU/TU} = -2.6490 \text{ km/s} \qquad (10.18)$$

where $\mathbf{v}_2$ is the aphelion velocity of the spacecraft on the Hohmann transfer (relative to the sun; Figure 10.5) and $\mathbf{v}_M$ is the circular orbital velocity of Mars relative to the sun. The magnitudes of these velocities are $v_2 = 0.72118$ AU/TU and $v_M = \sqrt{\mu_s/r_M} = 0.81012$ AU/TU, respectively. Figure 10.7 shows the Mars-arrival hyperbolic trajectory. Because the spacecraft's sun-relative velocity at aphelion (v_2) is *less* than Mars' sun-relative velocity (v_M), the spacecraft's hyperbolic trajectory approaches Mars along its "leading edge." In other words, the minus sign in Eq. (10.18) indicates that the Mars-relative velocity vector $\mathbf{v}_\infty^-$ is in the *opposite* direction as Mars' heliocentric velocity vector $\mathbf{v}_M$ as shown in Figure 10.7. It is very important to note that the instant the spacecraft crosses Mars'

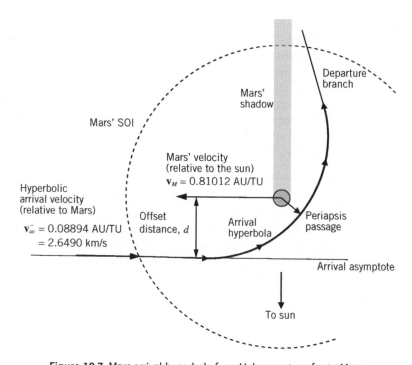

Figure 10.7 Mars-arrival hyperbola for a Hohmann transfer to Mars.

SOI, the patched-conic method demands that we switch to a two-body problem with an inertial frame fixed at the center of Mars. We may evaluate the energy of the two-body trajectory relative to Mars at its SOI boundary:

$$\xi_M = \frac{\left(v_\infty^-\right)^2}{2} - \frac{\cancelto{0}{\mu_M}}{r_\infty} = \frac{\left(v_\infty^-\right)^2}{2} \tag{10.19}$$

If the spacecraft has any Mars-relative velocity at the SOI boundary, then energy ξ_M is positive and hence the spacecraft *must* be on a hyperbolic trajectory relative to Mars.

Figure 10.7 shows the spacecraft traveling along the arrival branch of the hyperbola to periapsis and then departing Mars along the outgoing hyperbolic branch. A rocket burn or flight through Mars' atmosphere will alter the hyperbolic trajectory. Note that Figure 10.7 shows periapsis passage occurring on the sun-lit side of Mars. If the mission objectives call for a Mars orbiter, then the spacecraft fires an onboard rocket (typically at periapsis) to reduce energy and establish a closed orbit. If a direct landing is desired (such as the Mars Exploration Rover missions or Mars Science Laboratory), then the hyperbola's periapsis radius is targeted so that it is below the appreciable Martian atmosphere. The complete characteristics of the Mars-arrival hyperbola (such as its periapsis radius) can only be determined if the *offset distance d* is known. Figure 10.7 shows that d is the distance between the arrival asymptote and a parallel line passing through the center of Mars. We will discuss the planetary arrival trajectory in more detail in Section 10.4.

At this point, it is useful to review Figure 10.4 and identify the three distinct phases of the patched-conic method: (1) the Earth-departure hyperbola (Figure 10.6); (2) the heliocentric interplanetary cruise (Figure 10.5); and (3) the Mars-arrival hyperbola (Figure 10.7). Each phase consists of a two-body problem with respect to the appropriate central body. The reader should carefully note that Figure 10.4 shows a "general" patched-conic Earth–Mars mission where Mars' orbit is depicted accurately as an ellipse. Figures 10.5–10.7 present a specialized case: a coplanar Hohmann transfer between circular Earth and Mars orbits. It is important to remember that modeling the planets' orbits as coplanar circles about the sun simplifies the analysis at the cost of accurately calculating the heliocentric transfer time and launch energy. In Section 10.5, we will pose the heliocentric orbit transfer as Lambert's problem with the planetary positions defined by an accurate ephemeris (i.e., Table 10.3).

Example 10.1 Figure 10.8 shows a heliocentric Hohmann transfer from Earth to Venus. Assuming a coplanar transfer between circular planetary orbits, determine the following parameters: (1) flight time for the heliocentric interplanetary cruise phase; (2) launch energy C_3 for the Earth-departure phase; (3) velocity increment Δv_{TVI} for the "trans-Venus injection" (TVI); and (4) hyperbolic excess velocity v_∞^- at Venus' SOI. Assume that an upper stage performs the Δv_{TVI} burn in a 185-km altitude circular low-Earth orbit.

The semimajor axis of the Hohmann transfer is half of the sum of the circular radii of Earth and Venus:

$$a_t = \frac{1}{2}(r_E + r_V) = 0.8617 \text{ AU}$$

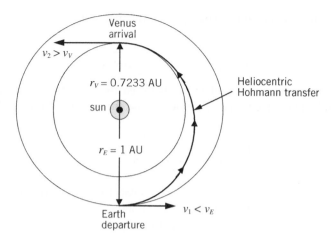

Figure 10.8 Earth–Venus heliocentric Hohmann transfer (Example 10.1).

The radius of Venus' orbit is 0.7233 AU. Flight time for the interplanetary cruise is half of the period of the Hohmann-transfer ellipse:

$$t_H = \frac{\pi}{\sqrt{\mu_s}} a_t^{3/2} = 2.5128 \text{ TU} = \boxed{146.1 \text{ days}}$$

Energy of the Hohmann transfer ellipse is

$$\xi_t = \frac{-\mu_s}{2a_t} = -0.58027 \text{ AU}^2/\text{TU}^2$$

Using the energy equation, we can compute the sun-relative velocity at Earth departure (aphelion on the Hohmann transfer):

$$\text{Aphelion:} \quad v_1 = \sqrt{2\left(\xi_t + \frac{\mu_s}{r_E}\right)} = 0.9162 \text{ AU/TU}$$

Immediately, we note that the spacecraft's sun-relative velocity v_1 is *less* than the Earth's orbital velocity ($v_E = 1$ AU/TU) as shown in Figure 10.8. Therefore, when the spacecraft crosses Earth's SOI, its velocity relative to the sun must be less than Earth's orbital velocity so that it begins an *inward* transfer to Venus (see Figure 10.8). The spacecraft's velocity relative to the sun as it crosses the Earth's SOI (at the first "patch point") is

$$\mathbf{v}_1 = \mathbf{v}_E + \mathbf{v}_\infty^+$$

The Earth-relative hyperbolic excess velocity is

$$\mathbf{v}_\infty^+ = \mathbf{v}_1 - \mathbf{v}_E = -0.0838 \text{ AU/TU} = -2.4954 \text{ km/s}$$

where $\mathbf{v}_1$ is the aphelion velocity of the spacecraft on the Hohmann transfer (relative to the sun; see Figure 10.8) and $\mathbf{v}_E$ is the circular orbital velocity of Earth relative to the sun. The negative sign on hyperbolic departure velocity indicates that $\mathbf{v}_\infty^+$ is in the opposite direction as $\mathbf{v}_1$. Figure 10.9 shows the Earth-departure hyperbola. The reader should note that the vector addition $\mathbf{v}_E + \mathbf{v}_\infty^+$ in Figure 10.9 yields $\mathbf{v}_1 = 0.9162$ AU/TU as required for the Hohmann transfer in Figure 10.8.

We can determine launch energy by squaring the magnitude of the hyperbolic excess velocity:

$$C_3 = \left(v_\infty^+\right)^2 = \boxed{6.2271 \text{ km}^2/\text{s}^2}$$

The TVI velocity impulse is the difference between the hyperbolic perigee velocity at radius r_{LEO} and LEO velocity. First, we determine the energy of the Earth-departure hyperbola using v_∞^+

$$\xi_E = \frac{\left(v_\infty^+\right)^2}{2} = \frac{v_p^2}{2} - \frac{\mu_E}{r_{\text{LEO}}} = 3.1135 \text{ km}^2/\text{s}^2$$

Using $r_{\text{LEO}} = 185 \text{ km} + R_E = 6{,}563 \text{ km}$ and $\mu_E = 3.986(10^5) \text{ km}^3/\text{s}^2$ (for an Earth-relative hyperbola), we find that perigee velocity is $v_p = 11.3003 \text{ km/s}$. Circular velocity in LEO is $v_{\text{LEO}} = \sqrt{\mu_E/r_{\text{LEO}}} = 7.7932 \text{ km/s}$. Hence, the TVI impulse is

$$\Delta v_{\text{TVI}} = v_p - v_{\text{LEO}} = \boxed{3.5070 \text{ km/s}}$$

The reader should compare the velocity impulses for trans-Mars injection and TVI: these two velocity impulses differ by a little more than 100 m/s! However, for a Mars mission, the departure asymptote is aligned with Earth's sun-relative velocity (Figure 10.6) whereas for a Venus mission, the departure asymptote is in the opposite direction as $\mathbf{v}_E$ (Figure 10.9).

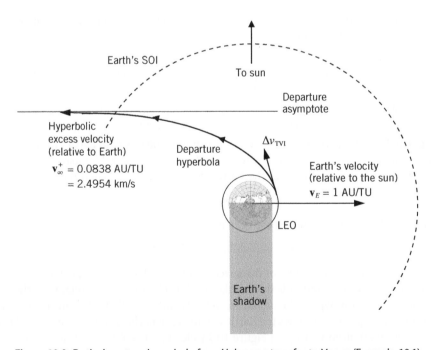

Figure 10.9 Earth-departure hyperbola for a Hohmann transfer to Venus (Example 10.1).

The final calculation involves the spacecraft's hyperbolic arrival velocity at Venus' SOI. First, we must compute the spacecraft's sun-relative velocity at Venus' orbit (perihelion on the Hohmann transfer):

$$\text{Perihelion:} \quad v_2 = \sqrt{2\left(\xi_t + \frac{\mu_s}{r_V}\right)} = 1.2667\,\text{AU/TU}$$

Venus' orbital velocity is $v_V = \sqrt{\mu_s/r_V} = 1.1758\,\text{AU/TU}$. Therefore, the spacecraft's sun-relative velocity is greater than Venus' orbital velocity at SOI arrival (see Figure 10.8). The Venus-relative hyperbolic arrival velocity is

$$v_\infty^- = v_2 - v_V = 0.0909\,\text{AU/TU} = \boxed{2.7066\ \text{km/s}}$$

We will revisit the Venus-arrival phase in Section 10.4.

10.3 Phase Angle at Departure

The previous section has presented an Earth–Mars Hohmann transfer in some detail. For simplicity, we have assumed that both planetary orbits are coplanar circles about the sun. Figure 10.10 shows the Earth–Mars Hohmann transfer previously depicted in Figure 10.5. Using "clock-face" coordinates, we see that the spacecraft departs Earth's SOI when Earth is at the 6 o'clock position and arrives at Mars' SOI when Mars is at 12 o'clock. We must now consider Mars' position when the spacecraft begins its heliocentric interplanetary cruise. Clearly, Mars must also be at the 12 o'clock position when the spacecraft arrives at Mars' orbit. Therefore, Mars must travel an angular displacement $\omega_M(t_2 - t_1)$ during the Hohmann transfer as shown in Figure 10.10 (note that ω_M is the angular velocity of Mars' circular orbit). Our previous analysis shows that

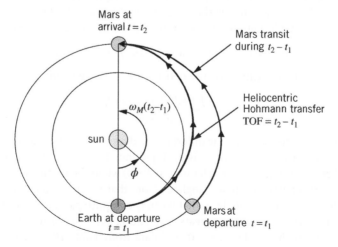

Figure 10.10 Phase angle at departure for an Earth–Mars Hohmann transfer between coplanar circular orbits.

the TOF of the heliocentric Hohmann transfer is $t_2 - t_1 = 258.9$ days. Recall that Table 10.4 presents a simple "2-D ephemeris" for Venus, Earth, and Mars, where the planets' orbits are coplanar and circular. Using Mars' angular velocity $\omega_M = 0.5240$ deg/day, we see that Mars travels about $135.7°$ during the interplanetary cruise (see Figure 10.10). We can now define the *phase angle at departure* (ϕ) as the angular separation between Earth and Mars at the start of the heliocentric transfer. For the Earth–Mars Hohmann transfer shown in Figure 10.10, the phase angle at departure is

$$\phi = 180° - \omega_M(t_2 - t_1) \tag{10.20}$$

Using the Mars transit angle of $135.7°$, we find that $\phi = 44.3°$. Therefore, Mars must "lead" Earth by $44.3°$ on the date when the spacecraft leaves Earth's SOI on a Hohmann transfer. The general expression for the phase angle at departure is

$$\phi = \Delta\theta - \omega_T(t_2 - t_1) \tag{10.21}$$

where $\Delta\theta$ is the transfer angle of the interplanetary cruise and ω_T is the angular velocity of the target planet. If the phase angle $\phi > 0$, then the target planet "leads" ahead of Earth (as shown in Figure 10.10); if $\phi < 0$ then the target planet "lags" behind Earth.

Figure 10.10 shows the proper Earth–Mars geometry at departure so that a 259-day Hohmann transfer delivers the spacecraft to Mars at the arrival date. However, for the sake of simplicity, Figure 10.10 shows Earth and Mars at the 6 and 12 o'clock positions, respectively. In practice, we must consider the actual positions of these planets in the heliocentric frame on a given date. Using the simplified 2-D ephemeris models (Table 10.4) for Earth and Mars, we can write the true longitude of both planets at any arbitrary time t as

$$\text{Earth:} \quad l_E(t) = l_{E0} + \omega_E(t - t_0) \tag{10.22}$$

$$\text{Mars:} \quad l_M(t) = l_{M0} + \omega_M(t - t_0) \tag{10.23}$$

where l_E and l_M are the true longitude angles of Earth and Mars (measured from the inertial $+X_H$ axis), respectively, l_{E0} and l_{M0} are the true longitudes at the epoch January 1, 2000 (listed in Table 10.4), and t_0 is the epoch date of January 1, 2000. The phase angle at departure is the true longitude of Mars minus the true longitude of Earth:

$$\phi = l_M(t) - l_E(t) = l_{M0} - l_{E0} + (\omega_M - \omega_E)(t - t_0) \tag{10.24}$$

Solving Eq. (10.24) for the time past the epoch date, $t - t_0$, we obtain

$$t - t_0 = \frac{(\phi \pm k360°) - l_{M0} + l_{E0}}{\omega_M - \omega_E} \tag{10.25}$$

Equation (10.25) expresses the time difference in days if all angles and angular rates are in degrees and degrees per day, respectively. Equation (10.25) allows us to compute the dates (either before or after the epoch January 1, 2000) when the Earth–Mars geometry at departure exhibits a particular phase angle ϕ. Note that we have added (or subtracted) multiples of $360°$ to the phase angle so that we may compute past and future dates when the proper Earth–Mars geometry repeats. Evaluating Eq. (10.25) for the Earth–Mars Hohmann transfer with $\phi = 44.3°$, $k = 0$, and data from Table 10.4, we obtain $t - t_0 = 456$ days (or 1.25 years). Therefore, the first feasible departure date for an Earth–Mars Hohmann transfer is 456 days after January 1, 2000, or April 1, 2001. The next feasible departure date is obtained with $k = -1$, or $t - t_0 = 1{,}236$ days (3.38 years) after January 1, 2000. The time difference $t - t_0 = 1{,}236$ days after the epoch translates to a calendar date of May 21, 2003. Note that we must use $k < 0$ in Eq. (10.25) in order to predict *future* departure dates because the denominator $\omega_M - \omega_E < 0$ (because Mars' period is greater than Earth's orbital period).

Table 10.6 presents feasible departure dates for an Earth–Mars Hohmann transfer for missions prior to and after January 1, 2000. In addition, column three of Table 10.6 lists actual Mars missions and their launch dates. Interestingly, the actual mission launch dates show a good match with the Hohmann-transfer departure dates predicted by Eq. (10.25). Space agencies (such as NASA) have taken advantage of every favorable launch date summarized in Table 10.6 with a mission to Mars except for the launch

Table 10.6 Predicted departure dates for an Earth–Mars Hohmann transfer and actual Mars mission launch dates.

Days past January 1, 2000	Hohmann-transfer departure date	Actual launch date
−1,103	December 24, 1996	December 4, 1996 (Mars Pathfinder)
−323	February 12, 1999	December 11, 1998 (Mars Climate Orbiter) January 3, 1999 (Mars Polar Lander)
456	April 1, 2001	April 7, 2001 (Mars Odyssey)
1,236	May 21, 2003	June 2, 2003 (Mars Express – ESA) June 10, 2003 (Mars Exploration Rover-A) July 8, 2003 (Mars Exploration Rover-B)
2,016	July 9, 2005	August 12, 2005 (Mars Reconnaissance Orbiter)
2,796	August 28, 2007	August 4, 2007 (Phoenix)
3,576	October 16, 2009	—
4,356	December 5, 2011	November 8, 2011 (Fobos-Grunt – Russia) November 26, 2011 (Mars Science Laboratory)
5,136	January 23, 2014	November 18, 2013 (MAVEN)
5,916	March 13, 2016	March 4–30, 2016 (InSight – postponed) March 14, 2016 (Schiaparelli – ESA)
6,696	May 2, 2018	May 5, 2018 (planned) (InSight – new launch date)

opportunity in late 2009. Space agencies utilize the efficiency of the Hohmann transfer for Mars missions in order to maximize the payload mass of the spacecraft.

If we scrutinize Table 10.6, we see that the Earth-departure dates corresponding to the desired phase angle of ϕ = 44.3° (for a Hohmann transfer) repeat every 780 days or 2.135 years. The time we must wait for the Earth–Mars phase angle to repeat is called the *synodic period*, and it can be determined by modifying Eq. (10.25)

$$T_{synodic} = \frac{360°}{|\omega_M - \omega_E|} \tag{10.26}$$

We must express the angular rates in Eq. (10.26) in degrees per day to obtain the synodic period in days. The synodic period is the time required for *any* phase angle ϕ to repeat, whether it is a Hohmann or non-Hohmann transfer. A more general equation for the synodic period is

$$T_{synodic} = \frac{360°}{|\omega_T - \omega_E|} \tag{10.27}$$

where ω_T is the angular velocity of the target planet. Using the data in Table 10.4, we see that the synodic period for a desired Earth–Venus phase angle is 584 days (1.60 years). Equation (10.27) shows that the synodic period becomes large for planetary orbits that have a small difference in their angular velocities (such as Earth and Mars); conversely, the synodic period becomes small for "slow-moving" target planets such as Uranus and Neptune.

Example 10.2   Let us return to Example 10.1 and the heliocentric Hohmann transfer from Earth to Venus. Use the coplanar 2-D ephemeris data in Table 10.4 to compute the Earth-departure date for the first mission opportunity after January 1, 2000. In addition, compute the date of the first Earth–Venus mission opportunity after January 1, 2016.

First, we use Eq. (10.21) to compute the required phase angle at departure for an Earth–Venus Hohmann transfer with transfer angle $\Delta\theta$ = 180°

$$\phi = 180° - \omega_V(t_2 - t_1)$$

where ω_V = 1.6021 deg/day is the angular velocity of Venus' orbit, and $t_2 - t_1$ = 146.1 days is the Earth–Venus flight time on the Hohmann transfer (see Example 10.1). Using these values, we find that the phase angle at departure is ϕ = −54.1°. Hence, Venus "lags" behind Earth at the SOI departure date (see the Mariner 2 heliocentric trajectory shown in Figure 10.1). Next, we use Eq. (10.25) to determine the first launch opportunity after the epoch date t_0 = January 1, 2000:

$$t - t_0 = \frac{(\phi \pm k360°) - l_{V0} + l_{E0}}{\omega_V - \omega_E}$$

The true longitudes of Earth and Venus at the epoch t_0 are l_{E0} = 100.47° and l_{V0} = 181.98° (Table 10.4). We must use k = +1 to obtain the first departure date *after* January 1, 2000 because the denominator $\omega_V - \omega_E$ is positive. Using these values in the above equation, the time difference is $t - t_0$ = 364 days. Therefore, the first departure opportunity is nearly 1 year after January 1, 2000. The departure date is December 30, 2000.

We can obtain future Earth-departure dates by adding multiples of the synodic period to December 30, 2000. Equation (10.27) determines the Earth–Venus synodic period:

$$T_{\text{synodic}} = \frac{360°}{|\omega_V - \omega_E|} = 583.9\,\text{days} = 1.60\,\text{years}$$

We must add 10 synodic periods (5,839 days ≈ 16 years) to December 30, 2000 to obtain a departure date in the year 2016. The Earth-departure date for a Venus mission in 2016 is Christmas Day: December 25, 2016.

As a final note, if we subtract 24 synodic periods (~38.4 years) from the December 30, 2000 departure date we obtain August 18, 1962 which is very close to the actual launch date of August 27, 1962 for Mariner 2 (see Figure 10.1).

10.4 Planetary Arrival

Our previous discussion has addressed the planetary arrival phase to some degree. For example, Figure 10.7 shows the hyperbolic flight phase within Mars' SOI. Because Figure 10.7 depicts an Earth–Mars heliocentric Hohmann transfer, the hyperbolic arrival velocity vector $\mathbf{v}_\infty^-$ is in the opposite direction as Mars' sun-relative orbital velocity vector $\mathbf{v}_M$. Therefore, the spacecraft approaches Mars from the "leading edge." Figure 10.7 shows the hyperbolic arrival asymptote on the sunlit side of Mars. The placement of the arrival asymptote in Figure 10.7 is arbitrary because the patched-conic method provides no information on exactly where the Mars-relative velocity $\mathbf{v}_\infty^-$ crosses the SOI (only precise numerical integration of the N-body problem will accurately determine the position and velocity of the spacecraft as it approaches Mars). In fact, we could have just as easily "flipped" the hyperbola in Figure 10.7 so that the leading-edge arrival asymptote passes through Mars' shadow. Furthermore, we could have placed $\mathbf{v}_\infty^-$ collinear with $\mathbf{v}_M$ so the offset distance $d = 0$ and the spacecraft follows a rectilinear impact trajectory with Mars! In practice, the spacecraft's onboard rocket performs very small orbital maneuvers (so-called "mid-course corrections") before planetary arrival in order to target the appropriate offset distance d shown in Figure 10.7. Computing these mid-course corrections is beyond the scope of this textbook. Therefore, we will assume that *any* offset distance d can be achieved by a small trajectory correction performed well before planetary arrival (of course, the offset distance d must be less than the radius of the planet's SOI otherwise the spacecraft misses the target planet altogether).

Figure 10.11 shows the *B-plane*, which is a reference plane for targeting the planetary arrival phase. By definition, the B-plane intersects the center of the target planet and is normal to the arrival asymptote. An orthogonal coordinate frame *TRS* is fixed to the center of the target body where the $+S$ axis is parallel to the arrival asymptote in the direction of $\mathbf{v}_\infty^-$. Axes T and R lie in the B-plane where $+T$ is in the target planet's ecliptic plane (note that so far we have assumed a coplanar solar system model and therefore the $+T$ axis is in a common ecliptic plane shared by all planets). The $+R$ axis is normal to the T and S axes and points toward the target planet's South Pole (however, $+R$ is not aligned with the South Pole unless the planet's equatorial plane coincides with its ecliptic plane). The vector $\mathbf{B}$ shown in Figure 10.11 lies in the B-plane and points from the planet's center to the arrival asymptote's intersection with the B-plane. Hence, the $\mathbf{B}$ vector only

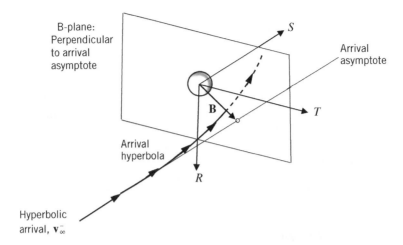

B-plane:
Perpendicular
to arrival
asymptote

S

Arrival
asymptote

B

T

Arrival
hyperbola

R

Hyperbolic
arrival, $\mathbf{v}_\infty^-$

Figure 10.11 Planetary arrival phase: B-plane target.

has *T* and *R* coordinates. Figure 10.11 shows a 3-D planetary arrival where the approach hyperbola crosses the B-plane below the planet's ecliptic plane. If we utilize our simplified coplanar solar-system model, then the **B** vector will only have a *T*-axis component.

Figure 10.12 shows a 2-D view of Figure 10.11 in the trajectory plane with an "edge-on" view of the B-plane. Note that the offset distance *d* is defined as the distance between the

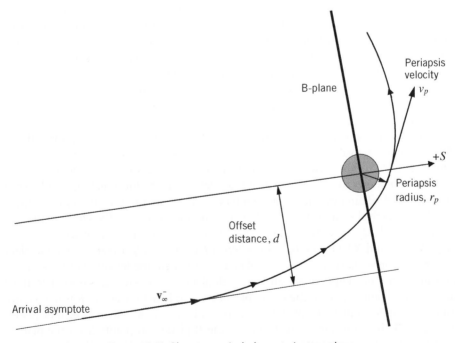

Periapsis
velocity
v_p

B-plane

$+S$

Periapsis
radius, r_p

Offset
distance, *d*

$\mathbf{v}_\infty^-$

Arrival asymptote

Figure 10.12 Planetary arrival phase: trajectory plane.

arrival asymptote and the $+S$ axis and hence it is the magnitude of the **B** vector shown in Figure 10.11. Comparing Figures 10.12 and 10.7, we see that the $+S$ axis of the B-plane frame will be collinear with the planet's sun-relative velocity *only* when the terminal apse of the heliocentric ellipse is tangent to the planet's orbit (i.e., a Hohmann transfer). Note that if we attached the B-plane to Figure 10.7, the $+S$ axis would be in the opposite direction as Mars' heliocentric velocity $\mathbf{v}_M$.

Now we may relate the offset distance d to a desired (or target) periapsis radius r_p for the hyperbolic trajectory. Using Figure 10.12, we can express the angular momentum of the hyperbola as

$$h = dv_\infty = r_p v_p \tag{10.28}$$

Note that both terms in Eq. (10.28) consist of the magnitude of the velocity multiplied by the respective "moment arm" perpendicular to an axis that passes through the center of the planet (in addition, note that we have dropped the superscript minus for hyperbolic arrival velocity $\mathbf{v}_\infty^-$). Next, use the energy equation evaluated at an infinite distance and at periapsis:

$$\xi = \frac{v_\infty^2}{2} = \frac{v_p^2}{2} - \frac{\mu_B}{r_p} \tag{10.29}$$

where μ_B is the gravitational parameter for the planetary body. Solving Eq. (10.29) for periapsis velocity, we obtain

$$v_p = \sqrt{v_\infty^2 + \frac{2\mu_B}{r_p}} \tag{10.30}$$

Substituting Eq. (10.30) into Eq. (10.28) and solving for the offset distance yields

$$d = \frac{r_p}{v_\infty} \sqrt{v_\infty^2 + \frac{2\mu_B}{r_p}} \tag{10.31}$$

Equation (10.31) determines the offset distance that is required to achieve a desired periapsis radius r_p. Remember that the offset distance is the magnitude of the **B** vector in the B-plane, or $d = \|\mathbf{B}\|$. For a 3-D planetary approach, we specify the TR coordinates of the **B** vector in order to target a desired periapsis radius and inclination (Figure 10.11).

The planetary encounter phase may simply be a "flyby" where the spacecraft passes through the targeted periapsis radius and then leaves the planet on the departure asymptote. If a mission calls for an orbiting spacecraft, then an onboard rocket burn (typically at periapsis) reduces the energy and creates a closed orbit. If a direct planetary landing is desired (such as the Mars Science Laboratory mission), then the hyperbolic trajectory must terminate at the proper conditions at the planet's atmosphere. The so-called "entry interface" (EI) is the altitude above a planet where atmospheric drag becomes appreciable. For example, EI for Mars is an altitude of 125 km. For an entry scenario, we may express the energy equation (10.29) in terms of the radius and velocity at EI

$$\xi = \frac{v_\infty^2}{2} = \frac{v_{EI}^2}{2} - \frac{\mu_B}{r_{EI}} \tag{10.32}$$

where r_{EI} is the radius of entry interface (e.g., Mars' radius plus 125 km) and v_{EI} is the velocity at EI. The velocity at EI is

$$v_{EI} = \sqrt{v_\infty^2 + \frac{2\mu_B}{r_{EI}}} \qquad (10.33)$$

which is simply Eq. (10.30) with r_p replaced by r_{EI}. The angular momentum of the hyperbolic trajectory at EI is

$$h = r_{EI} v_{EI} \cos\gamma_{EI} \qquad (10.34)$$

where γ_{EI} is the flight-path angle at EI. Finally, we may substitute Eqs. (10.33) and (10.34) into Eq. (10.28) and solve for the offset distance:

$$d = \frac{r_{EI} \cos\gamma_{EI}}{v_\infty} \sqrt{v_\infty^2 + \frac{2\mu_B}{r_{EI}}} \qquad (10.35)$$

Equation (10.35) determines the offset distance required to achieve a desired entry flight-path angle γ_{EI}. It is important to note that radius r_{EI} is not a free parameter; it is the largest radial distance where the planet's atmosphere causes detectable aerodynamic drag. The key targeting parameter in Eq. (10.35) is the EI flight-path angle. We must carefully select γ_{EI} so that the vehicle can withstand deceleration and heating loads during entry. Equation (10.35) determines the B-plane offset distance required to generate any entry flight-path angle from $\gamma_{EI} = 0$ (i.e., "skimming" the edge of the planet's atmosphere) to $\gamma_{EI} = -90°$ (i.e., entering the atmosphere with purely vertical velocity). Note that specifying $\gamma_{EI} = -90°$ in Eq. (10.35) leads to $d = 0$ as expected. We will discuss atmospheric entry in Chapter 11.

Example 10.3 Consider again the Venus-arrival hyperbola that resulted from the Earth–Venus Hohmann transfer described in Example 10.1. Determine the offset distance d (B-plane target) if the mission calls for a 350-km altitude periapsis at Venus. In addition, compute the impulsive velocity increment required to establish a 350-km altitude circular orbit about Venus.

In Example 10.1, we determined that the Venus-relative hyperbolic arrival velocity is $v_\infty = 2.7066$ km/s. The target periapsis radius is $r_p = 350$ km $+ R_V = 6{,}402$ km, where $R_V = 6{,}052$ km is the mean radius of Venus. Using Eq. (10.31), we can compute the offset distance

$$d = \frac{r_p}{v_\infty} \sqrt{v_\infty^2 + \frac{2\mu_V}{r_p}} = \boxed{24{,}673.5 \text{ km}}$$

where $\mu_V = 3.2486(10^5)$ km^3/s^2 is Venus' gravitational parameter. We may use conservation of angular momentum, Eq. (10.28), to compute the spacecraft's velocity at periapsis passage

$$v_p = \frac{d v_\infty}{r_p} = 10.4313 \text{ km/s}$$

Finally, the impulse required for insertion into a circular Venus orbit is

$$\Delta v = v_p - v_{LVO} = \boxed{3.3079 \text{ km/s}}$$

where $v_{LVO} = \sqrt{\mu_V/r_p} = 7.1234$ km/s is the circular orbital speed for the desired low-Venus orbit (LVO).

Example 10.4 The Mars Science Laboratory (MSL) mission used a direct entry, descent, and landing profile. The target state at Mars entry interface was $r_{EI} = 3{,}521$ km, $v_{EI} = 5.845$ km/s, and flight-path angle $\gamma_{EI} = -15.47°$. Determine the Mars-relative hyperbolic arrival speed and the offset distance d (i.e., the B-plane target).

Because the hyperbolic trajectory has constant energy, we can write

$$\xi = \frac{v_\infty^2}{2} = \frac{v_{EI}^2}{2} - \frac{\mu_M}{r_{EI}} = 4.9184 \text{ km}^2/\text{s}^2$$

where $\mu_M = 4.2828(10^4)$ km^3/s^2 is Mars' gravitational parameter. Therefore, the hyperbolic arrival velocity is

$$v_\infty = \sqrt{2\xi} = \boxed{3.136 \text{ km/s}}$$

We can use Eq. (10.35) to determine the offset distance

$$d = \frac{r_{EI}\cos\gamma_{EI}}{v_\infty}\sqrt{v_\infty^2 + \frac{2\mu_M}{r_{EI}}} = \boxed{6{,}324 \text{ km}}$$

10.5 Heliocentric Transfers Using an Accurate Ephemeris

All examples of the interplanetary cruise phase thus far have involved planetary motion modeled by coplanar circular orbits about the sun. When we utilize a Hohmann transfer between concentric-coplanar planetary orbits, the flight time and velocity increments remain the same even when we change the departure date by multiples of the synodic period. Therefore, all Earth–Mars Hohmann transfers listed in column two of Table 10.6 have a heliocentric flight time of 258.9 days and launch energy $C_3 = 8.672$ km^2/s^2. However, the *actual* Earth–Mars missions listed in column three of Table 10.6 do not utilize the same 180° Hohmann transfer because of the eccentric planetary orbits and the difference in the orbital planes (Table 10.3 shows that Mars' eccentricity and inclination are 0.093 and 1.85°, respectively).

We can improve the accuracy of the interplanetary cruise phase by defining each planet's orbit using the six orbital elements listed in Table 10.3. Now we may pose the heliocentric transfer between planets as *Lambert's problem* as described in Section 4.6: given two known position vectors and the flight time between them, determine the corresponding orbit. Posing the appropriate Lambert problem for the heliocentric phase is conceptually easy, and the basic steps for determining the interplanetary trajectory are as follows:

1) Given a guess for the Earth-departure date (t_1), determine Earth's heliocentric position and velocity vectors ($\mathbf{r}_E, \mathbf{v}_E$) in the heliocentric-ecliptic frame.
2) Given a guess for the planet-arrival date (t_2), determine the target planet's heliocentric position and velocity vectors ($\mathbf{r}_T, \mathbf{v}_T$).

3) Using the known heliocentric flight time $t_2 - t_1$, solve Lambert's problem and determine the heliocentric orbit that connects vectors $\mathbf{r}_E$ and $\mathbf{r}_T$. As a by-product of solving Lambert's problem, we have the heliocentric velocities $\mathbf{v}_1$ and $\mathbf{v}_2$ that correspond to the initial and terminal position vectors $\mathbf{r}_E$ and $\mathbf{r}_T$.

4) Determine the Earth-relative hyperbolic departure velocity vector at the SOI:

$$\mathbf{v}_\infty^+ = \mathbf{v}_1 - \mathbf{v}_E \tag{10.36}$$

5) Determine the launch energy from the magnitude $v_\infty = \left\| \mathbf{v}_\infty^+ \right\|$

$$C_3 = v_\infty^2 \tag{10.37}$$

Steps 1 and 2 require two solutions to *Kepler's problem* in order to determine the true anomalies of Earth and the target planet at times t_1 and t_2, respectively. Implicit in steps 1 and 2 are coordinate transformations from the six classical orbital elements to heliocentric vectors $(\mathbf{r}_E, \mathbf{v}_E)$ for Earth and $(\mathbf{r}_T, \mathbf{v}_T)$ for the target planet (see Section 3.5). Step 3 requires a numerical search algorithm that can solve Lambert's problem, such as the p-iteration method outlined in Section 4.6. Recall that we must specify either a "short-way" ($\Delta\theta < 180°$) or "long-way" ($\Delta\theta > 180°$) transfer when we solve Lambert's problem (again, see Section 4.6). Once we solve Lambert's problem, Eq. (10.36) determines the Earth-relative hyperbolic velocity vector (and in a similar manner, the hyperbolic velocity vector at the target planet). Finally, the launch energy C_3, computed by Eq. (10.37), provides a performance metric for the trial departure and arrival dates t_1 and t_2. We can systematically search across a range of departure and arrival dates until we determine the best combination that minimizes the launch energy C_3.

Some additional terminology is in order at this point. A short-way transfer with transfer angle $\Delta\theta < 180°$ is often called a "Type 1 transfer." A long-way transfer ($\Delta\theta > 180°$) is often called a "Type 2 transfer."

At this stage, the reader may (justifiably) believe that the Hohmann transfer for the concentric-coplanar solar-system model will approximate the best heliocentric transfer when we incorporate accurate 3-D planetary orbits. To some degree, this assumption is correct. However, it is worth noting that it is (usually) impossible to perform an exact 180° heliocentric transfer because of the difference in planetary orbital planes. Figure 10.13 shows Mars' orbit relative to Earth's ecliptic plane. The lightly shaded half of Mars' orbit in Figure 10.13 is "above" Earth's ecliptic plane, while the darker shaded half is "below" the ecliptic. Mars crosses Earth's ecliptic plane twice: at the ascending node ($\Omega = 49.56°$) and descending node. Figure 10.13 also shows a 180° heliocentric transfer that departs Earth on July 9, 2005 (i.e., a Hohmann-transfer departure date predicted in Table 10.6). The spacecraft's arrival at Mars (exactly one-half revolution later) is *infeasible* because at the arrival date Mars is *above* the ecliptic while the spacecraft is in the ecliptic plane. A 180° heliocentric transfer is possible *only* when the transfer terminates at Mars' ascending or descending node; that is, when Mars is crossing the ecliptic plane. Furthermore, heliocentric transfers that are slightly less than or greater than 180° become feasible only by "lofting" the spacecraft above (or below) the ecliptic so that its Z_H position component matches Mars' out-of-plane component. "Lofting" the heliocentric trajectory out of the ecliptic requires a large Z_H velocity component at Earth departure that is normal to Earth's heliocentric velocity vector $\mathbf{v}_E$. This additional Z_H velocity increment does not increase the energy of the heliocentric transfer; it represents an

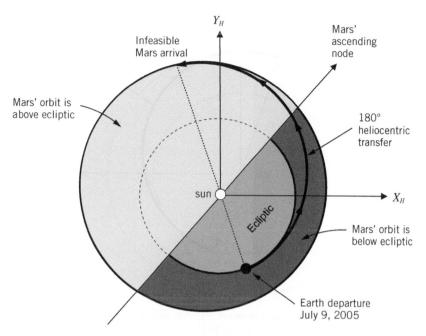

Figure 10.13 Orbital plane difference between Earth and Mars and an infeasible heliocentric transfer to Mars.

unnecessary plane change (the "lofting" maneuver) that ultimately penalizes heliocentric transfer angles that are near 180°.

The following example demonstrates the calculation of an accurate Earth–Mars heliocentric transfer by using the Lambert-problem method previously outlined. The example will highlight the effects of Mars' inclination on the launch energy required for the interplanetary cruise.

Example 10.5 Use the predicted departure and arrival dates for an Earth–Mars Hohmann transfer as a starting guess for a realistic Earth–Mars mission in mid 2005 (see column two of Table 10.6). Compare the launch energy (C_3) and Mars-arrival hyperbolic velocity v_∞^- for Earth–Mars transfers computed using concentric-coplanar planetary orbits and accurate (3-D) planetary orbits.

First, let us consider the Earth–Mars transfer using coplanar circular orbits for the two planets. Table 10.6 shows that on July 9, 2005 (or 2,016 days after January 1, 2000) Earth and Mars have the proper relative geometry for a Hohmann transfer. We may use Eqs. (10.22) and (10.23) and Table 10.4 to determine the true longitudes of Earth and Mars for the coplanar-concentric model. At departure time $t_1 = 2{,}016$ days (July 9, 2005), Earth's longitude is

$$\text{Earth:} \quad l_E(t_1) = l_{E0} + \omega_E(t_1 - t_0) = 2{,}087° = 287°$$

The Hohmann transfer flight time is 259 days. At arrival time $t_2 = 2{,}016 + 259 = 2{,}275$ days (March 25, 2006), Mars' longitude is

$$\text{Mars:} \quad l_M(t_2) = l_{M0} + \omega_M(t_2 - t_0) = 1{,}547° = 107°$$

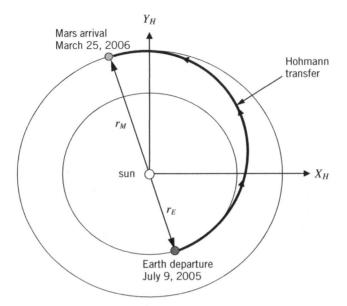

Figure 10.14 Heliocentric Earth–Mars Hohmann transfer with departure date July 9, 2005 using concentric-coplanar planetary orbits (Example 10.5).

which is 180° beyond the Earth-departure longitude. Remember that true longitude l is measured clockwise from the inertial $+X_H$ heliocentric axis. Figure 10.14 shows the Hohmann transfer between concentric-coplanar planetary orbits with departure date July 9, 2005. The arrival date at Mars is March 25, 2006 (i.e., 259 days after departure). The launch energy and Mars-arrival velocity for the Earth–Mars Hohmann transfer have been computed in Section 10.2, and these values are $C_3 = 8.672 \, \text{km}^2/\text{s}^2$ and $v_\infty^- = 2.649 \, \text{km/s}$, respectively.

Next, we compute an Earth–Mars transfer using an accurate ephemeris for the planetary orbits. To begin, we determine Earth's heliocentric state vector $(\mathbf{r}_E, \mathbf{v}_E)$ at departure date July 9, 2005. Although the process of computing vectors $(\mathbf{r}, \mathbf{v})$ in the heliocentric frame is systematic, we will not present the details here. An outline of the required steps follows. First, we compute the Earth's mean anomaly M_0 at epoch (January 1, 2000) using eccentricity and true anomaly θ_0 derived from the ephemeris data in Table 10.3. Next, we solve Kepler's problem, propagate Earth's orbit to the departure date (July 9, 2005), and obtain mean and true anomalies. Finally, using Earth's classical orbital elements and the true anomaly for July 9, 2005, we perform a transformation to the Cartesian heliocentric-ecliptic frame (see Section 3.5 for details regarding this transformation). After performing these steps, we obtain Earth's position and velocity vectors for departure date July 9, 2005:

$$\text{Earth departure (July 9, 2005):} \quad \mathbf{r}_E = \begin{bmatrix} 0.3035 \\ -0.9703 \\ 0 \end{bmatrix} \text{AU}, \quad \mathbf{v}_E = \begin{bmatrix} 0.9383 \\ 0.2948 \\ 0 \end{bmatrix} \text{AU/TU}$$

In order to set up Lambert's problem, we need Mars' heliocentric state vector $(\mathbf{r}_M, \mathbf{v}_M)$ at the end of the heliocentric transfer. The coplanar Hohmann transfer predicts that March 25, 2006 (i.e., 259 days after Earth departure) will provide a good option for the Mars-arrival date. Using Mars' orbital data in Table 10.3 and the steps previously outlined for the Earth departure, we obtain Mars' state vector for March 25, 2006:

$$\text{Mars arrival (March 25, 2006):} \quad \mathbf{r}_M = \begin{bmatrix} -0.6085 \\ 1.4998 \\ 0.0464 \end{bmatrix} \text{AU,} \quad \mathbf{v}_M = \begin{bmatrix} -0.7229 \\ -0.2368 \\ 0.0128 \end{bmatrix} \text{AU/TU}$$

Next, we pose Lambert's problem with initial position $\mathbf{r}_E$ (Earth), final position $\mathbf{r}_M$ (Mars), and flight time $t_2 - t_1 = 259$ days. We also determine the "long way" solution because $\Delta\theta = 185.0°$ for "counter-clockwise" (prograde) orbital motion from Earth position $\mathbf{r}_E$ to Mars position $\mathbf{r}_M$ (note the transfer angle is *not* 180° because we have used an accurate planetary ephemeris instead of concentric-coplanar circles). Solving Lambert's problem yields the spacecraft's heliocentric velocity vectors at each end of the transfer:

$$\text{Spacecraft's heliocentric velocity at Earth departure:} \quad \mathbf{v}_1 = \begin{bmatrix} 0.9670 \\ 0.3804 \\ -0.3611 \end{bmatrix} \text{AU/TU}$$

$$\text{Spacecraft's heliocentric velocity at Mars arrival:} \quad \mathbf{v}_2 = \begin{bmatrix} -0.6221 \\ -0.1983 \\ 0.2275 \end{bmatrix} \text{AU/TU}$$

Using Eq. (10.36), we determine the Earth-relative hyperbolic departure velocity vector:

$$\mathbf{v}_\infty^+ = \mathbf{v}_1 - \mathbf{v}_E = \begin{bmatrix} 0.0287 \\ 0.0857 \\ -0.3611 \end{bmatrix} \text{AU/TU}$$

Note the very large Z_H component for hyperbolic departure velocity that is required to "loft" the heliocentric transfer below the ecliptic plane for interception with Mars' orbit. The magnitude of $\mathbf{v}_\infty^+$ is 0.3722 AU/TU. Converting to SI units (1 AU/TU = 29.7847 km/s), yields $v_\infty^+ = 11.086$ km/s and the associated launch energy is $C_3 = v_\infty^2 = 122.898$ km^2/s^2. This launch energy is over 14 times greater than the Hohmann-transfer launch energy for concentric-coplanar planets! Obviously, this heliocentric transfer is unacceptable. Intercepting Mars when it is well above Earth's ecliptic plane causes the extremely high launch energy.

We can repeat the Lambert-solution process outlined above for the same Earth-departure date but with different heliocentric flight times. Figure 10.15 shows launch energy C_3 for heliocentric flight times ranging from 100 to 450 days. All heliocentric transfers leave Earth's SOI on July 9, 2005. Figure 10.15 shows that the short-way (or Type 1) transfer ($\Delta\theta < 180°$) is used when flight time is less than 249 days, and the long-way (or Type 2) transfer is used when flight time exceeds 249 days. When flight time is about 249 days, the transfer angle is 180° and the heliocentric transfer is not feasible due to the Z_H (out-of-plane) component of Mars' orbit. Hence, C_3 is extremely large for heliocentric transfer angles near 180° (C_3 is limited to 50 km^2/s^2 in Figure 10.15 for plotting purposes; the 259-day long-way transfer with $C_3 = 122.9$ km^2/s^2 is off the chart).

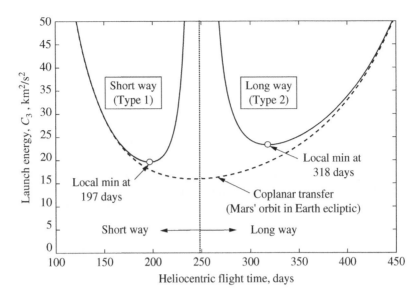

Figure 10.15 Earth–Mars launch energy vs. flight time for departure date July 9, 2005 (Example 10.5).

The dashed line in Figure 10.15 presents the launch energy for *coplanar* Earth–Mars transfers; that is, we have artificially removed Mars' inclination so that its orbit is in the ecliptic plane (Mars' orbit remains eccentric). Assuming coplanar (2-D) transfers removes the excessive launch energy for transfer angles near 180°. If we ignore out-of-plane motion, then a coplanar 242-day, short-way transfer ($\Delta\theta = 176.8°$) will minimize the launch energy for an Earth-departure date of July 9, 2005 (of course, a coplanar transfer is fictitious).

Figure 10.15 clearly shows the existence of two local minima for launch energy: flight time of 197 days (short way) and flight time of 318 days (long way). Let us analyze each transfer in some detail starting with the short-way transfer.

197-day short-way (Type 1) transfer

Mars' state vector at arrival date January 22, 2006 (197 days after Earth departure) is

$$\text{Mars arrival (January 22, 2006):} \quad \mathbf{r}_M = \begin{bmatrix} 0.2092 \\ 1.5364 \\ 0.0270 \end{bmatrix} \text{AU,} \quad \mathbf{v}_M = \begin{bmatrix} -0.7751 \\ 0.1788 \\ 0.0228 \end{bmatrix} \text{AU/TU}$$

Solving the Lambert problem using the July 9, 2005 departure date with a flight time of 197 days yields the initial and final velocity vectors:

$$\text{Spacecraft's heliocentric velocity at Earth departure:} \quad \mathbf{v}_1 = \begin{bmatrix} 1.0325 \\ 0.4000 \\ 0.0454 \end{bmatrix} \text{AU/TU}$$

$$\text{Spacecraft's heliocentric velocity at Mars arrival:} \quad \mathbf{v}_2 = \begin{bmatrix} -0.6964 \\ 0.2553 \\ -0.0242 \end{bmatrix} \text{AU/TU}$$

The Earth-relative hyperbolic departure velocity vector is

$$\mathbf{v}_\infty^+ = \mathbf{v}_1 - \mathbf{v}_E = \begin{bmatrix} 0.0943 \\ 0.1053 \\ 0.0454 \end{bmatrix} \text{AU/TU} = \begin{bmatrix} 2.8082 \\ 3.1354 \\ 1.3519 \end{bmatrix} \text{km/s}$$

The magnitude of $\mathbf{v}_\infty^+$ is 4.4209 km/s, and hence the launch energy is

$$\boxed{C_3 = v_\infty^2 = 19.545\,\text{km}^2/\text{s}^2 \quad (\text{197-day Type 1 transfer})}$$

This launch energy is the minimizing point for the short-way transfer in Figure 10.15. The launch energy for an Earth–Mars transfer using an accurate planetary ephemeris is significantly larger than $C_3 = 8.672\,\text{km}^2/\text{s}^2$ for the Hohmann transfer between concentric-coplanar planets.

Finally, the Mars-relative hyperbolic arrival velocity vector is

$$\mathbf{v}_\infty^- = \mathbf{v}_M - \mathbf{v}_2 = \begin{bmatrix} 0.0788 \\ 0.0765 \\ -0.0470 \end{bmatrix} \text{AU/TU} = \begin{bmatrix} 2.3467 \\ 2.2776 \\ -1.3990 \end{bmatrix} \text{km/s}$$

The magnitude of the Mars-relative velocity $\mathbf{v}_\infty^-$ is 3.5569 km/s which is significantly greater than the coplanar Hohmann transfer result ($v_\infty^- = 2.649\,\text{km/s}$).

318-day long-way (Type 2) transfer

Mars' state vector at arrival date May 23, 2006 (318 days after Earth departure) is

$$\text{Mars arrival (May 23, 2006):} \quad \mathbf{r}_M = \begin{bmatrix} -1.2437 \\ 1.0953 \\ 0.0535 \end{bmatrix} \text{AU}, \quad \mathbf{v}_M = \begin{bmatrix} -0.5069 \\ -0.5412 \\ 0.0011 \end{bmatrix} \text{AU/TU}$$

Solving the Lambert problem using the July 9, 2005 starting position $\mathbf{r}_E$ for a flight time of 318 days yields

$$\text{Spacecraft's heliocentric velocity at Earth departure:} \quad \mathbf{v}_1 = \begin{bmatrix} 1.0174 \\ 0.4185 \\ -0.0682 \end{bmatrix} \text{AU/TU}$$

$$\text{Spacecraft's heliocentric velocity at Mars arrival:} \quad \mathbf{v}_2 = \begin{bmatrix} -0.4260 \\ -0.5207 \\ 0.0350 \end{bmatrix} \text{AU/TU}$$

The Earth- and Mars-relative hyperbolic velocities are

$$\text{Earth departure:} \quad \mathbf{v}_\infty^+ = \mathbf{v}_1 - \mathbf{v}_E = \begin{bmatrix} 2.3562 \\ 3.6858 \\ -2.0310 \end{bmatrix} \text{km/s}$$

$$\text{Mars arrival:} \quad \mathbf{v}_\infty^- = \mathbf{v}_M - \mathbf{v}_2 = \begin{bmatrix} 2.4100 \\ 0.6116 \\ 1.0081 \end{bmatrix} \text{km/s}$$

The magnitude of departure $\mathbf{v}_\infty^+$ is 4.8230 km/s, and hence the launch energy is

$$\boxed{C_3 = v_\infty^2 = 23.2616 \text{ km}^2/\text{s}^2 \quad \text{(318-day Type 2 transfer)}}$$

which is the minimizing point for the long-way transfer in Figure 10.15. The magnitude of the Mars-relative velocity $\mathbf{v}_\infty^-$ is 2.6830 km/s, which is only slightly greater than the coplanar Hohmann transfer result.

Figure 10.16 displays the best short-way and long-way heliocentric transfers for an Earth-departure date of July 9, 2005. The transfer angles are 154.9° (short way) and 211.3° (long way), respectively. The short-way transfer is the best option for this fixed departure date because it has the lower launch energy and shorter transfer time. Note that the long-way transfer terminates at a radial distance of about 1.66 AU that is near Mars' aphelion radius of 1.67 AU.

Finally, we can compare the best short-way transfer in Figure 10.16 to the Mars Reconnaissance Orbiter heliocentric trajectory shown in Figure 10.2. Applying our Lambert-problem solver to the *actual* MRO Earth-departure date of August 12, 2005 and Mars-arrival date of March 10, 2006, we obtain $C_3 = 16.353 \text{ km}^2/\text{s}^2$ which shows a very close match with the actual launch energy of 16.4 km^2/s^2 [6]. The actual MRO launch energy is

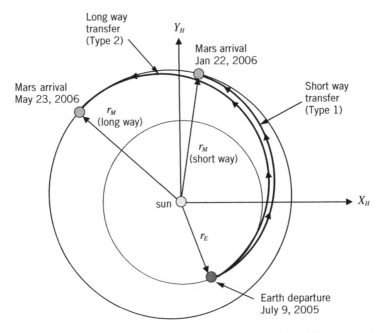

Figure 10.16 Heliocentric Earth–Mars transfers with departure date July 9, 2005 using 3-D planetary ephemeris (Example 10.5).

better than the launch energy corresponding to our optimal short-way transfer for a fixed departure date of July 9, 2005. Finding the optimal heliocentric transfer involves iterating on the flight time *and* the Earth-departure date.

10.5.1 Pork-Chop Plots

Example 10.5 presents a systematic process where we obtained the minimum-C_3 heliocentric transfer for a fixed departure date. In essence, we solved several Lambert problems for a range of flight times starting from a fixed Earth-departure position vector $\mathbf{r}_E$. Figure 10.15 is the result of this flight-time "sweep," which clearly shows the two optimal transfers that minimize launch energy for a particular departure date. In practice, mission designers find the best heliocentric transfer by varying *two* free parameters: the Earth-departure date and the flight time. Therefore, we may solve many Lambert problems for an array of *n* distinct departure dates and *m* flight times. By simply sorting the $n \times m$ array of Lambert-problem solutions, we can obtain the overall (or "global") optimal heliocentric transfer that minimizes launch energy C_3. This "brute force" method will determine the best combination of departure date and flight time.

It is useful to create contour plots of the $n \times m$ Lambert-problem solutions in order to visualize the design space for the interplanetary trajectory. Figure 10.17 shows a contour plot of launch energy C_3 vs. Earth-departure date and Mars-arrival date (of course, flight time is the difference between arrival and departure dates). Note that

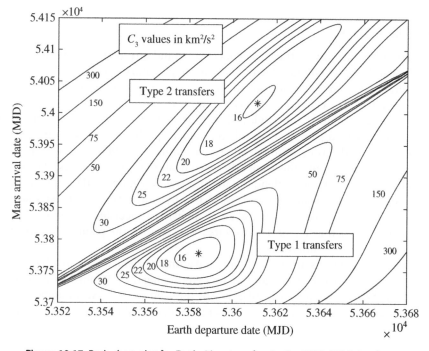

Figure 10.17 Pork-chop plot for Earth–Mars transfers in the 2005–2006 timeframe.

the Earth-departure and Mars-arrival dates (the x- and y-axis labels) in Figure 10.17 are presented as a *modified Julian date* (MJD) instead of a calendar date. Because the Julian date (JD) is a continuous count of days, the flight time (in days) is simply the difference between the arrival MJD and departure MJD (we will discuss conversions between the calendar date, JD, and MJD in the next subsection). Each contour line in Figure 10.17 depicts a heliocentric transfer with constant C_3. Figure 10.17 shows a cluster of two distinct heliocentric trajectories: the short-way (Type 1) and long-way (Type 2) transfers. The center of each grouping of contour lines is the optimal transfer that minimizes C_3. The steep "ridge" between the Type 1 and Type 2 transfers indicates the region where the transfer angle is nearly 180° and the corresponding C_3 is extremely high. Figure 10.17 is commonly known as a *pork-chop plot* because the contour lines resemble a pork chop (think of the "ridge" between the short- and long-way transfers as the bone). Mission designers use pork-chop plots to analyze the so-called *launch window*, that is, the effect that changes in the departure date has on launch energy and/or flight time. Note that Figure 10.15 is actually a "vertical slice" of the pork-chop plot contours in Figure 10.17 for the fixed departure date of July 9, 2005 (MJD = 53560). Figure 10.17 shows that the minimum-energy Type 1 transfer departs Earth on Aug 2, 2005 (MJD = 53584) with $C_3 = 14.94$ km²/s² and arrives at Mars' SOI 194 days later on February 12, 2006 (MJD = 53778). The minimum-energy Type 2 transfer departs Earth on Aug 29, 2005 (MJD = 53611) with $C_3 = 15.85$ km²/s² and arrives at Mars' SOI 406 days later on October 9, 2006 (MJD = 54017). For an Earth–Mars mission in 2005, the Type 1 (short-way) transfer requires the lowest launch energy. However, for some mission opportunities, the long-way transfer may be the minimum-C_3 transfer.

10.5.2 Julian Date

This section has introduced the basic concepts behind interplanetary mission design using the patched-conic method with an accurate planetary ephemeris. Determining the planetary position (i.e., true anomaly) at a specified date requires the solution of Kepler's problem. For example, suppose we want the true anomaly of Mars on September 23, 2020. Table 10.3 provides Mars' true anomaly at the epoch date of January 1, 2000. The next step involves solving Kepler's problem for a Mars transit time corresponding to the time difference between January 1, 2000 and September 23, 2020. How do we compute this time difference from the calendar dates? The easiest approach is to convert each calendar date into a *Julian date*. A JD is a continuous count of days starting from January 1, 4713 BC. By convention, a JD starts at noon so that astronomical observations (taken at night) are recorded under a single JD. After we convert each calendar date into its respective JD, the planetary transit time (in days) is simply the difference.

Many on-line programs are available for converting a calendar date to a JD and vice versa. We will present Vallado's algorithm for determining the JD from the calendar date [3; p. 183]. Let us represent the calendar date by three integers: y = year number (a four-digit number), m = month number (1–12), and d = day number (1–31). The associated JD is

$$JD = 367y - A + B + d + 1{,}721{,}013.5 \qquad (10.38)$$

where A and B are

$$A = \left\lfloor \frac{7\left\{ y + \left\lfloor \frac{m+9}{12} \right\rfloor \right\}}{4} \right\rfloor \tag{10.39}$$

$$B = \left\lfloor \frac{275m}{9} \right\rfloor \tag{10.40}$$

The operation $\lfloor x \rfloor$ is the *floor function*, which rounds to the nearest integer value towards minus infinity. In MATLAB syntax, this operation is `floor(x)`. For example, $\lfloor 2.6 \rfloor = 2$. As a quick example, the calendar date May 16, 2024 ($y = 2024$, $m = 5$, $d = 16$) can be converted to the Julian date JD = 2,460,446.5. Because the convention for calendar dates is to start a new day at midnight, the JD has a 0.5 fractional part (remember that a JD begins at noon).

   This conversion example leads to questions regarding the reference location for noon at the start of a JD. Where on Earth is noon the starting point for a new Julian day? The simple answer is that noon in Greenwich, UK is the basis for the start of a new JD (of course, this is also the location of the Greenwich meridian or $0°$ longitude). The local mean solar time relative to the Greenwich meridian is called Universal Time (UT), and is colloquially referred to as Greenwich Mean Time (GMT). For example, noon in Greenwich is 12:00:00 GMT (or 12 h, 0 min, and 0 s). The worldwide standard for time is "Coordinated Universal Time" (UTC), and is determined from an assembly of atomic clocks. If we know UTC (where $H = $ h, $M = $ min, and $S = $ s), we can add its fractional part to the JD; hence Eq. (10.38) becomes

$$\text{JD} = 367y - A + B + d + 1{,}721{,}013.5 + \frac{\dfrac{\left(\dfrac{S}{60} + M \right)}{60} + H}{24} \tag{10.41}$$

The JD of May 16, 2024 12:00:00 UTC (i.e., noon GMT) is 2,460,447.0, or the start of a new Julian date.

   Because a Julian date is a large number (greater than 2.4 million), it is common to use a JD with fewer digits. One variant is the *modified Julian date* (MJD), which has five digits to the left of the decimal point. We can compute the MJD by subtracting 2,400,000.5 from the JD. Thus, the MJD is a continuous count of days starting from midnight November 17, 1858. Recall that the pork-chop plot, Figure 10.17, presents the Earth-departure and Mars-arrival dates as MJDs.

   We also need to convert Julian dates to calendar dates. For example, Table 10.6 presents Earth–Mars Hohmann transfers between coplanar-concentric orbits that occur every synodic period. It is relatively easy to determine the JDs for launches relative to the January 1, 2000 epoch date (i.e., column one in Table 10.6); however, finding the corresponding calendar date is not so simple. Figure 10.18 presents the conversion algorithm, which is from the US Naval Observatory [7]. Note that every calculation involves integers and the floor function. The algorithm presented in Figure 10.18 can essentially be "cut-and-pasted" to create a MATLAB M-file for converting a given JD to the corresponding calendar date.

```
L = floor(JD + 68569.5)
N = floor( 4*L/146097 )
L = L - floor( (146097*N+3)/4 )
I = floor( 4000*(L+1)/1461001 )
L = L - floor(1461*I/4) + 31
J = floor(80*L/2447)
K = L - floor(2447*J/80)
L = floor(J/11)
J = J + 2 - floor(12*L)
I = floor(100*(N-49)) + I + L
year = I
month = J
day = K
```

Figure 10.18 US Naval Observatory algorithm for converting a Julian date (JD) to a calendar date [7].

We can append the Coordinated Universal Time to the calendar date computed by the algorithm contained in Figure 10.18. First, we compute the fractional part of the JD beyond noon GMT:

$$JD_{fraction} = JD - \lfloor JD \rfloor \tag{10.42}$$

The fractional part falls between the range $0 \le JD_{fraction} < 1$. Next, determine the hours past midnight:

$$\tau_{hour} = mod(12 + 24\,JD_{fraction}, 24) \tag{10.43}$$

where $mod(x,24)$ is the modulus function after division. For example, $mod(12.1,24) = 12.1$, $mod(23.6,24) = 23.6$, $mod(24.8,24) = 0.8$, and $mod(35.7,24) = 11.7$. Therefore, Eq. (10.43) ensures that $0 \le \tau_{hour} < 24$. The UTC hours H, minutes M, and seconds S are

$$H = \lfloor \tau_{hour} \rfloor \tag{10.44}$$
$$M = \lfloor 60(\tau_{hour} - H) \rfloor \tag{10.45}$$
$$S = \lfloor 3600(\tau_{hour} - H - M/60) \rfloor \tag{10.46}$$

Therefore, the UTC is [H:M:S]. As a quick example, if JD = 2,457,665.0, then the calendar date is October 3, 2016 UTC [12:00:00]. The reader can verify these calculations by using the algorithm in Figure 10.18 and Eqs. (10.42)–(10.46).

10.6 Gravity Assists

The final topic of this chapter is the *gravity assist* or so-called "flyby" or "swingby." In this scenario, a spacecraft enters a planet's SOI along a hyperbolic arrival asymptote, follows a coasting (unpowered) hyperbola through periapsis passage, and leaves the planet along a hyperbolic departure asymptote. The planetary flyby, however, changes the spacecraft's heliocentric energy *relative to the sun*. Therefore, a gravity assist is a maneuver where the energy of the spacecraft's heliocentric orbit is increased (or decreased) "for free" without expending any rocket propellant. Of course, setting up a planetary encounter and gravity assist often times requires "shaping" the heliocentric transfer by using mid-course trajectory corrections (propulsive burns) and/or by utilizing a sub-optimal departure date and launch energy. Consider again Figure 10.3, which shows the heliocentric phase of the

Cassini mission to Saturn. Note that the mission began with an *inward* transfer from Earth to Venus for the first gravity assist in late April 1998 – it appears that the spacecraft is initially moving in the wrong direction because it ultimately needs to move *outward* to Saturn's orbit! However, the first Venus gravity assist (VGA) significantly increased the energy of the spacecraft's heliocentric trajectory so that its aphelion was well beyond Earth's orbit. Figure 10.3 shows that the spacecraft performed a propulsive burn ("deep space maneuver") near aphelion on December 3, 1998. This rocket burn set up the second VGA on June 24, 1999. The second VGA further increased the spacecraft's heliocentric energy and set up an encounter with Earth for a third gravity assist less than 2 months later on August 18, 1999. The three gravity assists significantly increased the heliocentric energy so that the spacecraft was able to cross Jupiter's orbit 16 months after the Earth gravity assist. On December 30, 2000, the Cassini spacecraft encountered Jupiter for a fourth (and final) gravity assist that further increased the heliocentric energy for arrival at Saturn in early July 2004.

The preceding discussion indicates that a gravity assist can increase the heliocentric energy of the spacecraft's orbit. As we shall soon see, a gravity assist can also decrease the orbital energy of the spacecraft relative to the sun. Figure 10.19 shows a "trailing edge" gravity assist where the spacecraft's periapsis passage is "behind" the planet's motion or along the "trailing edge" of the planet's motion relative to the sun. Figure 10.19 shows the special case where the periapsis direction of the hyperbolic

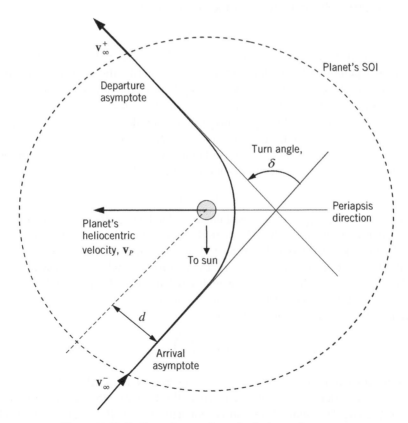

Figure 10.19 Trailing-edge gravity assist for increasing energy.

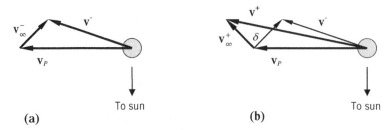

To sun

(a)

To sun

(b)

Figure 10.20 Trailing-edge gravity assist for increasing heliocentric energy: (a) heliocentric velocity $\mathbf{v}^-$ before flyby; and (b) heliocentric velocity $\mathbf{v}^+$ after flyby.

trajectory is collinear but opposite the planet's heliocentric velocity vector $\mathbf{v}_P$. The spacecraft's heliocentric velocity vector before the gravity assist is

$$\mathbf{v}^- = \mathbf{v}_P + \mathbf{v}_\infty^- \tag{10.47}$$

where $\mathbf{v}_\infty^-$ is the hyperbolic arrival velocity (at the SOI) of the spacecraft relative to the planet. Figure 10.20a shows the vector addition of Eq. (10.47); the reader should be able to correlate the arrival asymptote shown in Figure 10.19 with Figure 10.20a and Eq. (10.47). After the hyperbolic flyby, the spacecraft departs the planet and crosses the SOI with hyperbolic excess velocity vector $\mathbf{v}_\infty^+$ as shown in Figure 10.19. The spacecraft's heliocentric velocity vector *after* the gravity assist is

$$\mathbf{v}^+ = \mathbf{v}_P + \mathbf{v}_\infty^+ \tag{10.48}$$

Figure 10.20b depicts the vector addition after the flyby. It should be clear that the *magnitude* of the hyperbolic excess velocity is unchanged whether the spacecraft is approaching or leaving the planet, that is, $v_\infty = \left\| \mathbf{v}_\infty^- \right\| = \left\| \mathbf{v}_\infty^+ \right\|$. By using the patched-conic method, we have assumed that the two-body problem governs the spacecraft's orbit inside the SOI with the planet as the central body. Hence, the magnitude v_∞ is constant because the planet-relative energy of the hyperbolic trajectory is constant.

Returning to Eqs. (10.47) and (10.48) and Figure 10.20b, we see that the magnitude of the spacecraft's sun-relative velocity has increased after the gravity assist, that is

$$\left\| \mathbf{v}^+ \right\| > \left\| \mathbf{v}^- \right\| \tag{10.49}$$

Therefore, the trailing-edge gravity assist illustrated in Figures 10.19 and 10.20 *increases* the energy of the spacecraft's orbit relative to the sun. Figures 10.19 and 10.20b show that the length of the post-flyby velocity vector $\mathbf{v}^+$ increases because hyperbolic velocity vector $\mathbf{v}_\infty^-$ has been rotated counter-clockwise by the turning angle δ. It is easy to see from Figure 10.20b that increasing the turning angle δ will increase the magnitude of the post-flyby velocity vector $\mathbf{v}^+$. Recall from Chapter 2 that turning angle is solely a function of the eccentricity of the hyperbolic trajectory:

$$\delta = 2\sin^{-1}\left(\frac{1}{e}\right) \tag{10.50}$$

The turning angle δ increases as eccentricity e decreases. Recall that the energy of the hyperbola is solely determined by v_∞, whereas the B-plane offset distance d and v_∞ determine the angular momentum and eccentricity. Figure 10.21 shows two trailing-

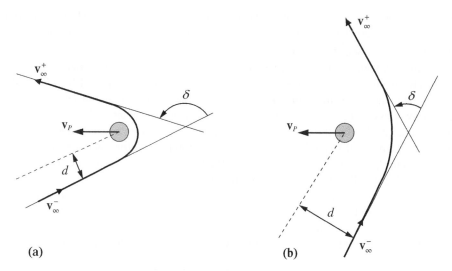

Figure 10.21 Effect of offset distance d on a trailing-edge gravity assist: (a) small offset d and large turning angle δ; and (b) large offset d and small turning angle δ.

edge gravity assists comparing "small" and "large" offset distances d. Decreasing offset d (for fixed v_∞) will decrease angular momentum, parameter p, and eccentricity e. Consequently, Eq. (10.50) shows that decreasing eccentricity (toward unity) *increases* δ as shown in Figure 10.21a. A large turning angle enhances the effect of the gravity assist (see Figure 10.20). Increasing the offset distance d increases the periapsis radius, angular momentum, and eccentricity and therefore "flattens out" the hyperbolic trajectory (i.e., decreasing δ) as shown in Figure 10.21b. Intuitively, Figure 10.21 makes sense: the closer the planetary flyby, the greater the change in heliocentric energy.

There are practical limits on the offset distance for gravity assists. If the offset d is too small, then the spacecraft will collide with the planet or enter its atmosphere. Flybys of Earth and Venus are typically limited to about 300 km altitude to avoid their atmospheres while a Jupiter flyby altitude is limited to roughly 500,000 km (about seven times the radius of Jupiter) in order to minimize exposure to Jupiter's hazardous radiation environment.

The spacecraft's heliocentric state vector before the gravity assist is $(\mathbf{r}^-, \mathbf{v}^-)$, whereas its new state vector is $(\mathbf{r}^+, \mathbf{v}^+)$ after the gravity assist. Remember that in our application of the patched-conic method, we have assumed that the spacecraft and planet share the same heliocentric position vector when the spacecraft's motion is within the planet's SOI. In this case, we have $\mathbf{r}^- = \mathbf{r}^+ = \mathbf{r}_P$ where $\mathbf{r}_P$ is the heliocentric position of the planet. This approximation is acceptable for gravity assists with small planets (Venus and Earth) but is less accurate for gravity assists with the so-called "gas giants" (Jupiter and Saturn). For example, Table 10.1 shows that the ratio of SOI radius and planetary semimajor axis is greater than 6% for Jupiter. For gravity assists with very large planets, we can improve the accuracy of the spacecraft's heliocentric position vector by accounting for the radius of the SOI:

$$\text{Before gravity assist:} \quad \mathbf{r}^- = \mathbf{r}_P + \mathbf{r}_\infty^-$$

$$\text{After gravity assist:} \quad \mathbf{r}^+ = \mathbf{r}_P + \mathbf{r}_\infty^+$$

The hyperbolic arrival radius $\mathbf{r}_\infty^-$ is the vector from the planet's center to the point where the spacecraft enters the SOI; similarly, $\mathbf{r}_\infty^+$ is the vector from the planet to the SOI departure point (of course, $\left\|\mathbf{r}_\infty^-\right\| = \left\|\mathbf{r}_\infty^+\right\| = r_{SOI}$).

In summary, Figures 10.19–10.21 show trailing-edge gravity assists that increase the spacecraft's heliocentric energy. Some mission scenarios (like the MESSENGER mission to Mercury) use one or more gravity assists to *decrease* the orbital energy. A "leading-edge" gravity assist (i.e., periapsis passage is in front of the planet) will reduce the heliocentric energy. To illustrate a leading-edge gravity assist, we can "flip" the hyperbolic trajectory in Figure 10.19 so that its periapsis direction points along heliocentric velocity $\mathbf{v}_P$. For an energy-loss gravity assist, Figure 10.20b depicts the velocity vector *before* the flyby while Figure 10.20a depicts $\mathbf{v}^+$ *after* the flyby.

Finally, we should note that the spacecraft's energy gain (or loss) is not "for free" without consequences. The spacecraft's kinetic energy change is balanced by an equal-and-opposite change in the planet's kinetic energy. However, because the spacecraft-to-planet mass ratio is extremely small, the planet's velocity change is immeasurable.

Example 10.6   Juno is a Jupiter-orbiting spacecraft launched on August 5, 2011 using an Atlas V 551. The Centaur upper stage provided a significant velocity change so that the Juno spacecraft initially followed a heliocentric trajectory with an aphelion distance of approximately 2.31 AU. The Juno spacecraft fired its onboard engines to perform two "deep space maneuvers" (DSM) near its aphelion on August 30, 2012 and September 3, 2012. These two successive DSMs corrected the trajectory to set up an Earth gravity assist on October 9, 2013. Figure 10.22 shows the heliocentric flight of the Juno spacecraft along with the DSMs, the Earth gravity assist (flyby), and the arrival at Jupiter.

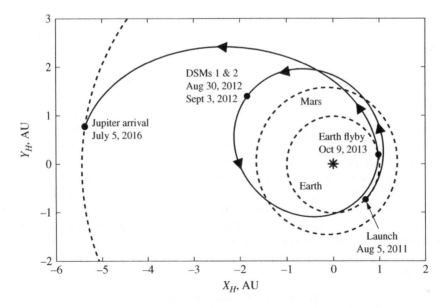

Figure 10.22 Juno spacecraft's heliocentric trajectory (Example 10.6).

The approximate heliocentric position and velocity vectors for Earth and Juno at the Earth gravity assist (October 9, 2013) are

$$\text{Heliocentric position:} \quad \mathbf{r}_E = \begin{bmatrix} 0.9577 \\ 0.2831 \\ 0 \end{bmatrix} \text{AU (Earth and Juno)}$$

$$\text{Heliocentric velocity (AU/TU):} \quad \mathbf{v}_E = \begin{bmatrix} -0.2998 \\ 0.9554 \\ 0 \end{bmatrix} \text{(Earth);} \quad \mathbf{v}^- = \begin{bmatrix} -0.0238 \\ 1.1691 \\ 0 \end{bmatrix} \text{(Juno)}$$

Determine:

a) Juno's hyperbolic arrival velocity, $\mathbf{v}_\infty^-$, before the Earth gravity assist.
b) Juno's heliocentric semimajor axis and eccentricity before and after the Earth gravity assist. The Earth gravity assist was a trailing-edge flyby with a perigee altitude of 560 km.

a) Figure 10.23 shows Juno's hyperbolic trajectory during the Earth gravity assist. Note that the Earth–sun direction shown in Figure 10.23 corresponds to the Earth–sun geometry presented in Figure 10.22 on the Earth-flyby date of October 9, 2013. Juno used a trailing-edge Earth flyby in order to increase its heliocentric energy. Its perigee passage on the hyperbolic flyby occurred a few minutes after entering the Earth's shadow. We can use Eq. (10.47) to compute the hyperbolic arrival velocity:

$$\mathbf{v}_\infty^- = \mathbf{v}^- - \mathbf{v}_E = \begin{bmatrix} -0.0238 \\ 1.1691 \\ 0 \end{bmatrix} - \begin{bmatrix} -0.2998 \\ 0.9554 \\ 0 \end{bmatrix} = \begin{bmatrix} 0.2760 \\ 0.2137 \\ 0 \end{bmatrix} \text{AU/TU}$$

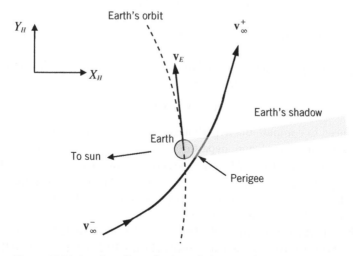

Figure 10.23 Juno's trailing-edge hyperbolic Earth flyby (Example 10.6).

Or, using 1 AU/TU = 29.785 km/s,

$$\mathbf{v}_\infty^- = \begin{bmatrix} 8.219 \\ 6.365 \\ 0 \end{bmatrix} \text{km/s}$$

The Cartesian coordinates of $\mathbf{v}_\infty^-$ make sense if we refer the hyperbolic arrival velocity vector shown in Figure 10.23 to the inserted X_H–Y_H frame. The magnitude of the hyperbolic arrival velocity is $v_\infty^- = 10.3955$ km/s.

b) First, let us determine the energy of Juno's trajectory relative to the sun *before* the Earth gravity assist using heliocentric position vector $\mathbf{r}_E$ and velocity vector $\mathbf{v}^-$:

$$\xi^- = \frac{(v^-)^2}{2} - \frac{\mu_s}{r_E} = -281.80 \text{ km}^2/\text{s}^2$$

where Earth's radial position is $r_E = 0.9987$ AU $= 149{,}404{,}653$ km, and $\mu_s = 1.32712(10^{11})$ km^3/s^2. Semimajor axis before the gravity assist is

$$a^- = -\frac{\mu_s}{2\xi^-} = \boxed{2.3547\,(10^8)\,\text{km}\,(=1.574\,\text{AU})}$$

The angular momentum (before the flyby) is

$$\mathbf{h}^- = \mathbf{r}_E \times \mathbf{v}^- = 5.01898\,(10^9)\,\mathbf{K}\,\text{km}^2/\text{s}$$

Therefore, the pre-flyby parameter is

$$p^- = \frac{(h^-)^2}{\mu_s} = 1.8981\,(10^8)\,\text{km}$$

Finally, the pre-flyby eccentricity is

$$e^- = \sqrt{1 - \frac{p^-}{a^-}} = \boxed{0.4404}$$

The reader should remember that these elements are with respect to the *sun* (heliocentric frame).

For the post-flyby elements, we need to compute Juno's sun-relative velocity vector *after* the gravity assist using Eq. (10.48)

$$\mathbf{v}^+ = \mathbf{v}_E + \mathbf{v}_\infty^+$$

Hence, we need to compute the hyperbolic departure velocity vector $\mathbf{v}_\infty^+$. Clearly, the magnitudes of the arrival and departure hyperbolic velocities are equal, that is, $v_\infty = \lVert \mathbf{v}_\infty^- \rVert = \lVert \mathbf{v}_\infty^+ \rVert$. Figure 10.24 shows the geometry of the vector addition for a trailing-edge flyby (cf. Figures 10.23 and 10.24a to identify the directions of the asymptotic velocity vectors $\mathbf{v}_\infty^-$ and $\mathbf{v}_\infty^+$). Note that the hyperbolic arrival velocity $\mathbf{v}_\infty^-$ is at an angle of 37.75° relative to the heliocentric $+X_H$ axis [i.e., $\tan^{-1}(6.365/8.219) = 37.75°$]. Consequently, the hyperbolic *departure* velocity vector $\mathbf{v}_\infty^+$ is at an angle 37.75° + δ measured counterclockwise from the $+X_H$ axis as shown in Figure 10.24a (recall that δ is the turning angle of the hyperbolic flyby – see Figure 10.21). We determine the asymptotic departure

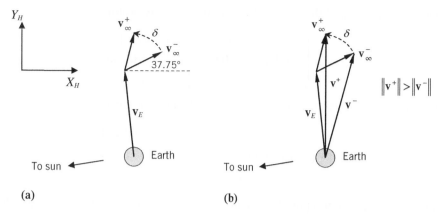

Figure 10.24 Juno's Earth gravity assist: (a) orientation of the asymptotic velocity vectors; and (b) Juno's heliocentric velocity vectors before and after the Earth flyby (Example 10.6).

vector in X_H–Y_H coordinates by multiplying v_∞ by the appropriate cosine and sine terms:

$$\mathbf{v}_\infty^+ = v_\infty \begin{bmatrix} \cos(37.75° + \delta) \\ \sin(37.75° + \delta) \\ 0 \end{bmatrix}$$

Equation (10.50) shows that the turning angle is

$$\delta = 2 \sin^{-1}\left(\frac{1}{e}\right)$$

The reader should take care to note that hyperbolic turning angle δ is a function of the *hyperbolic* trajectory eccentricity e; that is, the segment of the gravity assist *within* Earth's sphere of influence as shown in Figure 10.23. First, let us use Eq. (10.30) to compute Juno's perigee velocity during the hyperbolic flyby

$$v_p = \sqrt{v_\infty^2 + \frac{2\mu}{r_p}} = 14.932 \text{ km/s}$$

where $r_p = R_E + 560$ km $= 6{,}938$ km is the perigee radius, $v_\infty = 10.396$ km/s, and μ is the Earth's gravitational parameter. We compute the hyperbolic eccentricity using the succession of calculations summarized below (remember that the Earth is the central body in all two-body calculations):

Hyperbolic flyby semimajor axis: $\quad a = -\dfrac{\mu}{2\xi} = -\dfrac{\mu}{v_\infty^2} = -3{,}688 \text{ km}$

Hyperbolic flyby parameter: $\quad p = \dfrac{h^2}{\mu} = \dfrac{(r_p v_p)^2}{\mu} = 26{,}962 \text{ km}$

Hyperbolic flyby eccentricity: $\quad e = \sqrt{1 - \dfrac{p}{a}} = 2.8810$

Hence, $\delta = 40.62°$ and the hyperbolic departure velocity vector is

$$\mathbf{v}_\infty^+ = v_\infty \begin{bmatrix} \cos(37.75° + 40.62°) \\ \sin(37.75° + 40.62°) \\ 0 \end{bmatrix} = \begin{bmatrix} 2.095 \\ 10.182 \\ 0 \end{bmatrix} \text{ km/s}$$

Finally, Juno's heliocentric (sun-relative) velocity after the flyby is

$$\mathbf{v}^+ = \mathbf{v}_E + \mathbf{v}_\infty^+ = \begin{bmatrix} -8.928 \\ 28.455 \\ 0 \end{bmatrix} + \begin{bmatrix} 2.095 \\ 10.182 \\ 0 \end{bmatrix} = \begin{bmatrix} -6.833 \\ 38.637 \\ 0 \end{bmatrix} \text{ km/s}$$

The magnitude of Juno's post-flyby *heliocentric* velocity is $v^+ = 39.237$ km/s, which is greater than v^- as shown in Figure 10.24b. The post-flyby *heliocentric* calculations follow the same steps as the pre-flyby calculations:

Sun-relative energy: $\xi^+ = \dfrac{(v^+)^2}{2} - \dfrac{\mu_s}{r_E} = -118.496 \text{ km}^2/\text{s}^2$

Sun-relative semimajor axis: $a^+ = -\dfrac{\mu_s}{2\xi^+} = \boxed{5.5999(10^8) \text{ km } (= 3.743 \text{ AU})}$

Sun-relative angular momentum: $\mathbf{h}^+ = \mathbf{r}_E \times \mathbf{v}^+ = 5.82527(10^9)\mathbf{K} \text{ km}^2/\text{s}$

Therefore, the post-flyby parameter is

$$p^+ = \dfrac{(h^+)^2}{\mu_s} = 2.5569(10^8) \text{ km}$$

Finally, the post-flyby eccentricity is

$$e^+ = \sqrt{1 - \dfrac{p^+}{a^+}} = \boxed{0.7372}$$

The reader should remember that these elements are with respect to the *sun* (i.e., the heliocentric frame) *after* the Earth flyby. Clearly, the Juno spacecraft has gained energy (relative to the sun) after the gravity assist (i.e., $a^+ > a^-$), and its heliocentric trajectory has become more eccentric ($e^+ > e^-$). Figure 10.22 shows that Juno followed a highly elliptical transfer after the Earth-gravity assist in October 2013 that eventually reached Jupiter's orbit in early July 2016.

10.7 Summary

This chapter has presented interplanetary trajectories. We have focused on the *patched-conic method*, which divides the interplanetary mission into three separate two-body problems: (1) Earth departure; (2) heliocentric cruise; and (3) planetary arrival. We determine the characteristics of each mission segment by employing the two-body fundamentals (constants of motion, conic-section orbits, TOF, etc.) developed in Chapters 2 and 4.

For example, the spacecraft follows a hyperbolic trajectory for the Earth-departure phase because it must "escape" and reach "infinity" relative to a two-body system with Earth as the central body. Likewise, the planetary arrival phase is also a hyperbolic trajectory because the spacecraft approaches from an "infinite" distance. The heliocentric cruise, by far the longest flight segment, is an elliptical transfer orbit where the sun is the central body. The patched-conic method relies on properly "patching" together the various two-body conic sections at the fictitious "sphere of influence." In essence, the patching process is relatively simple vector addition: the spacecraft's inertial velocity in the heliocentric frame is determined by adding its Earth-relative hyperbolic excess velocity to the velocity of the moving frame (i.e., the Earth's velocity relative to the sun). This vector addition (and, the entire patched-conic method) becomes very easy to implement for a solar system modeled by coplanar, circular planetary orbits (the reader should note the similarity between the Hohmann transfer presented in Chapter 7 and the heliocentric cruise between concentric-coplanar planets). We also showed how to apply the patched-conic method to elliptical and inclined planetary orbits defined by their orbital elements (i.e., an *ephemeris*). Here we obtain the heliocentric cruise by solving Lambert's problem, and we can systematically search for the optimal (minimum-energy) cruise by obtaining multiple transfers for a range of Earth-departure and planet-arrival dates. We ended this chapter by discussing a *gravity assist*, a non-propulsive flight maneuver where a spacecraft's heliocentric orbital energy changes after a hyperbolic encounter with a planet.

References

1 https://saturn.jpl.nasa.gov/resources/1776/. Accessed August 17, 2017.
2 Battin, R.H., *An Introduction to the Mathematics and Methods of Astrodynamics, Revised Edition*, AIAA Education Series, Reston, VA, 1999.
3 Vallado, D.A., *Fundamentals of Astrodynamics and Applications*, 4th edn, Microcosm Press, Hawthorne, CA, 2013, pp. 945–948, 1040–1042.
4 Prussing, J.E. and Conway, B.A., *Orbital Mechanics*, 2nd edn, Oxford University Press, New York, 2013.
5 NASA Launch Services Program, https://elvperf.ksc.nasa.gov/Pages/Query.aspx. Accessed August 17, 2017.
6 Johnston, M.D., Graf, J.E., Zurek, R.W., Eisen, H.J., Jai, B., and Erickson, J.K., "The Mars Reconnaissance Orbiter Mission: From Launch to the Primary Science Orbit," IEEE Aerospace Conference, Paper 1001, April 2007.
7 http://aa.usno.navy.mil/faq/docs/JD_Formula.php. Accessed August 17, 2107.

Problems

Conceptual Problems

10.1 Determine an interplanetary Earth—Jupiter mission that uses a heliocentric Hohmann transfer. Assume coplanar concentric planetary orbits using the data in Table 10.3. Determine the following parameters:

a) Hyperbolic departure velocity, v_∞^+ in km/s
b) Hyperbolic arrival velocity, v_∞^- in km/s
c) Hohmann-transfer flight time in days
d) Phase angle at departure
e) Departure date for first launch opportunity after January 1, 2020.

10.2 Repeat Problem 10.1 for an Earth–Saturn mission.

10.3 Repeat Problem 10.1 for an Earth–Uranus mission.

10.4 Repeat Problem 10.1 for a Mars–Earth return mission.

10.5 Determine all feasible departure dates spanning the years 2020–2030 for a Mars–Earth return mission that uses a heliocentric Hohmann transfer.

10.6 Determine all feasible departure dates spanning the years 2020–2030 for an Earth–Venus mission that uses a heliocentric Hohmann transfer.

10.7 Determine all feasible departure dates spanning the years 2020–2030 for a Venus–Earth return mission that uses a heliocentric Hohmann transfer.

10.8 Figure P10.8 shows a leading-edge hyperbolic Venus flyby along with the heliocentric X_H–Y_H coordinate axes. Below are the spacecraft's hyperbolic arrival velocity vector (relative to Venus) and Venus' heliocentric velocity vector:

$$\text{Spacecraft: } \mathbf{v}_\infty^- = \begin{bmatrix} 0 \\ 2.71 \\ 0 \end{bmatrix} \text{ km/s}, \quad \text{Venus: } \mathbf{v}_V = \begin{bmatrix} 0 \\ 35.0208 \\ 0 \end{bmatrix} \text{ km/s}$$

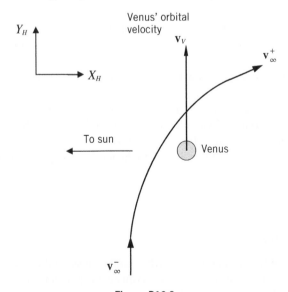

Figure P10.8

In addition, Venus' heliocentric position vector during the flyby is

$$\text{Venus:} \quad \mathbf{r}_V = \begin{bmatrix} 0.72333 \\ 0 \\ 0 \end{bmatrix} \text{AU}$$

The periapsis flyby altitude is 400 km. Determine:

a) The spacecraft's sun-relative (heliocentric) semimajor axis, eccentricity, perihelion radius, and aphelion radius before the flyby (express distances in AU).
b) The spacecraft's sun-relative (heliocentric) semimajor axis, eccentricity, perihelion radius, and aphelion radius after the flyby (express distances in AU).
c) The B-plane offset distance (in km) for the arrival hyperbola.

10.9 An interplanetary, Earth–Venus mission calls for a direct atmospheric entry at Venus. The targeted state at Venus' atmospheric entry interface (EI) is $r_{EI} = 6{,}252$ km, $v_{EI} = 11.25$ km/s, and flight-path angle $\gamma_{EI} = -9°$. What is the target offset distance d in the B-plane?

MATLAB Problems

10.10 Write an M-file that transforms an arbitrary calendar date to the corresponding Julian date. The inputs should be a 1×3 row vector corresponding to the calendar date, `cal_date = [month day year]`, and a 1×3 row vector corresponding to the Universal Time, `UTC = [hour minute second]`. The output should be the Julian date. Test your M-file against the US Naval Observatory's on-line Julian date calculator [7].

10.11 Write an M-file that transforms an arbitrary Julian date to the corresponding calendar date. The input should be the Julian date and the outputs should be a 1×3 row vector corresponding to the calendar date, `cal_date = [month day year]`, and a 1×3 row vector corresponding to the Universal Time, `UTC = [hour minute second]`. Test your M-file against the US Naval Observatory's on-line Julian date calculator [7].

10.12 Write an M-file that will compute the performance metrics of an interplanetary heliocentric Hohmann transfer between two arbitrary planets. Assume a concentric-coplanar solar-system model (use Table 10.3 for planetary radii). The inputs should be two integers for the departure and arrival planets (use 1=Mercury, 2=Venus, 3=Earth, etc.) and the desired launch year. The M-file should obtain the first feasible Hohmann transfer after January 1 of the desired launch year. The outputs should be asymptotic speeds on the departure and arrival hyperbolas, the transfer time (in days), and the departure date (as a calendar date). Test your M-file by re-solving Examples 10.1 and 10.2 and verifying the departure dates for concentric Earth–Mars transfers in Table 10.6.

Mission Applications

Problems 10.13–10.17 involve a future Earth–Mars mission. A trajectory design engineer is searching for possible Earth–Mars transfers in the 2024 time frame. Using an accurate planetary ephemeris and software that can optimize the patched-conic method, she obtains a good candidate Earth–Mars transfer with the following heliocentric coordinates:

Earth-departure date: October 1, 2024

$$\text{Earth (Oct 1, 2024):} \quad \mathbf{r}_E = \begin{bmatrix} 0.98879 \\ 0.15521 \\ 0 \end{bmatrix} \text{AU,} \quad \mathbf{v}_E = \begin{bmatrix} -0.17138 \\ 0.98430 \\ 0 \end{bmatrix} \text{AU/TU}$$

$$\text{Spacecraft (Oct 1, 2024):} \quad \mathbf{r} = \begin{bmatrix} 0.98879 \\ 0.15521 \\ 0 \end{bmatrix} \text{AU,} \quad \mathbf{v} = \begin{bmatrix} -0.18915 \\ 1.09683 \\ -0.00665 \end{bmatrix} \text{AU/TU}$$

Mars-arrival date: September 18, 2025

$$\text{Mars (Sept 18, 2025):} \quad \mathbf{r}_M = \begin{bmatrix} -1.13581 \\ -1.07875 \\ 0.00532 \end{bmatrix} \text{AU,} \quad \mathbf{v}_M = \begin{bmatrix} 0.59085 \\ -0.52030 \\ -0.02542 \end{bmatrix} \text{AU/TU}$$

$$\text{Spacecraft (Sept 18, 2025):} \quad \mathbf{r} = \begin{bmatrix} -1.13581 \\ -1.07875 \\ 0.00532 \end{bmatrix} \text{AU,} \quad \mathbf{v} = \begin{bmatrix} 0.56830 \\ -0.44095 \\ 0.00313 \end{bmatrix} \text{AU/TU}$$

10.13 The engineer wants to send a total spacecraft mass of at least 3,000 kg to Mars. Use NASA's on-line launch-vehicle performance calculator [5] to determine the smallest (i.e., cheapest!) possible launch vehicle that can deliver this spacecraft mass.

10.14 Is this candidate Earth–Mars mission a Type 1 or Type 2 transfer? Justify your answer.

10.15 The mission design team wants to consider a direct entry, descent, and landing (EDL) profile at Mars. The target state at Mars entry interface (EI) is radius $r_{EI} = 3{,}521$ km and flight-path angle $\gamma_{EI} = -15.4°$. Determine the spacecraft's velocity at Mars EI.

10.16 Using the Mars EI target presented in Problem 10.15, determine the B-plane offset distance for the Mars-arrival hyperbola.

10.17 The EDL engineers want to determine the sensitivity of the Mars EI flight-path angle γ_{EI} relative to dispersions in the B-plane offset distance. Plot EI flight-path angle vs. offset distance and determine the range for offset distance such that the flight-path angle is within its acceptable limits of $-15.8° \leq \gamma_{EI} \leq -15.0°$.

Problems 10.18–10.21 involve the Mars Science Laboratory (MSL) mission, which launched on November 26, 2011 and successfully landed on Mars on August 6, 2012. The patched-conic method and an accurate planetary ephemeris have determined the approximate heliocentric position and velocity vectors of the MSL spacecraft, Earth, and Mars for the departure and arrival dates listed below:

Earth-departure date: November 26, 2011

$$\text{Earth: } \mathbf{r}_E = \begin{bmatrix} 0.4362 \\ 0.8853 \\ 0 \end{bmatrix} \text{AU}, \quad \mathbf{v}_E = \begin{bmatrix} -0.9135 \\ 0.4382 \\ 0 \end{bmatrix} \text{AU/TU}$$

$$\text{MSL: } \mathbf{r} = \begin{bmatrix} 0.4362 \\ 0.8853 \\ 0 \end{bmatrix} \text{AU}, \quad \mathbf{v} = \begin{bmatrix} -0.9726 \\ 0.5372 \\ -0.0099 \end{bmatrix} \text{AU/TU}$$

Mars-arrival date: August 6, 2012

$$\text{Mars: } \mathbf{r}_M = \begin{bmatrix} -0.9375 \\ -1.2267 \\ -0.0027 \end{bmatrix} \text{AU}, \quad \mathbf{v}_M = \begin{bmatrix} 0.6770 \\ -0.4243 \\ -0.0255 \end{bmatrix} \text{AU/TU}$$

$$\text{MSL: } \mathbf{r} = \begin{bmatrix} -0.9375 \\ -1.2267 \\ -0.0027 \end{bmatrix} \text{AU}, \quad \mathbf{v} = \begin{bmatrix} 0.5716 \\ -0.4205 \\ 0.0062 \end{bmatrix} \text{AU/TU}$$

10.18 Determine the launch energy, C_3.

10.19 Determine the transfer angle and show that the MSL used a Type 1 trajectory.

10.20 Determine MSL's hyperbolic arrival velocity v_∞^- at Mars.

10.21 It turns out that a Type 2 transfer with a departure date of November 8, 2011 provided a lower launch energy than the Type 1 departure on November 26. MSL mission planners had hoped to use the November 8 launch, but it became unavailable because the Juno mission to Jupiter used the same launch pad in August 2011 and consequently the launch facilities were not ready in time. The predicted heliocentric positions and velocities for Earth and MSL on November 8, 2011 are

Earth-departure date: November 8, 2011

$$\text{Earth: } \mathbf{r}_E = \begin{bmatrix} 0.6929 \\ 0.7082 \\ 0 \end{bmatrix} \text{AU}, \quad \mathbf{v}_E = \begin{bmatrix} -0.7312 \\ 0.6957 \\ 0 \end{bmatrix} \text{AU/TU}$$

$$\text{MSL: } \mathbf{r} = \begin{bmatrix} 0.6929 \\ 0.7082 \\ 0 \end{bmatrix} \text{AU}, \quad \mathbf{v} = \begin{bmatrix} -0.7991 \\ 0.7647 \\ 0.02991 \end{bmatrix} \text{AU/TU}$$

Determine the launch energy, C_3, for the November 8, 2011 launch option.

10.22 In January 2006, an Atlas V 551 booster injected the New Horizons spacecraft into a hyperbolic departure trajectory with launch energy $C_3 = 158$ km^2/s^2. Assuming that the departure asymptote was aligned with Earth's heliocentric velocity vector, determine if the New Horizons spacecraft left the Earth's sphere of influence with enough energy to "escape" the solar system.

10.23 In February 2007 the New Horizons spacecraft approached Jupiter on a hyperbolic trajectory with an asymptotic arrival velocity of $v_\infty^- = 18.427$ km/s. If the probe's flyby speed at periapsis ("perijove") was 21.2 km/s, determine the offset distance d for the B-plane target.

10.24 In March 2006, the Mars Reconnaissance Orbiter (MRO) spacecraft approached Mars on a hyperbolic trajectory with eccentricity $e = 1.7804$ and asymptotic arrival speed $v_\infty^- = 2.9572$ km/s. Determine the offset distance d for the B-plane target.

10.25 The Juno spacecraft was launched on August 5, 2011. The approximate heliocentric position and velocity vectors for Earth and the Juno spacecraft on August 7, 2011 are

$$\text{Earth (August 7, 2011):} \quad \mathbf{r}_E = \begin{bmatrix} 0.71097 \\ -0.72325 \\ 0 \end{bmatrix} \text{AU}, \quad \mathbf{v}_E = \begin{bmatrix} 0.69660 \\ 0.69756 \\ 0 \end{bmatrix} \text{AU/TU}$$

$$\text{Juno (August 7, 2011):} \quad \mathbf{r} = \begin{bmatrix} 0.71415 \\ -0.71805 \\ 0 \end{bmatrix} \text{AU}, \quad \mathbf{v} = \begin{bmatrix} 0.79304 \\ 0.86108 \\ 0 \end{bmatrix} \text{AU/TU}$$

a) Determine the Juno spacecraft's distance from the Earth's SOI on August 7, 2011 (use the SOI radius defined by Table 10.1).

b) Compute the launch energy C_3 and use NASA's on-line launch-vehicle performance calculator [5] to determine Juno's launch mass (Juno was launched using an Atlas V 551). Compare this computed value to a published value of Juno's launch mass.

10.26 The Juno spacecraft approached Jupiter with hyperbolic arrival speed $v_\infty^- = 5.376$ km/s and B-plane offset distance $d = 816,600$ km. Determine the periapsis ("perijove") radius of the hyperbolic arrival trajectory.

11

Atmospheric Entry

11.1 Introduction

The principle focus of this textbook thus far has been on orbital motion in the vacuum of space. However, in our discussion of propulsion in Chapter 6, we briefly analyzed the launch or ascent phase where a booster rocket experiences atmospheric flight during its path to an orbital target. For missions with a crew, such as the Apollo lunar missions or visits to an orbiting space station, an entry flight phase is required to return the crew safely to the Earth's surface. For a robotic sample-return space probe (such as the Stardust spacecraft) or a planetary landing (such as Mars Science Laboratory), an atmospheric entry flight phase must be properly designed to bring the payload safely to the target planet's surface. This chapter presents an introduction to atmospheric entry: the flight mechanics of a spacecraft as it moves from orbital (Keplerian) motion to flight through a planetary atmosphere.

Whereas the launch phase involves increasing a vehicle's total energy as it attempts to reach an orbital target, *dissipating* the spacecraft's total energy is the primary challenge of the entry phase. The vehicle sheds its stored energy in the form of heat due to aerodynamic drag. An entry vehicle's total energy is, of course, composed of potential and kinetic energy, where its maximum potential energy (for the entry phase) occurs at the so-called "entry interface" (EI) or altitude where aerodynamic drag becomes perceptible. Specific kinetic energy at EI is $v_{EI}^2/2$, and can range from roughly 30 km^2/s^2 for a vehicle returning from low-Earth orbit (LEO) to 60 km^2/s^2 for a capsule returning from the moon.

Our primary performance metrics for atmospheric entry are peak deceleration (for structural and safety considerations), down-range distance (for targeting a landing position), and heating loads. As we shall soon see, the governing equations of motion for entry are nonlinear due to the inclusion of aerodynamic forces. We will present the so-called "first-order" analytical solutions for entry that were developed in the late 1950s by researchers such as Eggers *et al.* [1], Allen and Eggers [2], and Chapman [3]. Although these analytical solutions have shortcomings, they allow an engineer to quickly evaluate an entry vehicle or entry scenario, observe trends in trade studies, and generally gain an insight to the entry problem. While our focus is not on numerical solutions, we will present numerically integrated entry trajectories in order to make comparisons with the analytical results.

Before delving into the entry equations, let us briefly define different classes of space flight maneuvers that involve passage through a planet's atmosphere. Figure 11.1 shows an *aerobraking* maneuver where a spacecraft makes several passes through a planet's

Space Flight Dynamics, First Edition. Craig A. Kluever.
© 2018 John Wiley & Sons Ltd. Published 2018 by John Wiley & Sons Ltd.
Companion website: www.wiley.com/go/Kluever/spaceflightmechanics

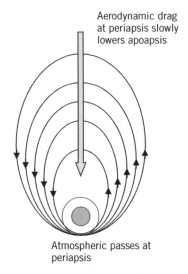

Aerodynamic drag
at periapsis slowly
lowers apoapsis

Atmospheric passes at
periapsis

Figure 11.1 Aerobraking maneuver.

atmosphere to lower its apoapsis altitude. Aerobraking begins by firing an onboard rocket to slow down the spacecraft's hyperbolic approach and establish a highly elliptical orbit about the target planet. A subsequent small propulsive burn applied at apoapsis positions the periapsis altitude at the planet's upper atmosphere. Consequently, the spacecraft experiences a small amount of atmospheric drag during each periapsis passage and the apoapsis is slowly reduced over a long time frame. Eventually, the elliptical orbit becomes nearly circular, whereupon a small propulsive burn raises periapsis out of the upper atmosphere. Aerobraking greatly reduces the propellant mass requirement for inserting a spacecraft into a low target orbit. The Mars Global Surveyor (MGS) underwent an extensive aerobraking maneuver from September 1997 to March 1999 that contracted its original 45-h elliptical orbit to a 2-h circular orbit about Mars. The MGS mission shows that while aerobraking enables propellant-mass savings, it extends the mission and delays satellite operations.

Figure 11.2 shows an *aerocapture* maneuver, where a single high-drag pass through a planet's atmosphere greatly reduces the spacecraft's velocity so that its hyperbolic approach trajectory becomes a closed (captured) orbit about the target body. Because aerocapture requires a great deal of kinetic energy dissipation during its single atmospheric pass, it requires a thermal protection system and hence is a much more aggressive atmospheric maneuver when compared with aerobraking. After the spacecraft exits the atmosphere, it fires a small propulsive burn at apoapsis to raise periapsis out of the planet's atmosphere and establish the target orbit. Although it offers significant propellant mass savings, no space mission to date has employed aerocapture.

11.2 Entry Flight Mechanics

Figure 11.3 presents a free-body diagram of the vertical forces that act on an entry vehicle during atmospheric flight. Two forces govern the vehicle's motion: gravitational force mg (acting from the vehicle to the planet's center), and aerodynamic force (resolved into lift

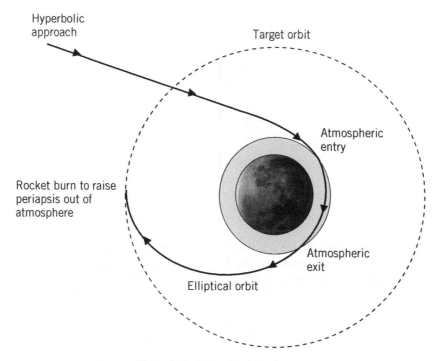

Figure 11.2 Aerocapture maneuver.

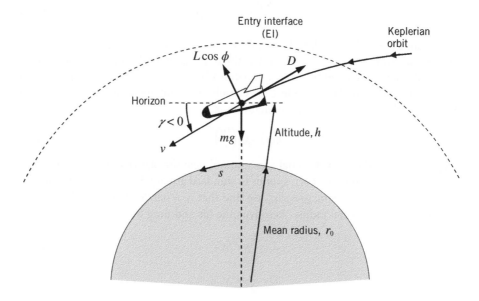

Figure 11.3 Atmospheric entry forces in a vertical plane.

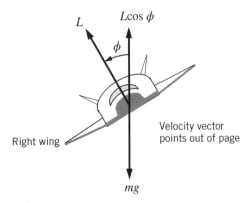

Figure 11.4 Bank angle ϕ and vertical-plane lift component $L\cos\phi$.

L and drag D). By definition, lift L is the aerodynamic force component normal to the flight path and drag D is tangent to the flight path but opposite the velocity vector. Figure 11.4 shows that the entry vehicle may rotate or "bank" about its velocity vector, thus rotating lift force L out of the vertical plane. Symbol ϕ denotes the bank angle, and Figure 11.4 shows a positive bank angle ($\phi > 0$, right wing down) where the velocity vector is pointing out of the page. Except for the elimination of the thrust force and the addition of the bank angle, Figure 11.3 is identical to Figure 6.4, the free-body diagram for the ascent phase of a launch vehicle. Hence, the entry-flight dynamical equations are obtained from the launch trajectory equations (6.17) and (6.18) by simply removing thrust T and replacing lift with the vertical-plane lift component, $L\cos\phi$:

$$\dot{v} = -\frac{D}{m} - g\sin\gamma \tag{11.1}$$

$$v\dot{\gamma} = \frac{L\cos\phi}{m} - \left(g - \frac{v^2}{r}\right)\cos\gamma \tag{11.2}$$

We maintain the familiar symbols here, where v is the vehicle's inertial velocity, r is the radial distance from the planet's center, γ is the flight-path angle, and m is the vehicle's mass. Recall that we derived the ascent equations by assuming a stationary planet, and therefore Eqs. (11.1) and (11.2) neglect the Coriolis and centripetal accelerations caused by a rotating planet. Gravitational acceleration is

$$g(r) = g_0\left(\frac{r_0}{r}\right)^2 \tag{11.3}$$

where g_0 is the "standard gravitational acceleration" near the planet's surface, and r_0 is the mean radius of the planet that corresponds to standard gravity. For Earth, standard gravity is $g_0 = 9.80665$ m/s^2. Using $g_0 = \mu/r_0^2$, we find that $r_0 = 6{,}375.42$ km, which is slightly less than Earth's equatorial radius. Aerodynamic lift and drag forces are

$$L = \frac{1}{2}\rho v^2 SC_L \tag{11.4}$$

$$D = \frac{1}{2}\rho v^2 SC_D \tag{11.5}$$

where ρ is the planet's atmospheric density, S is a vehicle reference area, and C_L and C_D are the aerodynamic lift and drag coefficients, respectively. Because we use inertial

Table 11.1 Planetary atmospheric constants.

Planet	Entry interface altitude, h_{EI} (km)	Atmospheric density at surface, ρ_0 (kg/m³)	Inverse scale height, β (km^{-1})
Earth	122	1.225	0.1378
Mars	125	0.016	0.0943
Venus	200	65.0	0.0629

velocity v in the aerodynamic force calculations, we have assumed that the planet's atmosphere is stationary. In practice, lift and drag coefficients will vary during entry as the flow-field dynamics transition between the hypersonic, supersonic, and subsonic regimes. However, because we are interested in analytical solutions, we simplify matters by treating the aerodynamic coefficients as constants (it turns out that C_L and C_D are approximately constant for hypersonic flight at constant angle-of-attack, which comprises the majority of the entry profile).

To complete the force models, we need to represent the variation of atmospheric density ρ with altitude. Because our goal is to obtain approximate analytical solutions, we use an exponential model of the planet's density

$$\rho = \rho_0 e^{-\beta h} \tag{11.6}$$

where ρ_0 is the atmospheric density at the planet's surface and β is the "inverse scale height." Although an altitude-dependent β provides the most accurate density model, we will use the simplest possible model and treat β as a constant. Table 11.1 presents the EI altitude h_{EI}, surface density ρ_0, and inverse scale height β for Earth, Mars, and Venus. Although the parameters presented in Table 11.1 are not definitive values, they are the planetary constants that we will utilize in this textbook.

We include two additional kinematic equations to determine the position of the entry vehicle:

$$\dot{h} = v \sin \gamma \tag{11.7}$$

$$\dot{s} = \frac{r_0}{r} v \cos \gamma \tag{11.8}$$

Equations (11.7) and (11.8) define the time-rates for altitude h and down-range distance s. Note that because radius is $r = r_0 + h$, its time-rate is $\dot{r} = \dot{h}$.

Equations (11.1), (11.2), (11.7), and (11.8) are the governing equations of motion for atmospheric entry in a vertical plane. Just as the launch trajectory equations, these ordinary differential equations (ODEs) are highly nonlinear and hence there is no analytical solution. While it is possible to obtain numerical solutions using a computer, we will develop approximate analytical solutions to gain insight into the behavior of entry flight mechanics.

A final note regarding planar motion is in order. Figure 11.4 shows that banking the lift vector will cause a side force, which produces turning or out-of-plane motion. We will neglect cross-track motion in this chapter because we are primarily concerned with the entry vehicle's motion in the vertical plane (i.e., velocity and along-track range). In practice, any cross-track excursions caused by banking are removed by so-called "bank reversals" (e.g., rotating the vehicle from $\phi = 40°$ to $-40°$). The Space Shuttle used bank reversals to eliminate cross-track deviations while maintaining a desired vertical lift component $L \cos \phi$ [4].

11.3 Ballistic Entry

Our first analytical entry solution involves a *ballistic entry* where the lift force is zero. An axially symmetric vehicle flown at zero angle-of-attack will follow a ballistic entry trajectory. The Stardust Sample Return Capsule was a blunt-body entry vehicle that followed a ballistic trajectory when it re-entered Earth's atmosphere in early 2006 after collecting dust samples from the tail of comet Wild 2.

In order to derive a first-order analytical solution for ballistic entry, we make two fundamental assumptions:

1) Drag acceleration D/m dominates the path-tangent acceleration equation (11.1) and therefore we can neglect $g \sin \gamma$.
2) The centrifugal term v^2/r cancels gravity g in the path-normal acceleration equation (11.2).

The first assumption is questionable when the vehicle reaches EI because drag is initially very small due to the extremely thin upper atmosphere. The first assumption is reasonable when the drag force is significant and the flight-path angle γ is small. The second assumption imposes limitations to the first-order ballistic solution. We may express the second assumption (centrifugal term cancels gravity) as

$$\frac{v^2}{r} = g$$

or,

$$v = \sqrt{gr} = \sqrt{\frac{\mu}{r}}$$

which holds because $g = \mu/r^2$. In other words, v^2/r cancels g only when the vehicle is entering the planet's atmosphere at near-circular speed. Therefore, the second assumption is reasonable for a ballistic entry after a de-orbit burn from LEO. However, by enforcing the second approximation, we cannot obtain analytical ballistic trajectories for vehicles entering the atmosphere at hyperbolic or near-hyperbolic speeds. Furthermore, even for entry at near-circular speed, the second condition ($v^2/r = g$) only holds during the initial high-altitude entry. Eventually, aerodynamic drag will slow the vehicle such that gravity g will begin to dominate the centrifugal term v^2/r. The path-normal equation (11.2) shows that in the absence of lift, the gravity force will eventually rotate the flight path downward toward the vertical direction.

Applying the first assumption to Eq. (11.1) yields

$$\dot{v} = -\frac{D}{m} = -\frac{1}{2}\rho v^2 \frac{S}{m} C_D \tag{11.9}$$

Applying the second assumption ($v^2/r = g$) to the path-normal equation (11.2) with zero lift ($L = 0$), we see that the time-rate of the flight-path angle is zero (i.e., $\dot{\gamma} = 0$). Therefore, flight-path angle remains constant and equal to its value at the entry interface, $\gamma = \gamma_{EI}$. Next, let us change the independent variable of Eq. (11.9) from time t to altitude h by dividing Eq. (11.9) by Eq. (11.7):

$$\frac{\dot{v}}{\dot{h}} = \frac{dv/dt}{dh/dt} = \frac{dv}{dh} = \frac{-\rho v^2 S C_D}{2mv \sin \gamma_{EI}} \tag{11.10}$$

Note that we have used $\dot{h} = v \sin \gamma_{EI}$ because flight-path angle remains constant. At this point, it is convenient to group the three constants S, C_D, and m into a single constant called the *ballistic coefficient*, C_B

$$C_B \equiv \frac{m}{S C_D} \tag{11.11}$$

The ballistic coefficient has units of kilograms per meter squared. Substituting Eq. (11.11) into Eq. (11.10) and separating variables yields

$$\frac{dv}{v} = \frac{-\rho}{2 C_B \sin \gamma_{EI}} dh \tag{11.12}$$

Next, substitute the exponential density model (11.6) into Eq. (11.12):

$$\frac{dv}{v} = \frac{-\rho_0}{2 C_B \sin \gamma_{EI}} e^{-\beta h} dh \tag{11.13}$$

Integrating both sides of Eq. (11.13) from the initial conditions at entry interface (v_{EI}, h_{EI}) to the final conditions (v, h) yields

$$\ln\left(\frac{v}{v_{EI}}\right) = \frac{\rho_0}{2\beta C_B \sin \gamma_{EI}} \left(e^{-\beta h} - e^{-\beta h_{EI}}\right) \tag{11.14}$$

or,

$$\ln\left(\frac{v}{v_{EI}}\right) = B\left(e^{-\beta h} - e^{-\beta h_{EI}}\right) \tag{11.15}$$

where the constant B is the first term on the right-hand side of Eq. (11.14)

$$B \equiv \frac{\rho_0}{2\beta C_B \sin \gamma_{EI}} \tag{11.16}$$

The constant B is dimensionless and negative because $\gamma_{EI} < 0$ (the reader should compute B with care; surface density ρ_0 has units of kg/m^3, ballistic coefficient C_B has units of kg/m^2, while inverse scale height β is typically given in units of km^{-1}). Finally, we can solve Eq. (11.15) for the vehicle's velocity along the ballistic entry trajectory:

$$v = v_{EI} \exp\left[B\left(e^{-\beta h} - e^{-\beta h_{EI}}\right)\right] \tag{11.17}$$

This expression for velocity can be simplified by neglecting the very small term $e^{-\beta h_{EI}}$, which becomes even more negligible compared with $e^{-\beta h}$ as the vehicle descends and altitude h decreases

$$v = v_{EI} \exp\left[Be^{-\beta h}\right] \tag{11.18}$$

Equation (11.18) is the spacecraft's velocity during ballistic entry as a function of altitude. The vehicle's entry-velocity history depends on the planet's inverse scale height β and the constant B, which in turn is a function of the planet's surface density, inverse scale height, ballistic coefficient, and entry flight-path angle. Of course, the vehicle's velocity also depends on its initial value, v_{EI}, which must be approximately equal to circular orbital speed to satisfy the second assumption for ballistic entry.

Vehicle deceleration is an important performance metric for an entry trajectory due to its correlation with limits for the vehicle's structural integrity. Furthermore, humans cannot tolerate sustained exposure to excessive "g-loads." Using the chain rule, we take the time derivative of the velocity equation (11.18):

$$\frac{dv}{dt} = v_{EI}B\exp\left[Be^{-\beta h}\right]\left(-\beta e^{-\beta h}\right)\frac{dh}{dt} \tag{11.19}$$

Substituting $dh/dt = v \sin\gamma_{EI} = v_{EI}\exp\left[Be^{-\beta h}\right]\sin\gamma_{EI}$ into Eq. (11.19) yields

$$\dot{v} = -\beta B v_{EI}^2 \sin\gamma_{EI} e^{-\beta h}\exp\left[2Be^{-\beta h}\right] \tag{11.20}$$

Equation (11.20) is the vehicle's acceleration along the ballistic flight path as a function of altitude h. Because $B < 0$ and $\sin\gamma_{EI} < 0$, Eq. (11.20) shows that acceleration dv/dt is always negative. We will define deceleration a as negative acceleration, $a = -\dot{v}$, and hence it is always positive for a ballistic entry. Deceleration will reach a peak (maximum) value at a particular altitude during the entry. We obtain maximum deceleration by setting the derivative of Eq. (11.20) with respect to altitude h to zero:

$$\frac{d}{dh}\dot{v} = \beta^2 B v_{EI}^2 \sin\gamma_{EI} e^{-\beta h}\exp\left[2Be^{-\beta h}\right] + \beta^2 B v_{EI}^2 \sin\gamma_{EI} e^{-\beta h}\left(2Be^{-\beta h}\right)\exp\left[2Be^{-\beta h}\right] = 0 \tag{11.21}$$

Factoring out the common term in Eq. (11.21) yields

$$\frac{d}{dh}\dot{v} = \beta^2 B v_{EI}^2 \sin\gamma_{EI} e^{-\beta h}\exp\left[2Be^{-\beta h}\right]\left(1 + 2Be^{-\beta h}\right) = 0 \tag{11.22}$$

Equation (11.22) is zero only when the parenthetical term is zero, or

$$1 + 2Be^{-\beta h} = 0 \tag{11.23}$$

We may express the above condition for peak deceleration as

$$e^{-\beta h} = \frac{-1}{2B} \tag{11.24}$$

We can solve for altitude by taking the natural logarithm of Eq. (11.24):

$$h_{crit} = \frac{1}{-\beta}\ln\left(\frac{-1}{2B}\right) \tag{11.25}$$

Or,

$$h_{crit} = \frac{\ln(-2B)}{\beta} \tag{11.26}$$

Equation (11.26) presents the *critical altitude* h_{crit} where peak deceleration occurs during ballistic entry. Finally, we can substitute Eq. (11.26) into the deceleration equation (11.20) to determine the peak deceleration. After some algebra, we obtain

$$a_{max} = \frac{-\beta \, v_{EI}^2 \sin\gamma_{EI}}{2e} \tag{11.27}$$

Equation (11.27) is the maximum deceleration along the ballistic flight path. Note that it is positive because $\sin\gamma_{EI} < 0$. Furthermore, Eq. (11.27) provides deceleration in units of kilometers per second squared if we express inverse scale height β in units of per kilometer and entry interface velocity v_{EI} in kilometers per second. Typically, we express vehicle deceleration in terms of an "Earth g-load" which we can easily obtain by dividing deceleration by Earth's standard gravitational acceleration g_0. The reader should note that peak deceleration only depends on entry speed v_{EI}, atmospheric inverse scale height β, and entry flight-path angle γ_{EI}. Furthermore, vehicle characteristics (i.e., ballistic coefficient C_B) *do not* influence the peak deceleration along a ballistic entry – this result is not intuitive. However, Eq. (11.26) shows that the vehicle's ballistic coefficient *does* influence the altitude where peak acceleration occurs.

Equation (11.27) has limitations: note that for a "grazing" entry with flight-path angle $\gamma_{EI} \approx 0$, Eq. (11.27) predicts zero deceleration. Chapman [3] developed and applied a second-order method to ballistic entry and showed that for entry from LEO, the peak deceleration is about $8g_0$ even at very shallow flight-path angles. The first-order method presented here (based on the research of Allen and Eggers [2]) produces peak decelerations that show a good match with Chapman's second-order method for entry flight-path angles steeper than about $-5°$.

The final piece of information regarding ballistic entry is the critical velocity where peak deceleration occurs. To find this velocity, we substitute the expression for critical altitude, Eq. (11.26), into the velocity equation (11.18)

$$v_{crit} = v_{EI} \exp\left\{ B \exp\left[\frac{-\beta \ln(-2B)}{\beta} \right] \right\} \tag{11.28}$$

Equation (11.28) reduces to the simple expression:

$$v_{crit} = v_{EI} e^{-0.5} \tag{11.29}$$

or,

$$v_{crit} = 0.6065 v_{EI} \tag{11.30}$$

The first-order ballistic entry analysis predicts that the peak deceleration *always* occurs at a critical velocity v_{crit} that is 60.7% of the velocity at entry interface regardless of entry flight-path angle γ_{EI}, ballistic coefficient C_B, or even the characteristics of the planet's atmosphere!

Ballistic entry offers a simple entry strategy. We may control peak deceleration for a ballistic entry by selecting the EI states, v_{EI} and γ_{EI}. However, as previously noted, the peak deceleration for a shallow entry from LEO exceeds $8g_0$. As we shall see in the next section, adding even a small amount of aerodynamic lift will greatly diminish the peak deceleration. The following example illustrates a ballistic entry trajectory and compares the first-order analytical solution with a numerically integrated trajectory.

Example 11.1 The Soviet Union's Vostok spacecraft was used for the first manned space flights in the early 1960s. It was spherical in shape (with drag coefficient $C_D = 2.0$), and had a mass of 2,460 kg and diameter of 2.3 m. The Vostok capsule departs a low-Earth orbit and reaches entry interface (EI) altitude $h_{EI} = 122$ km with velocity $v_{EI} = 7.74$ km/s and flight-path angle $\gamma_{EI} = -3.2°$. It follows a ballistic entry.

a) Compute the velocity, flight-path angle, and deceleration (in Earth g_0) at altitudes of 80, 60, and 40 km.
b) Compute the peak deceleration and associated critical altitude and velocity at peak deceleration.
c) Create plots of velocity and deceleration vs. altitude for the analytical ballistic entry and a numerically integrated ballistic entry profile.

a) The analytical ballistic entry equations require the constants C_B and B. Ballistic coefficient for the Vostok capsule is

$$C_B = \frac{m}{SC_D} = 296.05 \text{ kg/m}^2$$

where the Vostok reference area is $S = \pi(2.3/2)^2 = 4.1548 \text{ m}^2$. The dimensionless coefficient is

$$B = \frac{\rho_0}{2\beta C_B \sin \gamma_{EI}} = -268.9650$$

where the values for Earth's atmospheric constants ρ_0 and β are taken from Table 11.1 (note that β must be expressed in units of m^{-1} if surface density ρ_0 is in units of kg/m^3). Equation (11.18) provides velocity during the ballistic entry as a function of altitude:

$$v = v_{EI} \exp\left[Be^{-\beta h}\right]$$

The velocities at the required altitudes are

$$h = 80 \text{ km:} \quad \boxed{v = 7.7062 \text{ km/s}}$$

$$h = 60 \text{ km:} \quad \boxed{v = 7.2239 \text{ km/s}}$$

$$h = 40 \text{ km:} \quad \boxed{v = 2.6126 \text{ km/s}}$$

Note that the Vostok's velocity changes very little from EI to $h = 80$ km (due to low drag) but dramatically changes between 60 and 40 km altitude.

The analytical ballistic solution for flight-path angle assumes that it remains constant during the entire entry. Therefore, $\gamma = \gamma_{EI} = -3.2°$ at $h = 80$, 60, and 40 km.

Equation (11.20) determines the vehicle's acceleration

$$\dot{v} = -\beta B v_{EI}^2 \sin \gamma_{EI} e^{-\beta h} \exp\left[2Be^{-\beta h}\right]$$

We compute deceleration in units of meters per second squared by expressing β in per meter, v_{EI} in meters per second, and h in meters. The values of vehicle deceleration $(a = -\dot{v}, \text{ in } g_0)$ at the three altitudes are

$$h = 80 \text{ km:} \quad a = 2.003 \text{ m/s}^2 = \boxed{0.204 g_0}$$

$$h = 60 \text{ km:} \quad a = 27.704 \text{ m/s}^2 = \boxed{2.825 g_0}$$

$$h = 40 \text{ km:} \quad a = 57.021 \text{ m/s}^2 = \boxed{5.814 g_0}$$

where we have used $g_0 = 9.80665 \text{ m/s}^2$.

b) We use Eq. (11.27) to determine the peak deceleration of the Vostok capsule

$$a_{max} = \frac{-\beta \, v_{EI}^2 \sin \gamma_{EI}}{2e} = 84.763 \text{ m/s}^2 = \boxed{8.643 g_0}$$

Critical altitude for peak deceleration is computed using Eq. (11.26)

$$h_{crit} = \frac{\ln(-2B)}{\beta} = \boxed{45.629 \text{ km}}$$

We use Eq. (11.30) to predict the critical velocity for peak deceleration:

$$v_{crit} = 0.6065 v_{EI} = \boxed{4.694 \text{ km/s}}$$

c) Finally, we can compute velocity and deceleration (in g_0) using Eqs. (11.18) and (11.20) determined at altitudes ranging from h_{EI} to about zero. We compare the analytical ballistic entry results with a more precise entry trajectory obtained by numerically integrating Eqs. (11.1), (11.2), and (11.7) with constant C_D, $C_L = 0$ (zero lift), inverse-square gravity (11.3), and an exponential atmosphere modeled by Eq. (11.6). We employ MATLAB's M-file ode45.m to numerically integrate the three nonlinear ODEs. Figure 11.5 shows velocity vs. altitude for the analytical and numerical entry solutions. The Vostok entry trajectory begins in the upper right

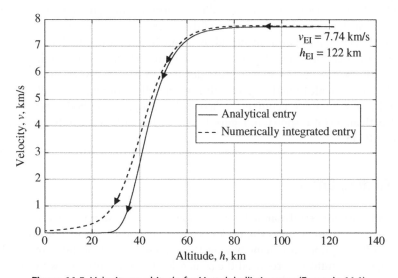

Figure 11.5 Velocity vs. altitude for Vostok ballistic entry (Example 11.1).

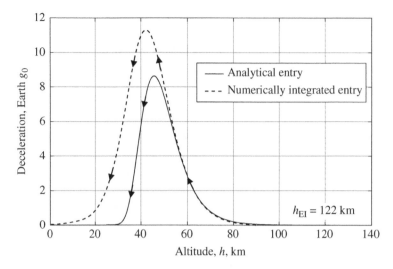

Figure 11.6 Deceleration vs. altitude for Vostok ballistic entry (Example 11.1).

corner of Figure 11.5 (at h_{EI} = 122 km), and progresses to the left as both altitude and velocity become smaller (note the arrows on the plots). The analytical velocity solution shows a good match with the "true" numerically integrated trajectory despite the major assumption that flight-path angle remains constant. Figure 11.6 presents the Vostok's deceleration for the analytical and numerical ballistic trajectories. The analytical method accurately predicts deceleration until altitude reaches 50 km, which is slightly before the critical altitude of 45.6 km. The peak deceleration of the analytical method is about $2.7g_0$ lower than the "true" peak deceleration from the numerically integrated trajectory.

   Figure 11.7 shows how the Vostok spacecraft's flight-path angle changes with altitude for the numerically integrated ballistic entry. Flight-path angle remains nearly constant for the high-altitude portion of the entry (i.e., the centrifugal force approximately cancels gravity) until an altitude of about 50 km. However, the Vostok capsule continues to decelerate (see Figure 11.5), and below 50 km gravity dominates the v^2/r term and consequently the flight-path angle turns downward toward the vertical direction. Below 6 km altitude the spacecraft is essentially dropping vertically toward the ground with $\gamma = -90°$. Of course, the first-order analytical method assumes constant flight-path angle (also shown in Figure 11.7) which is not an accurate approximation for $h < 50$ km.

11.4 Gliding Entry

An entry vehicle with aerodynamic lift capabilities offers two significant advantages over a ballistic entry: (1) lift reduces the peak deceleration; and (2) lift extends the vehicle's range and offers options for maneuvering the vehicle. The small lift provided by the Gemini capsule significantly reduced the peak g-load when compared with the ballistic entry of the Mercury capsule. The Space Shuttle's delta wing configuration provided

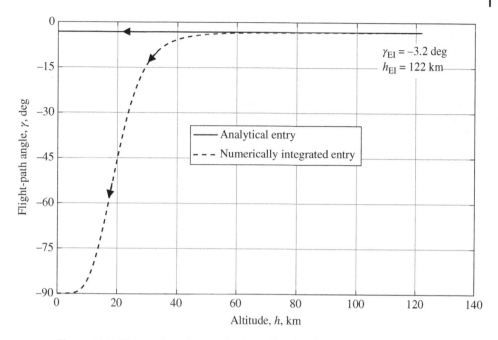

Figure 11.7 Flight-path angle vs. altitude for Vostok ballistic entry (Example 11.1).

more lift than any entry vehicle to date, which resulted in a long-range shallow glide with low deceleration loads.

As before, we will develop a first-order analytical solution for lifting entry; to do so, we employ two principle assumptions:

1) The lifting entry is a shallow glide at small flight-path angle and therefore we can set $\sin\gamma \approx \gamma$ and $\cos\gamma \approx 1$ in Eqs. (11.1) and (11.2).
2) The vertical lift and centrifugal terms, $L\cos\phi/m$ and v^2/r, cancel gravity g in the path-normal acceleration equation (11.2).

The second assumption is the so-called "equilibrium-glide" condition. Applying both approximations to Eq. (11.2), we obtain

$$v\dot{\gamma} = a_L \cos\phi - g + \frac{v^2}{r} = 0 \tag{11.31}$$

where $a_L \equiv L/m$ is the lift acceleration. The equilibrium glide is at constant (small) flight-path angle. Equation (11.31) is an algebraic equation that defines velocity during the gliding entry. However, lift acceleration a_L also depends on velocity. A convenient way to represent lift acceleration is

$$a_L = \frac{L/m}{D/m} a_D \tag{11.32}$$

where $a_D \equiv D/m$ is the drag acceleration, which can be expressed using Eq. (11.5):

$$a_D = \frac{1}{2}\rho v^2 \frac{S}{m} C_D = \frac{\rho v^2}{2C_B} \tag{11.33}$$

Note that we have used the ballistic coefficient, Eq. (11.11), to simplify Eq. (11.33). Substituting Eq. (11.33) into the lift acceleration equation (11.32) yields

$$a_L = (L/D)\frac{\rho v^2}{2C_B} \tag{11.34}$$

where L/D $(= a_L/a_D)$ is the dimensionless *lift-to-drag ratio* for the entry vehicle. As with ballistic coefficient, L/D is an important vehicle design parameter that greatly affects the gliding entry profile. We will treat L/D as a constant parameter. Substituting Eq. (11.34) into the equilibrium-glide condition (11.31) yields

$$(L/D)\frac{\rho v^2}{2C_B}\cos\phi - g + \frac{v^2}{r} = 0 \tag{11.35}$$

Multiplying all terms by radius r and rearranging, we obtain

$$v^2\left[1 + (L/D)\frac{r\rho\cos\phi}{2C_B}\right] = gr \tag{11.36}$$

Solving Eq. (11.36) for velocity yields

$$v = \sqrt{\frac{gr}{1 + (L/D)\dfrac{r\rho\cos\phi}{2C_B}}} \tag{11.37}$$

Equation (11.37) is the velocity along an equilibrium glide. Let us take a closer look at each term on the right-hand side of Eq. (11.37): we can assume that L/D and C_B are constants; bank angle ϕ can be modulated during the glide; and atmospheric density ρ changes dramatically with altitude. Because planets have "thin" atmospheres relative to their planetary radius, r varies less than 2% between EI and sea level for Earth. Similarly, Eq. (11.3) shows that Earth's gravity g varies less than 4% between EI and sea level. For these reasons, we use the surface values $r = r_0$ and $g = g_0$ in Eq. (11.37) for simplicity. Finally, we substitute the exponential density model (11.6) into Eq. (11.37) to obtain

$$v = \sqrt{\frac{g_0 r_0}{1 + (L/D)\dfrac{r_0\cos\phi}{2C_B}\rho_0 e^{-\beta h}}} \tag{11.38}$$

Equation (11.38) defines the velocity along an equilibrium glide as a function of altitude h. The assumptions leading to Eq. (11.38) have posed an additional limitation to its use. At entry interface $(h = h_{EI})$, the term $e^{-\beta h_{EI}}$ is very small, and hence the EI velocity predicted by Eq. (11.38) is approximately

$$v_{EI} \approx \sqrt{g_0 r_0} = v_s \tag{11.39}$$

Because $g_0 = \mu/r_0^2$, Eq. (11.39) shows that $v_s = \sqrt{\mu/r_0}$ is the circular orbital speed at the planet's surface. For Earth, $v_s = 7.907$ km/s. Therefore, imposing the equilibrium-glide condition restricts our first-order method to entry at near-circular speeds and shallow flight-path angle. It is important to reiterate that we *cannot* use Eq. (11.38) to compute velocity for an aerocapture maneuver where the spacecraft enters the atmosphere at hyperbolic or near-hyperbolic speed.

Deceleration along the flight path is another important metric. Neglecting the gravity component along the flight path in Eq. (11.1), we obtain

$$\dot{v} = -a_D = \frac{-a_L}{L/D} \tag{11.40}$$

where we have used Eq. (11.32) to replace drag acceleration a_D. The equilibrium-glide condition (11.31) dictates that the vertical lift component balances $g - v^2/r$

$$a_L \cos\phi = g_0 - \frac{v^2}{r_0} \tag{11.41}$$

Note that we have used the surface values $r = r_0$ and $g = g_0$ for simplicity. Substituting Eq. (11.41) for lift acceleration in Eq. (11.40) yields

$$\dot{v} = \frac{\frac{v^2}{r_0} - g_0}{(L/D)\cos\phi} \tag{11.42}$$

Equation (11.42) expresses vehicle acceleration along the flight path as a function of velocity and bank angle ϕ. Equation (11.38) determines velocity along the equilibrium glide. Let us observe deceleration ($a = -\dot{v}$) at the initial and terminal points of the trajectory. At entry interface $v_{EI} \approx \sqrt{g_0 r_0}$, and hence Eq. (11.42) shows that deceleration is nearly zero. The (theoretical) peak deceleration occurs at zero velocity, in which case Eq. (11.42) yields

$$a_{max} \approx \frac{g_0}{(L/D)\cos\phi} \tag{11.43}$$

For example, with $L/D = 1.1$ (e.g., the Space Shuttle) and zero bank angle, the peak deceleration is less than 9 m/s^2 or less than one g_0. Of course, the lifting vehicle cannot sustain the equilibrium-glide condition at low velocities as lift and centrifugal forces diminish and can no longer balance gravity.

Up to this point, the assumptions that lead to the equilibrium-glide condition for lifting entry might seem restrictive. As previously mentioned, we cannot use the first-order glide analysis for an aerocapture maneuver from super-circular entry speeds. However, a predominant segment of the Space Shuttle's entry profile followed an equilibrium glide. In fact, the Space Shuttle entry guidance algorithm used first-order analytical solutions to predict its down-range distance during the hypersonic entry. Therefore, it is instructive to develop an analytical expression for down-range during the equilibrium glide. To begin, we form the derivative ds/dv by dividing the down-range rate, Eq. (11.8), by the equilibrium-glide acceleration, Eq. (11.42):

$$\frac{\dot{s}}{\dot{v}} = \frac{ds}{dv} = \frac{v(L/D)\cos\phi}{\frac{v^2}{r_0} - g_0} \tag{11.44}$$

where we have used the approximations $r = r_0$ and $\cos\gamma = 1$ in the ds/dt equation. Separating variables yields

$$ds = \frac{v(L/D)\cos\phi}{\frac{v^2}{r_0} - g_0} dv \tag{11.45}$$

Integrating Eq. (11.45) with constant L/D and constant bank angle ϕ, we obtain

$$s = \frac{r_0}{2}(L/D)\cos\phi \ln\left(g_0 r_0 - v^2\right)\Big|_{v_1}^{v_2} \tag{11.46}$$

Evaluating the velocity at the boundary conditions and using $v_s^2 = g_0 r_0$ yields

$$s = \frac{r_0}{2}(L/D)\cos\phi \ln\left(\frac{v_s^2 - v_2^2}{v_s^2 - v_1^2}\right) \tag{11.47}$$

Equation (11.47) is the ground-track range for an equilibrium glide from initial velocity v_1 to final velocity v_2. Note that this range equation is sensitive to the value of initial velocity v_1 and becomes singular when $v_1 = v_s$ (this singularity makes sense: if initial velocity v_1 is *exactly* equal to surface-circular velocity v_s, then the vehicle remains in orbit and the ground-track is infinite). Equations (11.38) and (11.39) show that enforcing the equilibrium-glide condition results in an EI velocity that is less than but nearly equal to v_s. As previously mentioned, the Space Shuttle's onboard guidance computer used Eq. (11.47) to predict the ground-track range for the equilibrium-glide segment of its entry profile. We will discuss the Space Shuttle's entry guidance scheme and range prediction methods in Section 11.7.

Example 11.2 The Space Shuttle is in a 300-km altitude circular orbit when it performs a de-orbit burn with impulse $\Delta v = 90$ m/s (see Figure 11.8). The Space Shuttle has mass $m = 82{,}000$ kg, reference wing area $S = 250$ m^2, drag coefficient $C_D = 0.8$, $L/D = 1.1$, and constant bank angle $\phi = 50°$ during its equilibrium-glide entry.

a) Determine the Shuttle's velocity and flight-path angle at entry interface.
b) Compute the velocity and deceleration (in Earth g_0) at altitudes of 80, 60, and 40 km.
c) Plot velocity and deceleration for the analytical equilibrium-glide entry and a numerically integrated lifting entry profile. Let the final altitude be 30 km (i.e., the end of the entry phase).
d) Determine the total ground-track range from EI to an altitude of 30 km.

a) Figure 11.8 shows that the 90 m/s de-orbit burn (anti-tangent to LEO) creates a transfer orbit with an apogee that is tangent to LEO. The apogee velocity on the transfer orbit is

$$v_a = \sqrt{\frac{\mu}{r_{LEO}}} - \Delta v = 7.726 - 0.090 = 7.636 \text{ km/s}$$

where $r_{LEO} = 6{,}678$ km. The energy, semimajor axis, angular momentum, parameter, and eccentricity of the transfer orbit are

$$\text{Energy: } \xi = \frac{v_a^2}{2} - \frac{\mu}{r_{LEO}} = \frac{-\mu}{2a} = -30.536 \text{ km}^2/\text{s}^2$$

$\rightarrow$ semimajor axis is $a = 6{,}526.8$ km

Angular momentum and parameter: $h = r_{LEO} v_a = \sqrt{p\mu} = 50{,}992 \text{ km}^2/\text{s}$

$\rightarrow$ parameter is $p = 6{,}523.3$ km

$$\text{Eccentricity: } e = \sqrt{1 - \frac{p}{a}} = 0.0232$$

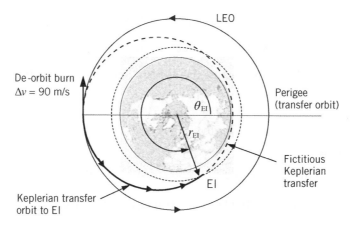

Figure 11.8 De-orbit burn from low-Earth orbit (LEO) and transfer to entry interface (EI) (Example 11.2).

The perigee radius of the transfer orbit is

$$r_p = \frac{p}{1+e} = 6{,}375.6 \text{ km} \approx r_0$$

We see that perigee radius of the Keplerian transfer is nearly equal to the Earth's mean radius as shown in Figure 11.8 (of course, the perigee radius is fictitious because the Shuttle does not follow a Keplerian orbit beyond EI). Velocity at EI is obtained from the energy equation using $r_{EI} = R_E + 122 \text{ km} = 6{,}500 \text{ km}$

$$\xi = \frac{v_{EI}^2}{2} - \frac{\mu}{r_{EI}} = -30.536 \text{ km}^2/\text{s}^2 \;\rightarrow\; \boxed{v_{EI} = 7.847 \text{ km/s}}$$

We can compute flight-path angle at EI from angular momentum (h), r_{EI}, and v_{EI}

$$\cos\gamma_{EI} = \frac{h}{r_{EI} v_{EI}} = 0.99974 \;\rightarrow\; \boxed{\gamma_{EI} = -1.306°}$$

Of course, flight-path angle at EI is negative because the Shuttle is approaching Earth. We can use the transfer orbit elements to compute the true anomaly of the Shuttle at EI: $\theta_{EI} = 278.9°$ as shown in Figure 11.8.

b) We use Eq. (11.38) to compute the velocity of the Shuttle on an equilibrium glide. We need the Shuttle's ballistic coefficient; using Eq. (11.11) we obtain

$$C_B = \frac{m}{SC_D} = 410 \text{ kg/m}^2$$

Using Earth's atmosphere parameters (β and ρ_0) from Table 11.1, $L/D = 1.1$, $\phi = 50°$, $r_0 = 6{,}375{,}416 \text{ m}$, and $g_0 = 9.80665 \text{ m/s}^2$, we may now use Eq. (11.38) to compute the Shuttle's velocity on the equilibrium glide at the three altitudes:

$$\text{At 80 km altitude:} \quad v = \sqrt{\frac{g_0 r_0}{1 + (L/D)\dfrac{r_0 \cos\phi}{2C_B}\rho_0 e^{-\beta h}}} = \boxed{7{,}506 \text{ m/s}}$$

At 60 km altitude: $\boxed{v = 4{,}787 \text{ m/s}}$

At 40 km altitude: $\boxed{v = 1{,}489 \text{ m/s}}$

Next, we use Eq. (11.42) to compute the deceleration $(a = -\dot{v})$ in Earth g_0

$$\frac{a}{g_0} = \frac{g_0 - \dfrac{v^2}{r_0}}{g_0(L/D)\cos\phi}$$

Deceleration is

At 80 km altitude $(v = 7{,}506 \text{ m/s})$: $\boxed{a = 0.140\,g_0}$

At 60 km altitude $(v = 4{,}787 \text{ m/s})$: $\boxed{a = 0.896\,g_0}$

At 40 km altitude $(v = 1{,}489 \text{ m/s})$: $\boxed{a = 1.364\,g_0}$

c) We can compute velocity and deceleration (in g_0) using Eqs. (11.38) and (11.42) evaluated at altitudes ranging from $h_{EI} = 122$ to 30 km. In addition, we determine a more precise entry trajectory by numerically integrating Eqs. (11.1), (11.2), (11.7), and (11.8) with constant $C_D = 0.8$, constant $C_L = (L/D)C_D = 0.88$, constant bank angle $\phi = 50°$, inverse-square gravity and an atmosphere modeled by Eq. (11.6). We use MATLAB's ode45.m for the numerical integration process. Figure 11.9 shows velocity vs. altitude for the analytical and numerical entry solutions. Note that the velocity of the numerically integrated gliding entry "oscillates" about the analytical equilibrium-glide solution. These oscillations occur because altitude increases and decreases during the "true" gliding entry with fixed bank $\phi = 50°$ (note that altitude oscillates left-to-right in Figure 11.9 because h is the x-axis). Therefore, the numerically integrated glide entry is not a "true" equilibrium glide: the vertical lift and centrifugal forces do not always balance gravity and the flight-path angle experiences both positive and negative rates during the gliding entry. Figure 11.10 shows the Shuttle's deceleration vs. velocity. As with Figure 11.9, deceleration from the numerically integrated glide exhibits oscillations about the analytical solution. Again, these oscillations occur because the numerically integrated glide experiences regions where the Shuttle rises above $(\dot{\gamma} > 0)$ and dips below $(\dot{\gamma} < 0)$ a true equilibrium glide. Below a velocity of about 3 km/s, the analytical equilibrium-glide solution overestimates the deceleration. Nevertheless, the analytical solutions illustrate the general trends of the gliding entry. It is interesting to note that the analytical solution predicts a final velocity of 758 m/s (at an altitude of 30 km), which shows a good match with the velocity at the end of the Space Shuttle's entry phase (about Mach 2.5).

d) We use Eq. (11.47) to predict the total ground-track range:

$$s = \frac{r_0}{2}(L/D)\cos\phi \ln\left(\frac{v_s^2 - v_2^2}{v_s^2 - v_1^2}\right)$$

The initial velocity v_1 is the velocity at EI, or $v_{EI} = 7{,}847$ m/s, and the final velocity at altitude $h = 30$ km is $v_2 = 758$ m/s [as determined by Eq. (11.38); see Figure 11.9]. The circular orbital velocity at the Earth's surface is

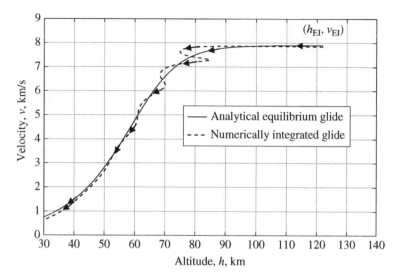

Figure 11.9 Velocity vs. altitude for equilibrium-glide entry (Example 11.2).

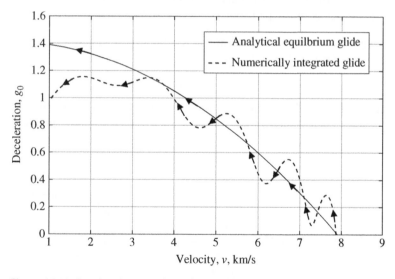

Figure 11.10 Deceleration vs. velocity for equilibrium-glide entry (Example 11.2).

$$v_s = \sqrt{\mu/r_0} = \sqrt{g_0 r_0} = 7{,}907 \text{ m/s}$$

where $g_0 = 9.80665 \text{ m/s}^2$ and $r_0 = 6{,}375{,}416$ m. Using these values, the total ground-track range is

$$\boxed{s = 9{,}425 \text{ km}}$$

The total ground-track range from the numerically integrated glide is 9,130 km, and therefore the range error is about 3%.

11.5 Skip Entry

A second category of lifting entry maneuvers is the *skip entry*. Figure 11.11 shows a schematic diagram of a skip entry. The spacecraft enters the planet's atmosphere with velocity v_{EI} and flight-path angle γ_{EI}. During passage through the atmosphere, the aerodynamic lift rotates the velocity vector upward ($\dot\gamma > 0$) until the flight-path angle is zero and the vehicle has reached the minimum altitude or "pull-up altitude." The lift force maintains the positive flight-path angle rotation and the vehicle ascends and exits the atmosphere with exit conditions $v_{exit} < v_{EI}$ and $\gamma_{exit} > 0$. After a second Keplerian flight phase, the vehicle re-enters the atmosphere down-range of atmospheric exit (not shown in Figure 11.11). A skip entry can extend and control the down-range distance for targeting purposes. The Apollo command module used a skip entry as an optional entry flight profile.

As before, we wish to develop analytical solutions for this particular entry flight profile. We can develop first-order analytical solutions for a skip entry by imposing two basic assumptions:

1) The vertical lift term $L\cos\phi/m$ dominates the combined effect of the centrifugal and gravity terms in the path-normal acceleration equation (11.2).
2) Gravity is negligible in the path-tangent acceleration equation (11.1).

We begin with the dynamical equation for the flight-path angle. Applying the first assumption, $L\cos\phi/m \gg v^2/r - g$, Eq. (11.2) becomes

$$v\dot\gamma = a_L\cos\phi \qquad (11.48)$$

Recall that $a_L \equiv L/m$ is the lift acceleration. This approximate ODE shows that the time rate of flight-path angle is always positive for $\cos\phi > 0$ (i.e., the lift vector points

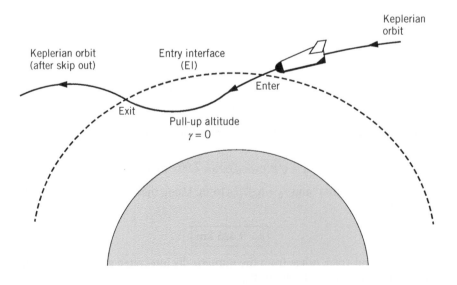

Figure 11.11 Skip entry.

"up"). We may use Eq. (11.32) to replace lift acceleration with $a_L = (L/D)a_D$. Equation (11.33) shows drag acceleration a_D expressed in terms of density, velocity, and ballistic coefficient. With these substitutions, Eq. (11.48) becomes

$$v\dot{\gamma} = \frac{\rho v^2}{2C_B}(L/D)\cos\phi \tag{11.49}$$

Next, divide both sides by velocity v

$$\dot{\gamma} = \frac{\rho v}{2C_B}(L/D)\cos\phi \tag{11.50}$$

We can change the independent variable from time t to altitude h by dividing Eq. (11.50) by the time-rate of altitude, $\dot{h} = v\sin\gamma$, defined by Eq. (11.7). The result is

$$\frac{d\gamma}{dh} = \frac{\rho}{2C_B\sin\gamma}(L/D)\cos\phi \tag{11.51}$$

Using the exponential density model (11.6), Eq. (11.51) becomes

$$\frac{d\gamma}{dh} = \frac{\rho_0 e^{-\beta h}}{2C_B\sin\gamma}(L/D)\cos\phi \tag{11.52}$$

Separating variables and integrating from EI conditions to an arbitrary state, we obtain

$$-\cos\gamma\Big|_{\gamma_{EI}}^{\gamma} = \frac{-\rho_0 e^{-\beta h}}{2\beta C_B}(L/D)\cos\phi\Big|_{h_{EI}}^{h} \tag{11.53}$$

Finally, we can simplify the right-hand side of Eq. (11.53) by neglecting the density term at EI (i.e., $\rho_0 e^{-\beta h_{EI}} \approx 0$). After rearranging Eq. (11.53), we obtain

$$\cos\gamma = \cos\gamma_{EI} + \frac{\rho_0 e^{-\beta h}}{2\beta C_B}(L/D)\cos\phi \tag{11.54}$$

Equation (11.54) is the variation in flight-path angle during the skip entry as a function of altitude. Because $\rho_0 e^{-\beta h_{EI}} \approx 0$ at EI altitude, Eq. (11.54) shows that at atmospheric entrance *and* exit (i.e., $h = h_{EI}$), we have $\cos\gamma_{exit} \approx \cos\gamma_{EI}$. Therefore, the exit flight-path angle is

$$\gamma_{exit} = -\gamma_{EI} \tag{11.55}$$

Of course, the exit flight-path angle must be positive because the vehicle is gaining altitude. We desire the minimum altitude during the skip entry (or, "pull-up altitude"), which is determined by setting $\gamma = 0$ in Eq. (11.54):

$$1 = \cos\gamma_{EI} + \frac{\rho_0 e^{-\beta h_{pullup}}}{2\beta C_B}(L/D)\cos\phi \tag{11.56}$$

We can solve Eq. (11.56) for the atmospheric density at pull-up altitude, h_{pullup}

$$\rho_{pullup} = \rho_0 e^{-\beta h_{pullup}} = \frac{2\beta C_B}{(L/D)\cos\phi}(1-\cos\gamma_{EI}) \tag{11.57}$$

Taking the natural logarithm of Eq. (11.57) provides the solution for pull-up altitude

$$h_{pullup} = \frac{-1}{\beta} \ln \left(\frac{\rho_{pullup}}{\rho_0} \right) \qquad (11.58)$$

where ρ_{pullup} is defined by the right-hand side of Eq. (11.57). We see that pull-up altitude depends on the planet's atmosphere (ρ_0 and β), the vehicle's characteristics (C_B and L/D), bank angle ϕ, and entry flight-path angle γ_{EI}. Entry speed v_{EI} does not influence pull-up altitude.

Next, we turn our attention to the velocity of the skip entry. Applying the second assumption (neglect gravity) to Eq. (11.1), we obtain

$$\dot{v} = -a_D = \frac{-a_L}{L/D} \qquad (11.59)$$

Because we have a first-order solution for flight-path angle, we divide Eq. (11.59) by $\dot{\gamma}$ to remove time as the independent variable. Dividing Eq. (11.59) by Eq. (11.48), we obtain

$$\frac{dv}{d\gamma} = \frac{-v}{(L/D)\cos\phi} \qquad (11.60)$$

We can separate variables and integrate Eq. (11.60) with constant L/D and constant bank angle ϕ. Integrating with the EI state as the initial condition, we obtain

$$\ln \left(\frac{v}{v_{EI}} \right) = \frac{-1}{(L/D)\cos\phi}(\gamma - \gamma_{EI}) \qquad (11.61)$$

Using the exponential function, Eq. (11.61) becomes

$$v = v_{EI} \exp \left[\frac{\gamma_{EI} - \gamma}{(L/D)\cos\phi} \right] \qquad (11.62)$$

Equation (11.62) provides the spacecraft's velocity during the skip entry as a function of flight-path angle. Recall that flight-path angle during the skip is defined by Eq. (11.54). It is easy to determine the vehicle's velocity at pull-up and exit conditions: at pull-up, set $\gamma = 0$ in Eq. (11.62) to obtain

$$v_{pullup} = v_{EI} \exp \left[\frac{\gamma_{EI}}{(L/D)\cos\phi} \right] \qquad (11.63)$$

At atmospheric exit, set $\gamma = -\gamma_{EI}$ in Eq. (11.62) to obtain the skip-out velocity

$$v_{exit} = v_{EI} \exp \left[\frac{2\gamma_{EI}}{(L/D)\cos\phi} \right] \qquad (11.64)$$

Equation (11.62) shows that ballistic coefficient and inverse scale height influence the velocity during the skip through the flight-path angle profile (11.54). However, Eqs. (11.63) and (11.64) show that the vehicle's pull-up and exit velocities do not depend on C_B or β.

Vehicle acceleration is the final performance metric of the skip entry. We focused on the path-tangent acceleration for ballistic and gliding flight. Because a skip entry can involve a large lift force, we will determine the path-normal acceleration. Equation (11.59) shows that vehicle acceleration along the flight path is due solely to drag (recall that we have neglected gravity). Using Eq. (11.33) for drag acceleration, the path-tangent vehicle *deceleration* is

$$a_t = -\dot{v} = \frac{\rho_0 e^{-\beta h} v^2}{2C_B} \qquad (11.65)$$

where the new symbol a_t is the deceleration along the flight path. Equation (11.49) defines the total acceleration normal to the flight path; we repeat it below in a modified form:

$$a_n = v\dot{\gamma} = \frac{\rho_0 e^{-\beta h} v^2}{2C_B}(L/D)\cos\phi \qquad (11.66)$$

Equation (11.66) shows that the path-normal acceleration, a_n, is solely due to the lift during a skip entry. Note that aerodynamic lift and drag are the only forces accounted for in these acceleration expressions because we have neglected gravity and centrifugal terms. Furthermore, the integration of these two reduced acceleration equations has produced the first-order solutions for flight-path angle and velocity for the skip entry. Therefore, we can think of the first-order skip entry as "zero gravity" solutions for a "flat-planet" dynamical model (no curvature or no centrifugal term).

Peak acceleration occurs just prior to the pull-up altitude. We may determine the approximate peak deceleration by substituting Eq. (11.57) for pull-up density in the path-tangent deceleration (11.65)

$$a_{t_{\max}} \approx \frac{2\beta C_B(1-\cos\gamma_{EI})}{2C_B(L/D)\cos\phi} v^2_{\text{pullup}} \qquad (11.67)$$

Next, substitute Eq. (11.63) for pull-up velocity v_{pullup} and cancel the $2C_B$ terms:

$$a_{t_{\max}} \approx \frac{\beta(1-\cos\gamma_{EI})}{(L/D)\cos\phi} v^2_{EI} \exp\left[\frac{2\gamma_{EI}}{(L/D)\cos\phi}\right] \qquad (11.68)$$

Following the same steps, the peak path-normal acceleration is

$$a_{n_{\max}} \approx \beta(1-\cos\gamma_{EI}) v^2_{EI} \exp\left[\frac{2\gamma_{EI}}{(L/D)\cos\phi}\right] \qquad (11.69)$$

Equations (11.68) and (11.69) show that peak acceleration components depend only on the atmospheric inverse scale height β, entry conditions (v_{EI} and γ_{EI}), lift-to-drag ratio, and bank angle ϕ. Ballistic coefficient C_B (i.e., reference area, vehicle mass, or drag coefficient) does not affect peak acceleration. Note that for fixed entry conditions, increasing the L/D ratio decreases the peak deceleration along the flight path [Eq. (11.68)] but *increases* the peak normal acceleration [Eq. (11.69)]. Furthermore, when $L/D \cos\phi = 1$, the peak tangential and normal accelerations have equal magnitudes regardless of the entry conditions. Equations (11.68) and (11.69) show that when $L/D \cos\phi > 1$ (i.e., a winged or lifting-body vehicle with small bank angle), the peak normal acceleration *always* exceeds the peak tangential deceleration during the skip entry.

We close this section with a brief discussion of the limitations of the first-order skip entry solutions. Recall that the fundamental assumption that allowed us to obtain the analytical skip solutions is that the lift force is much larger than the difference between the centrifugal and gravity forces. Hence, the path-normal acceleration, Eq. (11.48), only contains the vertical component of lift acceleration (the $v^2/r - g$ term is neglected). However, as the vehicle enters the atmosphere, the lift (and drag) force is nearly zero because density is very small. Therefore, the first-order skip entry solutions are only accurate for a near-circular entry speed where $v_{EI}^2/r \approx g$. For an entry speed significantly greater than circular speed, the centrifugal term is larger than gravity, and hence it dominates lift during the initial high-altitude entry phase. It is for this reason that we cannot use the first-order skip entry solution to analyze a high-speed entry such as an aerocapture maneuver. Furthermore, the analytical skip entry solution is only accurate for entry vehicles with a significant lift-to-drag ratio, such as $L/D > 1$. For example, the first-order method cannot accurately predict a skip entry with an Apollo-style capsule ($L/D \approx 0.3$) even if the spacecraft enters at near-circular speed; the capsule simply cannot generate enough lift to skip out of the atmosphere. The Apollo capsule entered Earth's atmosphere at about 11 km/s with flight-path angle $\gamma_{EI} \approx -6.5°$. Applying the first-order method to the Apollo entry predicts a deep, sustained atmospheric pass where the peak path-tangent deceleration exceeds $20g_0$ and the exit velocity is about 5 km/s. The actual Apollo skip trajectory (with full lift up) is much shallower and exits the atmosphere at a velocity of about 7 km/s (the peak deceleration is about $10g_0$). The centrifugal force v_{EI}^2/r for the Apollo capsule dominates the initial entry and neglecting it renders the first-order analytical skip solution useless.

In summary, the first-order method can accurately predict the performance of skip-entry profiles for vehicles with a significant lift-to-drag ratio (e.g., $L/D > 1$) entering the atmosphere at near-circular speeds. The following example illustrates the accuracy of the analytical skip entry solution by comparing it with a numerically integrated skip entry.

Example 11.3 A winged entry vehicle with lift-to-drag ratio $L/D = 1.5$ enters Earth's atmosphere with near-circular velocity $v_{EI} = 7.8$ km/s and flight-path angle $\gamma_{EI} = -3°$ and performs a skip entry. The vehicle has mass $m = 100{,}000$ kg, reference wing area $S = 350$ m^2, drag coefficient $C_D = 0.8$, and zero bank angle during its skip entry.

a) Determine the vehicle's pull-up altitude and pull-up velocity during the skip entry.
b) Compute the peak deceleration along the flight-path and the peak normal acceleration.
c) Compute the atmospheric exit velocity and flight-path angle.
d) Compute the maximum altitude of the Keplerian phase after the first skip-out exit.
e) Create plots of altitude vs. velocity and path-normal acceleration for the analytical skip entry and a numerically integrated skip entry profile.

a) First, we compute the ballistic coefficient for this winged vehicle:

$$C_B = \frac{m}{SC_D} = 357.143 \text{ kg/m}^2$$

Next, we use Eq. (11.57) to compute density at the pull-up altitude

$$\rho_{\text{pullup}} = \frac{2\beta C_B}{(L/D)\cos\phi}(1 - \cos\gamma_{EI}) = 8.9929\left(10^{-5}\right) \text{kg/m}^3$$

The reader should remember that inverse scale height β must be in units of per meter in the previous expression because it is multiplied by ballistic coefficient C_B (which has units of kg/m^2). Next, we use Eq. (11.58) (with surface density $\rho_0 = 1.225$ kg/m^3 and $\beta = 0.1378$ km^{-1}) to determine the pull-up altitude in km:

$$h_{pullup} = \frac{-1}{\beta} \ln\left(\frac{\rho_{pullup}}{\rho_0}\right) = \boxed{69.08 \text{ km}}$$

We use Eq. (11.63) to determine the pull-up velocity:

$$v_{pullup} = v_{EI} \exp\left[\frac{\gamma_{EI}}{(L/D)\cos\phi}\right] = \boxed{7.532 \text{ km/s}}$$

b) Peak deceleration along the flight path is determined using Eq. (11.68):

$$a_{t_{max}} = \frac{\beta(1-\cos\gamma_{EI})}{(L/D)\cos\phi} v_{EI}^2 \exp\left[\frac{2\gamma_{EI}}{(L/D)\cos\phi}\right] = \boxed{7.143 \text{ m/s}^2}$$

(it is useful here to use $\beta = 0.1378(10^{-3})$ m^{-1} and $v_{EI} = 7{,}800$ m/s). The peak deceleration in units of Earth g_0 (using $g_0 = 9.80665$ m/s^2) is

$$\boxed{a_{t_{max}} = 0.728 g_0}$$

We use Eq. (11.69) to determine the peak acceleration normal to the flight path:

$$a_{n_{max}} = \beta(1-\cos\gamma_{EI})v_{EI}^2 \exp\left[\frac{2\gamma_{EI}}{(L/D)\cos\phi}\right] = \boxed{10.715 \text{ m/s}^2}$$

Or, in Earth g_0: $\boxed{a_{n_{max}} = 1.093 g_0}$

c) Exit flight-path angle is simply $\boxed{\gamma_{exit} = -\gamma_{EI} = 3°}$

We use Eq. (11.64) to determine exit velocity:

$$v_{exit} = v_{EI} \exp\left[\frac{2\gamma_{EI}}{(L/D)\cos\phi}\right] = \boxed{7.274 \text{ km/s}}$$

d) The vehicle follows a Keplerian orbit after atmospheric skip-out. We compute the energy of the orbit using v_{exit} and $r_{EI} = 122$ km $+ R_E = 6{,}500$ km:

$$\xi = \frac{v_{exit}^2}{2} - \frac{\mu}{r_{EI}} = -34.8673 \text{ km}^2/\text{s}^2 = -\frac{\mu}{2a}$$

Therefore, semimajor axis is $a = 5{,}715.95$ km. Angular momentum is

$$h = r_{EI} v_{exit} \cos\gamma_{exit} = 47{,}216.4 \text{ km}^2/\text{s} = \sqrt{p\mu}$$

So, parameter is $p = 5{,}593.05$ km. Eccentricity of the orbit is

$$e = \sqrt{1 - \frac{p}{a}} = 0.1466$$

Finally, the apogee radius is

$$r_a = \frac{p}{1-e} = 6{,}554.13 \text{ km}$$

Hence, the peak altitude after the skip is $r_a - R_E = 176.13$ km, or 54 km above EI altitude.

e) We determine the analytical skip entry using the following steps: first, use Eq. (11.54) to compute flight-path angle for a given altitude (repeated below):

$$\cos \gamma = \cos \gamma_{EI} + \frac{\rho_0 e^{-\beta h}}{2\beta C_B}(L/D)\cos\phi$$

Flight-path angle for the "down" phase is computed for altitudes *descending* from h_{EI} (122 km) to pull-up altitude $h_{pullup} = 69.08$ km. The flight-path angle profile for the "up" phase ($h > h_{pullup}$) is a mirror image of the "down" phase except with a sign change because $\dot{h} > 0$. Next, the velocity profile is computed using Eq. (11.62) (repeated below)

$$v = v_{EI}\exp\left[\frac{\gamma_{EI} - \gamma}{(L/D)\cos\phi}\right]$$

where flight-path angle (determined by the previous step) ranges from $\gamma_{EI} < \gamma < \gamma_{exit}$. After competing these steps, we may plot velocity vs. altitude. Although velocity is the dependent variable and altitude is the independent variable here, it is (perhaps) easier to visualize their relationship by plotting altitude on the y-axis and velocity on the x-axis (after all, altitude is the vertical direction).

Equation (11.66) determines the normal acceleration (repeated below):

$$a_n = \frac{\rho_0 e^{-\beta h} v^2}{2C_B}(L/D)\cos\phi$$

Here we use the "down-up" altitude profile and corresponding analytical velocity profile from the previous calculations. Finally, we compare the analytical skip entry to a numerically integrated skip trajectory obtained by using MATLAB's ode45.m. As with Examples 11.1 and 11.2, we numerically integrate the governing nonlinear equations [Eqs. (11.1), (11.2), and (11.7)] using the given entry conditions and vehicle parameters.

Figure 11.12 shows the analytical and numerical skip trajectories (altitude vs. velocity). The analytical solution accurately portrays the first skip maneuver: both solutions show essentially the same pull-up altitude (about 69 km), pull-up velocity (about 7.55 km/s), and exit velocity (between 7.2 and 7.3 km/s). The winged vehicle re-enters the atmosphere at a sub-circular velocity (equal to the first exit velocity) and performs a second skip maneuver. Because the second entry speed is sub-circular, the vehicle does not achieve a full skip-out exit, and only reaches a peak altitude of about 95 km. However, the analytical solution (by definition) assumes a full skip-out to EI as shown in Figure 11.12. Subsequent "reduced skips" with diminishing peak altitudes follow (see the third numerically integrated skip in Figure 11.12), but the analytical first-order method does not accurately predict these skips because the centrifugal-gravity balance no longer holds. In addition, note that the velocity profile of the numerically

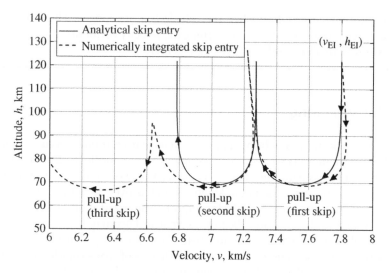

Figure 11.12 Altitude vs. velocity for a skip entry (Example 11.3).

integrated skip entry in Figure 11.12 shifts to the left (slows down) as altitude increases during skip out, and shifts to the right (speeds up) as altitude decreases during re-entry. This velocity shift vs. altitude is, of course, due to gravity. The analytical skip solution does not include the effect of gravity, and hence the skip-out and re-entry velocity vs. altitude profiles shown in Figure 11.12 appear as vertical lines.

Figure 11.13 shows normal acceleration computed by the analytical and numerical methods. The first-order method accurately predicts normal acceleration for the first two skips but thereafter becomes inaccurate for the reasons previously mentioned. The reader should note that the "true" numerically integrated skip entry shows small

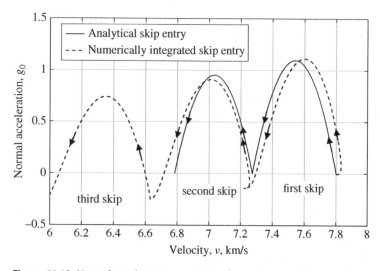

Figure 11.13 Normal acceleration vs. velocity for a skip entry (Example 11.3).

negative normal acceleration when the vehicle is at its apogee. This small negative normal acceleration is due to gravity turning the velocity vector downward. Recall that the first-order method neglects gravity altogether.

11.6 Entry Heating

As mentioned in Section 11.1, bringing a space vehicle from orbital speed to a safe landing on the surface of a planet requires dissipating an extraordinary amount of energy. Aerodynamic drag dissipates the vehicle's energy, and the result is aerodynamic heating. For entry at circular speed, the initial kinetic energy is over 30 MJ/kg. Several authors [5,6] point out that much of the vehicle would be vaporized if all of this tremendous energy was transferred to and absorbed by the vehicle. Fortunately, a significant portion of this thermal energy is shed to the flow field about the entry vehicle.

Because this textbook focuses on the flight mechanics of space vehicles, we will briefly discuss the heating problem without delving into the details of heat transfer and hypersonic aerodynamics. To begin our analysis, it is instructive to observe the energy dissipation rate during entry. Total mechanical energy per unit mass is

$$\xi = \frac{v^2}{2} - \frac{\mu}{r} \tag{11.70}$$

The time derivative of energy is

$$\dot{\xi} = v\dot{v} + \frac{\mu}{r^2}\dot{r} \tag{11.71}$$

Substituting Eq. (11.1) for $\dot{v}$, $\mu/r^2 = g$, and $\dot{r} = \dot{h} = v\sin\gamma$, the time rate of total energy becomes

$$\dot{\xi} = v\left(-\frac{D}{m} - g\sin\gamma\right) + gv\sin\gamma \tag{11.72}$$

The term $gv\sin\gamma$ cancels and the time rate of energy is

$$\dot{\xi} = -\frac{Dv}{m} \tag{11.73}$$

Equation (11.73) is a familiar result from basic mechanics: the time rate of energy (i.e., power) dissipated by a vehicle is the product of drag force D and velocity v. Using Eq. (11.5) for the drag force in Eq. (11.73), we obtain

$$\dot{\xi} = -\frac{1}{2}\rho v^3 \frac{SC_D}{m} = -\frac{\rho v^3}{2C_B} \tag{11.74}$$

Equation (11.73) shows that aerodynamic drag is the mechanism for energy dissipation during entry (of course, for Keplerian motion $D = 0$ and ξ is constant). Equation (11.74) shows that for a fixed vehicle (i.e., fixed ballistic coefficient C_B), the rate of energy loss is proportional to the product of atmospheric density and the *cube* of velocity. Several authors [5,6] have shown that the rate of heat transferred to the body of the entry vehicle is

$$\dot{Q} = \frac{1}{4}\rho v^3 S_w C_f \tag{11.75}$$

where S_w is the "body wall" or "wetted" area of the vehicle, and C_f is the "body-averaged" skin friction coefficient. Symbol Q is the heat or thermal energy transferred to the body in joules (J). Heat rate $\dot{Q}$ expressed in Eq. (11.75) has units of joules per second (J/s) or watts (W), whereas the total *specific* energy-rate expressed in Eq. (11.74) has units of watts per kilogram (W/kg). However, Eqs. (11.74) and (11.75) both illustrate that the energy-dissipation and heat-transfer rates are proportional to ρv^3. Aerodynamic drag (i.e., deceleration) is proportional to ρv^2.

We may obtain the total heat transferred to the body of the entry vehicle by integrating Eq. (11.75). As with the governing velocity and flight-path angle ODEs, we do not integrate the heat-transfer ODE with respect to time. Instead, we divide Eq. (11.75) by $\dot{v}$ to change the independent variable to velocity:

$$\frac{\dot{Q}}{\dot{v}} = \frac{dQ}{dv} = \frac{\frac{1}{4}\rho v^3 S_w C_f}{-\frac{1}{2}\rho v^2 \frac{S}{m}C_D} = -\frac{v}{2}\frac{S_w}{S}\frac{C_f}{C_D}m \tag{11.76}$$

Note that we neglected gravity in the along-path acceleration; that is, $\dot{v} \approx -D/m$. Separating variables in Eq. (11.76) leads to

$$dQ = -\frac{v}{2}\frac{S_w}{S}\frac{C_f}{C_D}m\,dv \tag{11.77}$$

Integrating Eq. (11.77) from entry speed (v_{EI}) to the at-rest condition ($v = 0$) yields the total heat load:

$$Q_{total} = \frac{1}{2}\frac{S_w}{S}\frac{C_f}{C_D}\frac{m\,v_{EI}^2}{2} \tag{11.78}$$

Equation (11.78) is the total thermal energy transferred to the vehicle during entry from EI conditions to zero velocity. Note that the last term on the right-hand side is the initial kinetic energy of the vehicle at atmospheric entry. Dividing both sides of Eq. (11.78) by initial kinetic energy yields

$$\frac{Q_{total}}{m\,v_{EI}^2/2} = \frac{1}{2}\frac{S_w}{S}\frac{C_f}{C_D} \tag{11.79}$$

Equation (11.79) expresses the fraction of initial kinetic energy transferred to the vehicle in the form of heat energy. The right-hand side of Eq. (11.79) is one-half of the ratio of skin-friction drag to total drag. Skin-friction drag coefficient C_f is a function of the tangential shear stress at the wall boundary, whereas the *total* drag coefficient C_D depends on the tangential shear stress *and* normal stress due to pressure imbalance. In other words, the total drag coefficient is

$$C_D = C_{D_p} + C_f \tag{11.80}$$

where C_{D_p} is the drag coefficient due to pressure imbalance and flow separation. Equation (11.79) shows that minimizing the ratio C_f/C_D will minimize the total thermal energy transferred to the vehicle. Therefore, a *blunt entry body*, where the majority of the total drag is "pressure drag" (i.e., $C_D \approx C_{D_p}$), will minimize the heat transferred to the vehicle's body. At the other extreme, "skin-friction drag" dominates the total drag for a slender, sleek body (i.e., $C_D \approx C_f$), and hence more heat is transferred to its body when compared with a blunt body. A blunt body will experience a very strong bow shock wave, and the resulting high-temperature gas behind the shock wave will flow past the vehicle. Hence, most of the kinetic energy heats the flow field and only a small portion reaches the vehicle. It is for these reasons that ballistic and near-ballistic entry vehicles such as the Mercury, Gemini, Apollo, and Stardust capsules utilized blunt-body aerodynamic designs (of course, the Gemini and Apollo capsules generated a small amount of lift to reduce the peak deceleration and provide maneuverability).

Let us return to Eq. (11.75), the instantaneous heating rate. The body-averaged heat-transfer rate per unit area is

$$q_{\text{avg}} = \frac{\dot{Q}}{S_w} = \frac{1}{4}\rho v^3 C_f \tag{11.81}$$

We may investigate the vehicle's average heat input rate for various entry strategies by substituting the appropriate first-order velocity solution into Eq. (11.81). Let us first obtain the heat input rate for a ballistic entry by substituting Eq. (11.18) for velocity in Eq. (11.81):

$$\text{Ballistic entry:} \quad q_{\text{avg}} = \frac{1}{4}\rho v_{\text{EI}}^3 C_f \exp\left[3Be^{-\beta h}\right] \tag{11.82}$$

Substituting the exponential atmosphere model into Eq. (11.82), we obtain

$$q_{\text{avg}} = \frac{1}{4}\rho_0 v_{\text{EI}}^3 C_f \exp\left[-\beta h + 3Be^{-\beta h}\right] \tag{11.83}$$

Equation (11.83) is the heat-transfer rate as a function of altitude. We are primarily interested in the critical altitude where the vehicle experiences the maximum heat-input rate. Therefore, we follow the same process as determining the peak deceleration: take the derivative of q_{avg} with respect to altitude h and set the result to zero. For the sake of brevity, we omit the details here; however, the process is very much like the steps surrounding Eqs. (11.21)–(11.26). Setting $dq_{\text{avg}}/dh = 0$ eventually leads to the condition

$$1 + 3Be^{-\beta h} = 0 \tag{11.84}$$

which is identical to Eq. (11.23) except that the factor 2 is replaced by 3. The critical altitude for peak heat-input rate is

$$\text{Ballistic entry:} \quad h_{\text{crit}} = \frac{\ln(-3B)}{\beta} \tag{11.85}$$

Recall that constant B is negative and depends on the atmospheric parameters, the vehicle characteristics, and the entry flight-path angle. Comparing Eqs. (11.26) and (11.85), we see that peak heating rate *always* occurs before peak deceleration during a ballistic entry. In fact, the difference in critical altitudes is $\ln(3/2)/\beta$, which is always 2.94 km for Earth (where $\beta = 0.1378$ km^{-1}).

Substituting the critical altitude for peak heating rate, Eq. (11.85), into Eq. (11.83) yields (after some algebra) the peak heat input:

$$\text{Ballistic entry:} \quad q_{\text{avg}_{\text{max}}} = -\frac{1}{6e}v_{\text{EI}}^3 C_f \beta C_B \sin \gamma_{\text{EI}} \tag{11.86}$$

Equation (11.86) is the peak heat-transfer rate to the vehicle for a ballistic entry. Because $\sin \gamma_{\text{EI}} < 0$, the peak heat rate is positive. Finally, the velocity at the peak heating rate can be determined by substituting Eq. (11.85) into the ballistic velocity profile, Eq. (11.18). The result is

$$\text{Ballistic entry:} \quad v_{\text{crit}} = \frac{v_{\text{EI}}}{e^{1/3}} = 0.7165 v_{\text{EI}} \tag{11.87}$$

Comparing Eq. (11.87) with Eq. (11.30) also confirms that the peak heat-transfer rate occurs before peak deceleration in a ballistic entry.

Example 11.4 Using the Vostok vehicle and entry data from Example 11.1, determine the critical altitude and velocity for peak heating rate during a ballistic entry.

Recall from Example 11.1 that the Vostok's ballistic coefficient is

$$C_B = \frac{m}{SC_D} = 296.05 \text{ kg/m}^2$$

and the dimensionless coefficient is

$$B = \frac{\rho_0}{2\beta C_B \sin \gamma_{\text{EI}}} = -268.9650$$

We use Eq. (11.85) to determine the critical altitude for peak heating rate:

$$h_{\text{crit}} = \frac{\ln(-3B)}{\beta} = \boxed{48.572 \text{ km}}$$

Referring back to Example 11.1, we see that the altitude for peak heating rate is about 2.94 km above the altitude for peak deceleration as expected.

We use Eq. (11.87) to compute the critical velocity for peak heating rate:

$$v_{\text{crit}} = \frac{v_{\text{EI}}}{e^{1/3}} = 0.7165 v_{\text{EI}} = \boxed{5.546 \text{ km/s}}$$

Recall that the critical velocity for peak deceleration is 4.694 km/s (see Example 11.1). Therefore, this result also shows that peak heating rate occurs before peak deceleration.

The previous discussion identified the critical heat-transfer parameters for a ballistic entry. We can follow the same basic steps and identify the critical parameters for an equilibrium-glide entry: (1) substitute the glide-entry velocity profile, Eq. (11.38), into the averaged heat-rate equation (11.81); (2) set the derivative with respect to altitude equal to zero; (3) solve the $dq_{\text{avg}}/dh = 0$ expression for the critical altitude; and (4) substitute the critical altitude into the heat-rate and velocity relations to obtain their critical values. We summarize the results as:

$$\text{Equilibrium gliding entry:} \quad q_{\text{avg}_{\text{max}}} = \sqrt{\frac{g_0^3 r_0}{27} \frac{C_f C_B}{(L/D)\cos\phi}} \tag{11.88}$$

$$h_{\text{crit}} = -\frac{1}{\beta}\ln\left(\frac{4C_B}{r_0\rho_0(L/D)\cos\phi}\right) \tag{11.89}$$

$$v_{\text{crit}} = \sqrt{\frac{g_0 r_0}{3}} \tag{11.90}$$

Equation (11.89) requires that $r_0\rho_0(L/D)\cos\phi > 4C_B$ so that the critical altitude for peak heating rate is positive [this is relatively easy to accomplish for an Earth entry trajectory because $r_0\rho_0 = 7.813(10^6)$ kg/m^2]. Equation (11.90) shows that the critical velocity for peak heating during an equilibrium glide has no dependence on the entry conditions or vehicle characteristics – it only depends on the mass and radius of the planet!

The method for computing the critical parameters associated with peak heat-rate input for a skip entry also follows the basic steps outlined for the gliding entry. However, the skip-entry velocity profile (11.62) is a function of flight-path angle. We can use Eq. (11.54) to express the atmospheric density in the heat-rate equation (11.81) in terms of flight-path angle and ultimately set the derivative $dq_{\text{avg}}/d\gamma = 0$ to find the critical flight-path angle where peak heating occurs during the skip. Vinh *et al.* [7] use a small-angle approximation for the critical flight-path angle γ_{crit} in order to obtain explicit results. We omit the details of the derivations here. The results are:

$$\text{Skip entry: } q_{\text{avg}_{\max}} = \frac{v_{\text{EI}}^3\,\gamma_{\text{EI}}^2 C_f \beta C_B}{4(L/D)\cos\phi}\exp\left[\frac{2\gamma_{\text{EI}}}{(L/D)\cos\phi}\right] \tag{11.91}$$

$$\gamma_{\text{crit}} \cong -\frac{3\,\gamma_{\text{EI}}^2}{2(L/D)\cos\phi} \tag{11.92}$$

$$h_{\text{crit}} = -\frac{1}{\beta}\ln\left(\frac{2\beta C_B(\cos\gamma_{\text{crit}} - \cos\gamma_{\text{EI}})}{\rho_0(L/D)\cos\phi}\right) \tag{11.93}$$

$$v_{\text{crit}} = v_{\text{EI}}\exp\left[\frac{\gamma_{\text{EI}} - \gamma_{\text{crit}}}{(L/D)\cos\phi}\right] \tag{11.94}$$

Of course, we must express the entry flight-path angle γ_{EI} in radians for use in Eqs. (11.91) and (11.92); the subsequent critical flight-path angle in Eq. (11.92) is also in radians.

In summary, we should note that computing *quantitative* values for the total heat load and/or peak heating rate requires knowledge of the body-averaged skin-friction coefficient C_f. Griffin and French suggest estimating the skin-friction coefficient by using flat-plate theory from classical aerodynamics [5]. As this textbook's primary focus is space flight mechanics, we will not pursue this topic any further. Instead, our analysis of the aerodynamic heating problem will focus on the critical altitude and velocity where the peak heat rate occurs. The following examples illustrate these concepts.

Example 11.5 Using the Space Shuttle vehicle data and entry conditions in Example 11.2, determine the critical altitude and velocity for peak aerodynamic heating rate for an equilibrium glide.

We determined the Shuttle's ballistic coefficient in Example 11.2:

$$C_B = \frac{m}{SC_D} = 410 \text{ kg/m}^2$$

Equation (11.89) is used to determine the critical altitude for peak heating rate on an equilibrium glide:

$$h_{crit} = \frac{-1}{\beta} \ln \left[\frac{4C_B}{r_0 \rho_0 (L/D) \cos \phi} \right]$$

Using $\beta = 0.1378$ km^{-1}, $r_0 = 6{,}375{,}416$ m, $\rho_0 = 1.225$ kg/m^3, $L/D = 1.1$, and bank angle $\phi = 50°$, we obtain

$$\boxed{h_{crit} = 58.94 \text{ km}}$$

We use Eq. (11.90) to determine the critical velocity for peak heating during the glide:

$$v_{crit} = \sqrt{\frac{g_0 r_0}{3}} = \boxed{4.565 \text{ km/s}}$$

where we have used $g_0 = 9.80665$ m/s^2 as the Earth's standard gravitational acceleration.

Example 11.6   Using the vehicle data and entry conditions from Example 11.3, determine the critical flight-path angle, altitude, and velocity corresponding to the peak aerodynamic heating rate for a skip entry.

The winged entry vehicle in Example 11.3 has $L/D = 1.5$, zero bank angle, and enters the atmosphere with flight-path angle $\gamma_{EI} = -3°$. Equation (11.92) allows us to compute the critical flight-path angle for peak heat rate during the skip entry:

$$\gamma_{crit} \cong \frac{-3\gamma_{EI}^2}{2(L/D)\cos\phi} = -0.002742 \text{ rad} = \boxed{-0.157°}$$

The critical flight-path angle is very shallow and hence very close to (but prior to) the pull-up altitude. The ballistic coefficient for the winged vehicle is

$$C_B = \frac{m}{SC_D} = 357.143 \text{ kg/m}^2$$

Using Eq. (11.93), the critical altitude for peak heating during the skip entry is

$$h_{crit} = \frac{-1}{\beta} \ln \left[\frac{2\beta C_B(\cos\gamma_{crit} - \cos\gamma_{EI})}{\rho_0 (L/D) \cos \phi} \right] = \boxed{69.10 \text{ km}}$$

Finally, we use Eq. (11.94) to find the critical velocity for peak heating:

$$v_{crit} = v_{EI} \exp \left[\frac{\gamma_{EI} - \gamma_{crit}}{(L/D) \cos \phi} \right] = \boxed{7.546 \text{ km/s}}$$

The pull-up altitude and velocity in Example 11.3 are 69.08 km and 7.532 km/s, respectively. The peak heating rate occurs just before the pull-up altitude where $\gamma = 0$.

Example 11.7   For the final entry-heating example, consider again the winged vehicle and entry conditions from Example 11.3. Compute the ratio of peak heating rates for an equilibrium glide and skip entry. Because the winged vehicle has relatively high L/D (=1.5), let the constant bank angle be $\phi = 60°$ for the equilibrium glide to avoid excessive skipping (the bank angle is zero for the skip entry).

Recall that the vehicle's ballistic coefficient is $C_B = 357.143$ kg/m^2. Using bank angle $\phi = 60°$ in Eq. (11.88), we find the maximum heating rate for the equilibrium-glide entry:

Equilibrium glide: $q_{avg_{max}} = \sqrt{\dfrac{g_0^3 r_0}{27} \dfrac{C_f C_B}{(L/D)\cos\phi}} = 7.106(10^6) C_f \text{ W/m}^2$

Using zero bank in Eq. (11.91) along with $\gamma_{EI} = -3°$ (= -0.05236 rad) and $v_{EI} = 7.8$ km/s, the maximum heating rate for the skip entry is

Skip entry: $q_{avg_{max}} = \dfrac{v_{EI}^3 \gamma_{EI}^2 C_f \beta C_B}{4(L/D)} \exp\left[\dfrac{2\gamma_{EI}}{L/D}\right] = 9.952(10^6) C_f \text{ W/m}^2$

The ratio of peak heating rates is

$$\frac{\text{Gliding peak heat rate}}{\text{Skip peak heat rate}} = \frac{7.106(10^6)C_f}{9.952(10^6)C_f} = 0.714$$

Therefore, the peak heating rate is significantly higher for a skip entry as compared with an equilibrium gliding entry.

We end this section with a brief discussion of the various methods for thermal protection. At present, three techniques exist for managing heat transferred to the vehicle's surface: (1) heat sink; (2) ablation; and (3) heat radiation. The heat sink (or "heat shield") method is a brute-force technique where a material with high thermal capacity and high melting point absorbs the total heat load. The early Mercury capsules used a massive beryllium blunt-body heat shield to absorb the heat load. The ablation method consists of a "heat shield" surface (typically a carbon fiber-phenolic resin matrix) that sheds the heat load by charring, melting, and subliming. Thus, the heat load is transported away in the flow field surrounding the vehicle. The Gemini, Apollo, and Stardust capsules used an ablative thermal protection system. The heat radiation method allows the vehicle's surface to reach a very high temperature until its radiated heat is in equilibrium with the incoming heat flux. Once at thermal equilibrium, the surface temperature can no longer increase. Of course, the heat radiation method requires an excellent insulating material such as the Space Shuttle's silica ceramic tiles.

11.7 Space Shuttle Entry

Section 11.4 and Example 11.2 indicated that the Space Shuttle used an equilibrium-glide entry. In reality, the Shuttle used four distinct flight phases during its entry from near-circular entry speed (~7.85 km/s) to a terminal speed of about 760 m/s (~Mach 2.5): (1) temperature control; (2) equilibrium glide; (3) constant drag; and (4) transition. Four different drag-acceleration (a_D) profiles define the four flight phases. We will briefly discuss the Shuttle's entry strategy in this section. Our primary objective here is to illustrate how the Space Shuttle's entry algorithm utilized some of the first-order analytical methods from this chapter. Harpold and Graves [4] present a detailed description of the Shuttle entry for the interested reader.

Figure 11.14 is a representation of the Space Shuttle's entry drag-acceleration profile as a function of velocity (note that we have reversed the x-axis so that the entry progresses left to right or from higher to lower velocity). A similar drag profile appears in Harpold and Graves [4] in British engineering system units. Figure 11.14 shows the Shuttle's four

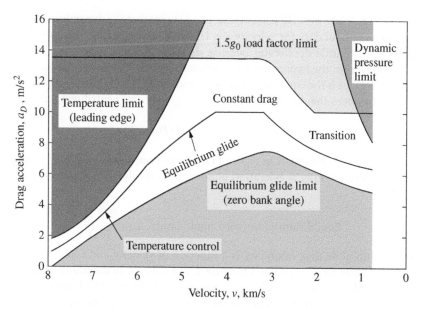

Figure 11.14 Drag-acceleration profile for Space Shuttle entry.

entry flight phases pieced together to form a continuous drag acceleration function that fits between four constraint boundaries. Before we discuss the flight phases, let us define the four constraints. The temperature boundary is the maximum allowable temperature on the surface of the Space Shuttle (typically the wing leading edge); this limit is transformed to the corresponding drag acceleration for a given velocity. The temperature boundary imposes a maximum (or upper) limit on drag acceleration for $5 < v < 8$ km/s as shown in Figure 11.14. The load-factor boundary defines the maximum allowable lift (or normal) acceleration. The non-dimensional load factor is lift divided by weight, that is, L/mg_0 or a_L/g_0. In order to express the load factor limit as a drag acceleration boundary, we use the lift-to-drag relationship $L/D = a_L/a_D$ with maximum $a_L = 1.5g_0$

$$\text{Load-factor limit} \left(\text{in m/s}^2\right): \quad a_D = \frac{1.5g_0}{L/D} \tag{11.95}$$

Using $L/D = 1.1$ in Eq. (11.95), the drag acceleration corresponding to the load-factor limit is 13.4 m/s² during the initial high-speed entry. However, the Shuttle's L/D begins to increase at velocity $v = 3.2$ km/s, which causes the load-factor limit to decrease as illustrated in Figure 11.14. The dynamic pressure boundary imposes an upper limit on drag acceleration at relatively low velocities (<2 km/s) as the vehicle decelerates to the low hypersonic (Mach 5) region. Dynamic pressure is defined as

$$\bar{q} = \frac{1}{2}\rho v^2 \tag{11.96}$$

and therefore drag acceleration is $a_D = \bar{q}SC_D/m$ [the reader should not confuse symbol $\bar{q}$ ("q-bar") with heating rate]. The drag-acceleration boundary for dynamic pressure is

$$\text{Dynamic-pressure limit:} \quad a_D = \bar{q}_{\max}\frac{SC_D}{m} \tag{11.97}$$

where $\bar{q}_{max} = 16{,}375$ N/m^2 (342 psf). Figure 11.14 shows that the dynamic-pressure limit is the upper drag boundary for velocity $v < 1.1$ km/s. The dynamic pressure drag limit decreases as the Shuttle decelerates because drag coefficient C_D decreases as the flow field transitions from the hypersonic to supersonic regime. The lower drag-acceleration limit in Figure 11.14 is the equilibrium-glide boundary, Eq. (11.41), where the vertical lift acceleration balances the difference between the gravity and centrifugal terms:

$$a_L \cos\phi = g_0 - \frac{v^2}{r_0} = a_D(L/D)\cos\phi \tag{11.98}$$

The equilibrium-glide boundary in Figure 11.14 corresponds to full lift "up" (bank $\phi = 0$). Using Eq. (11.98), the drag-acceleration boundary for equilibrium glide is

$$\text{Equilibrium-glide (zero-bank) limit:} \quad a_D = \frac{g_0 - \dfrac{v^2}{r_0}}{L/D} \tag{11.99}$$

The equilibrium-glide drag boundary is a function of velocity and L/D, and hence the boundary *decreases* for $v < 3.2$ km/s because L/D increases as the Shuttle decelerates.

Many of the Shuttle's drag acceleration boundaries in Figure 11.14 diminish at lower speeds because L/D increases. The Shuttle's dramatic change in angle-of-attack (α) explains the rise in L/D. Figure 11.15 shows the Shuttle's angle-of-attack profile as a function of velocity (again, note that we reversed the *x*- or velocity axis). For $v > 3.2$ km/s, the Shuttle glides at $\alpha = 40°$, which results in a high-drag configuration and degraded L/D ratio. The primary purpose of the high angle-of-attack profile during the high-speed entry phase is thermal control. As the Shuttle decelerates and its velocity drops below 3.2 km/s, the angle-of-attack ramps down as shown in Figure 11.15 and consequently C_D decreases and L/D increases.

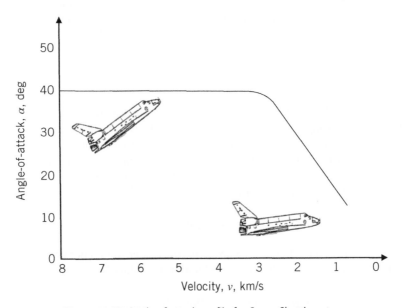

Figure 11.15 Angle-of-attack profile for Space Shuttle entry.

The narrow feasible region between the boundaries in Figure 11.14 is the Shuttle's *entry corridor*. The temperature, load factor, and dynamic pressure boundaries in Figure 11.14 impose an upper limit on drag known as the *undershoot boundary*. If the Shuttle exceeds the upper boundaries in Figure 11.14 (i.e., too much drag), it will violate heating, g-load, and structural limits. The equilibrium-glide limit in Figure 11.14 is the *overshoot boundary*, and it represents a lower limit on drag. If the Shuttle flies below this boundary (i.e., too little drag), it will not have proper control of its trajectory and may skip out of the atmosphere.

Now let us turn our attention to the Space Shuttle's drag-acceleration profile presented in Figure 11.14. The basic idea is to select a piecewise-continuous drag acceleration function $a_D(v)$ that fits within the entry corridor. In addition, the drag profile is selected so that the Shuttle's subsequent entry trajectory has the proper range to the landing site. To show the range–drag relationship, let us divide the down-range kinematic equation (11.8) by the along-path acceleration equation (11.1):

$$\frac{\dot{s}}{\dot{v}} = \frac{ds}{dv} = -\frac{v\cos\gamma}{a_D} \tag{11.100}$$

Note that we have assumed $r_0/r \approx 1$ for the down-range equation, and we have neglected the small gravity component ($g\sin\gamma \approx 0$) in the acceleration equation. Furthermore, because the glide is shallow, we set $\cos\gamma \approx 1$. After separating variables and integrating Eq. (11.100), we obtain

$$s_i = -\int_{v_A}^{v_B} \frac{v}{a_D} dv \tag{11.101}$$

where s_i is the down-track range flow by the Space Shuttle for a given drag-acceleration profile a_D between velocities v_A and v_B. The drag accelerations for the four entry flight phases are

$$\text{Temperature control:} \quad a_D(v) = c_0 + c_1 v + c_2 v^2 \tag{11.102}$$

$$\text{Equilibrium glide:} \quad a_D(v) = \frac{g_0 - \dfrac{v^2}{r_0}}{(L/D)\cos\phi} \tag{11.103}$$

$$\text{Constant drag:} \quad a_D(v) = c_3 \tag{11.104}$$

$$\text{Transition:} \quad a_D(\xi) = c_4 + c_5\xi \tag{11.105}$$

Substituting each drag-acceleration profile into Eq. (11.101) yields four analytical range integrals and their summation provides the predicted "range-to-go"

$$s_{go} = s_1 + s_2 + s_3 + s_4 \tag{11.106}$$

The temperature-control phase consists of a drag acceleration profile that is quadratic in velocity in order to maintain a constant heating rate. Harpold and Graves [4] present the detailed analytical range integrals (the range integral for the temperature-control phase is quite involved). Equation (11.47) presents the range integral for the equilibrium-glide phase (s_2) that we derived in Section 11.4. The constant-drag range integral is the easiest to derive because a_D is a constant, that is, $c_3 = 10$ m/s^2 (see Figure 11.14). Note that the drag acceleration profile for the transition phase is a linear function of energy ξ.

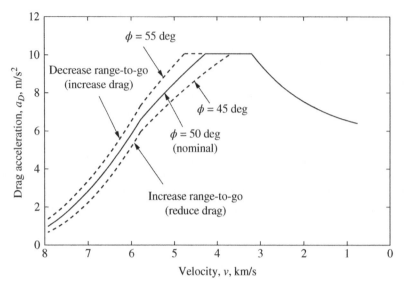

Figure 11.16 Adjusting the Shuttle's drag-acceleration profile for range control.

Therefore, we derive the appropriate range integral using $ds/d\xi = -1/a_D$; we obtain this expression by dividing Eq. (11.8) (with $r_0/r \approx 1$ and $\cos\gamma \approx 1$) by Eq. (11.73).

During its entry, the Space Shuttle's onboard guidance scheme computed the analytical predicted range-to-go s_{go} using Eqs. (11.101) and (11.106). The guidance algorithm then compared the analytical s_{go} with the *actual* remaining range to the landing site as determined by the navigation system. If the predicted range was *less* than the actual range, the a_D profile was shifted "down" (less drag) to extend the range flown. If s_{go} was *greater* than the actual range, the a_D profile was shifted "up" (more drag) to reduce the range. Figure 11.16 illustrates two scenarios early in the entry where the temperature-control and equilibrium-glide phases are shifted up or down in tandem to reduce or extend the range-to-go. Changing the polynomial coefficients c_i and bank angle ϕ in Eqs. (11.102)–(11.105) shifts the a_D profile. Of course, the shifted a_D profile must remain within the entry corridor shown in Figure 11.14. The Shuttle's guidance algorithm adjusted the free coefficients until the predicted s_{go} matched the actual range to the landing site. After establishing the reference drag-velocity function, the Space Shuttle used a closed-loop control law to modulate bank angle ϕ and keep the vehicle on the desired a_D profile. The entire process was repeated every 2 s so that the Shuttle constantly updated its drag-velocity profile to achieve the desired down-range distance. The interested reader should consult Harpold and Graves [4] to learn more about this application of first-order entry methods to the Space Shuttle entry problem.

11.8 Summary

In this chapter, we presented the so-called first-order methods for analyzing atmospheric entry. Our overall goal was to develop closed-form analytical solutions for three entry profiles: ballistic entry (no lift); gliding entry; and skip entry. The reader should go back and review the assumptions that are required for each first-order analytical solution. On a related note, the reader should understand the limitations of each method; for example, when he or she can and cannot use the first-order solutions. Later in the chapter, we illustrated how the tremendous kinetic energy at entry interface is dissipated by aerodynamic drag in the form of aerodynamic heating. The principle metrics for entry flight are deceleration and heating rate. We developed closed-form expressions for peak deceleration and peak heating rate and the corresponding critical altitudes and velocities where these peak values occur. Finally, we concluded this chapter with a discussion of the Space Shuttle's entry strategy. We illustrated how the various drag acceleration boundaries form the so-called entry corridor and how the first-order methods define a feasible entry profile that meets the down-range requirements.

References

1 Eggers, A.J., Allen, H.J., and Neice, S.E., "A Comparative Analysis of the Performance of Long-Range Hypersonic Vehicles," NACA TN-4046, 1957.
2 Allen, H.J. and Eggers, A.J., "A Study of the Motion and Aerodynamic Heating of Missiles Entering the Earth's Atmosphere at High Supersonic Speeds," NACA TR-1381, 1958.
3 Chapman, D.R., "An Approximate Analytical Method for Studying Entry into Planetary Atmospheres," NACA TN-4276, 1958.
4 Harpold, J.C. and Graves, C.A., "Shuttle Entry Guidance," *Journal of the Astronautical Sciences*, Vol. 27, 1979, pp. 239–268.
5 Griffin, M.D., and French, J.R., *Space Vehicle Design*, AIAA Educational Series, Washington, DC, 1991, pp. 255–264.
6 Regan, F.J. and Anandakrishnan, S.M., *Dynamics of Atmospheric Re-Entry*, AIAA Educational Series, Washington, DC, 1993, pp. 171–180, 216–221.
7 Vinh, N.X., Busemann, A., and Culp, R.D., *Hypersonic and Planetary Entry Flight Mechanics*, University of Michigan Press, Ann Arbor, MI, 1980, pp. 144–152.

Problems

Conceptual Problems

11.1 A capsule de-orbiting from a space station in low-Earth orbit (LEO) is entering the Earth's atmosphere (entry interface, EI) at velocity $v_{EI} = 7.85$ km/s. Assume that the vehicle's total specific energy is referenced to the surface of the Earth (i.e., $\xi = v^2/2 + g_0 h$). Note that with this definition, total energy is zero when the capsule is at rest on the surface of the Earth.

    a) Compute the total specific energy at entry interface.

    b) Determine the percentages of kinetic and potential energy at EI with respect to the total energy.

    c) Show that the time rate of the surface-referenced energy is identical to the result presented by Eq. (11.73).

11.2 A spent satellite is entering the Earth's atmosphere at velocity $v_{EI} = 7.75$ km/s and flight-path angle $\gamma_{EI} = -6°$. The satellite has a ballistic coefficient of 90 kg/m² and does not generate lift during entry.
   a) Estimate the satellite's velocity at an altitude of 65 km.
   b) What is the critical altitude for peak deceleration?
   c) Compute the peak deceleration of the satellite in units of Earth g_0.

11.3 A blunt-body capsule is at Earth's entry interface altitude with velocity $v_{EI} = 7.8$ km/s and flight-path angle $\gamma_{EI} = -5.5°$. The capsule has ballistic coefficient $C_B = 80$ kg/m² and follows a ballistic entry. Compute the drag acceleration D/m and centrifugal acceleration v^2/r at altitudes of 122 (EI), 90, 70, and 50 km. Compare these acceleration terms to Earth-surface g_0 and comment on the accuracy of the first-order ballistic entry solution at each altitude.

11.4 Using the blunt-body capsule data and entry conditions from Problem 11.3, determine the critical altitudes for peak deceleration and peak heat rate for the ballistic entry.

   Problems 11.5–11.8 involve an Apollo-style capsule with lift-to-drag ratio $L/D = 0.4$, mass $m = 6,000$ kg, reference area $S = 14$ m², and drag coefficient $C_D = 1.6$.

11.5 The capsule is at an altitude of 98 km with velocity $v = 9,750$ m/s and flight-path angle $\gamma = -5.5°$. If the bank angle is 180° (lift "down"), is the flight-path angle increasing or decreasing at this instant?

11.6 The capsule is on an entry trajectory at an altitude of 80 km with velocity $v = 8,250$ m/s and flight-path angle $\gamma = -5°$. Determine the bank angle required at this instant for an equilibrium glide.

11.7 Repeat Problem 11.6 if the capsule's altitude, velocity, and flight-path angle are 70 km, 6,930 m/s, and $-6.8°$, respectively.

11.8 Determine the drag acceleration (in Earth g_0) of the Apollo-style capsule at altitudes of 80 and 70 km. Use the information in Problems 11.6 and 11.7.

11.9 Spacecraft designers want to determine the effect of entry flight-path angle on peak deceleration for a ballistic entry. The spacecraft's entry interface velocity (for Earth) is $v_{EI} = 7.8$ km/s and its ballistic coefficient is 85 kg/m². Plot peak deceleration in units of Earth-surface g_0 (=9.80665 m/s²) for entry angles $-15° \leq \gamma_{EI} \leq -5°$ and determine the entry flight-path angle where peak deceleration is $20g_0$.

11.10 Spacecraft designers want to determine the effect of ballistic coefficient on the critical altitudes for peak deceleration and peak heating for a ballistic entry. The spacecraft's Earth entry interface velocity is $v_{EI} = 7.8$ km/s and its flight-path angle is $\gamma_{EI} = -5.8°$. Plot the critical altitudes for peak deceleration and peak heating for ballistic coefficients $50 < C_B < 200$ kg/m².

   Problems 11.11–11.14 involve a winged entry vehicle with lift-to-drag ratio $L/D = 1.35$, mass $m = 1,800$ kg, reference wing area $S = 6$ m², and drag coefficient $C_D = 0.8$.

11.11 The winged entry vehicle enters Venus' atmosphere (h_{EI} = 200 km) with near-circular velocity v_{EI} = 7.2 km/s and flight-path angle γ_{EI} = −2° and performs a skip entry. The vehicle has zero bank angle during its skip entry.
   a) Determine the vehicle's pull-up altitude and pull-up velocity during the skip entry.
   b) Compute the peak deceleration along the flight-path and the peak normal acceleration in units of Earth-surface g_0 (= 9.80665 m/s^2).
   c) Compute the atmospheric exit velocity and flight-path angle.
   d) Compute the peak altitude during the Keplerian "skip out" phase (use μ_V = 3.2486(10^5) km^3/s^2 for Venus' gravitational parameter and r_V = 6,052 km for the radius of Venus).

11.12 Using the Venus entry conditions from Problem 11.11, determine the critical flight-path angle, altitude, and velocity corresponding to the peak aerodynamic heating rate for a skip entry.

11.13 Repeat Problem 11.11 for a skip entry at Mars. Let the Mars entry conditions (h_{EI} = 125 km) be v_{EI} = 3.5 km/s and γ_{EI} = −2°.

11.14 Using the Mars entry conditions from Problem 11.13, determine the critical flight-path angle, altitude, and velocity corresponding to the peak aerodynamic heating rate for a skip entry.

11.15 A blunt-body capsule is at Earth's entry interface with entry speed v_{EI} = 7.8 km/s. The capsule has ballistic coefficient C_B = 65 kg/m^2. Plot the body-averaged heat-rate input (per unit skin-friction coefficient C_f) for a ballistic entry. Plot q_{avg}/C_f as a function of altitude for two entry flight-path angles: γ_{EI} = −5° and −7° (place both plots on the same figure). Compute the critical altitudes for peak heating and the peak q_{avg}/C_f values for both entry angles – do these values correspond to the peaks from the two plots?

Mission Applications

11.16 The Mercury spacecraft carried the first US astronauts to space and back to Earth in the early 1960s. The Mercury capsule used a ballistic entry and its parameters were mass m = 1,200 kg, reference area S = 2.812 m^2, and drag coefficient C_D = 1.6. Repeat parts (a) and (b) of Example 11.1 using the same EI states as the Vostok capsule. Compare the peak decelerations and critical altitudes of the two ballistic entry profiles.

Problems 11.17–11.24 involve the Space Shuttle's entry flight. Use mass m = 82,000 kg, reference wing area S = 250 m^2, drag coefficient C_D = 0.8, and L/D = 1.1 for all problems.

11.17 The Shuttle is currently in the equilibrium-glide phase and has velocity v = 5.4 km/s (the Shuttle defines the equilibrium-glide phase as the velocity range 4.27 < v < 5.8 km/s). Compute the constant bank angle ϕ so that the remaining down-range distance of the equilibrium glide is 700 km.

11.18 The Shuttle is currently in an equilibrium glide at velocity $v = 5.6$ km/s and bank angle $\phi = 52.5°$. The constant-drag phase has a drag acceleration of 10 m/s². The constant-drag phase ends at velocity $v = 3.2$ km/s.
a) Compute the Shuttle's current drag acceleration in m/s².
b) Determine the velocity that marks the beginning of the constant-drag phase.
c) Estimate the down-range distances of the remaining equilibrium-glide phase and the entire constant-drag phase.

11.19 The Shuttle is currently in the constant-drag phase and has velocity $v = 4$ km/s (the Shuttle defines the constant-drag phase as the velocity range $3.2 < v < 4.27$ km/s). The down-track range estimate for the transition phase ($0.76 < v < 3.2$ km/s) is 600 km. The Shuttle's *actual* range-to-go is 890 km. Compute the required constant drag acceleration so that the estimated range-to-go matches the actual remaining down-range distance (assume that the range for the transition phase does not change).

11.20 Estimate the down-track range of the Shuttle's *transition phase* where the drag acceleration profile is a linear function of energy. The drag acceleration and energy at the beginning and end of the transition phase are 10 m/s² and $-5.708(10^7)$ m²/s², and 6.3 m/s² and $-6.195(10^7)$ m²/s², respectively.

11.21 Plot drag acceleration vs. velocity for the Space Shuttle's equilibrium-glide boundary for three bank angles: $\phi = 0, 25°, 50°$ (place all three plots on the same figure). Use the following simple piecewise linear function for L/D vs. velocity:

$$L/D = 1.1 \quad \text{for} \quad v \ge 3{,}200 \text{ m/s}$$

$$L/D = 2.22 - 0.00035v \quad \text{for} \quad v \le 3{,}200 \text{ m/s}$$

11.22 Plot the critical altitude (in km) for peak heating rate during the Shuttle's equilibrium glide vs. bank angle for $0° \le \phi \le 80°$.

11.23 Determine the bank angle ϕ where the peak heat-rate input increases by 50% when compared to the peak heating using a zero-bank equilibrium glide.

11.24 Consider again the derivation surrounding Eq. (11.18), the spacecraft's velocity on a ballistic entry. The first-order ballistic entry method assumes (1) drag dominates gravity for entry at a shallow flight-path angle, and (2) flight-path angle is constant because the centrifugal force cancels gravity. Therefore, we can also use Eq. (11.18) to estimate velocity on a shallow equilibrium glide where flight-path angle is constant. The Space Shuttle's velocity at EI is $v_{EI} = 7{,}850$ m/s.
a) Starting from the state conditions $h = 85$ km and $\gamma = -1.2°$, compute the Shuttle's bank angle ϕ required to maintain an equilibrium glide [use Eq. (11.18) to compute velocity at $h = 85$ km].
b) At what altitude is it impossible for the Shuttle to maintain $\gamma = -1.2°$? Again, use Eq. (11.18) to compute the Shuttle's velocity as a function of altitude.
c) Plot bank angle ϕ vs. the altitude range where it is possible to maintain a true equilibrium glide (assume constant $L/D = 1.1$).

Problems 11.25–11.29 involve the entry flight of the Intermediate Experimental Vehicle (IXV), an unmanned, wingless, lifting-body entry vehicle designed and successfully flown by the European Space Agency (ESA) on February 11, 2015. Figure P11.25 shows the IXV. The IXV entered Earth's atmosphere (h_{EI} = 122 km) with velocity v_{EI} = 7.7 km/s and flight-path angle γ_{EI} = −1.2° and followed an equilibrium glide. Use mass m = 1,845 kg, reference area S = 7.26 m², drag coefficient C_D = 0.84, and L/D = 0.7 for all problems.

11.25 Determine the velocity and deceleration (in g_0) of the IXV during its equilibrium glide at altitudes of 80, 60, and 40 km if the bank angle is zero.

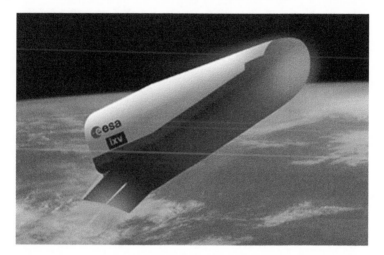

Figure P11.25 Intermediate Experimental Vehicle. *Source*: Courtesy of Huart.

11.26 Repeat Problem 11.25 (find the velocity and deceleration) if the bank angle is $\phi = 50°$.

11.27 Determine the constant bank angle of the IXV so that its range is 5,860 km on an equilibrium glide from EI velocity v_1 = 7.7 km/s to final velocity v_2 = 1 km/s.

11.28 Determine the critical altitude and velocity for peak aerodynamic heating rate for an equilibrium glide with zero bank angle.

11.29 Repeat Problem 11.28 (determine the critical altitude and velocity for peak heating) if the bank angle is $\phi = 50°$.

11.30 The Stardust capsule returned from deep space after sampling cosmic dust. It entered the Earth's atmosphere (EI) at velocity v_{EI} = 12.9 km/s. The capsule's mass is 46 kg and its reference and wetted areas are 0.52 and 1.2 m², respectively. Its drag coefficient is C_D = 1.5 and its skin-friction drag coefficient is 0.14. Determine the total thermal energy transferred to the Stardust capsule during its ballistic entry from EI conditions to zero velocity.

12

Attitude Dynamics

12.1 Introduction

All chapters up to this point have treated a satellite or spacecraft as a particle. In other words, we have only considered the motion of its center of mass. The satellite orbits, interplanetary trajectories, and atmospheric entry profiles analyzed in Chapters 2–11 are examples of *particle dynamics*: the application of Newton's laws to a single particle. In this chapter, we begin to consider a space vehicle's angular orientation and angular motion or *attitude dynamics*.

The distinction between particle dynamics and attitude dynamics is illustrated in Figure 12.1. Figure 12.1a shows a satellite's orbit about the Earth. The satellite is a particle where vectors **r** and **v** are the position and velocity of the satellite's center of mass as it moves along its orbit. All analysis of satellite motion thus far (such as orbital energy, orbital angular momentum, conic sections, etc.) has dealt with particle dynamics as depicted in Figure 12.1a. Figure 12.1b and 12.1c show two possible angular orientations of the satellite (vectors **r** and **v** are still the position and velocity of the satellite's center of mass). In Figure 12.1b, the satellite is oriented such that its antenna is pointed away from the Earth whereas in Figure 12.1c the antenna is pointed toward the Earth. The *attitude* is the satellite's angular orientation with respect to a reference frame. *Attitude dynamics* is the study of a satellite's rotational motion about its center of mass. Clearly, it is important to understand the attitude dynamics of a satellite so that it can be pointed in the correct direction for communication or scientific observations. In addition, a satellite powered by solar cells must maintain an attitude so that its solar arrays can collect the maximum amount of the sun's energy.

This chapter is an introduction to rigid body dynamics and satellite attitude dynamics. Some of this material should be familiar to readers who have taken a dynamics or mechanics course. Because most satellites operate in the vacuum of space where external moments or torques are zero, we will focus on a satellite's "torque-free" angular motion about its center of mass. This chapter will also address the stability of spinning satellites and disturbance torques. We will discuss methods for controlling a satellite's attitude in Chapter 13.

Space Flight Dynamics, First Edition. Craig A. Kluever.
© 2018 John Wiley & Sons Ltd. Published 2018 by John Wiley & Sons Ltd.
Companion website: www.wiley.com/go/Kluever/spaceflightmechanics

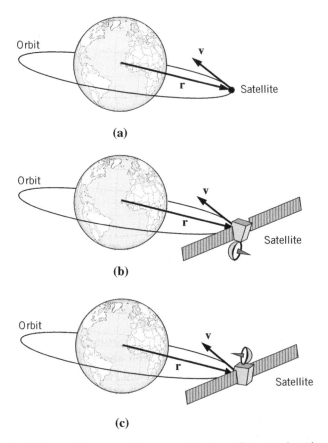

Orbit

v

r

Satellite

(a)

Orbit

v

r

Satellite

(b)

Orbit

v

r

Satellite

(c)

Figure 12.1 Earth-orbiting satellite: (a) satellite as a particle; (b) satellite attitude with antenna pointing away from Earth; and (c) satellite attitude with antenna pointing toward Earth.

12.2 Rigid Body Dynamics

Before we delve into the three-dimensional dynamics of a rotating body, let us consider the very simple case of a rotor or disk spinning about a *fixed* axis as shown in Figure 12.2. The disk in Figure 12.2 has moment of inertia I about its axis of symmetry defined by

$$I = \int r^2 dm \tag{12.1}$$

where dm is an infinitesimal mass with radial distance r from the symmetry axis (we have assumed that the rotation axis coincides with the disk's axis of symmetry). Figure 12.2 also shows an external torque (or moment) M applied to the disk. Positive angular displacement θ is counterclockwise (when viewed from above) as shown in Figure 12.2. We know from a dynamics course that the sum of the external torques is equal to the moment of inertia multiplied by the angular acceleration. For the rotor in Figure 12.2, we have

$$M = I\ddot{\theta} = I\dot{\omega} \tag{12.2}$$

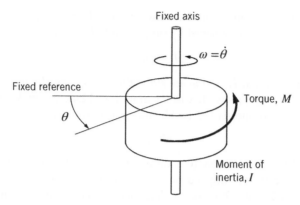

Figure 12.2 One degree-of-freedom system: disk rotating about a fixed axis.

Equation (12.2) is Newton's second law for a rotational mechanical system. We have assumed that the disk is rotating on frictionless bearings and therefore the applied torque M is the only moment acting on the disk. The disk shown in Figure 12.2 is called a *one degree-of-freedom* (1-DOF) system because it only requires one displacement coordinate (angle θ) to completely define its orientation.

Now let us derive the angular momentum of the symmetrical 1-DOF spinning disk. Figure 12.3 shows the disk rotating about the fixed z axis. The orthogonal xyz axes are rotating *body axes* fixed to the disk with its origin at the disk's center of mass (c.m.). The z body axis is an axis of symmetry. Furthermore, let us assume that the z axis is a fixed axis of rotation (e.g., a shaft as shown in Figure 12.2). Next, consider the incremental mass m_i shown in Figure 12.3. The incremental angular momentum associated with m_i is the moment of its linear momentum $(m_i v_i)$ about the z axis:

$$\mathbf{H}_i = \mathbf{r}_i \times m_i \mathbf{v}_i \tag{12.3}$$

where radial vector $\mathbf{r}_i$ is the perpendicular distance from the z axis to m_i. Because the disk is spinning about the fixed z axis, the velocity of mass m_i is $\mathbf{v}_i = \boldsymbol{\omega} \times \mathbf{r}_i$ which is always perpendicular to the spin axis and radius $\mathbf{r}_i$. It is easy to see that for this 1-DOF system

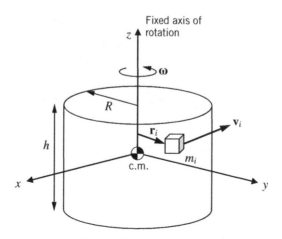

Figure 12.3 Homogeneous disk rotating about its fixed z body axis.

the incremental angular momentum $\mathbf{H}_i$ is always along the z axis and therefore the cross product in Eq. (12.3) can be replaced by simple multiplication:

$$H_i = r_i^2 m_i \omega \tag{12.4}$$

where it is understood that H_i and ω are along the z axis. Now, let us add up all of the angular momentum increments, that is, integrate Eq. (12.4), and make use of the definition of the moment of inertia (12.1). The integration result is

$$H = I\omega \tag{12.5}$$

For the simple 1-DOF rotational system, the angular momentum is the disk's moment of inertia I (about the z axis) multiplied by the angular velocity ω (also about the z axis). Comparing Eqs. (12.2) and (12.5), we see that the time derivative of angular momentum is equal to the applied torque (or moment).

12.2.1 Angular Momentum of a Rigid Body

Now let us extend the 1-DOF results to three dimensions. Figure 12.4 shows a satellite moving along its orbital path. Coordinate system XYZ is an inertial (fixed) frame fixed to the center of the gravitational body. Coordinate system xyz is a rotating *body* frame that is fixed to the spacecraft at its c.m. Therefore, frame xyz rotates relative to the inertial frame XYZ due to the satellite's angular motion. The satellite's angular velocity vector is $\boldsymbol{\omega}$, which is not necessarily aligned with a particular body axis. Let us consider the absolute (inertial) position of the incremental satellite mass m_i

$$\mathbf{R}_i = \mathbf{R}_{cm} + \mathbf{r}_i \tag{12.6}$$

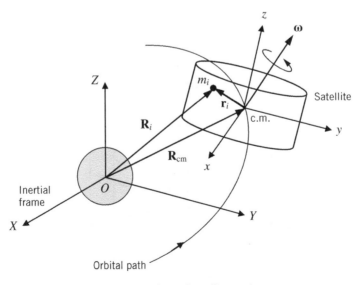

Figure 12.4 General satellite motion.

where $\mathbf{R}_{cm}$ is the absolute position of the satellite's center of mass and $\mathbf{r}_i$ is the position of m_i relative to the c.m. as expressed in the rotating xyz frame. The absolute velocity of incremental mass m_i is

$$\mathbf{v}_i = \dot{\mathbf{R}}_i = \dot{\mathbf{R}}_{cm} + \dot{\mathbf{r}}_i\big|_{rot} + \boldsymbol{\omega} \times \mathbf{r}_i \tag{12.7}$$

where $\dot{\mathbf{r}}_i\big|_{rot}$ is the velocity of m_i relative to the rotating xyz body-fixed frame. We assume that the satellite is a *rigid body* and therefore $\dot{\mathbf{r}}_i\big|_{rot} = \mathbf{0}$. Next, let us compute the incremental angular momentum of mass m_i relative to origin O, the center of the gravitational body:

$$\mathbf{H}_i = \mathbf{R}_i \times m_i \mathbf{v}_i$$
$$= \mathbf{R}_i \times m_i \left(\dot{\mathbf{R}}_{cm} + \boldsymbol{\omega} \times \mathbf{r}_i \right) \tag{12.8}$$

Summing all incremental masses over the entire satellite, we obtain the *total* angular momentum about the center of the gravitational body

$$\mathbf{H}_O = \sum \mathbf{R}_i \times m_i \mathbf{v}_{cm} + \sum \mathbf{R}_i \times m_i (\boldsymbol{\omega} \times \mathbf{r}_i) \tag{12.9}$$

where $\mathbf{v}_{cm} = \dot{\mathbf{R}}_{cm}$ is the absolute velocity of the satellite's c.m. The first summation on the right-hand side of Eq. (12.9) involves the inertial position of the satellite's c.m., or $\sum \mathbf{R}_i m_i = m \mathbf{R}_{cm}$ where m is the total mass of the satellite. The second summation must be expanded by using Eq. (12.6) for inertial position $\mathbf{R}_i$; the result is

$$\mathbf{H}_O = \mathbf{R}_{cm} \times m \mathbf{v}_{cm} + \sum (\mathbf{R}_{cm} + \mathbf{r}_i) \times m_i (\boldsymbol{\omega} \times \mathbf{r}_i)$$
$$= \mathbf{R}_{cm} \times m \mathbf{v}_{cm} + \sum \mathbf{R}_{cm} \times m_i (\boldsymbol{\omega} \times \mathbf{r}_i) + \sum \mathbf{r}_i \times m_i (\boldsymbol{\omega} \times \mathbf{r}_i) \tag{12.10}$$

The middle cross-product term in Eq. (12.10) is zero because $\sum m_i \mathbf{r}_i = \mathbf{0}$ is the definition of the c.m. Allowing m_i to become infinitesimally small permits us to replace the summation with an integral; hence Eq. (12.10) becomes

$$\mathbf{H}_O = \mathbf{R}_{cm} \times m \mathbf{v}_{cm} + \int \mathbf{r} \times (\boldsymbol{\omega} \times \mathbf{r}) dm \tag{12.11}$$

where $\mathbf{r}$ is the position of infinitesimal mass dm as expressed in the rotating body-fixed frame. We may express the satellite's *total* angular momentum as

$$\mathbf{H}_O = \mathbf{H}_{orbit} + \mathbf{H}_{body} \tag{12.12}$$

where $\mathbf{H}_{orbit} = \mathbf{R}_{cm} \times m \mathbf{v}_{cm}$ is the *orbital* angular momentum, or angular momentum due to the motion of the c.m. along its orbital path. The component $\mathbf{H}_{orbit}$ is the angular momentum we utilized in Chapters 2–10 when we treated the satellite as a point mass and it is identical to Eq. (2.13). The second component in Eq. (12.12), $\mathbf{H}_{body} = \int \mathbf{r} \times (\boldsymbol{\omega} \times \mathbf{r}) dm$, is the *body* angular momentum due to the satellite's rotation about its own c.m. Chapters 2–10 showed that orbital angular momentum is a key to analyzing orbital motion, or "orbital dynamics." Likewise, body angular momentum $\mathbf{H}_{body}$ is crucial for analyzing "attitude dynamics," or a satellite's rotational motion.

Let us express the body angular momentum in a simpler form. First, we can express the angular velocity vector in body-fixed (or *xyz*) coordinates:

$$\boldsymbol{\omega} = \begin{bmatrix} \omega_x \\ \omega_y \\ \omega_z \end{bmatrix} \tag{12.13}$$

Consider the cross product $\boldsymbol{\omega} \times \mathbf{r}$

$$\boldsymbol{\omega} \times \mathbf{r} = \begin{vmatrix} \mathbf{i} & \mathbf{j} & \mathbf{k} \\ \omega_x & \omega_y & \omega_z \\ x & y & z \end{vmatrix} \tag{12.14}$$

$$= \left(\omega_y z - \omega_z y\right)\mathbf{i} + \left(-\omega_x z + \omega_z x\right)\mathbf{j} + \left(\omega_x y - \omega_y x\right)\mathbf{k}$$

Equation (12.14) is a vector of the relative velocity components of mass *dm* as expressed in the body-fixed frame (unit vectors **ijk** are along the *xyz* coordinates of the rotating frame). Multiplying Eq. (12.14) by *dm* yields $(\boldsymbol{\omega} \times \mathbf{r})dm$, that is, the *linear* momentum components of mass *dm*. Next, cross the relative position vector **r** with $(\boldsymbol{\omega} \times \mathbf{r})dm$

$$\mathbf{r} \times (\boldsymbol{\omega} \times \mathbf{r})dm = \begin{vmatrix} \mathbf{i} & \mathbf{j} & \mathbf{k} \\ x & y & z \\ \omega_y z - \omega_z y & -\omega_x z + \omega_z x & \omega_x y - \omega_y x \end{vmatrix} dm$$

$$= \begin{bmatrix} \omega_x\left(y^2 + z^2\right) - \omega_y xy - \omega_z xz \\ -\omega_x xy + \omega_y\left(x^2 + z^2\right) - \omega_z yz \\ -\omega_x xz - \omega_y yz + \omega_z\left(x^2 + y^2\right) \end{bmatrix} dm \tag{12.15}$$

We can distribute the *dm* factor in Eq. (12.15) and write the 3×1 vector result as a matrix-vector product:

$$\mathbf{r} \times (\boldsymbol{\omega} \times \mathbf{r})dm = \begin{bmatrix} (y^2 + z^2)dm & -(xy)dm & -(xz)dm \\ -(xy)dm & (x^2 + z^2)dm & -(yz)dm \\ -(xz)dm & -(yz)dm & (x^2 + y^2)dm \end{bmatrix} \begin{bmatrix} \omega_x \\ \omega_y \\ \omega_z \end{bmatrix} \tag{12.16}$$

Note that the 3×3 matrix in Eq. (12.16) is symmetric with positive diagonal terms (recall that *x*, *y*, and *z* are the position coordinates of infinitesimal mass *dm* as expressed in the rotating, body-fixed frame). We can obtain body angular momentum $\mathbf{H}_{\text{body}}$ by integrating Eq. (12.16) over the satellite's mass distribution. The result of the integration may be succinctly written as

$$\mathbf{H}_{\text{body}} = \int \mathbf{r} \times (\boldsymbol{\omega} \times \mathbf{r})dm = \begin{bmatrix} I_x & -I_{xy} & -I_{xz} \\ -I_{xy} & I_y & -I_{yz} \\ -I_{xz} & -I_{yz} & I_z \end{bmatrix} \begin{bmatrix} \omega_x \\ \omega_y \\ \omega_z \end{bmatrix} \tag{12.17}$$

We may write Eq. (12.17) in a more compact form

$$\mathbf{H}_{body} = \mathbf{I}\boldsymbol{\omega} \tag{12.18}$$

where $\mathbf{I}$ is the 3×3 *inertia matrix*

$$\mathbf{I} = \begin{bmatrix} I_x & -I_{xy} & -I_{xz} \\ -I_{xy} & I_y & -I_{yz} \\ -I_{xz} & -I_{yz} & I_z \end{bmatrix} \tag{12.19}$$

The diagonal elements of the inertia matrix are the *moments of inertia*:

$$I_x = \int \left(y^2 + z^2 \right) dm \tag{12.20a}$$

$$I_y = \int \left(x^2 + z^2 \right) dm \tag{12.20b}$$

$$I_z = \int \left(x^2 + y^2 \right) dm \tag{12.20c}$$

Clearly, the moments of inertia are always positive. The off-diagonal elements of $\mathbf{I}$ are the *products of inertia*:

$$I_{xy} = \int xy\,dm \tag{12.21a}$$

$$I_{xz} = \int xz\,dm \tag{12.21b}$$

$$I_{yz} = \int yz\,dm \tag{12.21c}$$

The products of inertia may be positive, negative, or zero.

The moments and products of inertia are functions of the satellite's shape and mass distribution. As a relatively simple example, consider the disk shown in Figure 12.3. Let us assume that the disk is homogeneous (i.e., uniform mass distribution) with outer radius R, total mass m, and height h (see Figure 12.3). The moments of inertia for a homogeneous disk (cylinder) are

Homogeneous cylinder x, y axes: $\quad I_x = I_y = \dfrac{1}{12} m \left(3R^2 + h^2 \right) \tag{12.22}$

Homogeneous cylinder z axis: $\quad I_z = \dfrac{1}{2} mR^2 \tag{12.23}$

Because the homogenous cylinder is symmetric about the xy, xz, and yz planes, all three products of inertia are zero. Hence, $\mathbf{I}$ is a diagonal matrix for a homogeneous cylinder. In this simple example, the xyz body-fixed axes are the so-called *principal axes*. By definition, the principal axes are the orthogonal coordinates chosen so that the products of inertia are zero. We will discuss the principal axes in a bit more detail in the next subsection. The following example demonstrates the relationship between the rigid-body angular momentum $\mathbf{H}_{body}$ and the angular velocity vector $\boldsymbol{\omega}$ for symmetrical and asymmetrical rigid bodies.

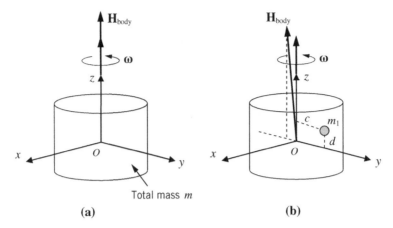

Figure 12.5 (a) Symmetrical homogeneous satellite and (b) asymmetrical satellite (Example 12.1).

Example 12.1 Consider the two cylindrically shaped satellites shown in Figure 12.5. Figure 12.5a shows a symmetrical satellite with homogeneous mass distribution, whereas Figure 12.5b shows an asymmetrical cylinder that is "unbalanced" due to the 50-kg mass m_1 offset from the origin. The offset distances for m_1 are $c = 0.5$ m (along the $+y$ body axis) and $d = 0.3$ m (along the $+z$ axis). Both cylindrical satellites have a radius of 1 m, height of 1.5 m, and total mass of 2,000 kg (without offset mass m_1). Both satellites are rotating about the $+z$ body axis with an angular velocity of 5 revolutions per minute (rpm). Compute the angular momentum of the symmetrical and asymmetrical rigid bodies.

We will consider the symmetrical satellite in Figure 12.5a first. Because the satellite is a homogeneous cylinder, its moments of inertia can be computed using Eqs. (12.22) and (12.23), radius $R = 1$ m, height $h = 1.5$ m, and mass $m = 2,000$ kg:

$$I_x = I_y = \frac{1}{12}m(3R^2 + h^2) = 875 \text{ kg-m}^2$$

$$I_z = \frac{1}{2}mR^2 = 1,000 \text{ kg-m}^2$$

The homogeneous satellite has zero products of inertia due to its symmetry. Therefore, its inertia matrix is

$$\text{Symmetrical satellite: } \mathbf{I} = \begin{bmatrix} I_x & 0 & 0 \\ 0 & I_y & 0 \\ 0 & 0 & I_z \end{bmatrix} = \begin{bmatrix} 875 & 0 & 0 \\ 0 & 875 & 0 \\ 0 & 0 & 1,000 \end{bmatrix} \text{kg-m}^2$$

The angular velocity is $\omega_z = (5 \text{ rpm})(2\pi \text{ rad}/1 \text{ rev})(1 \text{ min}/60 \text{ s}) = 0.5236$ rad/s. Expressed as a vector, the angular velocity is

$$\boldsymbol{\omega} = \begin{bmatrix} 0 \\ 0 \\ 0.5236 \end{bmatrix} \text{rad/s}$$

The angular momentum is computed using Eq. (12.18):

$$\text{Symmetrical satellite: } \mathbf{H}_{\text{body}} = \mathbf{I}\boldsymbol{\omega} = \begin{bmatrix} 875 & 0 & 0 \\ 0 & 875 & 0 \\ 0 & 0 & 1,000 \end{bmatrix} \begin{bmatrix} 0 \\ 0 \\ 0.5236 \end{bmatrix}$$

$$= \begin{bmatrix} 0 \\ 0 \\ 523.6 \end{bmatrix} \text{kg-m}^2/\text{s}$$

Therefore, all angular momentum is in the $+z$ body-axis direction. For a symmetrical satellite, the angular momentum $\mathbf{H}_{\text{body}}$ is aligned with the angular velocity $\boldsymbol{\omega}$ or the "spin axis" as shown in Figure 12.5a.

Next, we consider the asymmetrical (unbalanced) satellite shown in Figure 12.5b. We must re-compute the satellite's moments and products of inertia about the body-fixed *xyz* axes by adding the inertia elements for the homogeneous cylinder to the contribution due to the discrete offset mass m_1. The moments of inertia for the *unbalanced* satellite are

$$I_x = \bar{I}_x + \left(c^2 + d^2\right)m_1 = 875 + 17 = 892 \text{ kg-m}^2$$

$$I_y = \bar{I}_y + \left(0^2 + d^2\right)m_1 = 875 + 4.5 = 879.5 \text{ kg-m}^2$$

$$I_z = \bar{I}_z + \left(0^2 + c^2\right)m_1 = 1,000 + 12.5 = 1,012.5 \text{ kg-m}^2$$

Note that the moments of inertia for the homogeneous (symmetrical) satellite are $\bar{I}_x$, $\bar{I}_y$, and $\bar{I}_z$. The position components of the mass m_1 offset are $x = 0$, $y = c = 0.5$ m, and $z = d = 0.3$ m. The products of inertia for the unbalanced satellite are

$$I_{xy} = \bar{I}_{xy} + (0)(c)m_1 = 0$$

$$I_{xz} = \bar{I}_{xz} + (0)(d)m_1 = 0$$

$$I_{yz} = \bar{I}_{yz} + (c)(d)m_1 = 0 + 7.5 = 7.5 \text{ kg-m}^2$$

where $\bar{I}_{xy} = \bar{I}_{xz} = \bar{I}_{yz} = 0$ are the products of inertia associated with the symmetrical satellite. Because the *yz* plane remains a plane of symmetry for the unbalanced satellite, the two products of inertia associated with the *x* axis (I_{xy} and I_{xz}) are zero.

Assembling the inertia matrix for the unbalanced satellite yields

$$\text{Asymmetrical satellite: } \mathbf{I} = \begin{bmatrix} I_x & -I_{xy} & -I_{xz} \\ -I_{xy} & I_y & -I_{yz} \\ -I_{xz} & -I_{yz} & I_z \end{bmatrix} = \begin{bmatrix} 892 & 0 & 0 \\ 0 & 879.5 & -7.5 \\ 0 & -7.5 & 1,012.5 \end{bmatrix} \text{kg-m}^2$$

The angular momentum is the matrix-vector product:

$$\text{Asymmetrical satellite: } \mathbf{H}_{\text{body}} = \mathbf{I}\boldsymbol{\omega} = \begin{bmatrix} 892 & 0 & 0 \\ 0 & 879.5 & -7.5 \\ 0 & -7.5 & 1{,}012.5 \end{bmatrix} \begin{bmatrix} 0 \\ 0 \\ 0.5236 \end{bmatrix}$$

$$= \begin{bmatrix} 0 \\ -3.93 \\ 530.14 \end{bmatrix} \text{kg-m}^2/\text{s}$$

The angular momentum $\mathbf{H}_{\text{body}}$ for the asymmetrical satellite is *not* aligned with the angular velocity vector $\boldsymbol{\omega}$. The majority of $\mathbf{H}_{\text{body}}$ is along the $+z$ axis, but a small component is along the $-y$ body axis as shown in Figure 12.5b (not drawn to scale). The angle between $\mathbf{H}_{\text{body}}$ and the spin axis is $\tan^{-1}(3.93/530.14) = 0.42°$. Note that the magnitude (i.e., vector norm) of the asymmetrical $\mathbf{H}_{\text{body}}$ is greater than the magnitude of the symmetrical angular momentum due to the additional 50-kg mass m_1 that is offset from the axis of rotation.

12.2.2 Principal Axes

From our previous discussion of angular momentum, we see that the inertia matrix $\mathbf{I}$ consists of moments of inertia (diagonal elements) and products of inertia (off-diagonal elements). Example 12.1 shows that the products of inertia are zero with respect to body-fixed orthogonal axes that lie in the planes of symmetry. The coordinate axes are said to be the *principal axes* for the body if all products of inertia vanish. We will denote the principal axes as the 1, 2, and 3 axes (instead of x, y, and z axes). Therefore, a satellite's inertia matrix with respect to its principal axes is

$$\mathbf{I} = \begin{bmatrix} I_1 & 0 & 0 \\ 0 & I_2 & 0 \\ 0 & 0 & I_3 \end{bmatrix} \tag{12.24}$$

The diagonal elements I_1, I_2, and I_3 are the *principal moments of inertia*. It is important to remember that (1) the center of mass is the origin of the principal axes, and (2) the principal axes are body-fixed coordinates that rotate with the satellite.

A satellite's angular momentum with respect to the principal axes is

$$\mathbf{H} = \mathbf{I}\boldsymbol{\omega} = \begin{bmatrix} I_1 & 0 & 0 \\ 0 & I_2 & 0 \\ 0 & 0 & I_3 \end{bmatrix} \begin{bmatrix} \omega_1 \\ \omega_2 \\ \omega_3 \end{bmatrix} = \begin{bmatrix} I_1\omega_1 \\ I_2\omega_2 \\ I_3\omega_3 \end{bmatrix} \tag{12.25}$$

where ω_1, ω_2, and ω_3 are the angular velocity components along the principal axes (note that we have dropped the "body" subscript on $\mathbf{H}$ because we are only concerned with rotational motion). Another way to express Eq. (12.25) is

$$\mathbf{H} = I_1\omega_1\mathbf{u}_1 + I_2\omega_2\mathbf{u}_2 + I_3\omega_3\mathbf{u}_3 \tag{12.26}$$

where $\mathbf{u}_1$, $\mathbf{u}_2$, and $\mathbf{u}_3$ are unit vectors along the principal axes.

For a homogeneous body of revolution, it is fairly easy to select the principal axes. It is always possible to determine the principal axes where the products of inertia are zero. From linear algebra, we know that the eigenvectors of an arbitrary inertia matrix will determine the coordinates of the principal axes and the associated eigenvalues are the principal moments of inertia. We will not pursue the eigenvector problem and the determination of principal axes (see Kaplan [1; pp. 43–46] for details). Because using principal axes greatly simplifies the attitude dynamics and control problem, we will use them for the remainder of this chapter and Chapter 13.

12.2.3 Rotational Kinetic Energy

Just as the total energy of a point-mass spacecraft aided the solution of orbital mechanics problems, a satellite's *rotational kinetic energy* helps us solve attitude dynamics problems. The rotational kinetic energy of a rigid body consisting of particles m_i is

$$T_{\text{rot}} = \sum \frac{1}{2}m_i\mathbf{v}_i\cdot\mathbf{v}_i \tag{12.27}$$

where $\mathbf{v}_i$ is the velocity of incremental mass m_i due to the body's rotation:

$$\mathbf{v}_i = \boldsymbol{\omega}\times\mathbf{r}_i \tag{12.28}$$

Equation (12.28) is equal to Eq. (12.7) with zero absolute velocity of the body's center of mass, or $\dot{\mathbf{R}}_{\text{cm}} = \mathbf{0}$ (i.e., we are only considering *rotational* kinetic energy here and *not* the kinetic energy due to the speed of the body's center of mass). Substituting Eq. (12.28) for the *second* velocity vector term in Eq. (12.27) yields

$$T_{\text{rot}} = \frac{1}{2}\sum (m_i\mathbf{v}_i)\cdot(\boldsymbol{\omega}\times\mathbf{r}_i) \tag{12.29}$$

Using the scalar triple product [see Eq. (B.27) in Appendix B], we can swap the vectors so that Eq. (12.29) becomes

$$T_{\text{rot}} = \frac{1}{2}\sum \boldsymbol{\omega}\cdot(\mathbf{r}_i\times m_i\mathbf{v}_i) \tag{12.30}$$

Next, factor vector $\boldsymbol{\omega}$ out from the summation and substitute Eq. (12.28) for velocity $\mathbf{v}_i$ in Eq. (12.30):

$$T_{\text{rot}} = \frac{1}{2}\boldsymbol{\omega}\cdot\sum \mathbf{r}_i\times m_i(\boldsymbol{\omega}\times\mathbf{r}_i) \tag{12.31}$$

Note that the summation term in Eq. (12.31) is the *body* angular momentum; see the third term on the right-hand side of Eq. (12.10). Thus, Eq. (12.31) becomes

$$T_{\text{rot}} = \frac{1}{2}\boldsymbol{\omega}\cdot\mathbf{H} \tag{12.32}$$

Finally, we may substitute Eq. (12.18) for angular momentum **H** to yield

$$T_{\text{rot}} = \frac{1}{2}\boldsymbol{\omega} \cdot \mathbf{I}\boldsymbol{\omega} \tag{12.33}$$

It is important to note that rotational kinetic energy T_{rot} is a *scalar*, as illustrated by Eqs. (12.32) and (12.33) which show the dot product between angular velocity vector $\boldsymbol{\omega}$ and angular momentum vector $\mathbf{H} = \mathbf{I}\boldsymbol{\omega}$. Suppose we selected a body frame where one axis is instantaneously aligned with the angular velocity vector $\boldsymbol{\omega}$. In this case, the rotational kinetic energy is

$$T_{\text{rot}} = \frac{1}{2}I_{\text{spin}}\omega^2 \tag{12.34}$$

where I_{spin} is the satellite's moment of inertia about its "spin axis" or the axis of rotation (i.e., along $\boldsymbol{\omega}$), and ω is the magnitude of the angular velocity. Rotational kinetic energy is analogous to translational kinetic energy $mv^2/2$. However, note that a translating fixed-mass particle can change its kinetic energy only by changing its speed v, whereas a rotating body can change its kinetic energy by altering its angular velocity or its rotation axis (i.e., rotational inertia I_{spin}). Equation (12.34) is not very useful because I_{spin} is continuously changing due to the continuous change in the axis of rotation (i.e., vector $\boldsymbol{\omega}$). However, if a satellite is exhibiting a pure spin about a body axis (say the 3 axis), then the associated rotational kinetic energy is $T_{\text{rot}} = I_3\omega_3^2/2$.

To illustrate this concept, consider a thin, pencil-shaped cylinder called a *prolate body* shown in Figure 12.6. A prolate body is a solid of revolution about its *minor axis* or axis of smallest moment of inertia. For the prolate body in Figure 12.6, the 3 axis is the minor axis because $I_3 < I_1$ and $I_3 < I_2$ (of course, $I_1 = I_2$ for the cylinder in Figure 12.6). Figure 12.6a shows the prolate body spinning about its (minor) 3 axis with angular velocity ω_3 whereas Figure 12.6b shows the prolate body spinning end-over-end about its *major axis* (the 1 axis) with the same angular velocity, $\omega_1 = \omega_3$. For the minor-axis rotation (Figure 12.6a), we have angular momentum $H_3 = I_3\omega_3$ that is *less* than the angular momentum for the major-axis rotation (Figure 12.6b), $H_1 = I_1\omega_1$. Similarly, the rotational

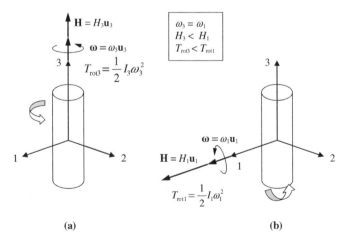

Figure 12.6 Prolate body: (a) spin about the minor 3 axis; and (b) spin about the major 1 axis.

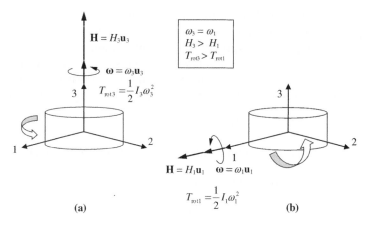

Figure 12.7 Oblate body: (a) spin about the major 3 axis; and (b) spin about the minor 1 axis.

kinetic energy for the minor-axis rotation (T_{rot3}) is *smaller* than the kinetic energy associated with the major-axis rotation (T_{rot1}) even though the angular velocities are the same.

Figure 12.7 shows a flat, squat cylinder called an *oblate body*. An oblate body is a solid of revolution about its *major axis* (axis of greatest moment of inertia; i.e., $I_3 > I_1$ and $I_3 > I_2$). Similar to Figure 12.6, the two scenarios in Figure 12.7 show rotations about the major and minor axes with the same angular velocity, $\omega_1 = \omega_3$. For the major-axis rotation of an oblate body, we have greater angular momentum ($H_3 > H_1$) and greater rotational kinetic energy ($T_{rot3} > T_{rot1}$) compared with the minor-axis rotation at the same angular velocity.

Figures 12.6 and 12.7 demonstrate that rotational kinetic energy depends on the angular velocity, the spin axis, and the body's inertia distribution (i.e., a prolate vs. oblate body). Unlike the total energy of a point-mass satellite moving in a two-body gravitational field, the *rotational* kinetic energy of a satellite may dissipate over time due to the motion of flexible appendages or internal damping devices. We will revisit rotational kinetic energy in a later section in order to characterize the stability of a spinning satellite.

12.2.4 Euler's Moment Equations

Thus far we have developed expressions for rotational angular momentum and rotational kinetic energy. These expressions are simplified by using the principal axes as the body-fixed coordinate frame. The next step is to characterize the motion of a rotating satellite; in other words, we want to solve the governing differential equations stemming from the rotational equivalent of Newton's second law.

Recall that Newton's second law may be stated as follows: the sum of external forces is equal to the time rate of linear momentum in an inertial frame. Similarly, for rotational motion, the sum of external torques (or moments) is equal to the time rate of angular momentum in an inertial frame, or

$$\mathbf{M} = \dot{\mathbf{H}}\big|_{\text{fix}} \tag{12.35}$$

where $\mathbf{M}$ is the summation of all torques or moments about the body's center of mass and the subscript "fix" indicates the time derivative with respect to an inertial (non-rotating) frame. However, our previous expressions for angular momentum, Eqs. (12.18) and (12.25), use a body-fixed frame that rotates with the satellite (i.e., $\mathbf{H} = \mathbf{I}\boldsymbol{\omega}$). Using a body-fixed frame is the only way to ensure a constant inertia matrix $\mathbf{I}$, a very desirable feature! Therefore, we can express the absolute time-rate of angular momentum as

$$\dot{\mathbf{H}}\Big|_{\text{fix}} = \dot{\mathbf{H}}\Big|_{\text{rot}} + \boldsymbol{\omega} \times \mathbf{H} \tag{12.36}$$

The time-rate of angular momentum with respect to the rotating (body) frame is

$$\dot{\mathbf{H}}\Big|_{\text{rot}} = \mathbf{I}\dot{\boldsymbol{\omega}} = I_1\dot{\omega}_1\mathbf{u}_1 + I_2\dot{\omega}_2\mathbf{u}_2 + I_3\dot{\omega}_3\mathbf{u}_3 \tag{12.37}$$

Note that we are using principal axes as the body-fixed frame. The cross-product term (due to the rotation of the principal axes) is

$$\boldsymbol{\omega} \times \mathbf{H} = \begin{vmatrix} \mathbf{u}_1 & \mathbf{u}_2 & \mathbf{u}_3 \\ \omega_1 & \omega_2 & \omega_3 \\ I_1\omega_1 & I_2\omega_2 & I_3\omega_3 \end{vmatrix} \tag{12.38}$$

$$= (I_3\omega_2\omega_3 - I_2\omega_2\omega_3)\mathbf{u}_1 - (I_3\omega_1\omega_3 - I_1\omega_1\omega_3)\mathbf{u}_2 + (I_2\omega_1\omega_2 - I_1\omega_1\omega_2)\mathbf{u}_3$$

Summing Eqs. (12.37) and (12.38) and setting the result equal to the external torque $\mathbf{M}$, yields the three component equations along the 1, 2, and 3 principal axes:

$$M_1 = I_1\dot{\omega}_1 + (I_3 - I_2)\omega_2\omega_3 \tag{12.39a}$$

$$M_2 = I_2\dot{\omega}_2 + (I_1 - I_3)\omega_1\omega_3 \tag{12.39b}$$

$$M_3 = I_3\dot{\omega}_3 + (I_2 - I_1)\omega_1\omega_2 \tag{12.39c}$$

Equations (12.39a), (12.39b), and (12.39c) are called *Euler's moment equations*. They are three nonlinear, coupled, first-order differential equations. The external moment vector $\mathbf{M}$ has components M_1, M_2, and M_3 along the principal axes. Solving Euler's moment equations will give us the evolution of the angular velocity components in the body-fixed frame (i.e., $\boldsymbol{\omega} = \omega_1\mathbf{u}_1 + \omega_2\mathbf{u}_2 + \omega_3\mathbf{u}_3$).

12.3 Torque-Free Motion

The primary objective of this chapter is to characterize a satellite's attitude dynamics or rotational motion about its center of mass. Euler's moment equations (12.39) are the governing dynamical equations of motion for a rotating satellite. However, they are nonlinear and a closed-form solution cannot be obtained for an arbitrary moment vector $\mathbf{M}$. A closed-form solution for the angular motion can be obtained in the absence of external torques.

Satellites are frequently "spin stabilized" by imparting an external torque until the rotating body achieves a desired angular velocity about a particular principal axis. For

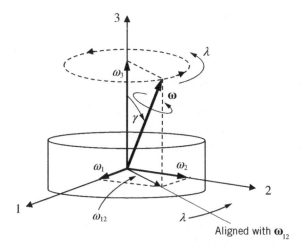

Figure 12.8 Spinning oblate satellite with zero external torques.

example, satellites are often spin stabilized before firing an onboard rocket for an orbital maneuver. Spinning the satellite provides "gyroscopic rigidity" or resistance to disturbance torques caused by misaligned engine thrust. We will investigate the advantages of spin stabilization in a later section. In this section, we will analyze the motion of a spinning satellite with zero external moments, or "torque-free motion."

Let us consider the spinning oblate satellite with principal axes 1, 2, and 3 as shown in Figure 12.8. Note that the angular velocity vector ω has "spin components" ω_1, ω_2, and ω_3 along the principal axes as shown in Figure 12.8. Because the satellite is axisymmetric, we have $I_1 = I_2$. Assuming that all external torque components are zero ($M_1 = M_2 = M_3 = 0$), Euler's moment equations (12.39) become

$$I_1\dot{\omega}_1 + (I_3 - I_2)\omega_2\omega_3 = 0 \tag{12.40}$$

$$I_2\dot{\omega}_2 + (I_1 - I_3)\omega_1\omega_3 = 0 \tag{12.41}$$

$$I_3\dot{\omega}_3 = 0 \tag{12.42}$$

Clearly, Eq. (12.42) shows that the angular velocity component about the 3 axis does not change (i.e., $\omega_3 = n =$ constant). Next, let us define the constant

$$\lambda = \frac{I_3 - I_2}{I_1}n = \frac{I_3 - I_1}{I_2}n \tag{12.43}$$

Equation (12.43) is satisfied because $I_1 = I_2$. Note that λ has units of angular velocity or radians per second. Using Eq. (12.43) in the 1- and 2-axis Euler equations (12.40) and (12.41), we obtain

$$\dot{\omega}_1 + \lambda\omega_2 = 0 \tag{12.44}$$

$$\dot{\omega}_2 - \lambda\omega_1 = 0 \tag{12.45}$$

Equations (12.44) and (12.45) are linear and coupled. One way to obtain a solution is to take the time derivative of Eq. (12.44) to yield $\ddot{\omega}_1 + \lambda\dot{\omega}_2 = 0$ and use Eq. (12.45) to substitute $\dot{\omega}_2 = \lambda\omega_1$. The result is an uncoupled second-order differential equation for angular velocity about the 1 axis:

$$\ddot{\omega}_1 + \lambda^2\omega_1 = 0 \tag{12.46}$$

The general solution to Eq. (12.46) is an undamped harmonic oscillator:

$$\omega_1(t) = C_1\cos\lambda t + C_2\sin\lambda t \tag{12.47}$$

where C_1 and C_2 are constants that depend on the initial conditions $\omega_1(0)$ and $\dot{\omega}_1(0)$. It is not difficult to see that the two constants are $C_1 = \omega_1(0)$ and $C_2 = \dot{\omega}_1(0)/\lambda$; therefore, the 1-axis spin component is

$$\omega_1(t) = \omega_1(0)\cos\lambda t + \frac{\dot{\omega}_1(0)}{\lambda}\sin\lambda t \tag{12.48}$$

Solving Eq. (12.44) for the 2-axis spin component, we obtain $\omega_2 = -\dot{\omega}_1/\lambda$. Thus, we can differentiate Eq. (12.48) and divide by $-\lambda$ to yield

$$\omega_2(t) = \omega_1(0)\sin\lambda t - \frac{\dot{\omega}_1(0)}{\lambda}\cos\lambda t \tag{12.49}$$

The projection of vector $\boldsymbol{\omega}$ onto the 1–2 plane is component ω_{12} (see Figure 12.8), and it is the hypotenuse of the right triangle formed by legs ω_1 and ω_2

$$\omega_{12}^2 = \omega_1^2 + \omega_2^2 \tag{12.50}$$

Squaring Eqs. (12.48) and (12.49) and adding yields

$$\omega_{12}^2 = \omega_1^2(0) + \frac{\dot{\omega}_1^2(0)}{\lambda^2} = \text{constant} \tag{12.51}$$

Therefore, the length of projection ω_{12} remains constant.

Another way to determine the spin components ω_1 and ω_2 is to project the vector $\boldsymbol{\omega}_{12}$ onto the 1 and 2 axes; the result is

$$\omega_1(t) = \omega_{12}\cos(\lambda t + \beta) \tag{12.52}$$

$$\omega_2(t) = \omega_{12}\sin(\lambda t + \beta) \tag{12.53}$$

where β is a phase angle. Clearly, Eq. (12.50) holds for solution set (12.52) and (12.53).

We can summarize our torque-free motion analysis for the oblate satellite shown in Figure 12.8:

1) Angular-velocity projection vector $\boldsymbol{\omega}_{12}$ has constant magnitude and rotates in the 1–2 plane at constant angular velocity λ.
2) Because $I_3 > I_1$ for an oblate cylinder, Eq. (12.43) shows that angular velocity $\lambda > 0$ and thus the vector $\boldsymbol{\omega}_{12}$ rotates counterclockwise about the 3 axis (see Figure 12.8).
3) The angular velocity vector $\boldsymbol{\omega}$ exhibits a counterclockwise *coning motion* about the symmetric 3 axis as shown by the dashed circle in Figure 12.8. An observer fixed to the satellite's 3 axis would see the vector $\boldsymbol{\omega}$ trace a cone with a period of $2\pi/\lambda$.

Figure 12.9 shows a spinning prolate body. Equations (12.48) and (12.49) or Eqs. (12.52) and (12.53) still provide the general solutions for the 1- and 2-axis spin components. Because $I_3 < I_1$ for a prolate cylinder, the angular velocity λ is negative and hence the

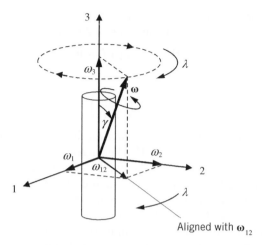

Figure 12.9 Spinning prolate satellite with zero external torques.

vector ω_{12} rotates clockwise as shown in Figure 12.9. Consequently, the coning motion of ω is clockwise for a prolate body.

Our torque-free motion analysis thus far has identified the coning motion of the angular velocity vector ω relative to the rotating body-fixed frame. We have not (yet) established an inertial vector. Equation (12.35) shows that when $\mathbf{M} = \mathbf{0}$, the angular momentum vector $\mathbf{H}$ is constant and fixed in inertial space. The angular momentum vector (in body-fixed coordinates) for an axisymmetric satellite is

$$\mathbf{H} = \mathbf{I}\omega = I_1\omega_1\mathbf{u}_1 + I_1\omega_2\mathbf{u}_2 + I_3\omega_3\mathbf{u}_3 \tag{12.54}$$

Note that we have substituted $I_1 = I_2$ in Eq. (12.54). It turns out that vector $\mathbf{H}$ is in a plane that contains the angular velocity vector ω and the 3-axis unit vector $\mathbf{u}_3$. We can use the scalar triple product to show that vectors $\mathbf{u}_3$, $\mathbf{H}$, and ω are coplanar:

$$\mathbf{u}_3 \cdot (\mathbf{H} \times \omega) = \begin{bmatrix} 0 \\ 0 \\ 1 \end{bmatrix} \cdot \begin{vmatrix} \mathbf{u}_1 & \mathbf{u}_2 & \mathbf{u}_3 \\ I_1\omega_1 & I_1\omega_2 & I_3\omega_3 \\ \omega_1 & \omega_2 & \omega_3 \end{vmatrix} \tag{12.55}$$

Because the third element of $\mathbf{u}_3$ is the sole non-zero element, we only need to compute the 3-axis component of the cross product $\mathbf{H} \times \omega$. Using Eq. (12.55), we see that the $\mathbf{u}_3$ component of the cross product is $I_1\omega_1\omega_2 - I_1\omega_1\omega_2 = 0$. Therefore, the scalar triple product is zero (regardless of the values of the spin components ω_1 and ω_2) and vectors $\mathbf{u}_3$, $\mathbf{H}$, and ω are *always* in the same plane. We can compute the angle between $\mathbf{u}_3$ (the symmetric body axis) and angular velocity ω using the trigonometric relationship:

$$\gamma = \tan^{-1}\left(\frac{\omega_{12}}{\omega_3}\right) = \tan^{-1}\left(\frac{\omega_{12}}{n}\right) \tag{12.56}$$

Note that although we are using symbol γ here it is *not* the flight-path angle. The angle γ (shown in Figures 12.8 and 12.9) is *constant* because ω_{12} and n are constants. In a similar fashion, the angle between the 3 axis and $\mathbf{H}$ can be determined by

$$\theta = \tan^{-1}\left(\frac{H_{12}}{H_3}\right) \tag{12.57}$$

Again, although we use symbol θ here it is *not* the true anomaly. The magnitude H_{12} can be computed from the projection of $\mathbf{H}$ onto the 1–2 plane:

$$\mathbf{H}_{12} = \mathbf{H}_1 + \mathbf{H}_2 = I_1\omega_1\mathbf{u}_1 + I_1\omega_2\mathbf{u}_2 = I_1\boldsymbol{\omega}_{12} \tag{12.58}$$

Using this result, Eq. (12.57) becomes

$$\theta = \tan^{-1}\left(\frac{I_1\omega_{12}}{I_3 n}\right) \tag{12.59}$$

The angle θ is also constant. It is called the *nutation angle*, and it determines the "tilt" of the 3 axis relative to the inertially fixed vector $\mathbf{H}$. The tangents of γ and θ differ by the inertia ratio I_1/I_3. Therefore, when $I_1 < I_3$ (oblate body), the nutation angle θ is less than γ and angular momentum $\mathbf{H}$ lies between the 3 axis and the angular velocity $\boldsymbol{\omega}$ (Figure 12.10).

Figure 12.11 attempts to show the complex "coning" torque-free motion for a spinning oblate satellite. First, the reader should note that the angular momentum vector $\mathbf{H}$ is *fixed*; it does not move or rotate relative to an inertial frame. Figure 12.11 shows $\mathbf{H}$ as a vertical vector so that the reader can more easily visualize its fixed direction. Secondly,

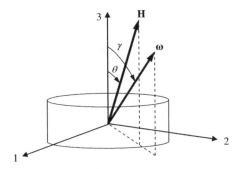

Figure 12.10 Spinning oblate satellite with zero external torques: angular momentum **H** and nutation angle θ.

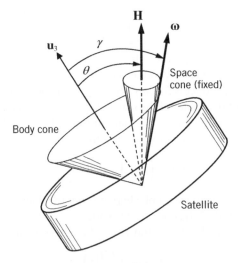

Figure 12.11 Spinning oblate satellite with zero external torques: body cone rolls along fixed space cone.

our analysis has shown that body axis $\mathbf{u}_3$, angular velocity $\boldsymbol{\omega}$, and angular momentum $\mathbf{H}$ always remain in the same plane. Because $\mathbf{H}$ is fixed in space, vectors $\mathbf{u}_3$ and $\boldsymbol{\omega}$ rotate about $\mathbf{H}$ with angular velocity λ. Furthermore, $\boldsymbol{\omega}$ traces a cone (with half angle γ) around the body axis $\mathbf{u}_3$ in a counterclockwise direction with angular velocity λ (see Figure 12.8). This conical trace is called the *body cone* and it is fixed to the satellite's body axes. The angular velocity vector $\boldsymbol{\omega}$ also traces a smaller cone (the *space cone* shown in Figure 12.11) about fixed vector $\mathbf{H}$. The space cone has half angle $\gamma - \theta$ and is fixed in inertial space. Thus, the body cone "rolls" along the fixed space cone in a counterclockwise direction such that $\boldsymbol{\omega}$ is the contact line between the two cones (for an oblate disk, the fixed skinny space cone is always inside the rotating broad body cone). An observer in a fixed frame would see the satellite "wobble" as the body-axes rotate about the angular momentum $\mathbf{H}$. In summary, vector $\mathbf{H}$ is fixed in inertial space, the plane containing $\mathbf{u}_3$, $\boldsymbol{\omega}$, and $\mathbf{H}$ rotates about fixed direction $\mathbf{H}$, and orientation angles γ and θ remain constant. This discussion should reinforce the notion that angular momentum $\mathbf{H}$ and angular velocity $\boldsymbol{\omega}$ are aligned only for a "pure spin" about the symmetric 3 axis. For a pure spin about the 3 axis, we have $\gamma = \theta = 0$ and the coning motion vanishes.

Figure 12.12 shows the coning torque-free motion for a spinning prolate (pencil-shaped) satellite. Because the prolate satellite's symmetry axis is the minor axis ($I_3 < I_1$), the nutation angle θ is greater than angle γ; see Eq. (12.59). Consequently, the angular velocity vector $\boldsymbol{\omega}$ is between $\mathbf{H}$ and the 3 axis as shown in Figure 12.12. For a prolate satellite, the body cone is *outside* the space cone (vector $\boldsymbol{\omega}$ is still the contact line between the two cones). In addition, Eq. (12.43) shows that the coning frequency λ is negative; in other words, the body cone rolls in a clockwise direction about the fixed space cone.

12.3.1 Euler Angle Rates

Our analysis thus far has not fully described the motion of the 123 body axes with respect to an inertial frame. Knowledge of the satellite's body-frame orientation is obviously important because its instruments (antennas, cameras, etc.) have known body coordinates.

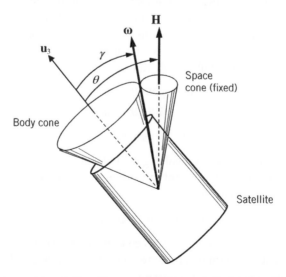

Figure 12.12 Spinning prolate satellite with zero external torques: body cone rolls along fixed space cone.

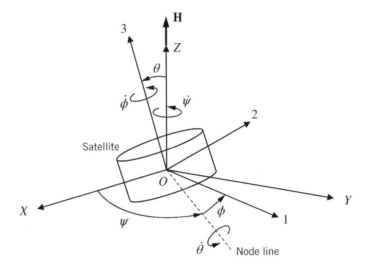

Figure 12.13 Euler angles and Euler angle rates.

It is customary to use *Euler angles* (ψ, θ, and ϕ) to define the angular orientation of the satellite's body axes relative to an inertial frame. Figure 12.13 shows the Euler angles and the 123 body axes. The $OXYZ$ coordinates are a non-rotating inertial frame where the $+Z$ axis is aligned with the angular momentum vector **H** (remember that for torque-free motion, **H** is constant in magnitude and direction). We can align the $OXYZ$ frame with the body-fixed 123 frame by employing three successive rotations: (1) rotate the $OXYZ$ frame by angle ψ about $+Z$ (or **H**); (2) rotate the intermediate frame about the "node line" by angle θ to establish the (symmetric) 3 axis; (3) rotate the intermediate frame about the $+3$ body axis by angle ϕ to establish the complete 123 body frame. The reader should note that this rotation sequence is *identical* to the set of rotations required to align the Earth-centered inertial (ECI) frame with the perifocal (**PQW**) frame for transforming orbital elements to Cartesian coordinates (see Section 3.5). Figure 12.13 also shows the directions of the Euler angle rates: $\dot{\psi}$ is along the $+Z$ axis; $\dot{\theta}$ is along the node line; and $\dot{\phi}$ is along the 3 axis.

Next, we can use Figure 12.13 to express the angular velocity components along the 123 body axes in terms of the Euler angle rates:

$$\omega_1 = \dot{\psi}\sin\theta\sin\phi \tag{12.60}$$

$$\omega_2 = \dot{\psi}\sin\theta\cos\phi \tag{12.61}$$

$$\omega_3 = \dot{\phi} + \dot{\psi}\cos\theta \tag{12.62}$$

Note that we did not include terms involving the Euler angle rate $\dot{\theta}$ because our previous analysis demonstrated that θ is constant for torque-free motion. Adding the squares of Eqs. (12.60) and (12.61) results in

$$\omega_1^2 + \omega_2^2 = \omega_{12}^2 = \dot{\psi}^2\sin^2\theta \tag{12.63}$$

Because component ω_{12} is constant (and θ is constant), we conclude that the Euler angle rate $\dot{\psi}$ is also constant. Furthermore, because ω_3 is constant [see Eq. (12.42)], Eq. (12.62) shows that the Euler angle rate $\dot{\phi}$ is constant.

Let us characterize the Euler angle rates $\dot{\phi}$ and $\dot{\psi}$. Equations (12.52) and (12.53) provide the 1- and 2-axis spin-component solutions, or the left-hand sides of Eqs. (12.60) and (12.61). Using Eq. (12.63), we can substitute $\dot{\psi}\sin\theta = \omega_{12}$ in the right-hand sides of Eqs. (12.60) and (12.61). The result is

$$\omega_1 = \omega_{12}\cos(\lambda t + \beta) = \omega_{12}\sin\phi \tag{12.64}$$

$$\omega_2 = \omega_{12}\sin(\lambda t + \beta) = \omega_{12}\cos\phi \tag{12.65}$$

which leads to the following relationship between the cosine and sine arguments in Eqs. (12.64) and (12.65).

$$\phi = \frac{\pi}{2} - \lambda t - \beta \tag{12.66}$$

Recall that (constant) phase angle β is determined by the initial conditions $\omega_1(0)$ and $\dot{\omega}_1(0)$. Taking the time derivative of Eq. (12.66) yields the Euler angle rate

$$\dot{\phi} = -\lambda \tag{12.67}$$

Therefore, the Euler angle rate $\dot{\phi}$ has the same magnitude but opposite direction as the coning angular velocity λ [Eq. (12.43) shows that λ depends on the 3-axis spin component n and moments of inertia I_1 and I_3]. Next, we can solve Eq. (12.62) for the Euler angle rate $\dot{\psi}$

$$\dot{\psi} = \frac{\omega_3 - \dot{\phi}}{\cos\theta}$$

$$= \frac{n - \dot{\phi}}{\cos\theta} \tag{12.68}$$

Recall that $\omega_3 = n$. Equation (12.43) shows that $n = I_1\lambda/(I_3 - I_1)$ for an axisymmetrical satellite with $I_1 = I_2$. Using Eq. (12.67), we can substitute $n = I_1\dot{\phi}/(I_1 - I_3)$ into Eq. (12.68) to obtain

$$\dot{\psi} = \frac{\dfrac{I_1}{I_1 - I_3}\dot{\phi} - \dot{\phi}}{\cos\theta}$$

Or,

$$\dot{\psi} = \frac{I_3\dot{\phi}}{(I_1 - I_3)\cos\theta} \tag{12.69}$$

Equation (12.69) is the *precession rate* $\dot{\psi}$ in terms of the inertial spin component $\dot{\phi}$. Referring again to Figure 12.13, we see that precession is the angular velocity of the node line. For an oblate satellite (as depicted in Figure 12.13), we have $I_3 > I_1$ and the precession rate is in the opposite direction of Euler angle spin rate $\dot{\phi}$; this scenario is called *retrograde precession*. For a prolate (pencil-shaped) satellite, $I_3 < I_1$ and the Euler angle rates $\dot{\psi}$ and $\dot{\phi}$ are in the same direction for *direct precession*. Precession is the apparent "wobbling" motion as seen by an inertial observer. Because

$$\left| \frac{I_3}{(I_1 - I_3)\cos\theta} \right| > 1 \tag{12.70}$$

always holds for an oblate satellite, the magnitude of the resulting precession rate $\dot{\psi}$ is greater than the coning angular velocity λ.

As a final note, let us consider the problem of determining the orientation of a spinning satellite in a torque-free environment. Suppose we know the initial Euler angles ψ_0 and ϕ_0 at time $t = 0$. The nutation angle θ_0 can be computed from Eq. (12.59) and knowledge of the moments of inertia and the instantaneous angular velocity vector $\boldsymbol{\omega} = [\omega_1 \ \omega_2 \ \omega_3]^T$ as measured in the body frame using sensors (gyroscopes) mounted along the 1, 2, and 3 axes. Of course, the nutation angle θ_0 remains constant during the ensuing coning/wobbling motion. Next, we can determine the *constant* Euler angle spin rate $\dot{\phi}$ from the coning frequency, Eq. (12.43), using the constant 3-axis spin component n. Equation (12.69) allows us to compute the constant precession rate $\dot{\psi}$. Using angular rate information, we can compute the Euler angles at any future time

$$\psi(t) = \psi_0 + \dot{\psi}t \tag{12.71}$$

$$\theta(t) = \theta_0 \tag{12.72}$$

$$\phi(t) = \phi_0 + \dot{\phi}t \tag{12.73}$$

Knowledge of the instantaneous Euler angles at any time t allows us to compute a rotation matrix that transforms an arbitrary vector in the inertial $OXYZ$ frame to the rotating 123 body frame. A more practical scenario involves transforming a vector expressed in the 123 body frame (such as the position vector of a camera or antenna) to the inertial $OXYZ$ frame. This transformation requires the inverse rotation matrix, or the rotation matrix $\widetilde{\mathbf{R}}$ defined by Eq. (3.39) where we substitute the Euler angles (ψ, θ, ϕ) for the orbital elements longitude of the ascending node, inclination, and argument of periapsis, respectively. The rotation matrix for the Euler angles is

$$\widetilde{\mathbf{R}} = \begin{bmatrix} c_\psi c_\phi - s_\psi s_\phi c_\theta & -c_\psi s_\phi - s_\psi c_\phi c_\theta & s_\psi s_\theta \\ s_\psi c_\phi + c_\psi s_\phi c_\theta & -s_\psi s_\phi + c_\psi c_\phi c_\theta & -c_\psi s_\theta \\ s_\phi s_\theta & c_\phi s_\theta & c_\theta \end{bmatrix} \tag{12.74}$$

We have used the short-hand notation $c_\alpha = \cos\alpha$ and $s_\alpha = \sin\alpha$ for the cosine and sine of the three rotation angles. Multiplying rotation matrix $\widetilde{\mathbf{R}}$ and *any* vector expressed in the rotating 123 body frame will transform it to inertial coordinates.

Example 12.2   The Intelsat II series of communication satellites were deployed in geostationary-equatorial orbit (GEO) in the late 1960s. Intelsat II was an oblate cylinder with a diameter of 1.42 m and height of 0.67 m. Before firing the apogee rocket engine for insertion into GEO, the Intelsat II was spun along its (symmetric) 3 axis at 12.57 rad/s (about 120 rpm). The mass of the Intelsat II was 162 kg before the apogee engine was fired. Let us assume that a pure spin about the 3 axis was *not* achieved and that at time $t = 0$ the angular velocity is $\boldsymbol{\omega}(0) = 0.3419\mathbf{u}_1 - 0.1974\mathbf{u}_2 + 12.5638\mathbf{u}_3$ rad/s along the 123 principal body axes as shown in Figure 12.14.

a) Compute the angular momentum vector $\mathbf{H}$ in body coordinates.
b) Determine the nutation angle θ and angle γ.
c) Compute the coning period and compare it with the spin period.
d) Compute the precession period and compare it with the spin period.

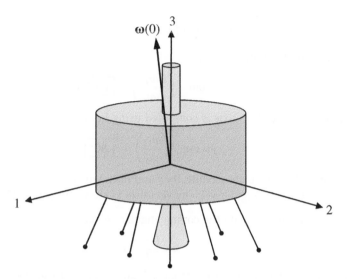

Figure 12.14 Intelsat II and its initial angular velocity (Example 12.2).

a) We know that angular momentum is $\mathbf{H} = \mathbf{I}\boldsymbol{\omega}$ and we are given the angular velocity vector $\boldsymbol{\omega}$ in 123 (or body) coordinates. Therefore, we need the inertia matrix of the Intelsat II. Assuming a homogeneous cylinder, the moment of inertia about the symmetric 3 axis is

$$I_3 = \frac{1}{2}mR^2 = 40.8321 \text{ kg-m}^2$$

where the cylinder radius is $R = (1/2)(1.42 \text{ m}) = 0.71 \text{ m}$ and mass is $m = 162$ kg. The moments of inertia about the transverse 1 and 2 axes are

$$I_1 = I_2 = \frac{1}{12}m\left(3R^2 + h^2\right) = 26.4762 \text{ kg-m}^2$$

where height $h = 0.67$ m. Because $I_3 > I_1$, the Intelsat II is an oblate ("flat") cylinder. Angular momentum is

$$\mathbf{H} = \mathbf{I}\boldsymbol{\omega} = \begin{bmatrix} 26.4762 & 0 & 0 \\ 0 & 26.4762 & 0 \\ 0 & 0 & 40.8321 \end{bmatrix} \begin{bmatrix} 0.3419 \\ -0.1974 \\ 12.5638 \end{bmatrix} = \begin{bmatrix} 9.052 \\ -5.226 \\ 513.006 \end{bmatrix} \text{ kg-m}^2/\text{s}$$

Or, we can express angular momentum as $\mathbf{H} = 9.052\mathbf{u}_1 - 5.226\mathbf{u}_2 + 513.006\mathbf{u}_3$ kg-m^2/s.

b) The nutation angle is determined from the body-axis components of $\mathbf{H}$; we may use either Eq. (12.57) or (12.59). Thus, we need to compute either the 1–2 projection H_{12} or ω_{12}. The projection of $\boldsymbol{\omega}$ onto the 1–2 plane is

$$\omega_{12} = \sqrt{\omega_1^2 + \omega_2^2} = 0.3948 \text{ rad/s}$$

Using Eq. (12.59), the nutation angle is

$$\theta = \tan^{-1}\left(\frac{I_1\omega_{12}}{I_3\omega_3}\right) = \boxed{1.17°}$$

The angle between ω and the 3 axis is computed using Eq. (12.56)

$$\gamma = \tan^{-1}\left(\frac{\omega_{12}}{\omega_3}\right) = \boxed{1.80°}$$

Because the Intelsat II is an oblate satellite, $\theta < \gamma$ and the angular momentum vector $\mathbf{H}$ is between the 3 axis and the angular velocity vector ω as shown in Figure 12.10.

c) We use Eq. (12.43) to compute the coning angular velocity

$$\lambda = \frac{I_3 - I_2}{I_1} n$$

where $I_1 = I_2 = 26.4762$ kg-m^2, $I_3 = 40.8321$ kg-m^2, and $n = \omega_3 = 12.5638$ rad/s (i.e., the constant spin component along the 3 axis). Using these values, we obtain $\lambda = 6.8123$ rad/s. The period of this coning frequency is

$$\tau_{\text{coning}} = \frac{2\pi}{\lambda} = \boxed{0.922 \text{ s}}$$

Therefore, the ω vector completes one cycle of coning motion in 0.922 s. The "spin period" is

$$\tau_{\text{spin}} = \frac{2\pi}{\omega} = \boxed{0.500 \text{ s}}$$

where $\omega = \|\omega\| = \sqrt{\omega_1^2 + \omega_2^2 + \omega_3^2} = 12.57$ rad/s. The spin period is the time required for the satellite to make one complete revolution (recall that the satellite makes two revolutions per second, or 120 rpm). Therefore, the ratio of the coning-to-spin period is $0.922/0.5 = 1.84$; in other words, the Intelsat II makes 1.84 revolutions during one coning cycle.

d) The precession rate is computed using Eq. (12.69)

$$\dot{\psi} = \frac{I_3\dot{\phi}}{(I_1 - I_3)\cos\theta}$$

where the Euler angle spin rate is $\dot{\phi} = -\lambda = -6.8123$ rad/s. Using this value and the moments of inertia and $\cos\theta = 0.99979$, we obtain

$$\dot{\psi} = 19.3802 \text{ rad/s}$$

The precession period is

$$\tau_{\text{precess}} = \frac{2\pi}{\dot{\psi}} = \boxed{0.324 \text{ s}}$$

As expected, the precession rate is greater than the coning angular velocity; it is also greater than the angular velocity magnitude. The ratio of the precession wobble-to-spin period is $0.324/0.5 = 0.65$ and the Intelsat II makes 0.65 revolutions during one "wobble" precession cycle.

As previously mentioned in this section, satellites are often "spin stabilized" before firing an onboard rocket. The Intelsat II was spin stabilized in order to provide "gyroscopic resistance" to any disturbance torques encountered during the second Hohmann-transfer burn.

Example 12.3 An axisymmetric oblate satellite has the principal moments of inertia $I_1 = I_2 = 50$ kg-m^2 and $I_3 = 90$ kg-m^2. Its initial angular velocity vector (in body-fixed 123 coordinates) is

$$\boldsymbol{\omega}(0) = \begin{bmatrix} 0 \\ 0.18 \\ 0.30 \end{bmatrix} = 0.18\mathbf{u}_2 + 0.30\mathbf{u}_3 \text{ rad/s}$$

The initial Euler angles are $\psi_0 = 0$ and $\phi_0 = 0$. The satellite is equipped with a camera with body-fixed position vector $\mathbf{r}_{cam} = 0.9\mathbf{u}_1$ m (i.e., the camera points along the satellite's 1 axis) as shown in Figure 12.15.

a) Determine the nutation angle θ and angle γ.
b) Compute the angular momentum vector **H** in inertial coordinates at time $t = 0$.
c) Determine the satellite's angular velocity vector $\boldsymbol{\omega}$ in 123 body coordinates at time $t = 25$ s.
d) Determine the position vector of the satellite's camera at time $t = 0$ and $t = 25$ s as expressed in the inertial $OXYZ$ frame.
e) Recalculate the angular momentum vector **H** in inertial coordinates at time $t = 25$ s using the satellite's angular velocity vector $\boldsymbol{\omega}$ (expressed in body coordinates) and show that it remains constant.

a) Figure 12.15 shows the initial configuration of the spinning oblate satellite. Note that there is no angular velocity component along the 1 axis at $t = 0$ and therefore $\omega_{12} = \omega_2(0) = 0.18$ rad/s. Using Eq. (12.56), we can determine the angle γ

$$\gamma = \tan^{-1}\left(\frac{\omega_{12}}{\omega_3}\right) = \tan^{-1}\left(\frac{0.18}{0.30}\right) = \boxed{30.964°}$$

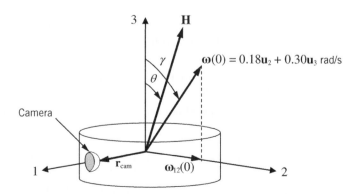

Figure 12.15 Spinning oblate satellite at time $t = 0$ (Example 12.3).

Using Eq. (12.59), the nutation angle is

$$\theta = \tan^{-1}\left(\frac{I_1\omega_{12}}{I_3\omega_3}\right) = \tan^{-1}\left(\frac{9}{27}\right) = \boxed{18.435°}$$

b) First, we compute the angular momentum vector in 123 body coordinates:

$$\mathbf{H}_{body} = \mathbf{I}\boldsymbol{\omega}_0 = \begin{bmatrix} 50 & 0 & 0 \\ 0 & 50 & 0 \\ 0 & 0 & 90 \end{bmatrix}\begin{bmatrix} 0 \\ 0.18 \\ 0.30 \end{bmatrix} = \begin{bmatrix} 0 \\ 9 \\ 27 \end{bmatrix} \text{kg-m}^2/\text{s}$$

To obtain $\mathbf{H}$ in the inertial frame, we need the rotation matrix $\tilde{\mathbf{R}}$ evaluated with the appropriate Euler angles. Using Eq. (12.74) with *initial* Euler angles $\psi_0 = 0$, $\phi_0 = 0$, and $\theta_0 = 18.435°$, we obtain

$$\tilde{\mathbf{R}}_0 = \begin{bmatrix} C_\psi C_\phi - S_\psi S_\phi C_\theta & -C_\psi S_\phi - S_\psi C_\phi C_\theta & S_\psi S_\theta \\ S_\psi C_\phi + C_\psi S_\phi C_\theta & -S_\psi S_\phi + C_\psi C_\phi C_\theta & -C_\psi S_\theta \\ S_\phi S_\theta & C_\phi S_\theta & C_\theta \end{bmatrix} = \begin{bmatrix} 1 & 0 & 0 \\ 0 & 0.9487 & -0.3162 \\ 0 & 0.3162 & 0.9487 \end{bmatrix}$$

We see that the initial rotation matrix is a pure rotation about the inertial X axis; intuitively this makes sense if we revisit Figure 12.13 with $\psi_0 = 0$ and $\phi_0 = 0$. Hence, the angular momentum in the inertial frame $OXYZ$ is

$$\mathbf{H}_{fix} = \tilde{\mathbf{R}}_0\mathbf{H}_{body} = \begin{bmatrix} 1 & 0 & 0 \\ 0 & 0.9487 & -0.3162 \\ 0 & 0.3162 & 0.9487 \end{bmatrix}\begin{bmatrix} 0 \\ 9 \\ 27 \end{bmatrix} = \begin{bmatrix} 0 \\ 0 \\ 28.4605 \end{bmatrix} \text{kg-m}^2/\text{s}$$

We know that for torque-free motion the satellite's angular momentum vector $\mathbf{H}_{fix}$ is constant; we will verify this fact in (e).

c) We will show two ways to determine the satellite's angular velocity at a future time. Because the satellite is axisymmetric ($I_1 = I_2$), we may use Eqs. (12.52) and (12.53) to project the 1–2 spin components to a future time:

$$\omega_1(t) = \omega_{12}\cos(\lambda t + \beta), \qquad \omega_2(t) = \omega_{12}\sin(\lambda t + \beta)$$

Because $\omega_{12} = \omega_2(0) = 0.18$ rad/s (as shown in Figure 12.15), the phase angle must be $\beta = 90°$ ($= \pi/2$ rad). Next, we use Eq. (12.43) to find the "coning-motion frequency"

$$\lambda = \frac{I_3 - I_2}{I_1}n = \frac{90 - 50}{50}(0.3\,\text{rad/s}) = 0.24\ \text{rad/s}$$

where $n = \omega_3$ is the constant spin rate about the 3 axis. Hence, the two spin components at $t = 25$ s are

$$\omega_1(25) = \omega_{12}\cos(0.24\cdot25 + \pi/2) = \boxed{0.0503\,\text{rad/s}\ (=2.88\ \text{deg/s})}$$

$$\omega_2(25) = \omega_{12}\sin(0.24\cdot25 + \pi/2) = \boxed{0.1728\,\text{rad/s}\ (=9.90\ \text{deg/s})}$$

The angular velocity vector at $t = 25$ s is $\boldsymbol{\omega}(25) = 0.0503\mathbf{u}_1 + 0.1728\mathbf{u}_2 + 0.3\mathbf{u}_3$ rad/s.

We can also determine the body-spin components from the Euler rates by using Eqs. (12.60)–(12.62):

$$\omega_1 = \dot{\psi}\sin\theta\sin\phi, \quad \omega_2 = \dot{\psi}\sin\theta\cos\phi, \quad \omega_3 = \dot{\phi} + \dot{\psi}\cos\theta$$

Recall that the nutation angle θ is constant for torque-free motion. Equation (12.67) shows that the Euler angle rate $\dot{\phi}$ is constant and opposite the coning frequency, that is, $\dot{\phi} = -\lambda = -0.24$ rad/s. The precession rate $\dot{\psi}$ is determined using Eq. (12.69):

$$\dot{\psi} = \frac{I_3\dot{\phi}}{(I_1 - I_3)\cos\theta} = 0.5692 \text{ rad/s} \ (= 32.61 \text{ deg/s})$$

The Euler angles ϕ and ψ at time $t = 25$ s are

$$\phi(25) = \phi_0 + \dot{\phi}t = 0 - (0.24)(25) = -6 \text{ rad} = -343.77° \ (= 16.23°)$$
$$\psi(25) = \psi_0 + \dot{\psi}t = 0 + (0.5692)(25) = 14.230 \text{ rad} = 95.32°$$

Using the Euler angles and their rates, the body-frame spin components at $t = 25$ s are

$$\omega_1 = \dot{\psi}\sin\theta\sin\phi = 0.0503 \text{ rad/s}$$
$$\omega_2 = \dot{\psi}\sin\theta\cos\phi = 0.1728 \text{ rad/s}$$
$$\omega_3 = \dot{\phi} + \dot{\psi}\cos\theta = 0.3 \text{ rad/s}$$

These alternate calculations match the previous solutions.

d) We can easily determine the inertial position of the satellite's camera at time $t = 0$ by multiplying matrix $\tilde{\mathbf{R}}_0$ with the camera's *body-fixed* coordinates $\mathbf{r}_{cam}$:

$$\mathbf{r}_{fix} = \tilde{\mathbf{R}}_0 \mathbf{r}_{cam} = \begin{bmatrix} 1 & 0 & 0 \\ 0 & 0.9487 & -0.3162 \\ 0 & 0.3162 & 0.9487 \end{bmatrix} \begin{bmatrix} 0.9 \\ 0 \\ 0 \end{bmatrix} = \begin{bmatrix} 0.9 \\ 0 \\ 0 \end{bmatrix} \text{ m}$$

Because the body 1 axis is initially aligned with the inertial X axis (recall that $\psi_0 = \phi_0 = 0$) and the camera points along the 1 axis, the inertial and rotating coordinates of the camera are coincident at time $t = 0$.

Determining the inertial position of the camera at 25 s requires the rotation matrix $\tilde{\mathbf{R}}$ evaluated with the Euler angles at $t = 25$ s, that is, $\psi(25) = 95.32°$, $\phi(25) = -343.77°$, and $\theta = 18.435°$:

$$\tilde{\mathbf{R}}_{25} = \begin{bmatrix} C_\psi C_\phi - S_\psi S_\phi C_\theta & -C_\psi S_\phi - S_\psi C_\phi C_\theta & S_\psi S_\theta \\ S_\psi C_\phi + C_\psi S_\phi C_\theta & -S_\psi S_\phi + C_\psi C_\phi C_\theta & -C_\psi S_\theta \\ S_\phi S_\theta & C_\phi S_\theta & C_\theta \end{bmatrix} = \begin{bmatrix} -0.3532 & -0.8810 & 0.3149 \\ 0.9314 & -0.3629 & 0.0294 \\ 0.0884 & 0.3036 & 0.9487 \end{bmatrix}$$

The *inertial* position of the camera at $t = 25$ s is

$$\mathbf{r}_{fix}(25) = \tilde{\mathbf{R}}_{25} \mathbf{r}_{cam} = \begin{bmatrix} -0.3532 & -0.8810 & 0.3149 \\ 0.9314 & -0.3629 & 0.0294 \\ 0.0884 & 0.3036 & 0.9487 \end{bmatrix} \begin{bmatrix} 0.9 \\ 0 \\ 0 \end{bmatrix} = \begin{bmatrix} -0.318 \\ 0.838 \\ 0.080 \end{bmatrix} \text{ m}$$

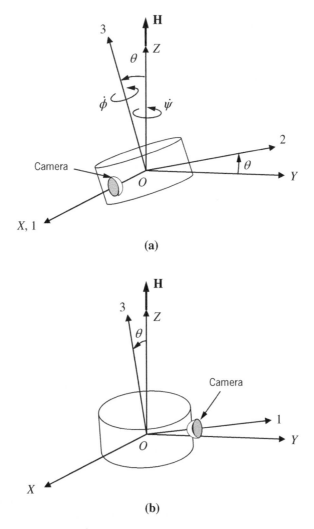

Figure 12.16 Inertial orientation of a spinning oblate satellite: (a) $t = 0$; and (b) $t = 25$ s (Example 12.3).

Figure 12.16 shows the orientation of the satellite with respect to the inertial $OXYZ$ frame at times $t = 0$ and $t = 25$ s. At $t = 0$ (Figure 12.16a), the satellite's camera is pointing along the inertial $+X$ axis which is collinear with the body 1 axis because $\psi_0 = \phi_0 = 0$. Figure 12.16b shows that at $t = 25$ s the camera is primarily pointing along the $+Y$ axis with a negative X-axis component.

e) The angular momentum vector in 123 body coordinates at $t = 25$ s is

$$\mathbf{H}_{body} = \mathbf{I}\boldsymbol{\omega}(25) = \begin{bmatrix} 50 & 0 & 0 \\ 0 & 50 & 0 \\ 0 & 0 & 90 \end{bmatrix} \begin{bmatrix} 0.0503 \\ 0.1728 \\ 0.30 \end{bmatrix} = \begin{bmatrix} 2.515 \\ 8.640 \\ 27 \end{bmatrix} \text{kg-m}^2/\text{s}$$

The angular momentum in the inertial frame $OXYZ$ is determined by multiplying this result by the rotation matrix at $t = 25$ s:

$$\mathbf{H}_{\text{fix}}(25) = \widetilde{\mathbf{R}}_{25}\mathbf{H}_{\text{body}} = \begin{bmatrix} -0.3532 & -0.8810 & 0.3149 \\ 0.9314 & -0.3629 & 0.0294 \\ 0.0884 & 0.3036 & 0.9487 \end{bmatrix} \begin{bmatrix} 2.515 \\ 8.640 \\ 27 \end{bmatrix}$$

$$= \begin{bmatrix} 0 \\ 0 \\ 28.4605 \end{bmatrix} \text{kg-m}^2/\text{s}$$

This calculation verifies that the satellite's angular momentum vector remains constant in a torque-free environment when referenced to the inertial $OXYZ$ frame (see Figures 12.16a and 12.16b).

12.4 Stability and Flexible Bodies

Spinning a rigid body gives it angular momentum and gyroscopic "stiffness" or resistance to a disturbance torque. A spinning toy top is a simple example of "spin stability." Spinning a satellite about a principal axis is a simple method for maintaining a fixed orientation in space. It is therefore natural to investigate the stability of a spinning satellite.

Exactly how do we characterize the "stability" of a spinning satellite? Here we define stability in the "bounded output" sense. Suppose we have a satellite in a "pure spin" about a particular principal body axis such as the 3 axis. If the satellite is perturbed such that the angular velocity vector $\boldsymbol{\omega}$ becomes misaligned with $\mathbf{u}_3$, the spin is said to be stable if vector $\boldsymbol{\omega}$ remains "close" to the 3 axis for all time (in other words, the spin perturbations remain bounded). Hence, the coning torque-free motion of an axisymmetric oblate or prolate satellite (illustrated in Figures 12.11 and 12.12) is stable.

One stability-analysis method is Poinsot's geometric approach based on two ellipsoids constructed from the angular momentum and rotational kinetic energy. The intersection of these two ellipsoids is called a *polhode* and it represents the path of the angular velocity vector $\boldsymbol{\omega}$ as seen in the body frame. We will not present the ellipsoid method here; instead we will discuss stability using a more heuristic approach. The interested reader may consult Kaplan [1; pp. 57–61], Thomson [2; pp. 121–126], or Wiesel [3; pp. 141–151] for details of the ellipsoid method.

12.4.1 Spin Stability about the Principal Axes

We will assess the stability of a satellite's torque-free motion when spinning about a principal axis by analyzing Euler's moment equations (12.39), repeated here with zero moments $M_1 = M_2 = M_3 = 0$:

$$I_1\dot{\omega}_1 + (I_3 - I_2)\omega_2\omega_3 = 0 \tag{12.75}$$

$$I_2\dot{\omega}_2 + (I_1 - I_3)\omega_1\omega_3 = 0 \tag{12.76}$$

$$I_3\dot{\omega}_3 + (I_2 - I_1)\omega_1\omega_2 = 0 \tag{12.77}$$

Let us consider a satellite with three distinct principal moments of inertia, that is, $I_1 \neq I_2 \neq I_3$. It is easy to see that a satellite will maintain a pure spin about *any* principal axis in the absence of any disturbance torques. For example, if $\omega_2 = 10$ rad/s and $\omega_1 = \omega_3 = 0$, we have a "pure spin" about the 2 axis and Eqs. (12.75)–(12.77) show that this spinning state does not change with time. This scenario holds for a pure spin about any principal axis. Our stability analysis will consider the consequences where the pure spin is slightly perturbed from its principal axis.

We characterize stability by employing the classic strategy of linearizing the governing equations of motion. Suppose the satellite is nominally spinning about the 3 axis with constant angular velocity $\boldsymbol{\omega}^* = \begin{bmatrix} 0 & 0 & n \end{bmatrix}^T$ where n is the reference spin rate. We use the superscript asterisk ($*$) to indicate the reference or nominal state just as we did in Chapter 8 when we developed the linear Clohessy–Wiltshire equations for relative orbital motion. Next, we define the satellite's *perturbation state* $\delta\boldsymbol{\omega} = \begin{bmatrix} \delta\omega_1 & \delta\omega_2 & \delta\omega_3 \end{bmatrix}^T$ as the difference between the actual and reference angular velocities:

$$\delta\boldsymbol{\omega} = \boldsymbol{\omega} - \boldsymbol{\omega}^* = \begin{bmatrix} \omega_1 \\ \omega_2 \\ \omega_3 - n \end{bmatrix} \tag{12.78}$$

The perturbations $\delta\omega_1$ and $\delta\omega_2$ are equal to the actual angular velocity components ω_1 and ω_2, respectively, because the reference spin rates about the 1 and 2 body axes are zero. Using Eq. (12.78) to substitute for the angular velocity components in Euler's torque-free equations (12.75)–(12.77) yields

$$I_1\delta\dot{\omega}_1 + (I_3 - I_2)\delta\omega_2(\delta\omega_3 + n) = 0 \tag{12.79}$$

$$I_2\delta\dot{\omega}_2 + (I_1 - I_3)\delta\omega_1(\delta\omega_3 + n) = 0 \tag{12.80}$$

$$I_3\delta\dot{\omega}_3 + (I_2 - I_1)\delta\omega_1\delta\omega_2 = 0 \tag{12.81}$$

Equations (12.79)–(12.81) are Euler's moment equations written in terms of the perturbation spin rates and the reference spin rate n. We can linearize these governing equations by neglecting the products of two (small) perturbations, for example, $\delta\omega_2\delta\omega_3 \approx 0$, $\delta\omega_1\delta\omega_3 \approx 0$, and $\delta\omega_1\delta\omega_2 \approx 0$:

$$I_1\delta\dot{\omega}_1 + (I_3 - I_2)\delta\omega_2 n = 0 \tag{12.82}$$

$$I_2\delta\dot{\omega}_2 + (I_1 - I_3)\delta\omega_1 n = 0 \tag{12.83}$$

$$I_3\delta\dot{\omega}_3 = 0 \tag{12.84}$$

Equation (12.84) is stable; any small spin perturbation from the 3 axis will not change over time. Equations (12.82) and (12.83) are linear and coupled. Taking the time derivative of Eq. (12.82) yields

$$I_1\delta\ddot{\omega}_1 + (I_3 - I_2)\delta\dot{\omega}_2 n = 0 \tag{12.85}$$

Substituting Eq. (12.83) for $\delta\dot{\omega}_2$, we obtain

$$\delta\ddot{\omega}_1 + \left[\frac{(I_3 - I_2)(I_3 - I_1)}{I_1 I_2} n^2\right]\delta\omega_1 = 0 \tag{12.86}$$

Following the same steps, Eq. (12.83) can also be written as a linear second-order differential equation:

$$\delta\ddot{\omega}_2 + \left[\frac{(I_3 - I_2)(I_3 - I_1)}{I_1 I_2} n^2\right]\delta\omega_2 = 0 \tag{12.87}$$

Note that Eq. (12.86) is identical to Eq. (12.46) for the axisymmetric satellite with equal moments of inertia, $I_1 = I_2$. For an axisymmetric satellite, the bracket term in Eqs. (12.86) and (12.87) is *always* positive, and hence the solutions for perturbation spins $\delta\omega_1$ and $\delta\omega_2$ are harmonic functions (sine and cosine functions) and we have the coning motion depicted in Figures 12.11 and 12.12. This scenario is stable.

For a satellite with three distinct moments of inertia, the bracket term in Eqs. (12.86) and (12.87) is positive for two cases: (1) $I_3 > I_2$ and $I_3 > I_1$; and (2) $I_3 < I_2$ and $I_3 < I_1$. In the first case, I_3 is the maximum moment of inertia and the 3 axis is the *major axis*. Therefore, if an oblate satellite is perturbed from a pure spin about its major axis, the subsequent motion is stable (e.g., Figure 12.11). In the second case, I_3 is the minimum moment of inertia and the 3 axis is the *minor axis*; therefore a prolate satellite perturbed from a spin about its minor axis will exhibit stable oscillations (e.g., Figure 12.12). Now consider the case where the 3 axis is the intermediate axis, where $I_1 < I_3 < I_2$ or $I_2 < I_3 < I_1$. For a perturbed spin about the intermediate axis, the bracket term is *negative* and the linear solutions for $\delta\omega_1$ and $\delta\omega_2$ are hyperbolic sine and cosine functions (i.e., the solutions contain a divergent exponential function with a real exponent). This scenario is unstable because the angular velocity perturbations $\delta\omega_1$ and $\delta\omega_2$ are unbounded functions with magnitudes that increase with time.

These stability concepts can be demonstrated by spinning a tennis racket about its three principal axes as seen in Figure 12.17. It is intuitive that a perturbed spin about the minor 3 axis or major 1 axis will simply "wobble" about the original spin axis; that is, the perturbed spin is stable. However, a perturbed spin about the intermediate 2 axis will produce a "tumbling" motion or an unstable spin.

12.4.2 Stability of Flexible Bodies

To this point, we have considered the rotating satellite to be a rigid body where its mass distribution remains fixed relative to the center of mass. In reality, all spacecraft are semi-rigid or flexible to some degree. For example, during a rotational maneuver the solar array panels may bend or twist, whip antennas may flex, and liquid fuel may slosh around inside its tanks. This internal relative motion will dissipate the satellite's kinetic energy in the form of friction (heat). However, because these friction forces are internal forces, the corresponding torques appear in equal-and-opposite pairs and hence angular momentum $\mathbf{H}$ is conserved. Consequently, flexible or semi-rigid spacecraft present the scenario where rotational kinetic energy decreases with time while angular momentum remains constant.

Let us investigate the consequences of dissipating kinetic energy by observing an axisymmetric satellite in a pure spin about a principal axis. The magnitude of angular momentum and the rotational kinetic energy are

$$H = I\omega = \text{constant} \tag{12.88}$$

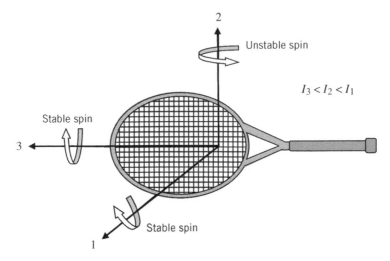

Figure 12.17 Spinning a tennis racket: stable spin about the minor 3 axis; stable spin about the major 1 axis and unstable spin about the intermediate 2 axis.

$$T_{rot} = \frac{1}{2}I\omega^2 \tag{12.89}$$

where it is understood that I is the moment of inertia about the principal axis that coincides with the spin axis. We can solve Eq. (12.88) for angular velocity ($\omega = H/I$) and substitute this result into the kinetic energy expression (12.89) to obtain

$$T_{rot} = \frac{H^2}{2I} \tag{12.90}$$

Suppose that the satellite is initially spinning about its *minor* axis or principal axis with the minimum moment of inertia. Equation (12.90) shows that a minor-axis spin results in the *maximum* kinetic energy for a constant angular momentum H. A flexible spacecraft will dissipate rotational kinetic energy, and therefore T_{rot} must decrease over time while H remains constant. Equation (12.90) shows that kinetic energy is *minimized* when the satellite is in a pure spin about the *major* axis (or principal axis with the maximum moment of inertia). This heuristic argument shows that a prolate (pencil-shaped) *flexible* satellite initially spinning about its longitudinal minor axis will eventually transition to a pure spin about its transverse major axis.

The first US satellite, Explorer 1, is a classic example of a spin-axis transition caused by energy dissipation. Explorer 1 was a prolate satellite with length and radius of approximately 2 m and 0.16 m, respectively. When Explorer 1 reached its orbit, it was initially spinning about its longitudinal minor 3 axis as shown in Figure 12.18a. During this initial spin, the four "whip antennas" on Explorer 1 began to flex and dissipate rotational kinetic energy. Consequently, the nutation angle steadily increased as Explorer 1 began to wobble and change its spin axis. Eventually, the kinetic energy dissipation ceased when Explorer 1 reached its minimum-energy state ($H^2/2I_{max}$) consisting of a "flat spin" about

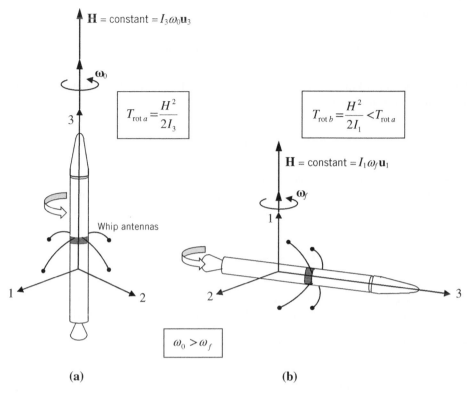

Figure 12.18 Explorer 1 energy dissipation: (a) initial spin about its minor axis; and (b) final "flat spin" about its major axis.

its transverse axis or end-over-end tumbling as shown in Figure 12.18b. The reader should note that the angular momentum vector **H** remained constant as Explorer 1 transitioned from a minor-axis spin (Figure 12.18a) to a major-axis spin (Figure 12.18b). The following example illustrates the spin-axis transition of the Explorer 1 satellite.

Example 12.4   The approximate mass properties of the Explorer 1 satellite are

Mass: $m = 14$ kg
Minor-axis moment of inertia: $I_3 = 0.17$ kg-m^2
Major-axis moment of inertia: $I_1 = I_2 = 5$ kg-m^2

Initially Explorer 1 was spinning at 750 rpm ($\omega_0 = 78.54$ rad/s) about its minor axis. Determine the initial and final rotational kinetic energies and the final spin rate ω_f.
We use Eq. (12.89) to determine the initial kinetic energy:

$$T_{\text{rot},0} = \frac{1}{2}I_3\omega_0^2 = \boxed{524.3252 \text{ kg-m}^2/\text{s}^2} \text{ (or units of joules, J)}$$

Alternatively, we can compute rotational kinetic energy from the constant angular momentum:

$$H = I_3\omega_0 = 13.3518 \text{ kg-m}^2/\text{s} = \text{constant}$$

Using Eq. (12.90)

$$T_{rot,0} = \frac{H^2}{2I_3} = 524.3252 \text{ kg-m}^2/\text{s}^2 \text{ (same result)}$$

The final rotational kinetic energy can be determined using Eq. (12.90) and the *maximum* moment of inertia I_1:

$$T_{rot,f} = \frac{H^2}{2I_1} = \boxed{17.8271 \text{ kg-m}^2/\text{s}^2}$$

Rotational kinetic energy has decreased by nearly a factor of 30.

The final spin rate about the major axis can be computed from the constant angular momentum H or final kinetic energy $T_{rot,f}$. Using angular momentum, the final spin rate is

$$\omega_f = \frac{H}{I_1} = \boxed{2.6704 \text{ rad/s}} \text{ (or 25.5 rpm)}$$

Thus, the angular velocity of the "flat spin" shown in Figure 12.18b is nearly 1/30 the initial spin rate along the minor axis. This 1/30 factor is the ratio of the minimum and maximum moments of inertia, or I_3/I_1.

The previous discussion of spinning flexible satellites and Example 12.4 illustrate the so-called "major-axis rule": a flexible spinning body will eventually transition to a pure spin about its principal axis of maximum moment of inertia. Unlike prolate satellites like Explorer 1, an *oblate* satellite can take advantage of the major-axis rule by intentionally including energy-dissipation devices to ensure a pure spin about its major axis. One such device is a *nutation damper*. In its simplest form, this device consists of a mass–spring mechanical system enclosed in a tube partially filled with a fluid. The tube is aligned with the spin axis so that any wobbling motion (nutation) causes the mass to vibrate and damp out the wobble. Hence, the nutation damper dissipates rotational kinetic energy and causes the major axis to realign with the angular momentum vector **H** (i.e., the nutation angle is removed). Nutation dampers provide a simple passive strategy for stabilizing a spinning oblate satellite. Of course, a spinning prolate satellite cannot use a passive damper. Other nutation damper designs exist; the interested reader may consult Hughes [4; pp. 391–400] for an excellent discussion of options for nutation dampers.

As we end this section, let us summarize the stability of spinning satellites:

1) A *rigid* spinning satellite can maintain a stable spin about its major or minor principal axes. Spin about the intermediate axis is unstable and leads to tumbling motion.
2) A semi-rigid (or flexible) satellite can maintain a stable spin about its major axis only (i.e., the so-called "major-axis rule"). An initial spin about the minor axis will lead to tumbling motion that eventually ends in a pure spin about the major axis. Nutation dampers may be used on oblate satellites to remove "wobble" and realign the major axis with the angular momentum vector **H**.

Example 12.5 An axisymmetric oblate satellite has principal moments of inertia $I_3 = 1,100 \text{ kg-m}^2$ (major axis) and $I_1 = I_2 = 700 \text{ kg-m}^2$ (minor axes). At time $t = 0$, the satellite has the angular velocity vector (in body-fixed coordinates)

$$\omega_0 = -0.2\mathbf{u}_1 + 0.1\mathbf{u}_2 + 8.3\mathbf{u}_3 \text{ rad/s}$$

The satellite is equipped with a nutation damper mounted along the 3 axis. Determine the following:

a) The initial nutation angle.
b) The initial rotational kinetic energy.
c) The final angular velocity vector in body-frame coordinates.
d) The final rotational kinetic energy.

Neglect the effect of the mass-spring nutation damper on the satellite's moments of inertia.

a) First, let us compute the angular momentum vector using Eq. (12.18)

$$\mathbf{H} = \mathbf{I}\omega_0 = \begin{bmatrix} 700 & 0 & 0 \\ 0 & 700 & 0 \\ 0 & 0 & 1,100 \end{bmatrix} \begin{bmatrix} -0.2 \\ 0.1 \\ 8.3 \end{bmatrix} = -140\mathbf{u}_1 + 70\mathbf{u}_2 + 9,130\mathbf{u}_3 \text{ kg-m}^2/\text{s}$$

Remember that the angular momentum vector $\mathbf{H}$ computed above is expressed in the *body frame* at time $t = 0$, and that the body axes are *not* inertially fixed due to the coning motion (however, $\mathbf{H}$ is a fixed vector relative to an inertial frame).

Equation (12.57) shows that the nutation angle is

$$\theta = \tan^{-1}\left(\frac{H_{12}}{H_3}\right)$$

where the component of $\mathbf{H}$ in the 1–2 plane of the body frame is

$$H_{12} = H_1 + H_2 = -140\mathbf{u}_1 + 70\mathbf{u}_2$$

The magnitude of H_{12} is $H_{12} = \sqrt{(-140)^2 + 70^2} = 156.5248 \text{ kg-m}^2/\text{s}$. Therefore, the initial nutation angle is

$$\theta = \tan^{-1}\left(\frac{H_{12}}{H_3}\right) = \tan^{-1}\left(\frac{156.5248}{9,130}\right) = \boxed{0.0171 \text{ rad} \, (= 0.9822°)}$$

The oblate satellite is spinning with a very slight wobble at time $t = 0$.

b) We can use Eq. (12.33) to compute the initial rotational kinetic energy

$$T_{\text{rot},0} = \frac{1}{2}\omega_0 \cdot \mathbf{I}\omega_0 = \frac{1}{2}[-0.2 \ \ 0.1 \ \ 8.3]\begin{bmatrix} -140 \\ 70 \\ 9,130 \end{bmatrix} = \boxed{37,907 \text{ kg-m}^2/\text{s}^2}$$

where the 3×1 column vector $\mathbf{H} = \mathbf{I}\omega_0$ was computed in (a). Because we are using principal axes, the rotational kinetic energy may be computed using

$$T_{rot,0} = \frac{1}{2}I_1\omega_1^2 + \frac{1}{2}I_2\omega_2^2 + \frac{1}{2}I_3\omega_3^2 = 37{,}907 \text{ kg-m}^2/\text{s}^2 \text{ (same result)}$$

c) The nutation damper will eventually remove the wobble and realign the major axis (the 3 axis) with the angular momentum vector. Because the nutation angle eventually goes to zero, the final angular velocity vector (and **H**) will only contain a component along the 3 axis (remember that the 3 axis is a coordinate of the rotating body frame and *not* a coordinate of an inertial frame). Because **H** is a constant vector, we can easily compute the final spin rate using

$$\omega_f = \frac{H}{I_3} = 8.3012 \text{ rad/s}$$

where the angular momentum magnitude is $H = \|\mathbf{H}\| = 9{,}131.34 \text{ kg-m}^2/\text{s}$. The final angular velocity vector expressed in body-fixed coordinates is

$$\boxed{\boldsymbol{\omega}_f = 8.3012\mathbf{u}_3 \text{ rad/s}}$$

Note that the angular momentum is $\mathbf{H} = \mathbf{I}\boldsymbol{\omega}_f = 9{,}131.34\mathbf{u}_3 \text{ kg-m}^2/\text{s}$, which has the same magnitude as $\mathbf{H} = \mathbf{I}\boldsymbol{\omega}_0$ as computed in (a).

d) The final rotational kinetic energy is easy to compute because the satellite is in a pure spin about its major axis:

$$T_{rot,f} = \frac{1}{2}I_3\omega_f^2 = \boxed{37{,}900.64 \text{ kg-m}^2/\text{s}^2}$$

Comparing this result to (b), we see that the nutation damper has dissipated a very small percentage of the initial rotational kinetic energy.

12.5 Spin Stabilization

We have alluded to the advantages of stabilizing a satellite by spinning it about a principal axis. The major-axis rule states that an oblate satellite equipped with a nutation damper will remove any wobbling motion and realign the major principal axis with the inertially fixed angular momentum **H**. Therefore, a properly designed "oblate spinner" will eventually point and hold its 3 axis in a fixed direction. This simple strategy is an example of *passive attitude control* (in Chapter 13 we will discuss *active attitude control*, i.e., using feedback from sensors to provide an actuating torque). Spin stabilization is often used to maintain a fixed direction during an onboard rocket burn for an orbit transfer. For example, the third stage of a launch vehicle is often a solid rocket intended for final orbit insertion (such as an apogee burn to establish GEO) or an injection burn to escape Earth and begin the interplanetary cruise phase. Before the burn, the mated third stage and spacecraft are "spun up" in order to provide gyroscopic stiffness and maintain a fixed direction during the burn (small thrusters mounted on the periphery of the stage are fired in pairs to produce a pure spin about the symmetric axis). Because the burn is relatively short (on the order of 1 min), it is possible to use this strategy for oblate and prolate satellites (the time scale for the spin divergence of a prolate minor-axis spinner is on the order of 1 h). Another example of spin stabilization is the Stardust capsule which collected samples from the comet Wild-2 and returned them to Earth. The oblate capsule was spun-up to 13.5 rpm prior to entering the Earth's atmosphere so that it could maintain a fixed attitude for the proper angle of attack at entry interface.

Let us demonstrate the "gyroscopic stiffness" caused by spinning a satellite. Figure 12.19 shows a symmetric spin-stabilized satellite performing a rocket burn. The satellite is initially spinning about the 3 axis when the burn is started. Suppose the thrust vector $\mathbf{F}$ has a slight angular misalignment β so that it does not perfectly point along the 3 axis. Consequently, the thrust misalignment produces an external disturbance moment $\mathbf{M}_d$ that is initially aligned with the 2 axis, that is

$$\mathbf{M}_d = Fd \sin \beta \mathbf{u}_2 \tag{12.91}$$

where d is distance from the center of mass to the thrust chamber along the 3 axis. Because moment $\mathbf{M}_d$ is initially perpendicular to $\mathbf{H}$, it causes a rotation of the angular momentum vector without changing its magnitude. The time-rate of angular momentum is the equal to the external moment:

$$\mathbf{M}_d = \frac{d\mathbf{H}}{dt} \approx \frac{\Delta\mathbf{H}}{\Delta t} \tag{12.92}$$

Expressing Eq. (12.92) in terms of magnitudes and using $\Delta H = H\Delta\theta$ (see Figure 12.19), we obtain

$$M_d = \frac{H\Delta\theta}{\Delta t} \tag{12.93}$$

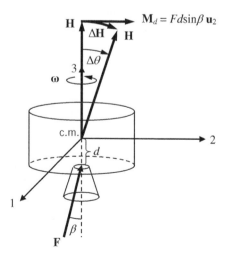

Figure 12.19 Spin-stabilized satellite with thrust misalignment.

Therefore, the change in nutation angle is

$$\Delta\theta = \frac{M_d \Delta t}{H} = \frac{M_d \Delta t}{I_3 \omega_3} \qquad (12.94)$$

Equation (12.94) shows that the effect of a given disturbance moment M_d on the nutation angle $\Delta\theta$ (i.e., the "wobble") can be diminished by making the satellite's initial angular momentum H large, or providing a high spin rate ω_3. In other words, a rapidly spinning satellite exhibits resistance (or "gyroscopic stiffness") to external moments that are perpendicular to the spin axis. Although this simple example involved a moment caused by thrust misalignment during a burn, M_d can represent any disturbance moment. A final note is in order here. We cannot use Eq. (12.94) to compute the nutation angle after burn time Δt because the subsequent satellite motion is governed by Euler's moment equations (12.39) which are nonlinear and coupled. This simple example is intended to illustrate the effect spinning a satellite has on resisting a disturbing moment.

12.5.1 Dual-Spin Stabilization

The first GEO communication satellites of the early 1960s were oblate satellites spinning about their major axis for stability. Although spinning the entire satellite provided directional stability, it greatly hindered the ability to point the satellite's antennas at a fixed location on Earth. One solution to this problem is the dual-spin satellite consisting of two sections that spin at different rates about a common axis. The Intelsat III series of GEO satellites of the late 1960s were oblate spinners with a "de-spun" antenna mounted on top that rotated at one revolution per sidereal day. Launched in the early 1970s, the Intelsat IV series of GEO communication satellites marked a dramatic achievement in dual-spin stabilization. Figure 12.20 shows the Intelsat IV in geostationary orbit. Intelsat IV consisted of a platform section and rotor section. The platform section (top half of satellite shown in Figure 12.20) contains the communication antennas and rotates at 1 revolution per day so that it always points at Earth. The rotor section (bottom half) spins at a higher rate in order to provide gyroscopic rigidity (i.e., the angular momentum vector **H** shown in Figure 12.20).

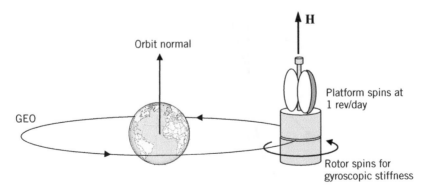

Figure 12.20 Schematic diagram of the dual-spinning satellite Intelsat IV in geostationary orbit.

The early GEO satellites were oblate spinners; the dual-spin Intelsat IV was a prolate spinner. Although Explorer 1 demonstrated the instability of prolate spinners, the geometric constraints imposed by the long, slender payload shroud of launch vehicles forced satellite designers to consider prolate configurations. The major-axis rule, however, dictated that only oblate satellites could utilize internal dampers to maintain a stable spin. Research in the mid-1960s revealed an exception to the major-axis rule: a prolate dual-spin satellite is stable about its minor axis if the majority of the energy dissipation is performed by the slowly spinning platform section. Demonstrating the stability of a prolate dual-spin satellite is beyond the scope of this textbook; the interested reader may consult Kaplan [1; pp. 178–188] or Wiesel [3; pp. 168–171] for details.

12.6 Disturbance Torques

Our analysis of attitude dynamics to this point has focused on torque-free motion. In general, the external torques caused by the satellite's environment are very small. However, all sources of disturbance torques must be considered in the design and operation of spacecraft. This section provides a brief overview of the disturbance torques that act on an orbiting satellite. The interested reader may consult Hughes [4; pp. 232–272] for an in-depth discussion of spacecraft torques.

12.6.1 Gravity-Gradient Torque

We know that gravitational force is proportional to $1/r^2$ where r is the radial distance from the center of the planet to the satellite. For particle dynamics (i.e., Chapters 2–11), the gravitational force acts on the center of mass. For attitude dynamics (i.e., rotational motion about the mass center), the gravitational force acts on the satellite's distributed mass. Figure 12.21 shows a cylindrical satellite in an Earth orbit where incremental mass dm is "below" the orbital path. The incremental gravitational force on dm is

$$dF = -dm\frac{\mu}{r^3}r \tag{12.95}$$

The position vector from the Earth's center to mass dm is

$$\mathbf{r} = \mathbf{r}_{cm} + \boldsymbol{\rho} \tag{12.96}$$

where $\mathbf{r}_{cm}$ is the position vector of the mass center and $\boldsymbol{\rho}$ is the relative position of dm with respect to the mass center. It should be clear to the reader that Eq. (12.95) is Newton's second law ($\mathbf{F} = m\mathbf{a}$) where the inverse-square gravitational acceleration is μ/r^2 and the unit vector $-\mathbf{r}/r$ points from dm to the center of the Earth. Equation (12.95) shows that the satellite will experience a larger attractive gravitation force on its "lower" side below the orbital path compared with its "upper side." Intuitively, the summation of $d\mathbf{F}$ on the satellite shown in Figure 12.21 will result in a net torque about the mass center that causes a counterclockwise rotation so that the long axis is aligned with the vertical direction. Mathematically we can sum all of the incremental gravitational torques $d\mathbf{M} = \boldsymbol{\rho} \times d\mathbf{F}$ over the entire satellite body to obtain

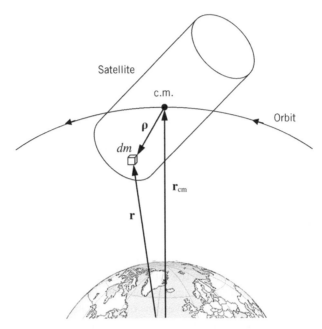

Figure 12.21 Geometry for determining the gravitational force on mass *dm*.

$$\mathbf{M}_{gg} = \int \boldsymbol{\rho} \times d\mathbf{F} \qquad (12.97)$$

Equation (12.97) is a general expression for the *gravity-gradient torque*. Substituting Eq. (12.95) for the incremental gravitational force on *dm*, Eq. (12.97) becomes

$$\mathbf{M}_{gg} = -\mu \int \frac{\boldsymbol{\rho} \times \mathbf{r}}{r^3} dm \qquad (12.98)$$

The gravity-gradient torque equation (12.98) can be expanded by carrying out the cross product and simplified by approximating the $1/r^3$ term with a binomial expansion. We will not show the complete results here; the interested reader may consult Hughes [4; pp. 233–238] for details. In the end, the gravity-gradient torque depends on the orbital radius r_{cm}, the satellite's inertia matrix $\mathbf{I}$, and the satellite's angular attitude. We will develop a complete expression for the gravity-gradient torque about one axis in the next section when we show how the gravitational torque can be used to provide a stable pointing direction.

12.6.2 Aerodynamic Torque

In Chapter 5, we briefly discussed how aerodynamic drag perturbs a satellite's orbit. The same drag force will result in a net torque if the aerodynamic center-of-pressure is offset from the satellite's center for mass. We know from basic mechanics that torque is the cross product of the position and force vectors; therefore a general equation for the aerodynamic torque is

$$\mathbf{M}_{\text{aero}} = \mathbf{r}_{\text{cp}} \times \mathbf{F}_{\text{aero}} \tag{12.99}$$

where $\mathbf{r}_{\text{cp}}$ is the position of the aerodynamic center-of-pressure (i.e., the location of the resultant aerodynamic force vector) in body-frame coordinates relative to the center of mass. The aerodynamic (drag) force vector is

$$\mathbf{F}_{\text{aero}} = -\frac{1}{2}\rho v_{\text{rel}}^2 S C_D \mathbf{u}_v \tag{12.100}$$

where ρ is the atmospheric density, v_{rel} is the satellite's velocity relative to the planet's atmosphere, S is the satellite's cross-sectional area, C_D is the drag coefficient, and $\mathbf{u}_v$ is a unit vector in the direction of the relative velocity. We insert a minus sign in Eq. (12.100) because aerodynamic drag is always opposite the vehicle's (atmospheric-relative) velocity vector. Equations (12.99) and (12.100) show that the aerodynamic torque $\mathbf{M}_{\text{aero}}$ depends on many variables such as orbital altitude (i.e., density ρ), satellite area and shape (S and C_D), and location of the aerodynamic center-of-pressure. Furthermore, a great deal of uncertainty exists because atmospheric density can vary dramatically due to solar activity and $\mathbf{r}_{\text{cp}}$ and C_D vary with the satellite's attitude as it moves along its orbit.

Let us demonstrate the magnitude of the aerodynamic torque using a satellite configuration presented in Chapter 5. Consider a satellite with ballistic coefficient $C_B = 85 \text{ kg/m}^2$ ($= m/SC_D$) in a 350-km altitude circular orbit. Table 5.4 shows that the drag acceleration F_{aero}/m is $2.13(10^{-6})$ m/s^2 for this satellite and orbit combination. If the satellite has mass $m = 1{,}000$ kg, then the magnitude of the aerodynamic force is 0.00213 N. For a center-of-pressure offset of 1 cm ($=0.01$ m), the aerodynamic torque acting on the satellite is $2.13(10^{-5})$ N-m.

12.6.3 Solar Radiation Pressure Torque

Chapter 5 also presented the perturbing force caused by solar radiation pressure (SRP). The resulting SRP torque is computed in a manner similar to the aerodynamic torque:

$$\mathbf{M}_{\text{SRP}} = \mathbf{r}_{\text{oc}} \times \mathbf{F}_{\text{SRP}} \tag{12.101}$$

where $\mathbf{r}_{\text{oc}}$ is the position of the optical center-of-pressure (i.e., the location of the resultant SRP force vector) in body-frame coordinates. The SRP force vector is

$$\mathbf{F}_{\text{SRP}} = C_R P_{\text{SRP}} A_s \mathbf{u}_s \tag{12.102}$$

where C_R is the satellite's surface reflectivity, P_{SRP} is the solar radiation pressure, A_s is the satellite's area normal to the sun vector, and $\mathbf{u}_s$ is a unit vector from the sun to the Earth (i.e., essentially the unit vector from the sun to the satellite) expressed in body-frame coordinates. Recall from Chapter 5 that the surface reflectivity C_R varies from zero (a translucent body) to a value of 2 (a perfectly reflective body, such as an ideal mirror). A "black body" that absorbs all sunlight has reflectivity $C_R = 1$. We also saw in Chapter 5 that the mean SRP is $P_{\text{SRP}} \approx 4.5(10^{-6})$ N/m^2, which is computed using the mean solar intensity.

We can demonstrate the magnitude of the SRP torque using the following representative values: let $A_s = 5$ m^2, $C_R = 1.5$, $P_{\text{SRP}} = 4.5(10^{-6})$ N/m^2, and $r_{\text{oc}} = 0.1$ m. The resulting SRP torque magnitude is $3.4(10^{-6})$ N-m, which is an order of magnitude smaller than the aerodynamic torque for a 400-km circular orbit. However, unlike aerodynamic torque, the SRP torque does not depend on the satellite's altitude. Therefore, the SRP torque is typically the dominant disturbance torque for geocentric satellites in high-altitude orbits such as a geostationary orbit.

12.6.4 Magnetic Torque

Planets that have a significant magnetic field produce a torque if the satellite has its own magnetic moment. This magnetic torque on the satellite is

$$\mathbf{M}_{mag} = \mathbf{M}_{sat} \times \mathbf{B} \tag{12.103}$$

where $\mathbf{M}_{sat}$ is the satellite's magnetic (dipole) moment and $\mathbf{B}$ is the magnetic flux density of the central gravitational body. The satellite's magnetic moment $\mathbf{M}_{sat}$ could be due to permanent magnets on the satellite and/or electrical current flowing through circuits on the satellite (i.e., Faraday's laws of induction). The planet's magnetic field $\mathbf{B}$ is proportional to $1/r^3$. Clearly, the orientation of the satellite's dipole moment $\mathbf{M}_{sat}$ relative to the Earth's magnetic field $\mathbf{B}$ is critical to the calculation of the magnetic torque.

Let us demonstrate the magnitude of the magnetic torque with a basic example. Suppose a satellite has several current-carrying loops with a total dipole moment $M_{sat} = 0.1$ A-m^2 (where A is one ampere or 1 amp). At 200-km altitude, the Earth's magnetic flux density is approximately $B = 3(10^{-5})$ Wb/m^2 (where Wb is one weber; note 1 Wb/m^2 = 1 tesla). If $\mathbf{M}_{sat}$ and $\mathbf{B}$ are at right angles, the magnetic torque is roughly $3(10^{-6})$ N-m which is on the order of the SRP torque.

12.7 Gravity-Gradient Stabilization

The previous section briefly presented and discussed disturbance torques that act on an orbiting satellite. It is possible to design a satellite so that the gravity-gradient torque provides a stable pointing direction. Intuitively, we may reason that a long, thin satellite with mass concentrated at each end will induce a gravitational torque when it is displaced from a vertical attitude. We will demonstrate gravity-gradient stabilization by considering a "dumbbell" satellite configuration (shown in Figure 12.22) consisting of two equal masses separated by a long slender rod of length $2L$. The dumbbell satellite in Figure 12.22 is in a circular orbit with radius r_0 and constant angular velocity ω_0. Its attitude angle θ is measured clockwise from the local vertical direction. Each mass is influenced by a gravitational force (F_{g1} and F_{g2}) and a centrifugal force (F_{cf1} and F_{cf2}) as shown in Figure 12.22. Although these forces act in radial directions, they are shown as parallel forces in Figure 12.22 because orbit radius r_0 is much larger than rod length $2L$. Let us compute the net vertical force (with positive convention toward the Earth) acting on each mass in Figure 12.22:

$$\text{Mass } m_1: \quad F_1 = +\downarrow \sum F = F_{g1} - F_{cf1}$$

$$= m_1 \left(\frac{\mu}{r_1^2} - \frac{v_1^2}{r_1} \right) \tag{12.104}$$

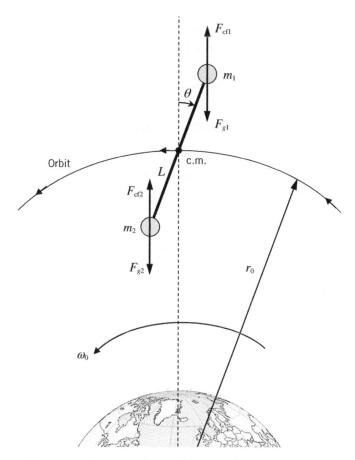

Figure 12.22 Gravity-gradient stabilization of a dumbbell satellite.

$$\text{Mass } m_2: \quad F_2 = +\!\downarrow \sum F = F_{g2} - F_{cf2}$$

$$= m_2\left(\frac{\mu}{r_2^2} - \frac{v_2^2}{r_2}\right) \tag{12.105}$$

The radial distance to each mass (for small attitude angle θ) is approximately

$$r_1 = r_0 + L\cos\theta \approx r_0 + L \tag{12.106}$$

$$r_2 = r_0 - L\cos\theta \approx r_0 - L \tag{12.107}$$

We approximate the inertial velocities of each mass using the product of the respective radial distance and the constant angular velocity of the circular orbit:

$$v_1 = r_1\omega_0 \tag{12.108}$$

$$v_2 = r_2\omega_0 \tag{12.109}$$

Any relative velocity components caused by attitude rotation (i.e., $L\dot{\theta}$) are extremely small compared with orbital speeds and are therefore neglected. Substituting Eqs. (12.106)–(12.109) into the net-force equations (12.104) and (12.105) yields

$$F_1 = m_1 \left[\frac{\mu}{(r_0 + L)^2} - (r_0 + L)\omega_0^2 \right] \tag{12.110}$$

$$F_2 = m_2 \left[\frac{\mu}{(r_0 - L)^2} - (r_0 - L)\omega_0^2 \right] \tag{12.111}$$

The net torque about the dumbbell satellite's center of mass is

$$M_{gg} = F_1 L \sin\theta - F_2 L \sin\theta \tag{12.112}$$

Recall that the net forces F_1 and F_2 are considered positive in the downward direction (i.e., toward Earth). After substituting Eqs. (12.110) and (12.111) and assuming a small attitude angle ($\sin\theta \approx \theta$), Eq. (12.112) becomes

$$M_{gg} = mL\theta \left\{ \mu \left[\frac{1}{(r_0 + L)^2} - \frac{1}{(r_0 - L)^2} \right] - \omega_0^2 [(r_0 + L) - (r_0 - L)] \right\} \tag{12.113}$$

Because the masses are equal, we use $m = m_1 = m_2$. The first bracket term on the right-hand side of Eq. (12.113) can be simplified to

$$\frac{1}{(r_0 + L)^2} - \frac{1}{(r_0 - L)^2} = \frac{-4r_0 L}{(r_0^2 - L^2)^2} \approx \frac{-4L}{r_0^3} \tag{12.114}$$

The final approximation in Eq. (12.114) is valid because $r_0 \gg L$. Using Eq. (12.114) in the torque equation (12.113), we obtain

$$M_{gg} = mL\theta \left(\frac{-4\mu}{r_0^3} L - 2\omega_0^2 L \right) \tag{12.115}$$

Recall from our previous chapters on orbital dynamics that the square of the orbital angular velocity is $\omega_0^2 = \mu/r_0^3$. Using this orbital relationship, Eq. (12.115) becomes

$$M_{gg} = -6mL^2\omega_0^2\theta \tag{12.116}$$

Equation (12.116) represents the gravity-gradient torque acting on the dumbbell satellite shown in Figure 12.22. Clearly, M_{gg} is a *restoring* torque because is it negative for a positive attitude rotation ($\theta > 0$) and positive for a negative rotation ($\theta < 0$). Hence, the gravity-gradient torque will cause a long, slender dumbbell satellite to rotate toward its local vertical attitude (i.e., $\theta = 0$). Because the moment of inertia of the dumbbell satellite is $I = 2mL^2$ (relative to an axis along the orbit normal), the gravity-gradient torque becomes

$$M_{gg} = -3\omega_0^2 I\theta \tag{12.117}$$

Next, let us include the gravity-gradient torque in Euler's moment equations. Recall that Eq. (12.39) utilizes the absolute angular velocity as expressed in a body-frame coordinates (the 123 principal axes). Let us assume that the dumbbell satellite in Figure 12.22 has its 2 axis pointing into the page; therefore the corresponding angular velocity component is

$$\omega_2 = \dot{\theta} - \omega_0 \tag{12.118}$$

Therefore, if the dumbbell satellite maintains a vertical attitude ($\theta = 0$ = constant), the spin component along the 2 axis is the negative orbital angular velocity (note that the vector ω_0 is *out* of the page in Figure 12.22). Recall Euler's equation (12.39b) about the 2 axis

$$M_2 = I_2\dot{\omega}_2 + (I_1 - I_3)\omega_1\omega_3 \tag{12.119}$$

Because the orbital rate ω_0 is constant, Eq. (12.118) shows that $\dot{\omega}_2 = \ddot{\theta}$. Assuming that the dumbbell satellite is not rotating about its 1 or 3 axes (i.e., $\omega_1 = \omega_3 = 0$) and moment M_2 is the gravity-gradient torque as expressed by Eq. (12.117), Eq. (12.119) becomes

$$I_2\ddot{\theta} = -3\omega_0^2 I\theta \tag{12.120}$$

Noting that $I_2 = I$, we obtain

$$\ddot{\theta} + 3\omega_0^2\theta = 0 \tag{12.121}$$

At this point we should summarize the results we have obtained thus far. Equation (12.121) represents the angular attitude dynamics of a dumbbell satellite influenced by the gravity-gradient torque. The only motion considered here is the attitude angle θ measured clockwise from the local vertical direction. Equation (12.121) is an undamped second-order differential equation, and hence its solution is a harmonic function with frequency $\sqrt{3}\omega_0$

$$\theta(t) = C_1 \sin\sqrt{3}\omega_0 t + C_2 \cos\sqrt{3}\omega_0 t \tag{12.122}$$

where constants C_1 and C_2 depend on the initial conditions. Therefore, a dumbbell satellite displaced from the local vertical will "rock" back and forth with frequency $\sqrt{3}\omega_0$. Unless onboard damping is present, the amplitude of this rocking or *libration* motion will remain constant. Furthermore, because the 2 axis is the maximum moment of inertia for the dumbbell satellite, the libration motion is in its minimum-energy configuration and hence it is stable. The libration frequency $\sqrt{3}\omega_0$ exhibits no dependence on the dumbbell satellite's moment of inertia (i.e., m and L) and instead solely depends on the radius of the circular orbit because $\omega_0 = \sqrt{\mu/r_0^3}$.

This example has focused on the 1-DOF attitude motion of a satellite subject to the gravity-gradient torque. We can describe the general 3-DOF attitude motion using angles relative to the so-called *local vertical/local horizontal* (LVLH) frame presented in Figure 12.23. The LVLH frame is defined by the local vertical axis (pointing from the satellite's mass center to the center of the Earth) and the local horizontal axis (perpendicular to the vertical axis and in the orbital plane in the direction of orbital motion). Roll angle ϕ is rotation about the horizontal axis, yaw angle ψ is rotation about the vertical axis, and pitch angle θ is rotation about an axis normal to the orbital plane but in the opposite direction as the orbital angular momentum **h**. Figure 12.23 should help the reader to see that the pitch axis in Figure 12.22 points into the page. Nominally, the 123 principal body axes point along the roll, pitch, and yaw axes shown in Figure 12.23. Because the origin of the LVLH frame is fixed at the satellite's center of mass (with its vertical axis pointing to the Earth), it is a moving and rotating frame.

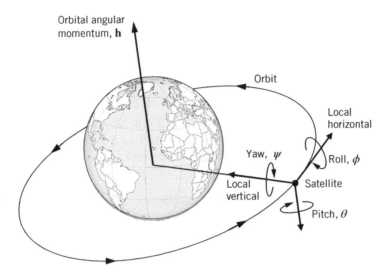

Figure 12.23 Local vertical/local horizontal (LVLH) frame.

It is possible to develop a general expression for the gravity-gradient torque vector in the body frame using the LVLH-frame angles and small-angle approximations; the result is

$$\mathbf{M}_{gg} = 3\omega_0^2 \begin{bmatrix} (I_3 - I_2)\phi \\ (I_3 - I_1)\theta \\ 0 \end{bmatrix} \tag{12.123}$$

The torque components in Eq. (12.123) (from top to bottom) are along the roll, pitch, and yaw axes, respectively. Deriving Eq. (12.123) is a bit tedious and requires carrying out the cross product in Eq. (12.98) and a coordinate transformation between the body and LVLH frames (the interested reader may consult Kaplan [1; pp. 199–204], Wiesel [3; pp. 162–168], or Hughes [4; pp. 233–238, 282–283] for details). Note that the 2-axis gravity-gradient torque component, $3\omega_0^2(I_3 - I_1)\theta$, is identical to the pitch-axis torque presented by Eq. (12.117) because $I = I_1 = I_2$ and $I_3 = 0$ for a dumbbell satellite with point masses at each end.

Inspection of Eq. (12.123) allows us to draw the following conclusions about the gravity-gradient torque $\mathbf{M}_{gg}$:

1) If $I_3 < I_2$, then the roll-axis gravity-gradient torque is restorative and stabilizing. In other words, a small positive roll angle ($\phi > 0$) will produce a negative roll torque that returns the satellite to the local vertical orientation.
2) If $I_3 < I_1$, then the pitch-axis gravity-gradient torque is restorative and stabilizing. A small positive pitch angle ($\theta > 0$) will produce a negative pitch torque that returns the satellite to the local vertical orientation.
3) A yaw angle rotation ψ does not affect the gravity-gradient torque. This result makes sense intuitively because a rotation along the vertical axis does not induce a gravitational torque.

4) The magnitude of the restoring gravity-gradient torques for the roll and pitch axes can be increased by designing a long, thin satellite configuration so that the moment of inertia differences $I_2 - I_3$ and $I_1 - I_3$ are "large." Therefore, the 3 axis is the long axis of the satellite.
5) Gravity does not induce a torque about the vertical (or yaw) axis.

Equation (12.123) provides the gravitational torques in the body frame; that is, the left-hand sides of Euler's moment equations (12.39). Instead of dealing with body-axis angular accelerations ($\dot{\omega}_1$, $\dot{\omega}_2$, and $\dot{\omega}_3$) and body-axis spin rates (ω_1, ω_2, and ω_3), we can use a linearized coordinate transformation to rewrite Euler's moment equations in terms of roll, pitch, and yaw angles and their derivatives. Neglecting the products of angular velocities, the linearized gravity-gradient roll, pitch, and yaw equations are

$$\text{Roll:} \qquad I_1\ddot{\phi} = 3\omega_0^2(I_3 - I_2)\phi \qquad (12.124)$$

$$\text{Pitch:} \qquad I_2\ddot{\theta} = 3\omega_0^2(I_3 - I_1)\theta \qquad (12.125)$$

$$\text{Yaw:} \qquad I_3\ddot{\psi} = 0 \qquad (12.126)$$

The pitch equation (12.125) is identical to the dumbbell satellite's dynamical equation (12.120); recall that $I_1 = I_2$ and $I_3 = 0$ for the dumbbell satellite. The general solutions for the roll and pitch motions are

$$\text{Roll:} \qquad \phi(t) = C_1 \sin\Omega_r t + C_2 \cos\Omega_r t \qquad (12.127)$$

$$\text{Pitch:} \qquad \theta(t) = C_3 \sin\Omega_p t + C_4 \cos\Omega_p t \qquad (12.128)$$

where the roll and pitch libration frequencies are

$$\text{Roll:} \qquad \Omega_r = \omega_0\sqrt{\frac{3(I_2 - I_3)}{I_1}} \qquad (12.129)$$

$$\text{Pitch:} \qquad \Omega_p = \omega_0\sqrt{\frac{3(I_1 - I_3)}{I_2}} \qquad (12.130)$$

Again, for stable libration motion, we have the conditions $I_1 > I_3$ and $I_2 > I_3$. A long, thin, rod-like satellite aligned along the local vertical axis will be stabilized by the gravitational torque.

Gravity-gradient stabilization has been used successfully on satellites that require "nadir pointing" where the satellite must always point "down" toward the center of the Earth. The long/thin inertia requirement for a sizeable gravitational torque is often achieved by extending a "tip mass" on a boom. An example of this strategy is GEOSAT, which was an Earth-observing satellite deployed by the US Navy in 1985. Figure 12.24 shows GEOSAT in its orbital configuration with its long, thin axis aligned with the local vertical direction. GEOSAT was equipped with a radar altimeter designed to measure the distance from the satellite to the ocean surface. Gravity-gradient stabilization was used to keep the long axis (and thus the radar) pointed downward in the vertical direction. GEO-SAT's long/thin inertia distribution was achieved by using a rigid boom to extend a tip mass away from the main satellite body. GEOSAT removed libration oscillations through a passive magnetic damper in its tip mass. As a result, GEOSAT was able to keep its libration angle amplitude to about $1°$.

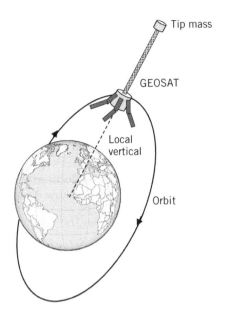

Figure 12.24 Schematic diagram of GEOSAT in orbit.

Example 12.6 Suppose GEOSAT (shown in Figure 12.24) has principal moments of inertia $I_3 = 100$ kg-m^2 (long minor axis) and $I_1 = I_2 = 2{,}600$ kg-m^2 (transverse major axes). If GEOSAT operated in an 800-km altitude circular orbit, compute the frequency and period of the pitch libration motion caused by the gravity-gradient torque.

Equation (12.130) provides the libration frequency for the pitching motion

$$\Omega_p = \omega_0 \sqrt{\frac{3(I_1 - I_3)}{I_2}}$$

We need the orbital angular velocity ω_0. The orbital radius is $r_0 = R_E + 800$ km $= 7{,}178$ km (recall that the Earth's equatorial radius is $R_E = 6{,}378$ km). The orbital angular velocity is

$$\omega_0 = \sqrt{\frac{\mu}{r_0^3}} = 0.001038 \text{ rad/s}$$

Using this value for ω_0 and the moments of inertia, the libration frequency is

$$\Omega_p = \omega_0 \sqrt{\frac{3(I_1 - I_3)}{I_2}} = \boxed{0.001763 \text{ rad/s}}$$

The period of the pitch libration motion is $\tau_{\text{lib}} = 2\pi/\Omega_p = \boxed{3{,}563.5 \text{ s} = 59.4 \text{ min.}}$
Note that GEOSAT's orbital period is $2\pi/\omega_0 = 6{,}053.2$ s $= 100.9$ min.

12.8 Summary

All prior chapters (Chapters 2–11) involved particle dynamics, that is, the motion of a space vehicle's center of mass along its flight path. This chapter analyzed a satellite's angular orientation (or *attitude*) and rotational motion about its center of mass. Attitude dynamics is an especially challenging discipline because it involves vectorial quantities with analysis based on rotating (i.e., non-inertial) coordinate frames. Our analysis methods rely on knowledge of the satellite's moments of inertia and angular velocity vector relative to a rotating, body-fixed coordinate system. We derived *Euler's moment equations* (i.e., the governing equations of motion for a rotating satellite) by equating the external torques (or moments) on a satellite to the time-rate of angular momentum relative to an inertial frame. Although these equations are nonlinear, we are able to obtain closed-form solutions for the torque-free motion of an axisymmetric satellite. We also investigated "spin stability" to show that a rigid spacecraft will maintain a stable (albeit "wobbly") spin about its axes of maximum and minimum moments of inertia. However, a spinning rigid body cannot maintain a stable spin about an axis with an intermediate moment of inertia. The situation is even more restrictive for flexible bodies, that is, spacecraft with moving internal parts, damping devices, or flexing appendages: flexible bodies can only maintain a stable spin about an axis of maximum moment of inertia (the so-called "major-axis rule"). It is important to remember that a satellite's angular momentum vector remains inertially *fixed* in space in the absence of external torques even though internal damping and flexure may cause the satellite to "tumble" and eventually spin about its major axis. We ended this chapter with a discussion of environmental disturbance torques that influence attitude dynamics. We showed that the so-called *gravity-gradient torque* causes a long, slender satellite to align its minor axis with the local vertical direction as it moves in orbit.

References

1 Kaplan, M.H., *Modern Spacecraft Dynamics and Control*, John Wiley & Sons, Inc., Hoboken, NJ, 1976.
2 Thomson, W.T., *Introduction to Space Dynamics*, Dover, New York, 1986.
3 Wiesel, W.E., *Spaceflight Dynamics*, 3rd edn, Aphelion Press, Beavercreek, OH, 2010.
4 Hughes, P.C., *Spacecraft Attitude Dynamics*, Dover, New York, 2004.

Problems

Conceptual Problems

12.1 A satellite has the following inertia matrix

$$\mathbf{I} = \begin{bmatrix} 550 & 0 & 0 \\ 0 & 550 & 0 \\ 0 & 0 & 280 \end{bmatrix} \text{kg-m}^2$$

and angular velocity $\boldsymbol{\omega} = 5\mathbf{u}_2 + 16\mathbf{u}_3$ rad/s in 123 body coordinates.
a) Is the satellite oblate or prolate? Explain your answer.
b) Determine the angular momentum vector in 123 body coordinates.
c) Determine the rotational kinetic energy.

12.2 A satellite's principal moments of inertia are $I_3 = 95$ kg-m^2 and $I_1 = I_2 = 60$ kg-m^2. The satellite is to be "spun up" from zero rotational kinetic energy to a pure spin of $\omega_3 = 0.7$ rad/s about its 3 axis using reaction jets that produce a total torque $M_3 = 5.5$ N-m about the 3 axis. Determine the total thruster time to complete the spin-up maneuver.

12.3 Show that the time-rate of rotational kinetic energy for an axisymmetric satellite (where $I_1 = I_2$) subjected to external torques $\mathbf{M} = M_1\mathbf{u}_1 + M_2\mathbf{u}_2 + M_3\mathbf{u}_3$ is

$$\dot{T}_{rot} = \boldsymbol{\omega} \cdot \mathbf{M}$$

where $\boldsymbol{\omega} = \omega_1\mathbf{u}_1 + \omega_2\mathbf{u}_2 + \omega_3\mathbf{u}_3$ and $\mathbf{u}_1$, $\mathbf{u}_2$, and $\mathbf{u}_3$ are unit vectors along the satellite's 123 body-fixed axes.

12.4 A rigid-body satellite has 123 body-fixed coordinates that correspond to principal axes. At a particular time instant, the satellite has angular velocity $\boldsymbol{\omega} = 0.1\mathbf{u}_1 - 0.2\mathbf{u}_2 + 0.6\mathbf{u}_3$ rad/s and angular momentum $\mathbf{H} = 3\mathbf{u}_1 - 6\mathbf{u}_2 + 57\mathbf{u}_3$ kg-m^2/s.
a) Determine the satellite's moment of inertia about its spin axis at this instant.
b) Determine the satellite's inertia matrix. Is this an oblate or prolate satellite?
c) Determine the nutation angle for torque-free motion.

12.5 A rigid axisymmetric satellite has principal moments of inertia $I_1 = I_2 = 90$ kg-m^2 and $I_3 = 140$ kg-m^2. At time $t = 0$, the satellite has angular velocity $\boldsymbol{\omega}_0 = 0.126040\mathbf{u}_1 - 0.072769\mathbf{u}_2 + 0.684703\mathbf{u}_3$ rad/s as expressed in 123 body-frame coordinates. Determine if the following angular velocity vectors at arbitrary time $t = t_1$ represent a *feasible solution* to *torque-free* motion. Explain your answers.
a) $\boldsymbol{\omega}(t_1) = 0.126040\mathbf{u}_1 + 0.072769\mathbf{u}_2 + 0.684703\mathbf{u}_3$ rad/s
b) $\boldsymbol{\omega}(t_1) = 0.126040\mathbf{u}_1 - 0.072769\mathbf{u}_2 + 0.598277\mathbf{u}_3$ rad/s
c) $\boldsymbol{\omega}(t_1) = -0.118275\mathbf{u}_1 + 0.084807\mathbf{u}_2 + 0.684703\mathbf{u}_3$ rad/s
d) $\boldsymbol{\omega}(t_1) = 0.017155\mathbf{u}_1 + 0.055209\mathbf{u}_2 + 0.684703\mathbf{u}_3$ rad/s.

12.6 An axisymmetric satellite has principal moments of inertia $I_1 = I_2 = 240$ kg-m^2 and $I_3 = 80$ kg-m^2. At time $t = 0$, the satellite has angular velocity $\boldsymbol{\omega}_0 = 0.9\mathbf{u}_3$ rad/s as expressed in 123 body-frame coordinates. The satellite has flexible antennas that dissipate energy over time. Determine if the following angular velocity vectors at arbitrary time $t = t_1$ represent a *feasible solution* to *torque-free* motion. Explain your answers.
a) $\boldsymbol{\omega}(t_1) = 0.051262\mathbf{u}_1 + 0.034882\mathbf{u}_2 + 0.870288\mathbf{u}_3$ rad/s
b) $\boldsymbol{\omega}(t_1) = 0.1\mathbf{u}_2 + 0.848528\mathbf{u}_3$ rad/s
c) $\boldsymbol{\omega}(t_1) = 0.245746\mathbf{u}_1 + 0.172073\mathbf{u}_2$ rad/s.

Problems 12.7–12.11 involve the torque-free motion of an axisymmetric rigid satellite. The satellite's moments of inertia about the 123 principal axes are $I_1 = I_2 = 150$ kg-m^2 and $I_3 = 30$ kg-m^2, and the 3-axis spin rate is $\omega_3 = 0.6$ rad/s. Figure P12.7 shows the 1- and 2-axis angular velocity components for the torque-free motion.

12.7 Determine the nutation angle θ.

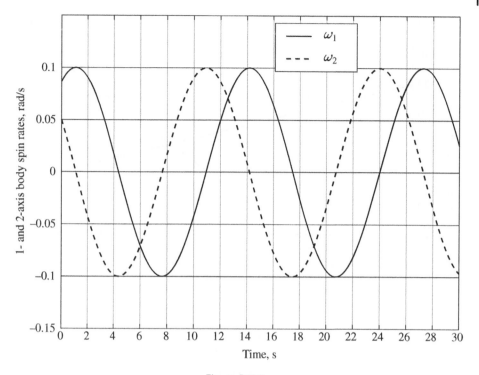

Figure P12.7

12.8 Determine the angle γ between $\mathbf{u}_3$ and $\boldsymbol{\omega}$.

12.9 Determine the precession rate $\dot{\psi}$ (in deg/s).

12.10 Determine the Euler angle rate $\dot{\phi}$ (in deg/s).

12.11 Determine the Euler angle ϕ at time $t = 0$.

12.12 A flexible satellite has the following inertia matrix

$$\mathbf{I} = \begin{bmatrix} 220 & 0 & 0 \\ 0 & 220 & 0 \\ 0 & 0 & 90 \end{bmatrix} \text{kg-m}^2$$

At time $t = 0$, its angular velocity is $\boldsymbol{\omega}_0 = 0.6\mathbf{u}_3$ rad/s in 123 body coordinates. The satellite's flexible appendages dissipate energy. Determine the satellite's spin rate (i.e., the magnitude of angular velocity vector $\boldsymbol{\omega}$) after the satellite has reached its minimum-energy state.

Problems 12.13–12.18 involve the torque-free motion of an axisymmetric rigid satellite. The satellite's moments of inertia about the 123 principal axes are $I_1 = I_2 = 90$ kg-m^2 and $I_3 = 140$ kg-m^2. Figure P12.13 shows the time histories of the satellite's Euler angles for the torque-free motion.

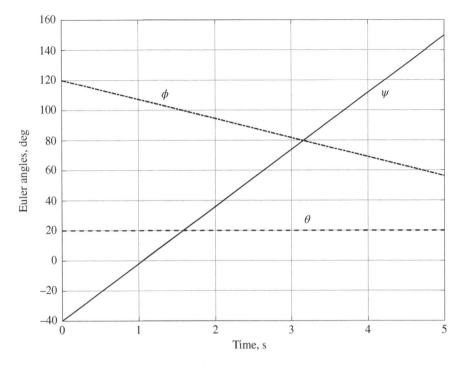

Figure P12.13

12.13 Determine the satellite's angular velocity vector $\boldsymbol{\omega}$ at $t = 0$ in the 123 body frame.

12.14 Determine the angle γ between the 3 axis and the angular velocity vector $\boldsymbol{\omega}$.

12.15 Determine the satellite's angular momentum vector **H** at $t = 0$ in the 123 body frame and in the inertial $OXYZ$ frame.

12.16 Determine the satellite's angular velocity vector $\boldsymbol{\omega}$ at $t = 20$ s in the 123 body frame.

12.17 Determine the satellite's angular momentum vector **H** at $t = 20$ s in the 123 body frame and in the inertial $OXYZ$ frame.

12.18 A sensor is located at body-frame coordinates $\mathbf{r} = 1.2\,\mathbf{u}_2$ m. Determine its inertial position vector (relative to the $OXYZ$ frame) at $t = 0$ and $t = 20$ s.

12.19 A satellite is spin stabilized by a pure spin about its 3 axis with angular velocity $\omega_3 = 2.1$ rad/s (≈ 20 rpm). Its principal moments of inertia are $I_3 = 90$ kg-m^2 and $I_1 = I_2 = 50$ kg-m^2. Suppose that a sinusoidal disturbance torque $M_d = 5\cos\omega_3 t$ N-m acts on the satellite such that it only changes the "tilt" of the satellite's angular momentum vector, that is, $dH/dt = Hd\theta/dt$, where angular momentum magnitude H is constant and θ is the nutation angle. Estimate the maximum amplitude of the nutation angle (in degrees).

Problems 12.20–12.21 involve a long, slender satellite with principal moments of inertia $I_3 = 200$ kg-m² (long minor axis) and $I_1 = I_2 = 1{,}800$ kg-m² (transverse major axes). The satellite is in a 600-km altitude circular orbit about the Earth.

12.20 The satellite has a roll angle of 5° (i.e., the minor axis is not aligned with the vertical direction). Compute the gravity-gradient torque and angular acceleration about the roll axis at this instant. Assume that the satellite has the angular velocity components $\omega_1 = -0.00043656$ rad/s and $\omega_2 = \omega_3 = 0$ at this instant.

12.21 Compute the maximum amplitude of the roll angle (in degrees) during the libration motion caused by the gravity gradient.

MATLAB Problem

12.22 Write an M-file that computes the torque-free motion of a rigid axisymmetric satellite. The inputs should be the principal moments of inertia about the axial (3 axis) and transverse axes (in kg-m²), the magnitude of the satellite's angular velocity, the angle γ between the 3 axis and the angular velocity vector, the initial Euler angles ψ_0 and ϕ_0, and the simulation time t_1. The outputs should be the satellite's initial angular momentum vector expressed in the inertial $OXYZ$ frame, the initial angular velocity vector (in 123 body-frame coordinates), the angular velocity vector at time t_1 (in body coordinates), the angular momentum vector at time t_1 expressed in the inertial $OXYZ$ frame, and the Euler angles ψ, θ, ϕ at time t_1. Test your M-file by verifying the solution of Example 12.3.

Mission Applications

12.23 Figure P12.23 shows the Juno space probe that reached Jupiter on July 5, 2016. The Juno spacecraft is powered by three very large solar array "wings." During launch, the solar arrays were folded up so that the spacecraft could fit within the payload shroud of the Atlas V launch vehicle. Prior to the upper-stage rocket burn that sent the Juno spacecraft on a hyperbolic escape trajectory, the probe was spin stabilized along its longitudinal axis. After the burn, the folded Juno probe was separated from the rocket stage. The spin rate of the *folded* (compact) Juno probe was 1.4 rpm and its momentum of inertia about the longitudinal (spin) axis was 7,500 kg-m². After stage separation, the solar arrays were deployed (unfolded) so that the Juno probe could achieve full power and charge its batteries. Compute the spin rate of the *unfolded* Juno probe if the spacecraft's moment of inertia is 22,300 kg-m² after the arrays are fully deployed as shown in Figure P12.23.

Figure P12.23 Juno spacecraft. *Source*: Courtesy of NASA.

Problems 12.24–12.26 involve the Galileo spacecraft. Galileo (Figure P12.24) was a spin-stabilized spacecraft that arrived at Jupiter in late 1995. Its approximate principal moments of inertia are $I_1 = I_2 = 3{,}450$ kg-m^2 and $I_3 = 5{,}370$ kg-m^2. Galileo was a flexible spacecraft that used a nutation damper mounted along its 3 axis so that it could maintain a spin rate of nearly 3 rpm. Suppose that Galileo's angular velocity vector is $\boldsymbol{\omega}_0 = -0.04\mathbf{u}_1 + 0.02\mathbf{u}_2 + 0.3\mathbf{u}_3$ rad/s in 123 body coordinates.

12.24 Determine the magnitude of the angular momentum vector **H** and the rotational kinetic energy $T_{\text{rot},0}$ at this instant.

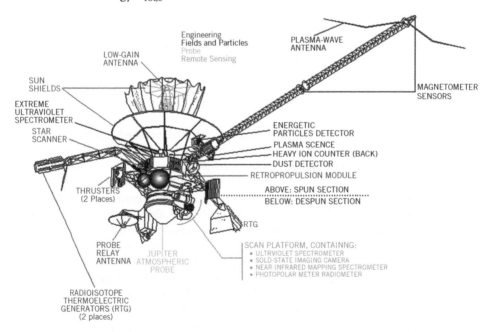

Figure P12.24 Galileo spacecraft. *Source*: Courtesy of NASA.

12.25 Determine the initial nutation angle of the Galileo spacecraft.

12.26 Determine the magnitude of the angular momentum vector **H** and the rotational kinetic energy $T_{rot,f}$ after the nutation damper has removed the "wobble" and returned the Galileo spacecraft to a pure spin about its 3 axis. What is the final spin rate?

Problems 12.27–12.29 pertain to the Stardust sample return capsule (SRC). The SRC was spin stabilized about its 3 axis during its Earth entry phase at 13.5 rpm. The approximate principal moments of inertia for the SRC are $I_1 = I_2 = 1.8$ kg-m^2 and $I_3 = 2.45$ kg-m^2. Suppose that the SRC angular velocity vector **ω** (with magnitude of 13.5 rpm) was slightly tilted away from its 3 axis by an angle of 1.5°.

12.27 Determine the nutation angle θ of the SRC.

12.28 Determine the SRC's precession rate $\dot{\psi}$ (in deg/s).

12.29 Determine the SRC's Euler angle rate $\dot{\phi}$ (in deg/s).

12.30 The Long Duration Exposure Facility (LDEF) was an unmanned orbiting platform for conducting scientific experiments and collecting information on the space environment (Figure P12.30). It was placed in a 475-km altitude circular low-Earth orbit by the US Space Shuttle Challenger in April 1984 and retrieved and returned to Earth by the Space Shuttle Columbia in January 1990. It was a large (9,700 kg), 12-sided, nearly cylindrical satellite that used gravity-gradient stabilization. Its principal moments of inertia are approximately $I_1 = 59,670$ kg-m^2, $I_2 = 60,890$ kg-m^2, and $I_3 = 22,100$ kg-m^2. Determine the roll and pitch libration frequencies for the LDEF.

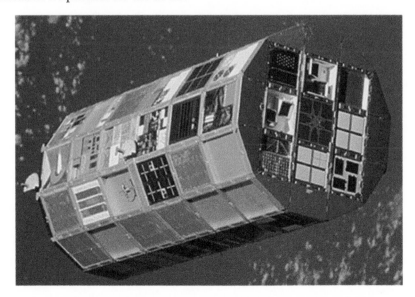

Figure P12.30 Long Duration Exposure Facility. *Source*: Courtesy of NASA.

13

Attitude Control

13.1 Introduction

The previous chapter analyzed a satellite's rotational motion or attitude dynamics. We primarily investigated torque-free rotational motion, spin stability, and the various disturbance torques that act on a satellite in its orbital environment. We briefly discussed two strategies for controlling a satellite's attitude or angular orientation: (1) spin stabilization; and (2) gravity-gradient stabilization. Both methods are used to maintain a fixed pointing direction; spin stabilization is often used to hold a fixed attitude for a Δv maneuver, whereas gravity-gradient stabilization is used to keep a long, thin satellite aligned with its local vertical direction for nadir pointing. As mentioned in Chapter 12, both strategies are examples of *passive attitude control*, where the satellite exploits the "natural dynamics" associated with gyroscopic stiffness or gravitational torque. This chapter introduces *active attitude control* techniques, where onboard sensors provide attitude feedback information to control algorithms that determine the appropriate actuator commands. Onboard actuators (such as thruster jets or spinning reaction wheels) are then operated to change the satellite's attitude until it reaches the desired orientation.

Feedback control theory is an expansive subject that is often introduced in a required undergraduate course in most mechanical and aerospace engineering curricula. Although we will discuss some feedback control concepts, this chapter is not intended to be an exhaustive treatment of control theory (References [1–3] provide a thorough study of control systems). Therefore, we will not focus on the traditional control-system analysis tools such as Laplace transforms, the root-locus method, and frequency-response analysis using Bode diagrams. Instead, we will use rudimentary closed-loop analysis methods coupled with the time-domain solutions of ordinary differential equations. This chapter will also focus on automatic attitude control strategies that use reaction jets (onboard thrusters) and momentum-exchange devices such as reaction wheels.

13.2 Feedback Control Systems

This section will present a brief overview of control systems sufficient for the purposes of this chapter. Figure 13.1 shows a block diagram of a general closed-loop feedback system for attitude control. The block labeled *satellite dynamics* essentially represents Euler's

Space Flight Dynamics, First Edition. Craig A. Kluever.
© 2018 John Wiley & Sons Ltd. Published 2018 by John Wiley & Sons Ltd.
Companion website: www.wiley.com/go/Kluever/spaceflightmechanics

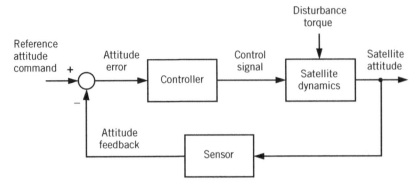

Figure 13.1 General closed-loop attitude control system.

moment equations (12.39). The block labeled *sensor* in Figure 13.1 denotes the physical measuring devices and algorithms that provide attitude feedback information. Most satellites use a combination of horizon sensors, sun sensors, star sensors, and gyroscopes coupled with an attitude determination scheme to discern the current rotational state of the spacecraft. The *controller* block in Figure 13.1 denotes the "control rules" and the physical actuating device that alters the satellite's attitude. The input to the controller block is usually an error signal which is the difference between the reference attitude command (a desired satellite orientation) and the actual satellite attitude (feedback signal) as measured by the sensor. The output of the controller is the control signal that drives the satellite dynamics. For example, the controller logic might decide to fire a pair of onboard thruster jets to create a torque that rotates the satellite in the desired direction. If the attitude control system is designed properly, the satellite's angular orientation will eventually match the reference attitude command. Finally, Figure 13.1 shows that the satellite may be subjected to disturbance torques due to solar radiation pressure, the gravity gradient, and so on.

13.2.1 Transfer Functions

The controller, satellite dynamics, and sensor blocks in Figure 13.1 are often represented by *transfer functions*. A transfer function of a linear, time-invariant differential equation is the Laplace transform of the output divided by the Laplace transform of the input with the assumption of zero initial conditions. Transfer functions provide a convenient method for representing and analyzing the input–output relationship of a dynamic system. Although they are formally defined by the Laplace transformation, we may develop a system's transfer function by utilizing the so-called differential operator (or "*D* operator"):

$$D \equiv \frac{d}{dt}$$

Therefore, time derivatives can be written as powers of operator D: for example, $Dy = \dot{y}$, $D^2 y = \ddot{y}$, and so on.

Let us demonstrate the *D*-operator by considering a simple second-order linear differential equation:

$$2\ddot{y} + 6\dot{y} + 20y = 3\dot{u} + 18u \tag{13.1}$$

where *y* is the output and *u* is the input. Applying the *D*-operator to Eq. (13.1), we obtain

$$\left(2D^2 + 6D + 20\right)y = (3D + 18)u \tag{13.2}$$

Next, we use Eq. (13.2) to form the ratio of output over input, or *y/u*

$$\frac{y}{u} = \frac{3D + 18}{2D^2 + 6D + 20} \tag{13.3}$$

In Laplace transform theory, the differentiation theorem states that the Laplace transform of $\dot{y}$ is equal to $sY(s) - y(0)$, where *s* is the complex Laplace variable. Furthermore, the Laplace transform of $\ddot{y}$ is $s^2Y(s) - sy(0) - \dot{y}(0)$, and so on for higher-order derivatives. Because all initial conditions [$y(0), \dot{y}(0)$, etc.] are assumed to be zero for a transfer function, we can conclude that multiplying by the *k*th power of *s* in the Laplace domain is equivalent to the *k*th derivative in the time domain. If we are primarily interested in *representing* dynamic systems, we can simply interchange the *D* and *s* symbols to produce transfer functions. Using this approach, Eq. (13.3) becomes

$$G(s) = \frac{3s + 18}{2s^2 + 6s + 20} \tag{13.4}$$

Function *G(s)* defined in Eq. (13.4) is the transfer function that represents the second-order differential equation (13.1). Figure 13.2 shows how the transfer function *G(s)* is used in a block diagram. We show the input and output signals in Figure 13.2 as time-domain functions *u(t)* and *y(t)*, respectively. The reader should note that multiplying input signal *u(t)* by the transfer function *G(s)* in Figure 13.2 yields the solution to the second-order differential equation (13.1) (remember that multiplying by *s* is equivalent to a time derivative). It is important to note that the transfer function *G(s)* represents the system dynamics and does *not* depend on the nature of the input function. Therefore, the block diagram shown in Figure 13.2 may be used to simulate the dynamic system (13.1) for *any* input *u(t)*; that is, the input could be a constant, a sinusoid, or a random signal.

Now let us turn our attention to Euler's moment equations and rotation about a single axis. Figure 13.3 shows a cylindrical satellite with a control torque M_c about the 3 axis. The control torque could be produced by firing a pair of thruster jets mounted on the periphery of the satellite (not shown in Figure 13.3). Euler's moment equation (12.39c) for the 3 axis is

$$M_3 = I_3\dot{\omega}_3 + (I_2 - I_1)\omega_1\omega_2 \tag{13.5}$$

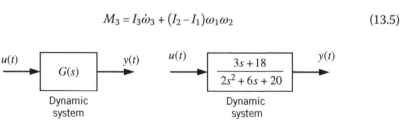

Figure 13.2 Transfer function representations of the input–output differential equation (13.1).

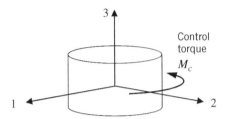

Figure 13.3 Satellite with a control torque and single-axis rotation.

For a pure spin about the 3 axis, we have $\omega_1 = \omega_2 = 0$ and Eq. (13.5) becomes

$$M_c = I_3\dot{\omega}_3 = I_3\ddot{\phi} \tag{13.6}$$

where the control torque M_c is the external torque along the 3 axis and ϕ is the satellite rotation angle about the 3 axis (therefore, $\dot{\phi} = \omega_3$ and $\ddot{\phi} = \dot{\omega}_3$). We should also note that Eq. (13.5) is reduced to Eq. (13.6) for general rotational motion of an axisymmetric satellite (i.e., $I_1 = I_2$) as shown in Figure 13.3. Applying the D-operator to Eq. (13.6) yields

$$M_c = I_3 D^2 \phi \tag{13.7}$$

Forming the ratio of output (rotation angle ϕ) to input (control torque M_c), we obtain

$$\frac{\phi}{M_c} = \frac{1}{I_3 D^2} \tag{13.8}$$

Finally, we may replace the D-operator with the Laplace variable s to obtain the satellite's transfer function

$$G(s) = \frac{1}{I_3 s^2} \tag{13.9}$$

Equation (13.9) is the transfer function that represents the satellite's attitude dynamics for a single-axis rotation about the 3 axis. Figure 13.4 shows two block diagram representations of the satellite dynamics (13.6). Figure 13.4a shows how transfer function $G(s)$ is used as a single block to represent the satellite dynamics $I_3\ddot{\phi} = M_c$. Figure 13.4b presents an equivalent block diagram of the satellite dynamics by using a chain of two integrator blocks. It should be clear to the reader that multiplying the three series blocks in Figure 13.4b produces transfer function $G(s)$ defined by Eq. (13.9). It should also be clear that the first block multiplication in Figure 13.4b represents Eq. (13.6); that is, torque divided by moment of inertia is equal to angular acceleration. Angular acceleration $\ddot{\phi}$

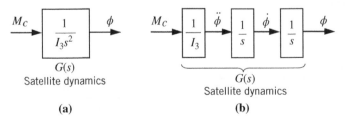

(a)   **(b)**

Figure 13.4 Equivalent block diagrams for single-axis satellite attitude dynamics: (a) single transfer function; and (b) "double-integrator" representation.

is then integrated once to produce angular velocity $\dot{\phi}$, which in turn is integrated to produce rotational angle ϕ. The $1/s$ blocks in Figure 13.4b represent integration (recall that multiplication by s is equivalent to a time derivative; therefore, multiplication by its inverse is integration with time).

13.2.2 Closed-Loop Control Systems

Next, we will briefly investigate a feedback system for satellite attitude control. Figure 13.5 shows a closed-loop control system for satellite rotation angle ϕ about its 3 axis. Note that the "double-integrator" representation of the satellite dynamics is employed. Comparing Figure 13.5 with Figure 13.1, we see that the sensor block in the feedback path is missing. Here we are assuming that the sensor (and the embedded attitude determination algorithm) provides perfect feedback information to the summing junction. Thus, we have used a *unity-feedback system* where the actual satellite attitude ϕ is compared with the reference attitude command ϕ_{ref} to form the attitude error. For real-world control systems, the feedback will be imprecise due to imperfect measurements and sensor noise. Finally, note that the controller algorithm in Figure 13.5 is represented by the transfer function $G_C(s)$.

The purpose of this subsection is to develop a *closed-loop* transfer function for the system in Figure 13.5. The satellite's attitude can be expressed as

$$\phi = G_C(s)G(s)\phi_e \tag{13.10}$$

which is obtained by multiplying the blocks in the forward path in Figure 13.5 from left to right. Figure 13.5 shows that the attitude error is simply the output of the summing junction, or $\phi_e = \phi_{ref} - \phi$. Substituting this result for ϕ_e in Eq. (13.10) yields

$$\phi = G_C(s)G(s)(\phi_{ref} - \phi) \tag{13.11}$$

Rearranging Eq. (13.11), we obtain

$$\phi[1 + G_C(s)G(s)] = G_C(s)G(s)\phi_{ref} \tag{13.12}$$

Finally, we can solve Eq. (13.12) for the ratio of the output ϕ to the input ϕ_{ref} to yield

$$\frac{\phi}{\phi_{ref}} = \frac{G_C(s)G(s)}{1 + G_C(s)G(s)} \tag{13.13}$$

Equation (13.13) is the *closed-loop transfer function* that relates the control system output (satellite attitude ϕ) to the control system input (reference attitude angle ϕ_{ref}). The satellite dynamics are represented by transfer function $G(s)$, that is, Eq. (13.9).

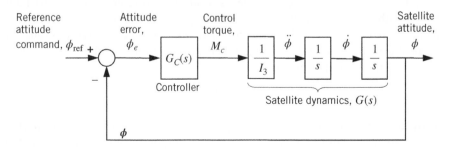

Figure 13.5 Closed-loop feedback control for single-axis satellite attitude dynamics.

The performance of the closed-loop control system (i.e., response speed, system damping, etc.) is dictated by the *poles* of the closed-loop transfer function (13.13). The closed-loop poles are the values of s that make the denominator of the closed-loop transfer function equal to zero, that is

$$1 + G_C(s)G(s) = 0 \tag{13.14}$$

The closed-loop poles determine the transient response of the closed-loop system. This result stems from the homogeneous solution of the governing differential equation that can be derived from Eq. (13.13). Hence, Eq. (13.14) is the closed-loop characteristic equation that determines the poles or *roots*. Readers with experience in feedback control systems may recall that Eq. (13.14) is the basis for the graphical root-locus method which determines the paths of the closed-loop poles (or roots) as a control gain varies.

13.2.3 Second-Order System Response

As stated in Section 13.1, we will not focus on traditional analysis tools such as the root-locus method or Bode diagram. Instead, we will concentrate on directly calculating the poles (or roots) of Eq. (13.14) for a given controller transfer function $G_C(s)$. For single-axis rotational maneuvers, the closed-loop characteristic equation (13.14) will be a second-order polynomial in s because the satellite dynamics $G(s)$ are second order; see Eq. (13.9). A general form of an *underdamped* second-order characteristic equation is

$$s^2 + 2\zeta\omega_n s + \omega_n^2 = 0 \tag{13.15}$$

where ζ is the *damping ratio* and ω_n is the *undamped natural frequency*. The two roots (or poles) of Eq. (13.15) are

$$s = -\zeta\omega_n \pm j\omega_n\sqrt{1-\zeta^2} \tag{13.16}$$

where $j = \sqrt{-1}$ is the imaginary number. Hence, the two roots are complex numbers with a negative real part and conjugate imaginary parts. The "free response" or homogeneous solution of an underdamped second-order system is

$$\phi_H(t) = Ke^{-\zeta\omega_n t}\cos(\omega_d t + \beta) \tag{13.17}$$

where $\omega_d = \omega_n\sqrt{1-\zeta^2}$ is the *damped frequency* or imaginary part of the complex conjugate roots defined by Eq. (13.16), and β is a phase angle that depends on the initial conditions. The coefficient K is the initial amplitude of the sinusoidal response, and it also depends on the initial conditions. Equation (13.17) shows that the transient response of an underdamped second-order system is a *damped sinusoid*; that is, a harmonic oscillation with an exponentially decaying envelope. Note that the transient response oscillates at frequency ω_d and eventually "dies out" at time $t = 4/\zeta\omega_n$ because the exponential term becomes $e^{-4} = 0.018$ (i.e., less than 2% of its initial value). Therefore, we can define the "settling time" as

$$t_S = \frac{4}{\zeta\omega_n} \tag{13.18}$$

Clearly, the damping ratio ζ and undamped natural frequency ω_n greatly affect the transient response. Figure 13.6 shows the transient responses of second-order systems with an undamped natural frequency $\omega_n = 2$ rad/s and damping ratios $\zeta = 0.4$, 0.5, 0.7, and 0.9. Decreasing the damping ratio increases the number of oscillations before

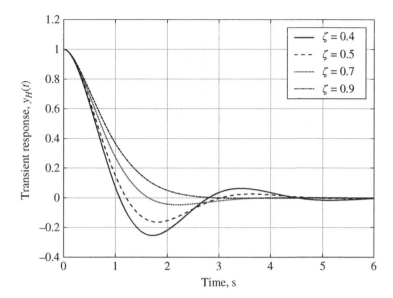

Figure 13.6 Transient responses for an underdamped second-order system with undamped natural frequency $\omega_n = 2$ rad/s.

the settling time and increases the under- and overshoot peaks of the damped sinusoidal response. The damped frequency of the response with $\zeta = 0.4$ is $\omega_d = \omega_n\sqrt{1-\zeta^2} = 1.833$ rad/s, and therefore its period is $2\pi/\omega_d = 3.43$ s as seen in Figure 13.6. Note that when $\zeta = 0.7$, the transient response seen in Figure 13.6 exhibits very little undershoot and essentially "dies out" at settling time $t_S = 2.86$ s as predicted by Eq. (13.18).

This subsection has presented a very brief overview of an underdamped second-order system response. Detailed analysis of second-order system response can be found in Kluever [4; pp. 213–230] and Ogata [5; pp. 388–398]. The following example involves a single-axis satellite attitude maneuver and is intended to illustrate the nature of a second-order system response for a closed-loop feedback system.

Example 13.1 Figure 13.7 shows a cylindrical-shaped satellite equipped with two pairs of reaction jets (thrusters). Firing the two jet pairs as shown in Figure 13.7b imparts a positive control torque M_c about the 3 axis (of course, firing the opposite jet pairs produces a negative torque). Suppose the satellite's moment of inertia about the 3 axis is $I_3 = 1,000$ kg-m². The jet pairs are located 1.5 m from the 3 axis. Assuming that each jet can be throttled to produce variable thrust up to a maximum value of 20 N, design a feedback controller that provides a fast, well-damped response to a constant attitude-rotation command of π rad (i.e., a 180° attitude maneuver). We will assume that the satellite is initially at rest $[\dot{\phi}(0) = 0]$ with zero attitude angle $\phi(0) = 0$.

Figure 13.8 shows the closed-loop attitude control system for this example (it is essentially the same as Figure 13.5 with the satellite dynamics represented by a single transfer function). Let us begin by investigating an extremely simple controller strategy known as *proportional control*, that is, the control torque is proportional to the attitude error:

$$\text{Proportional control: } M_c = K_P \phi_e \tag{13.19}$$

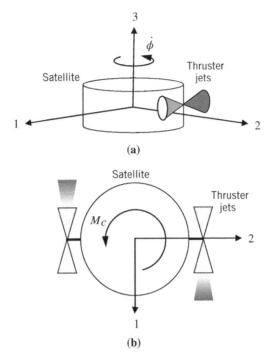

3

$\dot{\phi}$

Satellite

Thruster
jets

1   2

(a)

Satellite

Thruster
jets

M_C

2

1

(b)

Figure 13.7 Thruster jets for attitude control: (a) jet configuration; and (b) top-down view and positive control torque M_c (Example 13.1).

where K_P is called the *proportional gain*. In this case, K_P has units of torque/angular displacement or N-m/rad. For this fictitious example, we have assumed that each thruster can be throttled from zero to 20 N, and therefore the *maximum* control torque magnitude is $(2)(20 \text{ N})(1.5 \text{ m}) = 60$ N-m (recall that the jets are fired in pairs and that each jet has a moment arm of 1.5 m from the 3 axis). Because the initial attitude error is π rad for this example (i.e., a 180° rotation maneuver), the maximum proportional gain is $(60 \text{ N-m})/\pi$ rad $= 19.1$ N-m/rad.

Comparing Eq. (13.19) with Figure 13.8, we see that the controller is simply $G_C(s) = K_P$. Let us determine the closed-loop transfer function using Eq. (13.13):

$$\frac{\phi}{\phi_{\text{ref}}} = \frac{G_C(s)G(s)}{1 + G_C(s)G(s)}$$

$$= \frac{K_P \dfrac{1}{I_3 s^2} \dfrac{I_3 s^2}{I_3 s^2}}{1 + K_P \dfrac{1}{I_3 s^2} \dfrac{I_3 s^2}{I_3 s^2}} \tag{13.20}$$

$$= \frac{K_P}{I_3 s^2 + K_P}$$

Equation (13.20) is the closed-loop transfer function with a proportional controller $G_C(s) = K_P$. The poles (or roots) of the denominator polynomial determine the nature of the closed-loop transient response. Using $I_3 = 1,000$ kg-m^2 and $K_P = 19$ N-m/rad, the closed-loop characteristic equation is

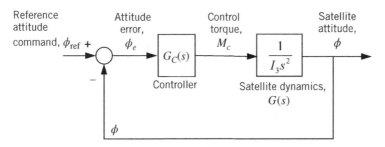

Figure 13.8 Closed-loop attitude control system (Example 13.1).

$$1{,}000s^2 + 19 = 0$$

or,

$$s^2 + 0.019 = 0 \qquad (13.21)$$

The closed-loop poles are $s = \pm j0.1378$, which are purely imaginary poles. Comparing the closed-loop characteristic equation for a system with proportional control with Eq. (13.15) (i.e., the "standard" second-order characteristic equation), we see that damping ratio $\zeta = 0$ and undamped natural frequency $\omega_n = \sqrt{0.019} = 0.1378$ rad/s. In other words, the closed-loop system has no damping mechanism and subsequently the satellite's angular response $\phi(t)$ is an undamped harmonic oscillation with a frequency of 0.1378 rad/s (or, period $= 2\pi/\omega_n = 45.6$ s).

Figure 13.9 shows the closed-loop satellite attitude angle $\phi(t)$ for the reference command $\phi_{\text{ref}} = \pi$ rad and proportional controller $K_P = 19$ N-m/rad. Here we see that the attitude angle oscillates without damping about the reference ($\phi_{\text{ref}} = \pi$ rad) with a period of 45.6 s. The satellite attitude continually overshoots the reference attitude by 180° (or one-half of a revolution). Figure 13.10 shows the control torque history for a proportional

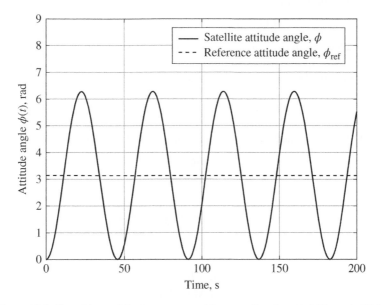

Figure 13.9 Closed-loop attitude response with proportional control (Example 13.1).

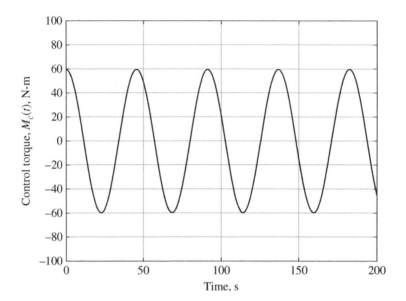

Figure 13.10 Control torque with proportional control (Example 13.1).

controller with $K_P = 19$ N-m/rad. The control torque oscillates between its extreme values of ± 60 N-m with a period of 45.6 s. Clearly, the proportional control scheme provides very poor closed-loop performance because it cannot drive the satellite attitude to the desired reference angle.

A proportional controller will not work because the resulting closed-loop system does not possess a damping term [i.e., the denominator of Eq. (13.20) does not contain a first-order s term]. A well-known solution to this problem is to employ *proportional-derivative* (PD) *control* defined by

$$\text{PD control:} \quad M_c = K_P \phi_e + K_D \dot{\phi}_e \qquad (13.22)$$

where K_D is called the *derivative gain* (with units of N-m-s/rad in this case). Now the control torque is proportional to the attitude error and the derivative of the attitude error. The PD controller transfer function is

$$\text{PD control:} \quad \frac{M_c}{\phi_e} = G_C(s) = K_P + K_D s \qquad (13.23)$$

Using the PD transfer function, the closed-loop transfer function is

$$\frac{\phi}{\phi_{\text{ref}}} = \frac{G_C(s)G(s)}{1 + G_C(s)G(s)}$$

$$= \frac{(K_P + K_D s)\dfrac{1}{I_3 s^2}}{1 + (K_P + K_D s)\dfrac{1}{I_3 s^2}} \frac{I_3 s^2}{I_3 s^2} \qquad (13.24)$$

$$= \frac{K_P + K_D s}{I_3 s^2 + K_D s + K_P}$$

The characteristic equation of the closed-loop system using PD control is

$$I_3 s^2 + K_D s + K_P = 0$$

or,

$$s^2 + \frac{K_D}{I_3} s + \frac{K_P}{I_3} = 0 \tag{13.25}$$

Comparing Eq. (13.25) with the standard second-order system (13.15), we see that the zeroth-order term in Eq. (13.25) determines the closed-loop undamped natural frequency, that is, $\omega_n = \sqrt{K_P/I_3}$. The first-order term in Eq. (13.25) determines the damping ratio:

$$\frac{K_D}{I_3} = 2\zeta\omega_n$$

These expressions show that we can achieve any desired natural frequency ω_n and damping ratio ζ by proper selection of control gains K_P and K_D. However, we cannot "over gain" the system because control torque is limited to 60 N-m. Therefore, let us use $K_P = 19$ N-m/rad (as before) and set $\zeta = 0.7$ for good closed-loop damping. Using these values, we find that the derivative gain must be $K_D = 192.98$ N-m-s/rad.

Figure 13.11 shows the closed-loop attitude response using a PD controller. Note that the satellite attitude converges to the desired reference attitude $\phi_{ref} = \pi$ rad in less than 45 s. In fact, the settling time can be estimated from the undamped natural frequency and damping ratio as $t_S = 4/\zeta\omega_n = 41.5$ s. Furthermore, the attitude response shows very good damping with a slight overshoot of the desired attitude angle. If the overshoot is too large, we can reduce it by increasing the derivative gain K_D. Figure 13.12 shows the control

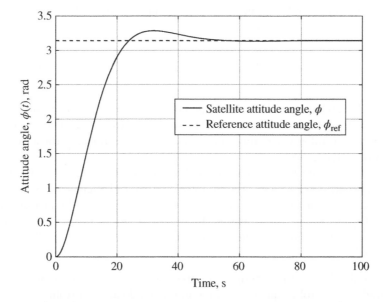

Figure 13.11 Closed-loop attitude response with proportional-derivative control (Example 13.1).

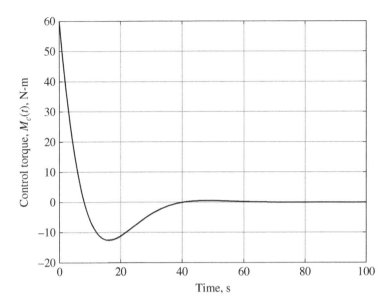

Figure 13.12 Control torque with proportional-derivative control (Example 13.1).

torque as commanded by the PD controller. The control torque is near its maximum value of 60 N-m at time $t = 0$ because the proportional gain was chosen to be $K_P = 19$ N-m/rad and the initial angular error is 3.1416 rad. The positive control torque causes the satellite to initially rotate toward the reference attitude. However, after time $t = 8$ s, the PD controller switches the control torque to a negative value (i.e., jet reversal) in order to provide a "braking torque" to decelerate the rotation as the satellite approaches its target attitude angle. This braking action is attributed to the derivative-control term. The control torque steadily diminishes to zero as the attitude angle matches the reference and the attitude error goes to zero.

Example 13.1 illustrates some basic feedback control concepts that are summarized as follows:

1) Because the satellite attitude dynamics do not possess any natural damping, a simple proportional controller will not work. The solution is to add "artificial" damping by using a PD controller where the control torque is proportional to the attitude error and its time derivative.
2) Because the satellite dynamics are second order, the closed-loop transfer function using PD control will also be second order. The control engineer may achieve any desired closed-loop natural frequency ω_n and damping ratio ζ by proper selection of the PD gains K_P and K_D. However, the system cannot be over gained; if the gains are too large, the control torque will exceed physical limitations (i.e., "control saturation" will occur).

Finally, we should note that Example 13.1 presented a satellite with variable-thrust jets that allowed the control torque to be continuously modulated. In reality, the relatively small thrusters used for attitude control cannot be throttled, and instead operate in an "on–off" switching or pulsed mode. Hence, the magnitude of the control torque is a single value and the control scheme depicted in Figure 13.8 is not realistic or feasible. The reader should remember that the purpose of Example 13.1 (and Section 13.2) is to illustrate the basic attributes of a closed-loop control system. We will discuss a realistic attitude control system that uses pulsed (on–off) jets in Section 13.5.

13.3   Mechanisms for Attitude Control

In this section, we will discuss the devices or actuators that are used to control a satellite's attitude. The three major mechanisms for active attitude control are: (1) reaction jets; (2) momentum-exchange devices; and (3) magnetic torquers. We will briefly discuss each device.

13.3.1   Reaction Jets

Reaction jets (or *thrusters*) produce an external torque on the satellite by expelling mass. They may be "hot-gas" or "cold-gas" thrusters. Hydrazine is often the propellant of choice for hot-gas thrusters that produce a force via a chemical reaction. Cold-gas thrusters produce relatively small forces by opening a valve and expelling a compressed neutral gas such as nitrogen. In either case, a reaction-jet control system must use consumables (propellant or compressed gas) that obviously contribute to the usable lifetime of the satellite. Reaction jets are typically on–off devices that deliver a fixed thrust magnitude (and hence fixed control torque) when activated. This pulsed mode of operation results in a discontinuous (or nonlinear) control signal and complicates the control design. We present control systems that use reaction jets in Sections 13.5 and 13.6.

13.3.2   Momentum-Exchange Devices

Momentum-exchange devices operate on the principle of conservation of angular momentum in a torque-free environment. These devices consist of spinning wheels attached to the satellite. Reaction wheels, momentum wheels, and control moment gyros (CMGs) are all momentum-exchange devices.

   Reaction wheels are small rotating disks attached to the satellite and driven by an electric motor. The basic operation of a reaction wheel is fairly intuitive: spinning a wheel mounted along the 3 axis in a clockwise direction will cause a counterclockwise satellite spin about the 3 axis so that total angular momentum is conserved. Figure 13.13 illustrates this concept. In Figure 13.13, a person is standing on a platform mounted on (theoretical) frictionless bearings. Suppose the person has a hand-held electric drill connected to a flywheel and is pointing the drill's spin axis along his or her longitudinal (vertical) axis. In Figure 13.13a, the person is motionless and the drill's flywheel is not rotating, and hence the total angular momentum of the system (person + platform + flywheel) is zero. In Figure 13.13b, the person turns on the drill so that the rotating flywheel adds positive ("upward") angular momentum $\Delta \mathbf{H}_w$. Consequently, the person and platform rotate in the *opposite* direction as the flywheel. The opposite spin of the person and platform contributes "downward" angular momentum change $\Delta \mathbf{H}_p$ so that the total

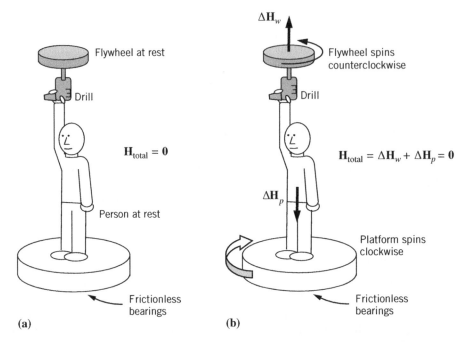

ΔH_w

Flywheel at rest

Drill

$H_{total} = 0$

Person at rest

Frictionless
bearings

(a)

Flywheel spins
counterclockwise

Drill

$H_{total} = \Delta H_w + \Delta H_p = 0$

ΔH_p

Platform spins
clockwise

Frictionless
bearings

(b)

Figure 13.13 Reaction wheel concept and the conservation of angular momentum: (a) person and flywheel are at rest; and (b) spinning flywheel causes person and platform to spin in the opposite direction.

angular momentum is conserved, that is $H_{total} = \Delta H_w + \Delta H_p = 0$ as shown in Figure 13.13b. After the person has rotated to the desired attitude, he or she turns off the drill and comes to rest so that total angular momentum remains zero. The reader should note that the drill's torque acting on the flywheel is an *internal* torque that does not change the total angular momentum of the system.

Figure 13.13 illustrates how one rotating flywheel can control one rotational axis. A minimum of three orthogonally mounted reaction wheels are required to fully control the rotational motion of a satellite. Figure 13.14 shows a satellite with three independent reaction wheels aligned with the 123 principal axes. Each reaction wheel is driven by a

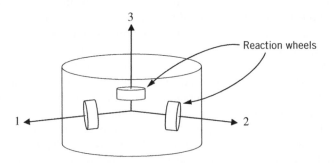

3

Reaction wheels

1

2

Figure 13.14 Three reaction wheels mounted along the 123 principal axes.

separate electric motor, and hence the satellite's rotational motion about any axis can be controlled.

Momentum wheels are essentially reaction wheels that maintain a nominal spin rate for gyroscopic rigidity. Because reaction wheels are used to change the angular orientation of a satellite whose normal state is zero angular velocity, the wheels also normally have zero angular velocity (hence the total angular momentum of a satellite with reaction-wheel control is zero). Conversely, a satellite equipped with one or more momentum wheels has non-zero angular momentum due to the wheel's nominal spin rate. The satellite's attitude may be controlled by changing the spin rate of the momentum wheel. Attitude control systems that use momentum wheels are sometimes called *momentum-bias* systems because they maintain a constant non-zero angular momentum in the absence of external torques.

Control moment gyros are relatively large spinning momentum wheels mounted on movable gimbals. The gimbal mounts can be rotated by electric motors in order to change the direction of the wheel's spin axis. Consequently, the directional change of the CMG's angular momentum vector causes the satellite's angular momentum vector to undergo an opposite change for momentum conservation. Whereas reaction wheels have variable spin rates along a fixed axis, CMGs typically have constant spin rates and a variable-direction axis.

Figure 13.15 illustrates the momentum-exchange concept of a CMG. Consider again a person standing on a frictionless platform with a hand-held electric drill capable of rotating a flywheel. In Figure 13.15a, the person is stationary and pointing the drill axis in a direction perpendicular to his or her longitudinal (or vertical) axis. The drill is turned on and rotating the flywheel so that it has angular momentum $\mathbf{H}_w$ along a transverse axis pointing away from the person as shown in Figure 13.15a. Hence the flywheel's angular momentum is the total angular momentum of the system (person + platform + flywheel), or $\mathbf{H}_{total} = \mathbf{H}_w$. In Figure 13.15b, the person rotates the drill axis downward (using an internal torque) to change the direction of $\mathbf{H}_w$. If the flywheel's change in angular momentum is vector $\Delta \mathbf{H}_w$ (downward), then the person and platform must rotate counterclockwise to produce an opposite momentum change $\Delta \mathbf{H}_p$ (upward) so that $\mathbf{H}_{total}$ remains constant. Of course, pointing the drill upward will cause the person to rotate clockwise. The person's rotational motion can be stopped by pointing the drill axis along its original transverse direction shown in Figure 13.15a.

Figure 13.16 shows a satellite with a CMG aligned with its 1 axis. The single gimbal shown in Figure 13.16 allows the CMG to be rotated about the 2 axis, which would create a satellite spin about its 3 axis. Adding a second (outer) gimbal (not shown in Figure 13.16) would allow the CMG to be rotated about the 3 axis. Consequently, a dual-gimbaled CMG can control the satellite's spin about the 2 and 3 axes. Two dual-gimbaled CMGs will provide attitude control about three orthogonal axes.

Momentum-exchange devices allow smooth and continuous control signals because the wheel speed (or CMG axis) can be continuously modulated between upper and lower bounds. As a result, the feedback control system is linear and easier to analyze and design when compared with nonlinear control systems.

There is one significant disadvantage to using momentum-exchange devices for active attitude control. Consider a scenario where we wish to keep a satellite in an inertially fixed attitude with zero angular momentum. Suppose that an external disturbance torque (possibly due to solar radiation pressure) acts on the satellite and causes a small but

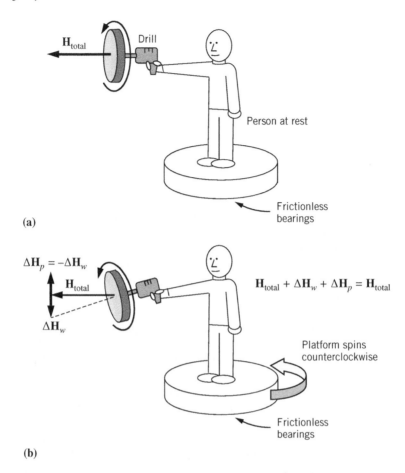

(a)

(b)

Figure 13.15 Control moment gyro concept and the conservation of angular momentum: (a) person is at rest and flywheel is spinning; and (b) rotating axis of spinning flywheel downward causes person and platform to spin.

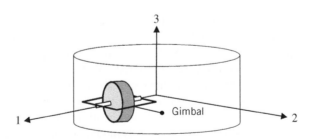

Figure 13.16 Single-gimbal control moment gyro initially spinning along the 1 axis.

steady increase in angular momentum (remember that external torque is equal to the time-rate of angular momentum). A reaction wheel can absorb the added angular momentum by spinning in the opposite direction of the disturbance torque. Eventually, the wheel will spin at its maximum angular velocity; at this point, it has become *saturated* and cannot absorb additional angular momentum. The satellite needs to "dump" the wheel's stored momentum by using an external torque provided by a separate actuation device. Typically, the "momentum dump" is performed by a reaction-jet system. Requiring a second attitude-control device for the momentum dump complicates the spacecraft design.

13.3.3 Magnetic Torquers

Magnetic torquers (or magnetic torque rods) interact with a planet's magnetic field to create an external torque on the satellite. This external torque is the cross product of the magnetic dipole (produced by passing current through a wire coil) and the planet's magnetic field vector. Hence, the external torque on the satellite is always perpendicular to the planet's magnetic field. Magnetic torquers are wire-coil electromagnetic rods that are mounted along a satellite's body axis. The direction of the dipole moment may be reversed by switching the flow of current in the coils. Because the planet's magnetic field strength varies with altitude, the effectiveness of magnetic torquers diminishes with altitude. Furthermore, magnetic torquers produce relatively small external torques.

13.4   Attitude Maneuvers Using Reaction Wheels

The previous discussion of momentum-exchange devices introduces the concept of utilizing conservation of angular momentum as a mechanism for controlling a satellite's attitude. Altering the spin rate of an internal reaction wheel causes a counteracting satellite spin rate, which can control the satellite's attitude. In this section, we will only consider attitude control about a single axis (the 3 axis).

Figure 13.17 shows a cylindrical satellite with 123 principal axes with a single reaction wheel mounted along the 3 axis. The satellite's moment of inertia about the 3 axis is I_{sat}, while the wheel's moment of inertia is I_w. We must stress two very important points here before continuing:

1) For the analysis to follow, the satellite's moment of inertia I_{sat} does *not* include the inertia contribution from the reaction wheel.
2) The reaction wheel spins on an axis that is *fixed* to the satellite and aligned with the 3 axis.

The satellite's principal moments of inertia are I_1, I_2, and I_3. Using our prior definitions, the 3-axis principal moment of inertia is $I_3 = I_{sat} + I_w$.

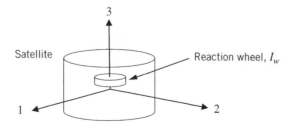

Figure 13.17 Reaction wheel mounted along the satellite's 3 axis.

Next, let us define the independent spin rates about the body-fixed 3 axis. The satellite's (inertial) angular velocity (or spin) component along its 3 axis is defined as $\dot{\phi} = \omega_3$. The reaction wheel's angular velocity ω_w is defined as the wheel's spin rate *relative* to the satellite. Therefore, the *absolute* or inertial angular velocity of the wheel is

$$\omega_{w,\text{abs}} = \omega_3 + \omega_w \tag{13.26}$$

Equation (13.26) shows that if the wheel is not spinning relative to the satellite (i.e., $\omega_w = 0$), then the wheel's absolute angular velocity is equal to the absolute angular velocity of the satellite, ω_3.

Let us only consider rotational motion about the 3 axis. The satellite's angular momentum is

$$H = (I_{\text{sat}} + I_w)\omega_3 + I_w\omega_w \tag{13.27}$$

Substituting $I_{\text{sat}} + I_w = I_3$, the angular momentum is

$$H = I_3\omega_3 + I_w\omega_w \tag{13.28}$$

Equations (13.27) and (13.28) both show that the satellite's angular momentum is the sum of the momentum contributions from the spinning satellite ($I_3\omega_3$) and the spinning wheel ($I_w\omega_w$). Next, take the time derivative of Eq. (13.28)

$$\dot{H} = I_3\dot{\omega}_3 + I_w\dot{\omega}_w = 0 \tag{13.29}$$

Angular momentum is constant because there are no external torques. Substituting $\ddot{\phi} = \dot{\omega}_3$ into Eq. (13.29), we obtain

$$I_3\ddot{\phi} = -I_w\dot{\omega}_w$$

or,

$$\ddot{\phi} = \frac{-I_w}{I_3}\dot{\omega}_w \tag{13.30}$$

Equation (13.30) is the governing dynamical equation if the satellite only rotates about its 3 axis. The satellite's absolute angular acceleration is always opposite the wheel's relative angular acceleration. Because the inertia ratio $I_w/I_3 = I_w/(I_{\text{sat}} + I_w)$ is always less than unity, the magnitude of the satellite's angular acceleration is less than the wheel's acceleration. When the wheel stops accelerating and spins at a constant rate, the satellite must also spin at a constant rate so that Eq. (13.29) is satisfied.

Next, we need to understand the governing dynamics of the reaction wheel. Figure 13.18 shows a free-body diagram of the external torques that act on the reaction wheel: motor torque T_m and friction torque T_f. Summing torques on the wheel and applying Newton's second law yields

$$I_w\dot{\omega}_{w,\text{abs}} = T_m - b\omega_w \tag{13.31}$$

Note that we must use the reaction wheel's *absolute* angular acceleration in Newton's second law. We can obtain the wheel's absolute acceleration by taking the time derivative of Eq. (13.26):

$$\dot{\omega}_{w,\text{abs}} = \dot{\omega}_3 + \dot{\omega}_w \tag{13.32}$$

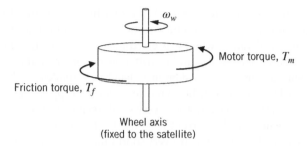

Figure 13.18 Free-body diagram of the reaction wheel.

Equation (13.32) shows that the wheel's absolute angular acceleration is the sum of the satellite's absolute acceleration ($\ddot{\phi} = \dot{\omega}_3$) and the wheel's acceleration relative to the satellite ($\dot{\omega}_w$). The right-hand side of Eq. (13.31) is the sum of the motor torque T_m and the opposing linear friction torque $b\omega_w$, where b is the viscous friction coefficient (in units of N-m-s/rad). Substituting Eq. (13.32) into Eq. (13.31) and solving for the wheel's acceleration relative to the satellite yields

$$\dot{\omega}_w = \frac{1}{I_w}(T_m - b\omega_w) - \ddot{\phi} \tag{13.33}$$

We can substitute Eq. (13.30) for the satellite's acceleration $\ddot{\phi}$ in Eq. (13.33) to yield

$$\dot{\omega}_w = \frac{1}{I_w}(T_m - b\omega_w) + \frac{I_w}{I_3}\dot{\omega}_w \tag{13.34}$$

Or,

$$(I_3 - I_w)\dot{\omega}_w = \frac{I_3}{I_w}(T_m - b\omega_w) \tag{13.35}$$

Finally, using $I_{\text{sat}} = I_3 - I_w$, we obtain

$$\dot{\omega}_w = \frac{I_3}{I_{\text{sat}}I_w}(T_m - b\omega_w) \tag{13.36}$$

Equation (13.36) is our model of the reaction-wheel dynamics.

Torque for a direct current (DC) motor is a linear function of motor current i_m

$$T_m = K_m i_m \tag{13.37}$$

where K_m is called the *motor-torque constant* (in N-m/A), and it depends primarily on the strength of the DC motor's magnetic field, the total length of wire wrappings about the rotating armature, and the radius of the rotor. The DC motor's electrical circuit consists of a voltage source e_{in} (in volts, V), electrical resistance R (in ohms, Ω), and inductance L (in henries, H) of the wire wrappings. We can apply Kirchhoff's voltage loop law to the DC motor circuit to obtain

$$L\frac{di_m}{dt} + Ri_m = e_{\text{in}} - K_b\omega_w \tag{13.38}$$

where K_b is called the *back-emf constant* (in V-s/rad), which arises from Faraday's laws of induction. It turns out that the back-emf constant K_b and motor-torque constant K_m have identical numerical values when expressed in SI units [i.e., the units N-m/A are equivalent to V-s/rad because volts can be written as power (in watts, 1 W = 1 N-m/s) divided by current (in amps, A)]. Additional details for modeling electromechanical systems (such as a DC motor) can be found in Kluever [4; pp. 57–62]. Typically, the "time constant" $\tau = L/R$ for a DC motor is very small relative to the mechanical (rotor) time constant. Consequently, the current response to the input voltage e_{in} is extremely rapid and reaches a steady-state value in a very short time. We can obtain the steady-state current response of Eq. (13.38) by neglecting the derivative term di_m/dt. In this case, the DC motor current is

$$i_m = \frac{1}{R}\left(e_{in} - K_b\omega_w\right) \tag{13.39}$$

We may think of the back-emf term $K_b\omega_w$ as a parasitic voltage that reduces the input voltage e_{in} when the motor is rotating. Equation (13.39) is Ohm's law: current is equal to the net voltage divided by resistance. Finally, we may multiply the motor current, Eq. (13.39), by K_m to obtain the motor torque:

$$T_m = \frac{K_m}{R}\left(e_{in} - K_b\omega_w\right) \tag{13.40}$$

Figure 13.19 presents a block diagram that illustrates the interaction between the DC motor, reaction wheel, and satellite. The reader should be able to trace the block diagram signal paths and identify the corresponding modeling equations. For example, the first summing junction in Figure 13.19 produces the voltage difference $(e_{in} - e_b)$, which is multiplied by K_m/R to produce the motor torque. The inner loop in Figure 13.19 shows the summation of the motor torque and friction torque in order to determine the reaction wheel's angular acceleration $\dot{\omega}_w$ in accordance with Eq. (13.36). Note that the wheel acceleration is integrated to produce wheel velocity, which is needed to compute the

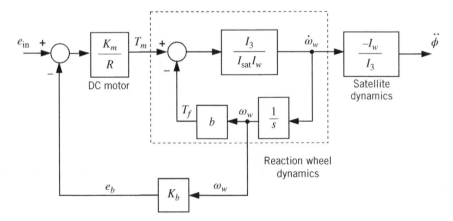

Figure 13.19 Open-loop control system: direct current motor, reaction wheel, and satellite dynamics.

friction torque and back-emf terms. The final block shows the conservation of angular momentum, Eq. (13.30), where multiplying the wheel's relative acceleration by the inertia ratio $-I_w/I_3$ produces the satellite's angular acceleration. Figure 13.19 is an *open-loop* control system because the control signal (DC motor input voltage e_{in}) does *not* depend on the satellite's attitude or angular velocity. In other words, we could apply a constant input voltage to the DC motor, and it will produce a motor torque that accelerates the reaction wheel, which ultimately causes an opposite angular acceleration of the satellite. However, the ensuing satellite motion will not necessarily meet our desired objective. The following examples illustrate the operation of a DC motor and reaction wheel in an open-loop setting.

Example 13.2   Satellite operators want to accurately determine a spacecraft's principal moment of inertia after it has fired an onboard rocket and expended propellant mass. The spacecraft is equipped with a reaction wheel, and both the spacecraft and wheel are initially at rest (i.e., zero angular momentum). The satellite operators command the reaction wheel to spin up to a constant angular velocity vector of $\boldsymbol{\omega}_w = 3{,}000\ \mathbf{u}_3$ rpm by using the DC motor's speed control mode. After the wheel reaches its constant spin rate, the onboard gyroscope measures the satellite's angular velocity vector as $\boldsymbol{\omega}_3 = -0.182\ \mathbf{u}_3$ rpm. If the reaction wheel's moment of inertia is $I_w = 0.055\ \text{kg-m}^2$, determine the satellite's moment of inertia about its 3 axis *without* the inertia contribution from the reaction wheel.

Because the satellite and wheel are initially at rest, the angular momentum is zero. We use Eq. (13.28) to determine the angular momentum

$$H = I_3\omega_3 + I_w\omega_w = 0$$

where I_3 is the *total* moment of inertia (satellite + wheel) about the 3 axis. Solving this expression for the total moment of inertia, we obtain

$$I_3 = -I_w\frac{\omega_w}{\omega_3} = \left(-0.055\,\text{kg-m}^2\right)(3{,}000\,\text{rpm})/(-0.182\,\text{rpm}) = 906.5934\ \text{kg-m}^2$$

Note that we set the satellite's spin rate ω_3 as a negative value because it spins in the opposite direction as the reaction wheel. Finally, the satellite's moment of inertia *without* the reaction wheel is

$$I_{sat} = I_3 - I_w = \boxed{906.5384\ \text{kg-m}^2}$$

Example 13.3   The DC motor associated with the reaction wheel and spacecraft from Example 13.2 has the following parameters: motor torque constant $K_m = 0.043$ N-m/A, back-emf constant $K_b = 0.043$ V-s/rad, resistance $R = 2.8\ \Omega$, and friction coefficient $b = 6(10^{-5})$ N-m-s/rad. Determine the input voltage to the DC motor required to maintain a constant wheel spin rate of 3,000 rpm.

We can manipulate Eq. (13.40) to obtain the DC motor's input voltage in terms of motor torque and the wheel's angular velocity:

$$e_{in} = \frac{R}{K_m}T_m + K_b\omega_w$$

Equations (13.33) or (13.36) show that the motor torque T_m is balanced by the wheel's friction torque $b\omega_w$ when the wheel reaches a constant angular velocity. Hence, the motor torque is

$$T_m = b\omega_w = \left[6\left(10^{-5}\right) \text{N-m-s/rad}\right](314.1593 \text{ rad/s}) = 0.018850 \text{ N-m}$$

Note that the wheel's angular velocity ω_w must be expressed in rad/s (i.e., 3,000 rpm = 314.1593 rad/s). Therefore, the input voltage is

$$e_{\text{in}} = \frac{R}{K_m} T_m + K_b \omega_w$$

$$= (2.8 \ \Omega)(0.018850 \text{ N-m})/0.043 \text{ N-m/A} + (0.043 \text{ V-s/rad})(314.1593 \text{ rad/s})$$

$$= \boxed{14.736 \text{ V}}$$

The back-emf voltage from the motor rotation is $e_b = K_b \omega_w = 13.509$ V, and the motor current is $i_m = (e_{\text{in}} - e_b)/R = 0.438$ A.

Figure 13.20 presents a *closed-loop* block diagram for single-axis attitude control with a reaction wheel. Note that the desired satellite attitude angle ϕ_{ref} (rotation angle about the 3 axis) is compared with the satellite's *actual* attitude ϕ to form the attitude error ϕ_e, which is the input to the controller transfer function $G_C(s)$. Electrical voltage input e_{in} to the DC motor is produced by the controller block $G_C(s)$. Our open-loop model of the motor, wheel, and satellite dynamics (Figure 13.19) resides in the forward path of Figure 13.20.

At this point, we would like to design the controller $G_C(s)$ using similar methods that were employed in Example 13.1. In particular, we will determine the feedback gains based on the desired closed-loop response characteristics. We can greatly simplify the controller design by ignoring the back-emf and friction torque terms in Figure 13.19. Figure 13.21 shows the simplified open-loop models of the DC motor, reaction wheel, and satellite dynamics when back-emf and friction torque are eliminated from

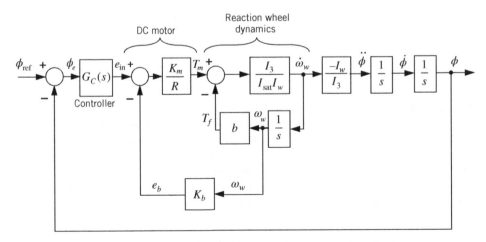

Figure 13.20 Closed-loop attitude control using a reaction wheel.

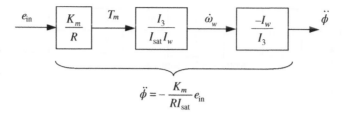

$$\ddot{\phi} = -\frac{K_m}{RI_{sat}}e_{in}$$

Figure 13.21 Simplified models of the direct current motor, reaction wheel, and satellite dynamics for controller design.

Figure 13.19. Multiplying the three gains in the forward path, we obtain the approximate relationship between input voltage and satellite acceleration:

$$\ddot{\phi} = -\frac{K_m}{RI_{sat}}e_{in} \qquad (13.41)$$

Figure 13.22 shows the closed-loop attitude control system with Eq. (13.41) approximating the combined models for the DC motor, reaction wheel, and satellite dynamics. Using Eq. (13.13), the closed-loop transfer function is

$$\frac{\phi}{\phi_{ref}} = \frac{G_C(s)\frac{a}{s^2}}{1 + G_C(s)\frac{a}{s^2}} = \frac{G_C(s)a}{s^2 + G_C(s)a} \qquad (13.42)$$

where $a = -K_m/RI_{sat}$. Using a PD controller $G_C(s) = K_P + K_D s$, the denominator of closed-loop transfer function is

$$s^2 + (K_P + K_D s)a = s^2 + aK_D s + aK_P \qquad (13.43)$$

$$= s^2 + 2\zeta\omega_n s + \omega_n^2 \qquad (13.44)$$

Comparing Eq. (13.43) with the "standard" second-order characteristic equation (13.44), we can express the first- and zeroth-order coefficients in terms of the undamped natural frequency ω_n and damping ratio ζ:

$$\text{First-order term: } aK_D = 2\zeta\omega_n$$

$$\text{Zeroth-order term: } aK_P = \omega_n^2$$

The undamped natural frequency ω_n and damping ratio ζ affect the response of a second-order system. Because the PD controller has two free gains (K_P and K_D), we can achieve *any* desired ω_n and ζ for the closed-loop system [however, we must remember that the closed-loop system depicted in Figure 13.22 and Eq. (13.42) is an approximation of the actual system because we have ignored the back-emf and friction torque]. The proportional and derivative gains are

$$\text{Proportional gain: } K_P = \frac{\omega_n^2}{a} = \frac{-RI_{sat}\,\omega_n^2}{K_m} \qquad (13.45)$$

$$\text{Derivative gain: } K_D = \frac{2\zeta\omega_n}{a} = \frac{-2RI_{sat}\zeta\omega_n}{K_m} \qquad (13.46)$$

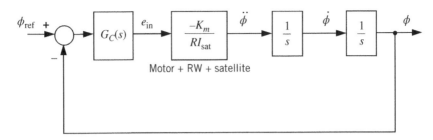

Figure 13.22 Closed-loop attitude control with approximate models for the direct current motor, reaction wheel, and satellite dynamics.

Equations (13.45) and (13.46) can be used to "tune" the PD controller gains in order to achieve a desired closed-loop system response. However, we must emphasize that a candidate PD controller designed using Eqs. (13.45) and (13.46) must ultimately be tested with a closed-loop simulation that includes the *accurate* system dynamics, that is, the system depicted in Figure 13.20. The following example illustrates a single-axis attitude control system using a reaction wheel.

Example 13.4 Figure 13.17 shows a satellite with a reaction wheel mounted along its 3 axis. The satellite's moment of inertia about the 3 axis is $I_{sat} = 500$ kg-m^2 (not including the inertia of the reaction wheel), and the wheel's inertia about its spin axis is $I_w = 0.06$ kg-m^2. The DC motor that drives the reaction wheel has the following parameters: motor torque constant $K_m = 0.04$ N-m/A, back-emf constant $K_b = 0.04$ V-s/rad, resistance $R = 3\ \Omega$, and friction coefficient $b = 6(10^{-5})$ N-m-s/rad. The DC motor has voltage limits of ± 28 V. Design a PD controller for the reaction wheel that can provide a single-axis attitude maneuver with good damping and fast response speed. Demonstrate the closed-loop control system with an attitude maneuver starting from $\phi_0 = 0$, $\dot{\phi}_0 = 0$ with a target attitude $\phi_{ref} = \pi/3$ (i.e., 60°). Because the satellite begins and ends with zero angular velocity, this maneuver is known as a "rest-to-rest maneuver."

Figure 13.20 shows the closed-loop attitude control system using the reaction wheel. The PD controller transfer function is

$$G_C(s) = K_P + K_D s$$

The attitude error ϕ_e (in rad) is the input to the controller, and its output is the voltage signal to the DC motor:

$$e_{in} = K_P \phi_e + K_D \dot{\phi}_e$$

Clearly, the PD controller gains have units of V/rad (K_P) and V-s/rad (K_D). Equations (13.45) and (13.46) show that the gains can be obtained by specifying the desired undamped natural frequency ω_n and damping ratio ζ. However, we cannot "over gain" the controller because the motor's input limit is ± 28 V. Let us select a "good" damping ratio $\zeta = 0.7$ and a settling time of $t_S = 250$ s. Using the definition of settling time, Eq. (13.18), we can compute the undamped natural frequency:

$$\omega_n = \frac{4}{t_S \zeta} = 4/(250\,\text{s})(0.7) = 0.0229 \text{ rad/s}$$

We use Eqs. (13.45) and (13.46) to obtain the PD controller gains:

$$\text{Proportional gain: } K_P = \frac{-RI_{\text{sat}} \, \omega_n^2}{K_m} = -19.5918 \text{ V/rad}$$

$$\text{Derivative gain: } K_D = \frac{-2RI_{\text{sat}} \zeta \omega_n}{K_m} = -1{,}200 \text{ V-s/rad}$$

Recall that Eqs. (13.45) and (13.46) were derived using a simplified model of the reaction-wheel system that ignored the friction torque and back-emf. At time $t = 0$, the attitude error is $\phi_e(0) = \phi_{\text{ref}} - \phi_0 = \pi/3$ rad, and hence the motor's initial voltage input is $e_{\text{in}}(0) = K_P \phi_e(0) + K_D \dot{\phi}_e(0) = -20.52$ V which is within the ± 28 V limits (recall that the satellite's initial angular velocity is zero).

We can use MATLAB's Simulink to numerically simulate the closed-loop attitude control system presented in Figure 13.20 (we should again emphasize that although the PD controller was designed using approximate models, its performance is tested using the accurate modeling equations for the DC motor and reaction wheel). Figure 13.23 shows the satellite's attitude response. Note that the satellite achieves the desired 60° reference angle in about 230 s without any discernable overshoot. Figure 13.24 shows the angular momentum contributions due to the angular velocity of the complete satellite ($I_3\omega_3$) and spinning reaction wheel ($I_w\omega_w$). Note that both angular momentum components start at zero (the satellite and wheel are initially at rest), and that they are "mirror images" of each other because the total angular momentum is conserved and therefore remains zero, that is, $H = I_3\omega_3 + I_w\omega_w = 0$. The wheel spins with negative angular velocity (i.e., opposite the 3 axis) so that the satellite spins with positive angular velocity. Because the reaction wheel's moment of inertia is very small ($I_w = 0.06$ kg-m^2) relative to the moment of inertia of the complete satellite and wheel

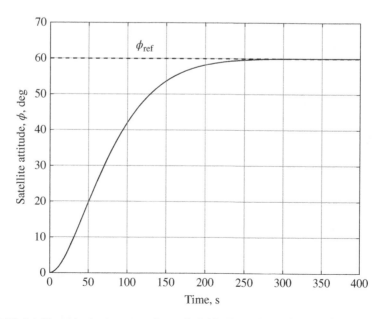

Figure 13.23 Satellite attitude response using a closed-loop reaction wheel controller (Example 13.4).

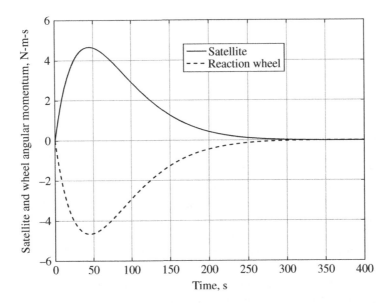

Figure 13.24 Satellite and reaction-wheel angular momentum (Example 13.4).

(I_3 = 500.06 kg-m²), it must spin at a much higher rate in order to counteract the satellite's angular momentum. For example, the wheel's peak angular velocity is approximately −742 rpm at about t = 45 s, whereas the satellite's peak angular velocity is about 0.09 rpm (≈0.5 deg/s). Finally, Figure 13.25 shows the input voltage to the DC

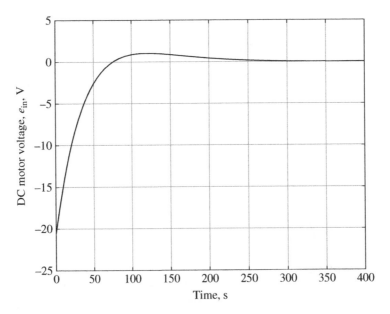

Figure 13.25 Input voltage to the DC motor (Example 13.4).

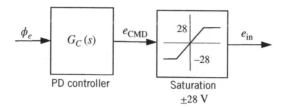

Figure 13.26 Saturation block for limiting input voltage to the DC motor.

motor during the closed-loop response. The initial voltage of −20.5 V (due to the initial attitude error) is observed, and the motor voltage (and consequently the motor torque) go to zero as the satellite comes to rest at the desired 60° attitude.

We can speed up the closed-loop attitude response by increasing the PD controller gains. However, we must keep the motor's input voltage within the ± 28 V limits. We do this by inserting a *saturation block* (or "limiter block") immediately after the PD controller block as shown in Figure 13.26. The saturation block will allow a voltage command (e_{CMD}) that is between ± 28 V to reach the DC motor unaltered. However, the saturation block will limit the motor voltage e_{in} to +28 V or −28 V if the controller commands an excessive voltage (i.e., $|e_{CMD}| > 28$ V). We can repeat this example with a settling time of 60 s and damping ratio $\zeta = 0.7$; the resulting PD gains are $K_P = -340.1361$ V/rad and $K_D = -5,000$ V-s/rad. Figure 13.27 shows that the satellite's attitude reaches its 60° reference in about 100 s (less than half the time of the previous case) and with a slight overshoot. Figure 13.28 shows that the satellite and reaction wheel both store more than twice as much angular momentum for this high-gain case when compared with the initial low-gain case. For example, the peak angular velocity of the reaction wheel is −2,090 rpm and the satellite's peak spin rate is 0.25 rpm (1.5 deg/s). While the wheel's peak spin rate

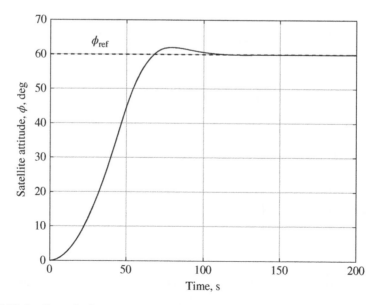

Figure 13.27 Satellite attitude response using the high-gain wheel controller and saturation block (Example 13.4).

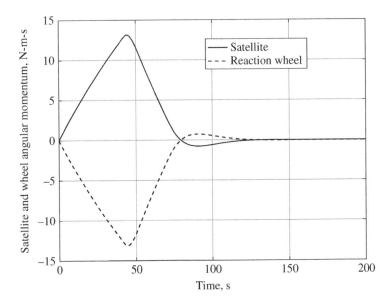

Figure 13.28 Satellite and reaction-wheel angular momentum using the high-gain wheel controller and saturation block (Example 13.4).

is quite large, many reaction wheels have maximum angular velocities as high as 5,000 rpm. Figure 13.29 presents the input voltage to the DC motor. Here we see that the high controller gains and saturation block initially produce the largest possible (negative) input voltage (−28 V) to the motor in order to provide the greatest (negative) torque

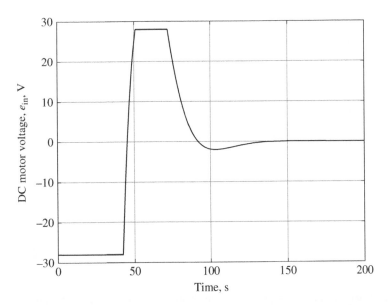

Figure 13.29 Input voltage to the DC motor using the high-gain wheel controller and saturation block (Example 13.4).

and angular acceleration for the wheel. At $t = 42$ s, the controller's damping term $K_D \dot{\phi}_e$ causes the input voltage to switch toward +28 V so that the motor torque slows down the wheel's spin rate as seen in Figure 13.28. For $t > 72$ s, the motor's input voltage is no longer saturated at its maximum value.

This example has demonstrated that we can adjust the PD controller gains to speed up the closed-loop attitude response. Consequently, many closed-loop controllers have "low-gain" and "high-gain" settings for added flexibility. Furthermore, the presence of a saturation block for limiting the control signal is a common feature in most operational systems. However, an active saturation block results in a nonlinear system, and therefore its closed-loop response can no longer be predicted by analytical formulas.

13.5 Attitude Maneuvers Using Reaction Jets

This section involves closed-loop attitude control using on–off reaction jets. Just as reaction-wheel control, we will demonstrate the concept by considering single-axis attitude maneuvers. The resulting reaction-jet control law will be nonlinear due to its discontinuous (on–off) pulsed operation.

13.5.1 Phase-Plane Analysis of Satellite Attitude Dynamics

Let us begin our analysis with the single degree-of-freedom attitude dynamics for satellite rotation about a principal axis as shown in Figure 13.30

$$M_c = I\ddot{\phi} \tag{13.47}$$

where it is understood that I is the satellite's moment of inertia about the axis normal to the page. Firing the reaction jets in pairs produces the control torque M_c

$$M_c = \pm 2Fr \tag{13.48}$$

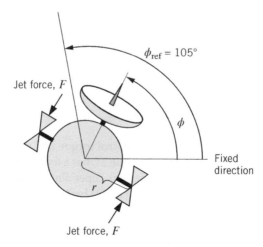

Figure 13.30 Attitude control using reaction jets (positive control torque is shown).

where F is the thrust of a single jet and r is the radial moment-arm distance from the jet location to the spin axis. Equation (13.48) shows that the control torque may be positive or negative depending on which pair of jets is fired. Figure 13.30 shows reaction jets fired to create a positive control torque on the satellite; obviously, firing the opposite jets causes a negative control torque. Because the jet force is produced by opening the valve to a hot-gas or cold-gas thruster, the force F is constant and consequently the control torque M_c has a constant magnitude.

In Figure 13.30, the satellite's attitude angle ϕ is the angular position of an antenna-pointing direction relative to a fixed direction. Our attitude-control problem might demand that the satellite rotates so that its antenna points in a new direction as specified by the desired attitude angle ϕ_{ref}. Figure 13.30 and Eq. (13.47) show that a positive control torque causes positive angular acceleration in the counterclockwise direction.

We will develop a control logic based on the attitude *error* dynamics. Therefore, let us define the attitude error as

$$\text{Attitude error:} \quad \phi_e = \phi - \phi_{ref} \tag{13.49}$$

Referring again to Figure 13.30, the satellite's current attitude is $\phi = 60°$. If the reference (desired) attitude is fixed at $\phi_{ref} = 105°$, then the current attitude error is $-45°$. Because the reference attitude is fixed, successive time derivatives of the attitude error are equal to the satellite's angular velocity and angular acceleration; that is

$$\text{Attitude error rate:} \quad \dot{\phi}_e = \dot{\phi} - \overset{0}{\cancel{\dot{\phi}_{ref}}} = \dot{\phi} \tag{13.50}$$

$$\text{Attitude error acceleration:} \quad \ddot{\phi}_e = \ddot{\phi} - \overset{0}{\cancel{\ddot{\phi}_{ref}}} = \ddot{\phi} \tag{13.51}$$

Note that our definition of attitude error, Eq. (13.49), is the opposite of attitude error shown in Figure 13.8 (Example 13.1) and Figure 13.20 (Example 13.4). We choose to define attitude error as $\phi_e = \phi - \phi_{ref}$ so that the time derivatives of attitude error and satellite attitude angle are equal as shown in Eqs. (13.50) and (13.51).

We can obtain a closed-form solution of the satellite's angular motion by using a state-variable approach. Our analysis will eliminate the motion's dependency on time. Let us begin by defining two state variables based on the attitude error: $x_1 = \phi_e$ and $x_2 = \dot{\phi}_e$. The two state-variable equations are simply the time derivatives of the two states:

$$\dot{x}_1 = \dot{\phi}_e = x_2 \tag{13.52}$$

$$\dot{x}_2 = \ddot{\phi}_e = \ddot{\phi} = \frac{M_c}{I} = u \tag{13.53}$$

where the control term u is the ratio of the control torque and moment of inertia (note that u has units of rad/s^2). Because control torque M_c has a fixed magnitude, control term u also has a fixed magnitude, and can switch signs depending on which jet pairs are fired. Next, we can eliminate time by dividing Eq. (13.52) by Eq. (13.53):

$$\frac{\dot{x}_1}{\dot{x}_2} = \frac{dx_1/dt}{dx_2/dt} = \frac{dx_1}{dx_2} = \frac{x_2}{u} \tag{13.54}$$

Separating variables, we obtain

$$udx_1 = x_2 dx_2 \qquad (13.55)$$

Equation (13.55) is easily integrated with respect to state variables x_1 and x_2 to yield

$$ux_1 = \frac{1}{2}x_2^2 + C \qquad (13.56)$$

where C is a constant of integration. Let us express Eq. (13.56) in terms of the attitude error $(x_1 = \phi_e)$ and the attitude error rate $(x_2 = \dot{\phi}_e)$:

$$\frac{1}{2}\dot{\phi}_e^2 - u\phi_e + C = 0 \qquad (13.57)$$

Equation (13.57) is the satellite's angular motion in terms of attitude error, attitude error rate, and torque acceleration u (remember that the control acceleration can only take on two values, i.e., $u = \pm 2Fr/I$). We can create a plot of attitude error rate $(\dot{\phi}_e)$ vs. attitude error (ϕ_e) for *positive* jet torque $u = 2Fr/I$ and different values of constant C. Figure 13.31 shows solutions to Eq. (13.57) for $u = +0.25$ rad/s^2 and $C = -0.5$, 0, and 0.5 rad/s^2. Equation (13.57) and Figure 13.31 show that a "sideways parabola" represents the attitude error rate plotted against the attitude error for a given value of C. The arrows on the parabolic curves in Figure 13.31 show the direction of the satellite's angular motion for positive jet torque. Equation (13.53) shows that $\ddot{\phi} = \ddot{\phi}_e > 0$ when the satellite fires the positive torque jets $(u > 0)$, and therefore attitude error rate $\dot{\phi}_e$ is always increasing as illustrated in Figure 13.31.

Figure 13.31 is called a *phase portrait* and it represents the trajectories of a dynamic system in the so-called *phase plane*, where the x axis is the attitude error and the y axis is its time derivative. Because our goal is to drive the attitude error and attitude error rate to

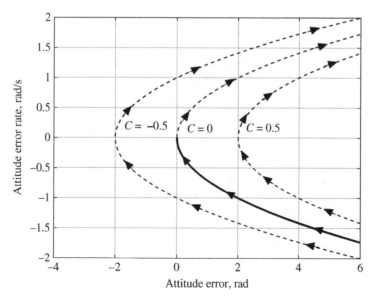

Figure 13.31 Attitude motion in the phase plane with positive jet torque, $u = 0.25$ rad/s^2.

zero, we desire the sideways parabolic path that intersects the origin of the phase plane. Figure 13.31 shows that the bottom half of the path with $C = 0$ (the solid curve) is the *only* possible path to the origin when using positive jet torque. The bottom half of the path with $C = 0$ describes a satellite with positive attitude error ($\phi_e > 0$) and negative angular velocity ($\dot{\phi}_e < 0$), and hence the positive jet torque is performing a *braking maneuver*. Consequently, the phase-plane trajectory to the origin with positive torque satisfies

$$\text{Path to origin with } u > 0: \quad \frac{1}{2}\dot{\phi}_e^2 = u\phi_e \quad \text{with} \quad \phi_e > 0, \quad \dot{\phi}_e < 0 \tag{13.58}$$

For example, a satellite with torque acceleration $u = +0.25$ rad/s^2, $\phi_e = 2$ rad (114.6°), and $\dot{\phi}_e = -1$ rad/s (−57.3 deg/s), will follow a parabolic path to the origin in the phase plane as shown in Figure 13.31. The reader should note that a satellite with attitude error $\phi_e = 2$ rad and *positive* error rate $\dot{\phi}_e = +1$ rad/s with $u = +0.25$ rad/s^2 will follow the upper-half parabolic path with $C = 0$; however, the positive torque will cause the satellite's attitude to drift away from the origin.

Figure 13.32 presents the phase portrait for negative torque jets, or $u = -0.25$ rad/s^2. The arrows on the phase-plane trajectories show that attitude error rate $\dot{\phi}_e$ is always decreasing along a parabolic path when $u < 0$. Figure 13.32 shows that the top half of the parabola with $C = 0$ (the solid curve) is the *only* path to the origin with negative torque. In this case, attitude error rate must be positive and attitude error must be negative:

$$\text{Path to origin with } u < 0: \quad \frac{1}{2}\dot{\phi}_e^2 = u\phi_e \quad \text{with} \quad \phi_e < 0, \quad \dot{\phi}_e > 0 \tag{13.59}$$

As with the positive-torque scenario, the reaction jets are producing a torque in the opposite direction as the satellite's angular velocity (again, a braking maneuver).

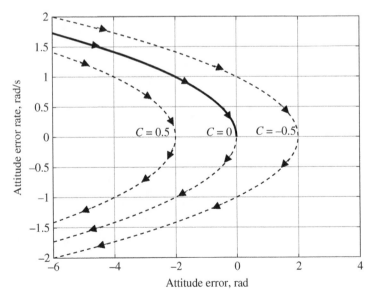

Figure 13.32 Attitude motion in the phase plane with negative jet torque, $u = -0.25$ rad/s^2.

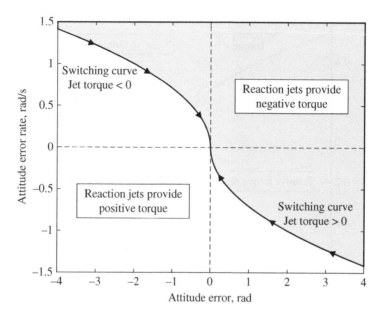

Figure 13.33 Switching curve in the phase plane.

The previous discussion shows that a satellite has only two paths to the origin of the phase plane (i.e., $\phi_e = 0$ and $\dot{\phi}_e = 0$) when using a constant-torque magnitude: the bottom-half parabola with positive jet torque (Figure 13.31) or the top-half parabola with negative jet torque (Figure 13.32). Figure 13.33 shows these two parabolic paths (with $C = 0$) pieced together in the phase plane. The composite of these two half-parabolic paths joined at the origin is called the *switching curve*. The top-half parabola is the negative-torque switching curve and the bottom-half parabola is the positive-torque switching curve. We may use Figure 13.33 to describe an attitude-control strategy for driving the satellite to the desired reference attitude (i.e., $\phi_e = 0$ and $\dot{\phi}_e = 0$) starting from *any* initial conditions. If the satellite's state $(\phi_e, \dot{\phi}_e)$ is "below" the switching curve in Figure 13.33, fire the *positive* torque jets in order to follow an "upward" parabolic path as shown in Figure 13.31. When the phase-plane path intersects the upper-half switching curve in Figure 13.33, the positive jets are turned off and the *negative* torque jets are fired. Consequently, the satellite follows the upper-half switching curve to the origin. If the satellite's state $(\phi_e, \dot{\phi}_e)$ is "above" the switching curve in Figure 13.33, the opposite strategy is employed: fire the negative torque jets until the parabolic path reaches the bottom-half switching curve, and then switch to positive torque jets. Figure 13.33 shows the "positive torque jet" region below the switching curve and the "negative torque jet" region above the switching curve.

Let us consider a satellite with an initial attitude error of 0.4 rad ($\approx 23°$) and initial angular velocity of 0.2 rad/s (≈ 11.5 deg/s). This initial state is shown in the phase plane in Figure 13.34. Because this initial state is "above" the switching curve, the *negative* torque jets are fired, and the phase-plane trajectory follows a "downward" parabola as shown in Figure 13.34. The satellite's phase-plane state reaches the bottom-half switching curve at attitude error $\phi_e = 0.24$ rad ($\approx 13.8°$) and attitude error rate $\dot{\phi}_e = -0.346$ rad/s

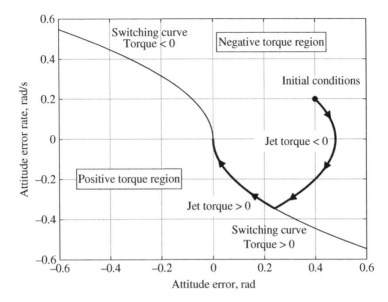

Figure 13.34 Attitude motion in the phase plane from an arbitrary initial condition.

($\approx$ –19.8 deg/s). At this point, the positive torque jets are fired and the phase-plane trajectory follows the switching curve to the origin as shown in Figure 13.34.

Figure 13.34 shows that the error states can be driven to zero with one switch in the
reaction jets (of course, the initial phase-plane conditions could lie *exactly* on a switching
curve and hence either positive or negative torque will drive the errors to zero without
the need for switching). This one-switch scenario represents the idealized case. In a real
attitude control system, the decision to switch jets is based on sensors that measure the
attitude angle and angular velocity. If the measurement (or sampling) rate is too slow or
inaccurate, the switching may be delayed as illustrated in Figure 13.35. A combination of
slow measurement rates or slow thruster dynamics causes the phase-plane trajectories to
overshoot the switching curve. Eventually, the error states ($\phi_e, \dot{\phi}_e$) become "close" to the
origin and enter a *limit cycle* where the reaction jets "chatter" between positive and negative at high frequency. Consequently, the attitude error and its rate are never driven to
zero and instead "loop" around the origin.

13.5.2 Reaction Jet Control Law

Our phase-plane analysis of satellite attitude dynamics with on–off reaction jets has led
to the switching-curve concept illustrated in Figures 13.33 and 13.34. In this subsection,
we will present an explicit feedback control equation that mechanizes the switching curve.

Equations (13.58) and (13.59) define the top-half and bottom-half switching curves
depicted in Figure 13.33. The basic switching-curve equation is

$$\frac{1}{2}\dot{\phi}_e^2 = u\phi_e \qquad (13.60)$$

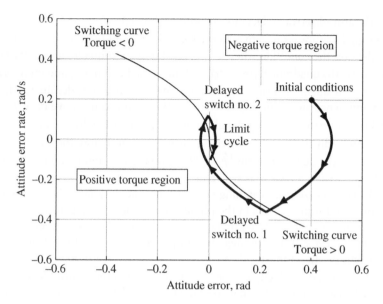

Figure 13.35 Attitude motion in the phase plane with delayed switching.

where $u = \pm 2Fr/I$ is the torque acceleration. The top-half switching curve in Figure 13.33 uses negative torque ($u = -2Fr/I$), while the bottom-half curve uses positive torque ($u = +2Fr/I$). We can manipulate Eq. (13.60) to develop a feedback control equation that determines when to use positive or negative reaction jets. Because we desire an equation that will determine the correct sign for the control torque, let us rewrite Eq. (13.60) and eliminate the torque acceleration u

$$\frac{1}{2}|\dot{\phi}_e|\dot{\phi}_e = -\alpha\phi_e \tag{13.61}$$

where the *magnitude* of the torque acceleration (or, torque-to-inertia ratio) is defined as

$$\alpha \equiv \frac{2Fr}{I} > 0 \tag{13.62}$$

Equation (13.61) uses $|\dot{\phi}_e|\dot{\phi}_e$ instead of $\dot{\phi}_e^2$ in order to retain the correct sign for attitude rate when the attitude error ϕ_e is positive. Note that Eq. (13.61) holds for the top-half switching curve (i.e., the second quadrant of the phase plane where $\phi_e < 0$ and $\dot{\phi}_e > 0$) and the bottom-half switching curve (i.e., the fourth quadrant where $\phi_e > 0$ and $\dot{\phi}_e < 0$). Next, rearrange Eq. (13.61) to define the *switching signal* σ

$$\sigma = -\alpha\phi_e - \frac{1}{2}|\dot{\phi}_e|\dot{\phi}_e \tag{13.63}$$

We will use the switching signal σ to determine which reaction jets to fire for attitude control. It should be clear to the reader that $\sigma = 0$ on the switching curve by simply comparing Eqs. (13.60), (13.61), and (13.63).

Let us check the sign of the switching signal σ in all four quadrants of the phase plane shown in Figure 13.33 (remember that constant α is always positive). In the first quadrant, the attitude error and its rate are both positive ($\phi_e > 0$, $\dot{\phi}_e > 0$), and therefore the switching signal σ is always negative. In the second quadrant ($\phi_e < 0$, $\dot{\phi}_e > 0$), the switching curve denotes the condition where $\sigma = 0$. For the second-quadrant region *above* the switching curve, the quadratic term $|\dot{\phi}_e|\dot{\phi}_e/2$ dominates the term $\alpha\phi_e$ and therefore $\sigma < 0$. Below the second-quadrant switching curve the reverse is true: $\alpha\phi_e$ dominates the quadratic term and hence switching signal $\sigma > 0$. In the third quadrant, we have $\phi_e < 0$ and $\dot{\phi}_e < 0$, and therefore the switching signal σ is always positive. Finally, the fourth quadrant ($\phi_e > 0$, $\dot{\phi}_e < 0$) is divided by the switching curve ($\sigma = 0$). Below the fourth-quadrant switching curve, the quadratic term $|\dot{\phi}_e|\dot{\phi}_e/2$ dominates $\alpha\phi_e$ and therefore $\sigma > 0$; above the switching curve, the reverse is true and $\sigma < 0$.

Table 13.1 summarizes the switching signal results and correlates them with the negative and positive torque regions in the phase plane depicted in Figure 13.33. It is clear that the switching signal σ defined by Eq. (13.63) dictates the reaction jet selection for attitude control: if $\sigma < 0$ fire the negative-torque reaction jets, if $\sigma > 0$ fire the positive-torque jets. Because the attitude error is $\phi_e = \phi - \phi_{\text{ref}}$, the attitude error rate is equal to the satellite's angular velocity (i.e., $\dot{\phi}_e = \dot{\phi}$) when the reference attitude angle is constant. Using this assumption, we may use Eq. (13.63) to develop the following feedback equation for control torque M_c

$$M_c = \begin{cases} +2Fr & \text{if } \sigma = -\alpha\phi_e - 0.5|\dot{\phi}|\dot{\phi} > 0 \\ -2Fr & \text{if } \sigma = -\alpha\phi_e - 0.5|\dot{\phi}|\dot{\phi} < 0 \end{cases} \tag{13.64}$$

Equation (13.64) is a *nonlinear feedback control law* that determines the control torque M_c. It is nonlinear because the control signal (torque M_c) is discontinuous; that is, the control torque will instantly jump from positive to negative (or vice versa) depending on the sign of the switching signal σ. Equation (13.64) dictates that control torque can only be two values ($\pm 2Fr$) that correspond to firing the positive or negative jets.

Figure 13.36 shows a closed-loop reaction-jet control system that utilizes Eq. (13.64) for satellite torque. The reader should be able to match the signal-path labels in Figure 13.36 with the respective equations: attitude error ϕ_e [Eq. (13.49)], switching signal σ [Eq. (13.63)], and control torque M_c [Eq. (13.64)]. The output w of the "relay with dead zone" block in Figure 13.36 is

Table 13.1 Switching signal and corresponding reaction-jet torque.

Phase-plane region	Switching signal	Reaction-jet torque
First quadrant	$\sigma < 0$	Negative
Second quadrant (above switching curve)	$\sigma < 0$	Negative
Second quadrant (below switching curve)	$\sigma > 0$	Positive
Third quadrant	$\sigma > 0$	Positive
Fourth quadrant (below switching curve)	$\sigma > 0$	Positive
Fourth quadrant (above switching curve)	$\sigma < 0$	Negative

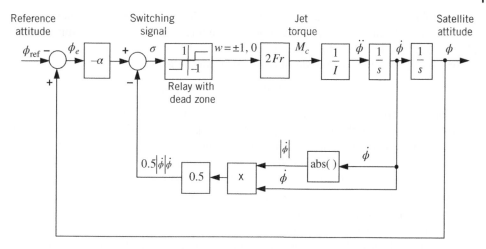

Figure 13.36 Closed-loop attitude control using reaction jets and switching signal.

$$w = \begin{cases} +1 & \text{if} \quad \sigma > \varepsilon \\ 0 & \text{if} \quad -\varepsilon \leq \sigma \leq \varepsilon \\ -1 & \text{if} \quad \sigma < -\varepsilon \end{cases} \qquad (13.65)$$

where $\pm \varepsilon$ is the width of the "dead zone." Therefore, the series connection of this relay block and the gain $2Fr$ mechanizes the jet torque M_c prescribed by Eq. (13.64). Setting $\varepsilon = 0$ (no dead zone) turns the relay in Figure 13.36 into a pure signum (or sign) function, that is, $w = \text{sgn}(\sigma)$. Using a pure signum switching function in an operational setting is not desirable because the switching signal σ will never *exactly* equal zero as the satellite approaches zero attitude error and zero angular velocity. Hence, a relay without dead zone will result in "chattering" control between positive and negative jets when the satellite attitude error approaches zero. This control strategy is obviously undesirable because it wastes onboard fuel even though the satellite's attitude error may be infinitesimally small.

The following examples illustrate the performance of a closed-loop reaction-jet system. We will show what effect the dead zone has on the control torque and system response.

Example 13.5 Consider again a single-axis attitude maneuver depicted in Figure 13.30. In this case, each reaction jet can supply a force of 400 N when fired (the jet thrust is constant and cannot be throttled). The satellite has moment of inertia $I_3 = 6{,}400$ kg-m^2 and the moment arm of each jet is 2 m (these satellite parameters are based on the roll axis of the Apollo capsule). Suppose the satellite's current attitude angle and angular rate are $\phi_0 = 60°$ and $\dot{\phi}_0 = -10$ deg/s, respectively, and that the reference (desired) attitude is $\phi_{ref} = 105°$ as shown in Figure 13.30. Obtain the closed-loop attitude response using the nonlinear feedback control law that utilizes the switching signal σ.

Figure 13.36 shows the reaction-jet control system using the switching signal. There is no way to obtain the system response using analytical methods because the control system is nonlinear; we must use a numerical simulation to obtain the response. We can create a block-diagram simulation of Figure 13.36 using MATLAB's Simulink software. The Simulink details are not presented here; the interested reader may consult Kluever [4; pp. 162–183, 469–484] for a primer on Simulink.

For this satellite, the magnitude of the control torque is

$$|M_c| = 2Fr = 2(400\,\text{N})(2\,\text{m}) = 1,600\,\text{N-m}$$

The constant torque acceleration (magnitude) is $\alpha = |M_c|/I_3 = 0.25\,\text{rad/s}^2$.

First, let us obtain the closed-loop attitude response with dead-zone $\varepsilon = 0.0004$ $(\text{rad/s})^2$. The switching-signal equation (13.63) shows that when the angular velocity is very small ($\dot{\phi} \approx 0$), the switching signal σ is in the dead zone when the magnitude of the attitude error is less than ε/α. For $\alpha = 0.25\,\text{rad/s}^2$ and $\varepsilon = 0.0004$ $(\text{rad/s})^2$, the jets are inactive when $|\phi_e| < 0.0016\,\text{rad}$ (or, $|\phi_e| < 0.1°$). Figures 13.37 and 13.38 show the time histories of the closed-loop attitude error (ϕ_e) and angular velocity ($\dot{\phi}$), respectively. Figure 13.37 shows that the attitude error begins at −45° and eventually goes to zero at about 4.1 s. Figure 13.38 shows that satellite's angular velocity starts at −10 deg/s and steadily increases at a linear rate with time due to the positive torque jets. Positive torque jets are initially used because $\sigma > \varepsilon$ at time $t = 0$ [see Eq. (13.63)]. At time $t = 2.5$ s, the switching signal σ becomes less than $-\varepsilon$, and control switches to negative torque jets. At approximately $t = 4.4$ s, Figures 13.37 and 13.38 show that the attitude error and attitude rate become small enough so that the switching signal is in the dead zone (i.e., $-\varepsilon \leq \sigma \leq \varepsilon$) and the jets are turned off. Figure 13.39 shows the time history of the reaction-jet control torque: the switching at $t = 2.5$ s is apparent in this figure. At $t = 4.4$ s, the jets are turned off because the attitude error and attitude rate are small enough so that $|\sigma| \leq \varepsilon$. However, at time $t = 4.9$ s, the satellite's attitude angle has drifted away from the

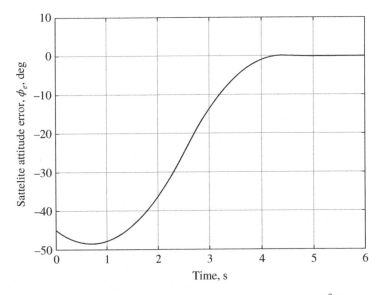

Figure 13.37 Closed-loop attitude error ϕ_e for dead-zone $\varepsilon = 0.0004$ $(\text{rad/s})^2$ (Example 13.5).

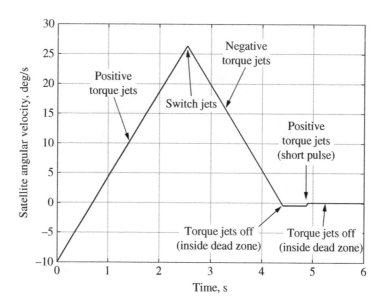

Figure 13.38 Closed-loop attitude rate for dead-zone $\varepsilon = 0.0004$ (rad/s)^2 (Example 13.5).

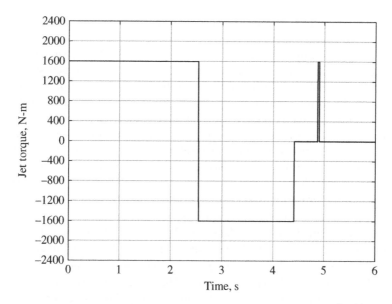

Figure 13.39 Closed-loop reaction-jet torque for dead-zone $\varepsilon = 0.0004$ (rad/s)^2 (Example 13.5).

reference attitude (note the very small negative attitude rate in Figure 13.38 for $t > 4.4$ s), and consequently $\sigma > \varepsilon$ and a very short positive torque pulse is applied to null the small negative attitude rate. Figure 13.40 shows the closed-loop attitude motion in the phase plane. The initial error state, $\phi_e = -45°$ and $\dot{\phi}_e = -10$ deg/s, is in the positive-torque

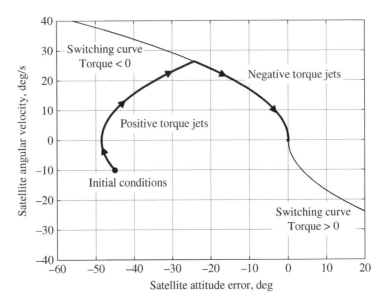

Figure 13.40 Closed-loop attitude motion in the phase plane (Example 13.5).

region, and therefore the switching-signal control law (13.64) correctly calls for positive torque at $t = 0$. Figure 13.40 also shows that the feedback control system correctly switched to negative torque jets when the error states reached the switching curve.

Finally, let us obtain the closed-loop attitude response with a much smaller dead-zone $\varepsilon = 5(10^{-8})$ (rad/s)^2. The closed-loop attitude error and error rate responses are essentially the same as the responses shown in Figures 13.37 and 13.38; that is, the satellite initially exhibits positive angular acceleration (positive torque) between $t = 0$ and $t = 2.5$ s, whereupon the torque jets switch to a negative value for $2.5 \le t \le 4.4$ s. The major effect of the tiny dead zone is displayed by the torque history *after* the satellite has essentially reached its zero-error state. Figure 13.41 shows the reaction-jet control torque history for the case with $\varepsilon = 5(10^{-8})$ (rad/s)^2. The torque profile matches the control torque for dead zone $\varepsilon = 0.0004$ (rad/s)^2 (Figure 13.39) until $t = 4.4$ s. For $t > 4.4$ s, the attitude error ϕ_e and attitude rate $\dot{\phi}$ are very small (i.e., these states are essentially at the origin of the phase plane), and hence the switching signal σ is also very small. However, because ε is tiny, the switching signal rapidly "drifts" in and out of the dead zone due to very small changes in ϕ_e and $\dot{\phi}$. The result is the high-frequency jet switching shown in Figure 13.41 for $t > 4.4$ s. This "chattering" control is undesirable because it wastes reaction-jet fuel as it attempts to keep the attitude errors within an unreasonable error threshold.

A final note is in order. In this example, the fixed step size of the numerical integration scheme is $\Delta t = 0.002$ s, which implies that the feedback switching signal σ is computed 500 times each second (i.e., the sample rate is $1/\Delta t = 500$ samples per second $= 500$ Hz). If the sampling (or feedback measurement) rate is too slow, then the jet switching will be delayed as shown in Figure 13.35. Increasing the sampling rate reduces the switching delay and extraneous control pulses when the error states

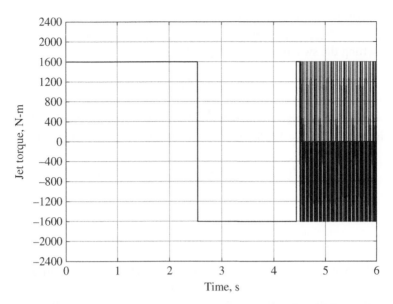

Figure 13.41 Closed-loop reaction-jet torque with dead-zone $\varepsilon = 5(10^{-8})$ (rad/s)^2 and high-frequency chatter at steady state (Example 13.5).

approach zero. These issues are associated with digital control systems and are beyond the scope of this textbook.

Example 13.6 Consider again the satellite and its initial conditions presented in Example 13.5. Determine the point in the phase plane where the first jet switching occurs (assuming a zero dead-zone for the relay), and estimate the total maneuver time to bring the satellite to rest at the desired attitude.

The initial state in the phase plane is attitude error $\phi_e(0) = \phi_0 - \phi_{\text{ref}} = -45°$ ($= -0.7854$ rad) and $\dot{\phi}_e(0) = -10$ deg/s ($= -0.1745$ rad/s). Because the initial state is "below" the switching curve (Figure 13.40), we select the positive torque jets [or, we can show that the initial switching signal determined by Eq. (13.64) is positive]. The phase plane profile is determined by Eq. (13.57):

$$\frac{1}{2}\dot{\phi}_e^2 - u\phi_e + C = 0 \tag{13.66}$$

where $u = +2Fr/I = 0.25$ rad/s^2 is the positive torque acceleration. Substituting the initial error state into Eq. (13.66), we find that $C = -0.2116$ (rad/s)^2. The switching point in the phase plane is the intersection of the *positive*-torque parabolic curve [Eq. (13.66] and the *negative*-torque switching curve (with $C = 0$; see Figure 13.32). The intersection of these two parabolic curves is

$$\underbrace{\frac{1}{2}\dot{\phi}_e^2 - 0.25\phi_e + C}_{\text{Positive jet-torque}} = \underbrace{\frac{1}{2}\dot{\phi}_e^2 + 0.25\phi_e}_{\substack{\text{Negative-torque} \\ \text{switching curve}}}$$

Using $C = -0.2116$ (rad/s)^2, we find that intersection occurs at the attitude error $\phi_e = -0.4232$ rad ($= -24.25°$). The attitude error rate at the switching curve can be determined by setting the switching signal, Eq. (13.63), to zero:

$$\sigma = -\alpha\phi_e - \frac{1}{2}|\dot{\phi}_e|\dot{\phi}_e = 0 \qquad (13.67)$$

Using $\phi_e = -0.4232$ rad and $\alpha = 0.25$ rad/s^2 in Eq. (13.67), we find that the attitude rate at the switching curve is $\dot{\phi}_e = 0.4560$ rad/s. In summary, the switching point is

$$\boxed{\phi_e(t_1) = -0.4232 \text{ rad}, \quad \dot{\phi}_e(t_1) = 0.4560 \text{ rad/s}}$$

Or,

$$\boxed{\phi_e(t_1) = -24.25°, \quad \dot{\phi}_e(t_1) = 26.35 \text{ deg/s}}$$

Time t_1 is the switching time. This switching point matches the numerical simulation results presented in Example 13.5 (see Figure 13.40).

The second part of this problem involves estimating the total maneuver time. We neglect the effect of the relay's dead-zone and assume "perfect switching" when $\sigma = 0$. Because the torque acceleration is constant, it is easy to integrate the angular acceleration of the attitude error by combining Eqs. (13.51), (13.47), and (13.48):

$$\ddot{\phi}_e = \frac{\pm 2Fr}{I} = \pm 0.25 \text{ rad/s}^2$$

The integral is

$$\dot{\phi}_e(t) = \dot{\phi}_e(0) \pm 0.25t \text{ rad/s} \qquad (13.68)$$

For the initial *positive*-torque segment, we use positive torque acceleration, $+0.25$ rad/s^2. Using the initial attitude rate [$\dot{\phi}_e(0) = -0.1745$ rad/s] and the attitude rate on the switching curve [$\dot{\phi}_e(t_1) = 0.4560$ rad/s], we determine the switching time:

$$t_1 = \frac{\dot{\phi}_e(t_1) - \dot{\phi}_e(0)}{0.25} = 2.538 \text{ s}$$

This switching time corresponds to the numerical simulation results from Example 13.5, and can be seen in Figures 13.38, 13.39, and 13.41. The remaining maneuver time to the origin (time on the negative-torque switching curve) can be computed using a modified version of Eq. (13.68):

$$\Delta t = \frac{\dot{\phi}_e(t_f) - \dot{\phi}_e(t_1)}{-0.25} = 1.840 \text{ s}$$

where $\dot{\phi}_e(t_f) = 0$ is the final attitude rate (the origin of the phase plane). The total maneuver time is $t_f = t_1 + \Delta t = 4.38$ s. This calculation shows a good match with the simulation results of Example 13.5 shown in Figures 13.38 and 13.39.

13.6 Nutation Control Using Reaction Jets

Sections 13.4 and 13.5 discussed single-axis attitude maneuvers using reaction wheels and reaction jets. In both cases, we developed feedback control systems that automatically performed a rotation about a single axis to a desired attitude angle. In this section, we will present a feedback control scheme for removing the nutation angle (or "wobble") of a spinning satellite.

Recall that the nutation angle θ is the "tilt" of the satellite's body-fixed 3 axis relative to the inertially fixed angular momentum vector $\mathbf{H}$ (see Section 12.3). Figure 13.42 (Figure 12.12 from Chapter 12) shows the nutation angle θ and angle γ for a prolate satellite. For torque-free motion, the space cone (i.e., vector $\mathbf{H}$) is fixed and the body cone (containing the 3 axis) rolls along the space cone as shown in Figure 13.42. A satellite is often spin stabilized before firing an onboard rocket for a large-scale orbital maneuver. Recall that a semi-rigid *oblate* satellite is stable about its major axis (the "major-axis rule"), and hence a passive nutation damper may be used to remove the nutation angle over time. However, the geometry of a launch vehicle's payload shroud favors prolate ("pencil-shaped") satellites. Therefore, spinning a semi-rigid prolate satellite about its minor axis will eventually lead to instability. An active feedback control scheme for removing nutation can be used on prolate satellites. We will briefly investigate a nutation control using on–off reaction jets.

Figure 13.43 shows one possible reaction-jet configuration for controlling rotational motion about all three axes. The six jets are in two clusters of three, where each cluster is offset from the 3 axis by radial distance r, and has vertical height z above the 1–2 plane. All six reaction jets are identical and produce thrust force F when fired. Jets a and d each produce force F in the $+\mathbf{u}_1$ direction, and jets c and f produce force F in the $-\mathbf{u}_1$ direction. Firing jet b produces a force in the $+\mathbf{u}_2$ direction, while firing jet e produces a force in the $-\mathbf{u}_2$ direction. No jets produce a force component along the 3 axis. Table 13.2 summarizes the jet combinations that produce positive and negative torques about the body axes. Note that the control torque about the 1 axis is half the magnitude of the 2-axis

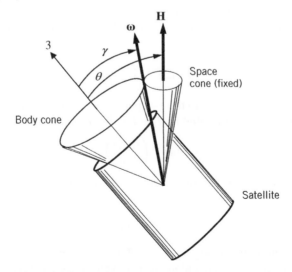

Figure 13.42 Spinning prolate satellite with zero external torques: body cone rolls along fixed space cone.

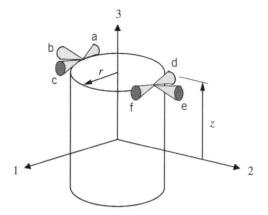

Figure 13.43 Reaction jet placement for three-axis control.

Table 13.2 Reaction jet torques corresponding to the jet configuration in Figure 13.43.

Jets fired	Control torque in 123 body frame
a, f	$2Fr\mathbf{u}_3$
c, d	$-2Fr\mathbf{u}_3$
e	$Fz\mathbf{u}_1$
b	$-Fz\mathbf{u}_1$
a, d	$2Fz\mathbf{u}_2$
c, f	$-2Fz\mathbf{u}_2$

control torque because a 1-axis control torque involves a single jet and a 2-axis torque involves firing two jets (jets a, d or c, f must be fired in pairs for a pure 2-axis torque, otherwise a 3-axis torque component exists).

Let us develop a control law for reducing a satellite's nutation angle θ. The satellite's nutation angle is

$$\theta = \tan^{-1}\left(\frac{H_{12}}{H_3}\right) \tag{13.69}$$

where H_{12} is the projection of $\mathbf{H}$ onto the 1–2 plane, and H_3 is the component of $\mathbf{H}$ along the 3 axis. For an axisymmetric satellite ($I_1 = I_2$), the nutation angle is

$$\theta = \tan^{-1}\left(\frac{I_1\omega_{12}}{I_3\omega_3}\right) \tag{13.70}$$

where ω_{12} is the component of the angular velocity vector $\boldsymbol{\omega}$ projected onto the 1–2 plane

$$\omega_{12} = \sqrt{\omega_1^2 + \omega_2^2} \tag{13.71}$$

and ω_3 is the spin component along the 3 axis. Therefore, a control law that drives both 1- and 2-axis spin components ω_1 and ω_2, respectively, to zero will remove the nutation angle and align the 3 axis and angular velocity vector $\boldsymbol{\omega}$ with the angular momentum vector $\mathbf{H}$. Because control torques exist, the angular momentum $\mathbf{H}$ is *not* constant.

Recall Euler's moment equations, Eq. (12.39), for general rotational motion with respect to principal 123 body axes:

$$M_1 = I_1\dot{\omega}_1 + (I_3 - I_2)\omega_2\omega_3 \tag{13.72}$$

$$M_2 = I_2\dot{\omega}_2 + (I_1 - I_3)\omega_1\omega_3 \tag{13.73}$$

$$M_3 = I_3\dot{\omega}_3 + (I_2 - I_1)\omega_1\omega_2 \tag{13.74}$$

where M_1, M_2, and M_3 are the control torques about the principal body axes. For an axisymmetric satellite ($I_1 = I_2$), Eq. (13.74) shows that $I_3\dot{\omega}_3 = 0$ if $M_3 = 0$, or $\omega_3 = n =$ constant. Using $I_1 = I_2$, $\omega_3 = n$, and the constant $a = (I_3 - I_2)n$, the first two Euler moment equations become

$$M_1 = I_1\dot{\omega}_1 + a\omega_2 \tag{13.75}$$

$$M_2 = I_1\dot{\omega}_2 - a\omega_1 \tag{13.76}$$

The analysis becomes difficult from this point because the jet torques M_1 and M_2 can only take the positive or negative values presented in Table 13.2; that is, $M_1 = \pm Fz$ and $M_2 = \pm 2Fz$. Hence, any feedback scheme for control torques M_1 and M_2 will be nonlinear because it can only command three values: positive, negative, or zero torque. We will arrive at a feasible nonlinear control law by using a heuristic argument that is based on the fictional *linear* control laws:

$$M_1 = -K_1\omega_1, \quad M_2 = -K_2\omega_2 \tag{13.77}$$

where K_1 and K_2 are feedback gains. Using these (fictional) control laws in Eqs. (13.75) and (13.76), we obtain

$$I_1\dot{\omega}_1 = -a\omega_2 - K_1\omega_1 \tag{13.78}$$

$$I_1\dot{\omega}_2 = a\omega_1 - K_2\omega_2 \tag{13.79}$$

The next step involves state-variable methods. We may define the two-element state vector

$$\mathbf{x} = \begin{bmatrix} x_1 \\ x_2 \end{bmatrix} = \begin{bmatrix} \omega_1 \\ \omega_2 \end{bmatrix} \tag{13.80}$$

Hence, Eqs. (13.78) and (13.79) may be written as

$$\dot{\mathbf{x}} = \begin{bmatrix} \dot{x}_1 \\ \dot{x}_2 \end{bmatrix} = \begin{bmatrix} \dot{\omega}_1 \\ \dot{\omega}_2 \end{bmatrix} = \begin{bmatrix} -K_1/I_1 & -a/I_1 \\ a/I_1 & -K_2/I_1 \end{bmatrix} \mathbf{x} \tag{13.81}$$

The stability of this linear system can be determined by computing the so-called *eigenvalues* of the 2×2 matrix in Eq. (13.81). A detailed discussion of linear system analysis is beyond the scope of this section (besides, the linear control law is fictional and the actual system is *not* linear). The eigenvalues are the roots of the following quadratic equation:

$$s^2 + \left(\frac{K_1 + K_2}{I_1}\right)s + \frac{K_1 K_2 + a^2}{I_1} = 0 \tag{13.82}$$

Closed-loop stability of the linear system is ensured if the first-order and zeroth-order polynomial coefficients in Eq. (13.82) are positive (the interested reader may consult Kluever [4; pp. 232–234] for a discussion of eigenvalues and stability). Therefore, the linear system is stable if both gains K_1 and K_2 are positive. Recall that the *fictional* linear control laws are $M_1 = -K_1\omega_1$ and $M_2 = -K_2\omega_2$. Both linear controls command jet torques in directions that are the *opposite* of the respective spin components. Because the control torques can only take on a fixed magnitude (Table 13.2), we can use the linear-analysis results to determine the *sign* of the controls:

$$M_1 = -Fz\,\text{sgn}(\omega_1) \tag{13.83}$$

$$M_2 = -2Fz\,\text{sgn}(\omega_2) \tag{13.84}$$

Equations (13.83) and (13.84) are nonlinear nutation control laws for the reaction-jet torques about the 1 and 2 principal axes. Recall that the *linear* system is stable when the controls are $M_1 = -K_1\omega_1$ and $M_2 = -K_2\omega_2$ (with positive gains), but this feedback law requires reaction jets with variable thrust. The nonlinear control laws (13.83) and (13.84) use the *signs* of the angular velocity feedback (ω_1 and ω_2) to determine which jets to fire (i.e., positive or negative torque).

Because the feedback torque laws (13.83) and (13.84) are nonlinear, there is no way to develop analytical expressions for the ensuing satellite motion with nutation control. Therefore, we must resort to numerical integration methods to determine the satellite's response. The following example illustrates the nonlinear nutation control system.

Example 13.7   Consider an axisymmetric prolate satellite with a reaction-jet configuration as shown in Figure 13.43. The principal moments of inertia are $I_1 = I_2 = 4{,}317$ kg-m^2 and $I_3 = 2{,}800$ kg-m^2 (these moments of inertia correspond to a cylindrical satellite with a mass of 1,400 kg, radius of 2 m, and height of 5 m). Each reaction jet provides force $F = 50$ N when fired, and the vertical height from the jet clusters to the 1–2 plane is $z = 2.5$ m. At time $t = 0$, the satellite's angular velocity is $\omega = 6$ rad/s. However, the angular velocity vector $\boldsymbol{\omega}$ is initially tilted 8° from the 3 axis (i.e., $\gamma = 8°$), which causes a wobbling motion. Demonstrate the effectiveness of the nonlinear nutation control laws (13.83) and (13.84) by using a numerical simulation of the closed-loop system.

First, we should note that because the satellite is axisymmetric ($I_1 = I_2$) and no torque is produced about the 3 axis ($M_3 = 0$), Euler's moment equations are reduced to Eqs. (13.75) and (13.76). Next, let us substitute the nonlinear nutation control laws (13.83) and (13.84) for control torques M_1 and M_2 in Eqs. (13.75) and (13.76). The result is

$$I_1\dot{\omega}_1 = -a\omega_2 - Fz\,\text{sgn}(\omega_1) \tag{13.85}$$

$$I_1\dot{\omega}_2 = a\omega_1 - 2Fz\,\text{sgn}(\omega_2) \tag{13.86}$$

where $a = (I_3 - I_2)n$ and $n = \omega_3$. Because $M_3 = 0$ and $I_1 = I_2$, the spin component along the 3 axis is constant, that is,

$$n = \omega_3 = \omega\cos\gamma = 5.9416\ \text{rad/s (constant)}$$

Using the 3-axis spin component n and moments of inertia I_3 and I_2, we determine the constant $a = -9{,}013.42$ kg-m^2/s. Figure 13.44 shows the closed-loop nutation control system for the axisymmetric satellite. Although it requires carefully tracing the signal

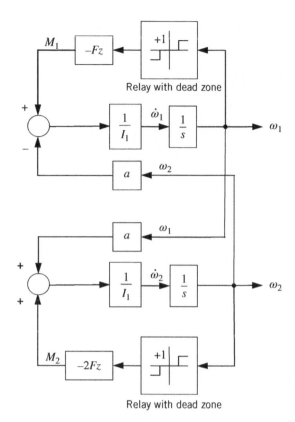

Figure 13.44 Nutation control system for an axisymmetric satellite (Example 13.7).

paths, the reader should be able to identify the governing equations (13.85) and (13.86) in Figure 13.44. Note also that a relay with dead zone is used in the feedback path instead of a pure "sign" or signum function (again, the dead zone will reduce the control "chatter" when spin components ω_1 and ω_2 become acceptably small). In an operational setting, the jet selection is simply determined by checking the signs of the feedback from rate gyroscopes mounted along the 1 and 2 axes. For example, if $\omega_1 > 0$ and $\omega_2 < 0$, fire the reaction jet b (for $M_1 = -Fz$) and jets a, d (for $M_2 = +2Fz$). If either spin component is smaller than the threshold of the dead zone, the jets are not fired.

Simulink was used to create and simulate the closed-loop system shown in Figure 13.44. The initial value of the component of the angular velocity vector ω projected onto the 1–2 plane is

$$\omega_{12} = \omega \sin \gamma = 0.8350 \text{ rad/s}$$

Suppose the ω_{12} vector is initially 30° (counterclockwise) from the 1 axis (recall for *torque-free* motion, the ω_{12} vector will rotate clockwise for a prolate satellite; see Figure 12.9). The initial spin components are $\omega_1(0) = \omega_{12} \cos(30°) = 0.7232$ rad/s, and $\omega_2(0) = \omega_{12} \sin(30°) = 0.4175$ rad/s. Each dead zone threshold is set to 0.001 rad/s (≈ 0.06 deg/s). Figure 13.45 shows the time histories of the spin components $\omega_1(t)$ and $\omega_2(t)$, respectively. The results of the nutation control effort are apparent in these figures as both off-axis spin components ω_1 and ω_2 are driven to zero in about 15 s.

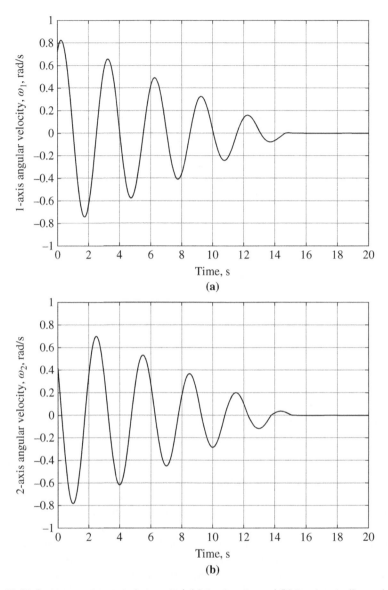

Figure 13.45 Response using nutation control: (a) 1-axis spin; and (b) 2-axis spin (Example 13.7).

After 15 s, the satellite's angular velocity vector $\boldsymbol{\omega}$ is essentially aligned with the 3 axis [recall that $\omega_3 = 5.9416$ rad/s (constant) at all times because $M_3 = 0$ and $I_1 = I_2$; therefore, the magnitude of the final angular velocity is equal to ω_3]. Figure 13.46 shows the 1- and 2-axis control torques, M_1 and M_2, respectively. Each control switching corresponds to a sign change of the appropriate spin component. Note that the magnitude of torque $M_1 = Fz = 125$ N-m (one jet), whereas the 2-axis torque magnitude is $M_2 = 2Fz = 250$ N-m (two jets). Both control torques go to zero for $t > 15$ s because each spin component is within the dead zone and the nutation has been removed.

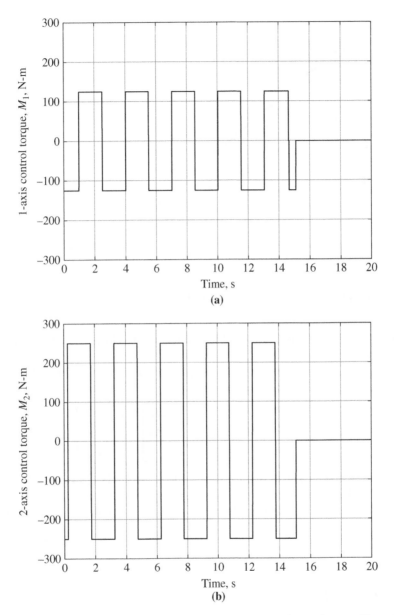

Figure 13.46 Nutation control: (a) 1-axis control torque; and (b) 2-axis control torque (Example 13.7).

Finally, Figure 13.47 shows the time histories of the nutation angle θ and "tilt" angle γ. The initial nutation angle is

$$\theta = \tan^{-1}\left(\frac{I_1\omega_{12}}{I_3\omega_3}\right) = 12.23°$$

Figure 13.47 shows that angles θ and γ steadily decrease as the nutation control torques remove the off-axis spin components ω_1 and ω_2. At $t = 15$ s, both angles have been driven

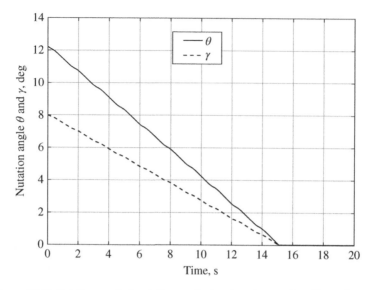

Figure 13.47 Nutation control: nutation angle θ and angle γ vs. time (Example 13.7).

to zero, and hence the 3 axis is aligned with the angular velocity vector $\boldsymbol{\omega}$ and angular momentum vector $\mathbf{H}$.

As a final note, we can determine the change in angular momentum for the nutation control. At $t = 0$, the initial angular momentum is

$$\mathbf{H}_0 = \mathbf{I}\boldsymbol{\omega}_0 = \begin{bmatrix} 4{,}317 & 0 & 0 \\ 0 & 4{,}317 & 0 \\ 0 & 0 & 2{,}800 \end{bmatrix} \begin{bmatrix} 0.7232 \\ 0.4175 \\ 5.9416 \end{bmatrix} = \begin{bmatrix} 3{,}122 \\ 1{,}802 \\ 16{,}637 \end{bmatrix} \text{kg-m}^2/\text{s}$$

The magnitude is $H_0 = 17{,}023$ kg-m^2/s. For $t > 15$ s, the satellite is essentially in a pure spin about its 3 axis, and therefore the final angular momentum is $H_f = I_3\omega_3 = 16{,}637$ kg-m^2/s. The external torque from firing reaction jets has reduced the magnitude of the angular momentum by about 2%.

13.7 Summary

This chapter has presented examples of feedback schemes for satellite attitude control. We began with a short introduction to feedback control concepts, including closed-loop transfer functions and simple design rules for tuning a PD controller. This chapter has focused on single-axis rotational maneuvers for satellites equipped with a reaction wheel or on–off reaction jets. Reaction wheels offer smooth, continuous control by exchanging the angular momentum between the satellite and the spinning wheel. Hence, reaction-wheel controllers can be designed using linear control concepts. Reaction jets, on the other hand, are on–off devices that deliver a fixed positive or negative control torque when fired. Therefore, the control torque is discontinuous and the control system is nonlinear. We developed a switching-curve control law by using phase-plane analysis of the satellite dynamics with on–off reaction jets. Finally, we investigated a nutation control scheme for removing the "wobbling" motion about the primary spin axis. Our nutation controller also used pulsed reaction jets and a nonlinear control law. Although this chapter does not provide an exhaustive treatment of three-axis attitude control, it does offer an introduction to realistic attitude control schemes.

References

1 Phillips, C.L., and Parr, J.M., *Feedback Control Systems*, 5th edn, Prentice Hall, Upper Saddle River, NJ, 2011.
2 Franklin, G.F., Powell, J.D., and Emami-Naeini A., *Feedback Control of Dynamic Systems*, 4th edn, Prentice Hall, Upper Saddle River, NJ, 2002.
3 Dorf, R.C., and Bishop, R.H., *Modern Control Systems*, 12th edn, Prentice Hall, Upper Saddle River, NJ, 2011
4 Kluever, C.A., *Dynamic Systems: Modeling, Simulation, and Control*, John Wiley & Sons, Inc., Hoboken, NJ, 2015.
5 Ogata, K., *System Dynamics*, 4th edn, Pearson Prentice Hall, Upper Saddle River, NJ, 2004.

Further Reading

Hughes, P.C., *Spacecraft Attitude Dynamics*, Dover, New York, 2004.
Kaplan, M.H., *Modern Spacecraft Dynamics and Control*, John Wiley & Sons, Inc., New York, 1976.
Thomson, W.T., *Introduction to Space Dynamics*, Dover, New York, 1986.
Wiesel, W.E., *Spaceflight Dynamics*, 3rd edn, Aphelion Press, Beavercreek, OH, 2011.

Problems

Conceptual Problems

13.1 Given the ordinary differential equation relating input u to output y

$$3\dot{y} + 8y = 0.2u$$

determine the corresponding transfer function $G(s)$.

13.2 Given the ordinary differential equation relating input u to output y

$$2\ddot{y} + 7\dot{y} + 50y = 3u$$

determine the corresponding transfer function $G(s)$.

13.3 Given the ordinary differential equation relating input u to output y

$$0.2\ddot{y} + 4\dot{y} + 12y = 0.2\dot{u} + u$$

determine the corresponding transfer function $G(s)$.

13.4 Determine the undamped natural frequency ω_n and damping ratio ζ for the system defined by the differential equation in Problem 13.2.

Problems 13.5–13.8 involve the closed-loop control system shown in Figure P13.5.

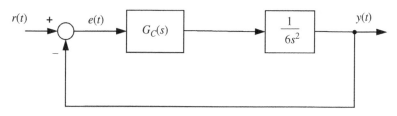

Figure P13.5

13.5 Determine the closed-loop transfer function if the controller is a simple gain, $G_C(s) = K_P$.

13.6 Determine the closed-loop transfer function for a proportional-derivative (PD) controller, $G_C(s) = K_P + K_D s$.

13.7 Determine the closed-loop system's undamped natural frequency ω_n and damping ratio ζ for the proportional controller $G_C(s) = 10$.

13.8 Determine the closed-loop system's undamped natural frequency ω_n and damping ratio ζ for the PD controller $G_C(s) = 10 + 12s$.

13.9 A satellite is equipped with a reaction wheel. The wheel is mounted along the 3 axis and has moment of inertia $I_w = 0.1$ kg-m². The satellite's *total* moment of inertia about the 3 axis (including the wheel) is $I_3 = 852.4$ kg-m². At time $t = 0$, the satellite is spinning with angular velocity $\omega_3 = 5$ revolutions per minute (rpm) about the 3 axis (i.e., $\omega_3 = 5\,\mathbf{u}_3$ rpm). At $t = 0$, the reaction wheel's angular velocity relative to the satellite is zero.
   a) Determine the angular momentum of the satellite.
   b) A direct current (DC) motor brings the reaction wheel to the spin rate $\omega_w = 3{,}500\,\mathbf{u}_3$ rpm. Determine the satellite's angular velocity vector after the wheel reaches this constant spin rate.
   c) A DC motor brings the reaction wheel to the spin rate $\omega_w = -3{,}500\,\mathbf{u}_3$ rpm. Determine the satellite's angular velocity vector after the wheel reaches this constant spin rate.

13.10 A satellite is equipped with a reaction wheel mounted along the 3 axis. The satellite's moment of inertia about its 3 axis (*not* including the wheel) is $I_{sat} = 1{,}030$ kg-m². At time $t = 0$, the satellite is spinning with angular velocity vector $\omega_3 = 0.35\,\mathbf{u}_3$ rpm and the reaction wheel's angular velocity vector (relative to the satellite) is $\omega_w = -2{,}751\,\mathbf{u}_3$ rpm. The DC motor driving the wheel is turned off and eventually the reaction wheel comes to rest. The satellite's spin rate also goes to zero when the wheel stops spinning relative to the satellite. Determine the moment of inertia of the reaction wheel.

Problems 13.11–13.14 involve a satellite equipped with on–off reaction jets for attitude control about all three axes. The satellite's three principal moments of inertia are $I_1 = 1{,}210$ kg-m^2, $I_2 = 1{,}080$ kg-m^2, and $I_3 = 1{,}730$ kg-m^2. Pairs of reaction jets can provide the following external torques to the satellite: $M_1 = \pm 95$ N-m, $M_2 = \pm 190$ N-m, and $M_3 = \pm 260$ N-m. Figure 13.36 presents the basic structure of the three feedback control laws for each body-fixed axis (for simplicity assume perfect switching with a zero dead zone). In all cases the attitude error is the satellite's *actual* attitude angle minus the reference attitude for a particular body axis. Assume single-axis maneuvers in all problems.

13.11 The satellite's attitude error about the 1 axis is 25° and its 1-axis gyroscope is sensing zero angular velocity. Determine the control-torque sequence and the total maneuver time to reach zero attitude error and zero attitude rate.

13.12 The satellite's attitude error about the 3 axis is −95° and its angular velocity about the 3 axis is 2.5 deg/s. Determine the control-torque sequence and the total maneuver time to reach zero attitude error and zero attitude rate.

13.13 The satellite's 2-axis gyroscope is sensing −15 deg/s at time $t = 0$. The closed-loop control system selects positive reaction-jet torque at $t = 0$ and the satellite reaches zero attitude error and zero attitude rate without switching to negative torque jets. Determine the satellite's attitude error at $t = 0$ and the maneuver time.

13.14 At time $t = 0$, the satellite's attitude error about the 3 axis is 11° and its 3-axis angular velocity is 20 deg/s.
  a) Determine the point in the phase plane where torque switching occurs.
  b) Determine the maneuver time to the torque-switching point.
  c) What is the satellite's maximum attitude error during the closed-loop maneuver?

13.15 Consider again the demonstration of nutation control presented in Example 13.7. Figure 13.45 shows that the 1- and 2-axis spin rates during nutation damping are periodic with an amplitude that appears to be linear with time. Therefore, we can approximate their time histories as

$$\omega_1(t) = (c_0 + c_1 t)\cos(\lambda t + \beta) \tag{13.P1}$$

$$\omega_2(t) = (c_0 + c_1 t)\sin(\lambda t + \beta) \tag{13.P2}$$

where c_0 and c_1 are constants that define the linear "amplitude envelop" seen in Figures 13.45a and 13.45b. The "coning motion" angular velocity λ is defined by the principal moments of inertia and the 3-axis spin rate ω_3 [see Eq. (12.43) in Chapter 12] and β is a phase angle that depends on initial conditions. Use Eqs. (13.P1) and (13.P2) to estimate the time history of the angular velocity projection onto the 1–2 plane, that is, $\omega_{12}(t)$. Take the time-derivative of the nutation angle

$$\theta = \tan^{-1}\left(\frac{I_1\omega_{12}}{I_3\omega_3}\right)$$

and use the approximate solution for $\omega_{12}(t)$ to estimate the time rate of nutation removal. Does the approximate nutation rate match the numerical simulation results as seen in Figure 13.47?

Mission Applications

Problems 13.16–13.19 involve the ARTEMIS spacecraft which was launched by the European Space Agency (ESA) in July 2001. Due to a partial failure of the Ariane 5 launch vehicle, the ARTEMIS spacecraft used its electric propulsion (EP) ion thrusters to complete the final phase of the transfer to geostationary-equatorial orbit (GEO). During the EP thrusting phase, the spacecraft must maintain a negative rotation about its pitch axis at the orbital angular velocity so that the body-fixed thrust vector remains aligned with the local horizon and the velocity vector (see Figure P13.16). The satellite's thrust vector intersects the center of mass and therefore does not impart an external torque. Assume that the ARTEMIS spacecraft had zero angular momentum when it was placed in sub-geostationary orbit by the launch vehicle.

The spacecraft and wheel properties are I_{sat} = 590 kg-m^2 (satellite moment of inertia about its pitch axis *without* the reaction wheel) and I_w = 0.03 kg-m^2 (reaction wheel moment of inertia mounted along the pitch axis). The DC motor properties are K_m = 0.025 N-m/A (motor torque constant), K_b = 0.025 V-s/rad (back-emf constant), resistance R = 2 Ω, and b = 4(10^{-5}) N-m-s/rad (friction-torque coefficient).

13.16 The ARTEMIS spacecraft began its orbit transfer to GEO in a circular orbit with an altitude of 31,000 km. Determine the reaction wheel's angular velocity (magnitude and direction) in order to maintain the satellite's angular velocity required to "steer" the EP thrust vector.

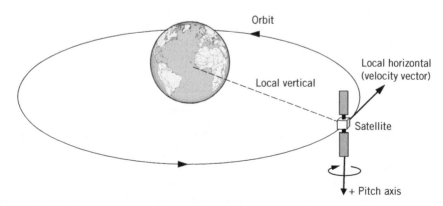

Figure P13.16

13.17 Using the orbital conditions in Problem 13.16, determine the DC motor torque T_m, the armature current i_m, and the input voltage to the DC motor e_{in}.

13.18 Repeat Problems 13.16 and 13.17 for the case where the ARTEMIS spacecraft is in a quasi-circular orbit at an altitude of 35,000 km (i.e., near GEO).

13.19 When the ARTEMIS spacecraft reached its GEO target, it needed to point its communication antenna toward Earth (along the local vertical direction shown in Figure P13.16). Therefore, the mission operators want the ARTEMIS space-craft to achieve a spin rate of 0.02 rad/s (about 1 deg/s) about its negative pitch axis. Determine the reaction wheel's spin rate ω_w (in rpm) and input voltage e_{in} required for this attitude maneuver about the pitch axis.

13.20 A communication satellite in GEO must keep its antenna pointed in the local vertical direction (i.e., pointing to Earth). Therefore, the satellite must maintain a constant (negative) angular velocity about its pitch axis (Figure P13.16 shows that the positive pitch axis is in the opposite direction as the *orbital* angular momentum vector). Suppose that the communication satellite is disturbed by the following solar radiation pressure (SRP) torque about its pitch axis:

$$M_{SRP} = 10^{-4} \cos\omega_0 t \text{ N-m}$$

where ω_0 is the orbital angular velocity for GEO and time t is measured relative to local sunrise so that zero SRP torque occurs at noon and midnight. Control engineers propose a pitch-axis reaction wheel for absorbing the SRP torque. The wheel has the following angular velocity profile:

$$\omega_w(t) = \bar{\omega}_w + \tilde{\omega}_w(t)$$

where $\bar{\omega}_w$ is the wheel's constant (nominal) spin rate and $\tilde{\omega}_w(t)$ is the wheel's time-varying spin rate. If the satellite's moment of inertia about its pitch axis is 560 kg-m^2 (without the reaction wheel) and the wheel's moment of inertia is $I_w = 0.06$ kg-m^2, determine the time-varying wheel spin rate $\tilde{\omega}_w(t)$ (in rpm) required to absorb the SRP torque.

Appendix A

Physical Constants

We need the physical parameters for planetary bodies to numerically calculate two-body orbits. Table A.1 presents the equatorial radius and gravitational parameter for celestial bodies in our solar system. Table A.2 presents the physical characteristics of the Earth, including equatorial and polar radii, standard gravitational acceleration, and the second zonal harmonic coefficient.

Table A.1 Physical parameters for celestial bodies.

Body	Equatorial radius (km)	Gravitational parameter, μ (km^3/s^2)
Sun	695,700	$1.32712440(10^{11})$
Mercury	2,439.7	$2.2033(10^4)$
Venus	6,051.8	$3.24860(10^5)$
Earth	6,378.137	$3.98600442(10^5)$
Earth's moon	1,737.5	$4.902801(10^3)$
Mars	3,396.19	$4.28283(10^4)$
Jupiter	71,492	$1.266865(10^8)$
Saturn	60,268	$3.793119(10^7)$
Uranus	25,559	$5.793940(10^6)$
Neptune	24,764	$6.836530(10^6)$
Pluto	1,151	$8.719(10^2)$

Space Flight Dynamics, First Edition. Craig A. Kluever.
© 2018 John Wiley & Sons Ltd. Published 2018 by John Wiley & Sons Ltd.
Companion website: www.wiley.com/go/Kluever/spaceflightmechanics

Table A.2 Physical characteristics of the Earth.

Characteristic	Value
Equatorial radius, R_E	6,378.137 km
Polar radius, R_P	6,356.752 km
Gravitational parameter, μ_E	$3.98600442(10^5)$ km^3/s^2
Standard acceleration due to gravity, g_0	9.80665 m/s^2
Rotation rate, ω_E	$7.29211576(10^{-5})$ rad/s
Second zonal harmonic coefficient (oblateness), J_2	0.0010826267

Appendix B

Review of Vectors

B.1 Introduction

This appendix presents a very brief review of vectors and operations involving vectors. Much of the material discussed here will likely be familiar to engineering students.

B.2 Vectors

We will denote vectors with bold-face typeset, such as position vector **r**. A vector is a quantity that has magnitude and direction. Figure B.1 shows vector **a** in Cartesian coordinate system $OXYZ$. Vector **a** points from origin O to particle P. The length of vector **a** is its magnitude, whereas its direction can be defined by using the directions of the three orthogonal axes X, Y, and Z.

A *unit vector* has a magnitude of one (unity) and a direction. A unit vector $\mathbf{u}_a$ that points in the same direction as vector **a** is defined by

$$\mathbf{u}_a = \frac{\mathbf{a}}{a} \tag{B.1}$$

where a is the magnitude (or length) of vector **a**. We can define Cartesian unit vectors **I**, **J**, and **K** along the $+X$, $+Y$, and $+Z$ axes, respectively. Thus, we may express vector **a** in terms of its Cartesian components shown in Figure B.1:

$$\mathbf{a} = a_X \mathbf{I} + a_Y \mathbf{J} + a_Z \mathbf{K} \tag{B.2}$$

The magnitude of vector **a** is

$$a = \|\mathbf{a}\| = \sqrt{a_X^2 + a_Y^2 + a_Z^2} \tag{B.3}$$

Space Flight Dynamics, First Edition. Craig A. Kluever.
© 2018 John Wiley & Sons Ltd. Published 2018 by John Wiley & Sons Ltd.
Companion website: www.wiley.com/go/Kluever/spaceflightmechanics

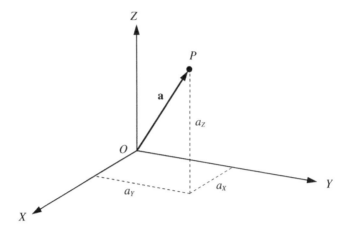

Figure B.1 Vector in a Cartesian coordinate system.

B.3   Vector Operations

B.3.1   Vector Addition

Figure B.2 illustrates vector addition: $\mathbf{c} = \mathbf{a} + \mathbf{b}$. Resultant vector $\mathbf{c}$ denotes the position of particle Q. We can obtain the **IJK** coordinates of vector $\mathbf{c}$ by individually adding the **IJK** components of vectors $\mathbf{a}$ and $\mathbf{b}$ as follows:

$$\mathbf{c} = \mathbf{a} + \mathbf{b} = (a_X + b_X)\mathbf{I} + (a_Y + b_Y)\mathbf{J} + (a_Z + b_Z)\mathbf{K} \tag{B.4}$$

Or, we may rewrite Eq. (B.4) as a three-element column vector:

$$\mathbf{c} = \mathbf{a} + \mathbf{b} = \begin{bmatrix} a_X + b_X \\ a_Y + b_Y \\ a_Z + b_Z \end{bmatrix} \tag{B.5}$$

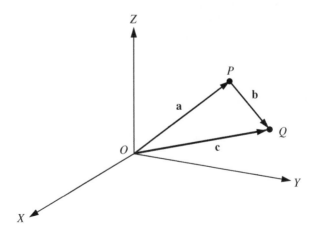

Figure B.2 Vector addition.

where it is understood that the top, middle, and bottom elements of vector $\mathbf{c}$ are the $\mathbf{I}$, $\mathbf{J}$, and $\mathbf{K}$ components. The time derivative of the vector addition operation is

$$\frac{d}{dt}(\mathbf{a} + \mathbf{b}) = \dot{\mathbf{a}} + \dot{\mathbf{b}} \tag{B.6}$$

B.3.2 Cross Product

The cross (or vector) product of vectors $\mathbf{a}$ and $\mathbf{b}$ is vector $\mathbf{c}$, written as

$$\mathbf{c} = \mathbf{a} \times \mathbf{b} \tag{B.7}$$

and read as "vector $\mathbf{c}$ equals $\mathbf{a}$ cross $\mathbf{b}$." The direction of resultant vector $\mathbf{c}$ can be determined by the "right-hand rule," where the fingers of the right hand are curled *from* vector $\mathbf{a}$ to vector $\mathbf{b}$, and the right-hand thumb points in the direction of $\mathbf{c}$. Hence, resultant vector $\mathbf{c}$ is perpendicular to the plane containing vectors $\mathbf{a}$ and $\mathbf{b}$. The *magnitude* of the cross product $\mathbf{c} = \mathbf{a} \times \mathbf{b}$ is

$$c = ab\sin\theta \tag{B.8}$$

where θ is the smallest angle between vectors $\mathbf{a}$ and $\mathbf{b}$ if we place their "tails" together (therefore, angle θ is between zero and 180°). It is important to note that the cross product is *not* commutative, that is, $\mathbf{a} \times \mathbf{b} \neq \mathbf{b} \times \mathbf{a}$. However, the cross product does follow

$$\mathbf{a} \times \mathbf{b} = -\mathbf{b} \times \mathbf{a} \tag{B.9}$$

The distributive rule for the cross product is

$$\mathbf{a} \times (\mathbf{b} + \mathbf{c}) = \mathbf{a} \times \mathbf{b} + \mathbf{a} \times \mathbf{c} \tag{B.10}$$

and the associative law for the cross product is

$$\alpha(\mathbf{a} \times \mathbf{b}) = (\alpha\mathbf{a}) \times \mathbf{b} = \mathbf{a} \times (\alpha\mathbf{b}) = (\mathbf{a} \times \mathbf{b})\alpha \tag{B.11}$$

where α is any (scalar) constant.

The cross product of two parallel vectors is the zero or null vector. For example,

$$\mathbf{a} \times \mathbf{a} = \mathbf{0} \tag{B.12}$$

It is important to note that although the right-hand side of Eq. (B.12) is zero, it is a 3×1 column vector with three zero elements. Hence, we can write the cross products of the IJK unit vectors for the Cartesian frame as

$$\mathbf{I} \times \mathbf{I} = \mathbf{J} \times \mathbf{J} = \mathbf{K} \times \mathbf{K} = \mathbf{0} \tag{B.13}$$

Because the cross product $\mathbf{c}$ is perpendicular to vectors $\mathbf{a}$ and $\mathbf{b}$, we can express the Cartesian frame unit vectors as cross products:

$$\mathbf{I} \times \mathbf{J} = \mathbf{K}$$
$$\mathbf{J} \times \mathbf{K} = \mathbf{I} \tag{B.14}$$
$$\mathbf{K} \times \mathbf{I} = \mathbf{J}$$

We can compute the cross product of vectors **a** and **b** expressed in Cartesian coordinates by using the determinant expansion:

$$\mathbf{c} = \mathbf{a} \times \mathbf{b} = \begin{vmatrix} \mathbf{I} & \mathbf{J} & \mathbf{K} \\ a_X & a_Y & a_Z \\ b_X & b_Y & b_Z \end{vmatrix} \tag{B.15}$$

$$= (a_Y b_Z - a_Z b_Y)\mathbf{I} - (a_X b_Z - a_Z b_X)\mathbf{J} + (a_X b_Y - a_Y b_X)\mathbf{K}$$

The time derivative of the cross product is

$$\frac{d}{dt}(\mathbf{a} \times \mathbf{b}) = \dot{\mathbf{a}} \times \mathbf{b} + \mathbf{a} \times \dot{\mathbf{b}} \tag{B.16}$$

B.3.3 Dot Product

The dot (or scalar) product of vectors **a** and **b** is a scalar, and is written as

$$\mathbf{a} \cdot \mathbf{b} = ab\cos\theta \tag{B.17}$$

and read as "**a** dot **b**." As with the cross product, θ is the smallest angle between vectors **a** and **b** when we place their tails together. The order of the dot product does *not* make a difference, and hence

$$\mathbf{a} \cdot \mathbf{b} = \mathbf{b} \cdot \mathbf{a} \tag{B.18}$$

The distributive law for the dot product is

$$\mathbf{a} \cdot (\mathbf{b} + \mathbf{c}) = \mathbf{a} \cdot \mathbf{b} + \mathbf{a} \cdot \mathbf{c} \tag{B.19}$$

Multiplication by scalar α can be performed in any order:

$$\alpha(\mathbf{a} \cdot \mathbf{b}) = (\alpha \mathbf{a}) \cdot \mathbf{b} = \mathbf{a} \cdot (\alpha \mathbf{b}) = (\mathbf{a} \cdot \mathbf{b})\alpha \tag{B.20}$$

The dot product of two orthogonal vectors is zero. For example, consider the **IJK** unit vectors for the Cartesian coordinate system:

$$\mathbf{I} \cdot \mathbf{J} = \mathbf{J} \cdot \mathbf{K} = \mathbf{K} \cdot \mathbf{I} = 0 \tag{B.21}$$

Note that the right-hand side of Eq. (B.21) is a scalar zero. If vectors **a** and **b** are expressed in **IJK** (Cartesian) coordinates, their dot product is

$$\mathbf{a} \cdot \mathbf{b} = a_X b_X + a_Y b_Y + a_Z b_Z \tag{B.22}$$

The dot product of a vector with itself is its magnitude squared:

$$\mathbf{a} \cdot \mathbf{a} = a_X a_X + a_Y a_Y + a_Z a_Z = a^2 \tag{B.23}$$

The time derivative of the dot product is

$$\frac{d}{dt}(\mathbf{a} \cdot \mathbf{b}) = \dot{\mathbf{a}} \cdot \mathbf{b} + \mathbf{a} \cdot \dot{\mathbf{b}} \tag{B.24}$$

We can determine the time derivative of the dot product of a vector with itself by differentiating both sides of Eq. (B.23) and making use of Eq. (B.24):

$$\frac{d}{dt}(\mathbf{a}\cdot\mathbf{a}) = (\dot{\mathbf{a}}\cdot\mathbf{a}) + (\mathbf{a}\cdot\dot{\mathbf{a}}) = 2a\dot{a} \tag{B.25}$$

Dividing Eq. (B.25) by 2, we obtain

$$\mathbf{a}\cdot\dot{\mathbf{a}} = a\dot{a} \tag{B.26}$$

We used the above dot-product property in Section 2.3 to demonstrate conservation of energy for the two-body problem.

B.3.4 Scalar Triple Product

The scalar triple product is an operation involving three vectors. As the name implies, the result is a scalar. The scalar triple product is defined by

$$\mathbf{a}\cdot(\mathbf{b}\times\mathbf{c}) = \mathbf{b}\cdot(\mathbf{c}\times\mathbf{a}) = \mathbf{c}\cdot(\mathbf{a}\times\mathbf{b}) \tag{B.27}$$

We used the scalar triple product in Section 2.4 in the derivation of the trajectory equation.

B.3.5 Vector Triple Product

The vector triple product involves two cross products and thus the result is a vector. It is defined by

$$\mathbf{a}\times(\mathbf{b}\times\mathbf{c}) = \mathbf{b}(\mathbf{a}\cdot\mathbf{c}) - \mathbf{c}(\mathbf{a}\cdot\mathbf{b}) \tag{B.28}$$

Note that the two parenthetical dot-product terms on the right-hand side of Eq. (B.28) are scalar multiplying factors for vectors **b** and **c**. The vector triple product was utilized in the derivation of the trajectory equation in Section 2.4.

Appendix C

Review of Particle Kinematics

C.1 Introduction

Many of the scenarios we encounter in space flight dynamics involve motion of a particle relative to a "fixed" or inertial frame (*absolute motion*) or relative to a moving and/or rotating frame (*relative motion*). Often it is advantageous to express the particle's motion in a moving frame. This appendix presents position, velocity, and acceleration of a point-mass particle in a variety of coordinate systems.

C.2 Cartesian Coordinates

Figure C.1 shows a fixed, non-rotating (inertial) Cartesian frame $O'XYZ$ and a moving Cartesian frame $Oxyz$. Cartesian frame $Oxyz$ can translate and rotate. The absolute position of particle P is

$$\mathbf{r} = \mathbf{r}_O + \boldsymbol{\rho} \tag{C.1}$$

where $\mathbf{r}_O$ is the absolute position vector of the moving frame $Oxyz$ relative to fixed frame $O'XYZ$, and $\boldsymbol{\rho}$ is the relative position of P expressed in the moving coordinates. Let us express vectors in the inertial frame using the unit vectors $\mathbf{IJK}$ aligned with the Cartesian axes XYZ. Therefore, the absolute position of the moving frame is

$$\mathbf{r}_O = X_O\mathbf{I} + Y_O\mathbf{J} + Z_O\mathbf{K} \tag{C.2}$$

In a similar fashion, the position of P relative to the moving/rotating frame is

$$\boldsymbol{\rho} = x\mathbf{u}_x + y\mathbf{u}_y + z\mathbf{u}_z \tag{C.3}$$

where unit vectors $\mathbf{u}_x \mathbf{u}_y \mathbf{u}_z$ are aligned with the moving/rotating axes xyz. The absolute velocity of P is obtained by taking the time derivative of Eq. (C.1):

$$\mathbf{v} = \dot{\mathbf{r}} = \dot{\mathbf{r}}_O + \dot{\boldsymbol{\rho}} \tag{C.4}$$

where the time derivatives of vectors $\mathbf{r}_O$ and $\boldsymbol{\rho}$ are

$$\dot{\mathbf{r}}_O = \frac{d}{dt}(X_O\mathbf{I} + Y_O\mathbf{J} + Z_O\mathbf{K}) = \dot{X}_O\mathbf{I} + \dot{Y}_O\mathbf{J} + \dot{Z}_O\mathbf{K} \tag{C.5}$$

Space Flight Dynamics, First Edition. Craig A. Kluever.
© 2018 John Wiley & Sons Ltd. Published 2018 by John Wiley & Sons Ltd.
Companion website: www.wiley.com/go/Kluever/spaceflightmechanics

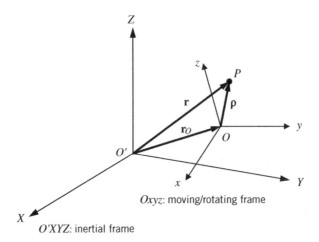

Figure C.1 Inertial and moving Cartesian coordinate systems.

$$\dot{\boldsymbol{\rho}} = \frac{d}{dt}\left(x\mathbf{u}_x + y\mathbf{u}_y + z\mathbf{u}_z\right) = \dot{x}\mathbf{u}_x + \dot{y}\mathbf{u}_y + \dot{z}\mathbf{u}_z + x\dot{\mathbf{u}}_x + y\dot{\mathbf{u}}_y + z\dot{\mathbf{u}}_z \qquad (C.6)$$

Note that the derivatives of unit vectors $\mathbf{u}_x\mathbf{u}_y\mathbf{u}_z$ must be included in Eq. (C.6) because their *directions* change due to the rotation of frame $Oxyz$. The derivatives of unit vectors **IJK** are zero because their lengths and directions remain constant because frame $O'XYZ$ is fixed in inertial space. Let $\boldsymbol{\omega}$ be the angular velocity vector of the rotating frame $Oxyz$ relative to the fixed frame. The time rates of the unit vectors of the moving frame are due to angular velocity $\boldsymbol{\omega}$ and can be expressed as cross products:

$$\dot{\mathbf{u}}_x = \boldsymbol{\omega} \times \mathbf{u}_x \qquad (C.7)$$

$$\dot{\mathbf{u}}_y = \boldsymbol{\omega} \times \mathbf{u}_y \qquad (C.8)$$

$$\dot{\mathbf{u}}_z = \boldsymbol{\omega} \times \mathbf{u}_z \qquad (C.9)$$

Using Eqs. (C.7)–(C.9) in Eq. (C.6), we obtain

$$\dot{\boldsymbol{\rho}} = \dot{\boldsymbol{\rho}}\big|_{\mathrm{rot}} + \boldsymbol{\omega} \times \boldsymbol{\rho} \qquad (C.10)$$

where

$$\dot{\boldsymbol{\rho}}\big|_{\mathrm{rot}} = \dot{x}\mathbf{u}_x + \dot{y}\mathbf{u}_y + \dot{z}\mathbf{u}_z \qquad (C.11)$$

is the time derivative of position vector $\boldsymbol{\rho}$ with respect to the moving/rotating frame. Substituting Eq. (C.10) in Eq. (C.4) yields

$$\dot{\mathbf{r}} = \dot{\mathbf{r}}_O + \dot{\boldsymbol{\rho}}\big|_{\mathrm{rot}} + \boldsymbol{\omega} \times \boldsymbol{\rho} \qquad (C.12)$$

Equation (C.12) is absolute velocity of particle P.

We obtain the absolute acceleration of particle P by taking the time derivative of every term in Eq. (C.12):

$$\mathbf{a} = \ddot{\mathbf{r}} = \ddot{\mathbf{r}}_O + \frac{d}{dt}\dot{\boldsymbol{\rho}}\bigg|_{\mathrm{rot}} + \frac{d}{dt}(\boldsymbol{\omega} \times \boldsymbol{\rho}) \qquad (C.13)$$

Using Eq. (C.11), the time derivative of $\dot{\boldsymbol{\rho}}|_{\text{rot}}$ is

$$\frac{d}{dt}\dot{\boldsymbol{\rho}}|_{\text{rot}} = \frac{d}{dt}\left(\dot{x}\mathbf{u}_x + \dot{y}\mathbf{u}_y + \dot{z}\mathbf{u}_z\right)$$

$$= \ddot{x}\mathbf{u}_x + \ddot{y}\mathbf{u}_y + \ddot{z}\mathbf{u}_z + \dot{x}\dot{\mathbf{u}}_x + \dot{y}\dot{\mathbf{u}}_y + \dot{z}\dot{\mathbf{u}}_z \qquad (C.14)$$

$$= \ddot{\boldsymbol{\rho}}|_{\text{rot}} + \boldsymbol{\omega} \times \dot{\boldsymbol{\rho}}|_{\text{rot}}$$

The time derivative of the third right-hand side term in Eq. (C.13) is

$$\frac{d}{dt}(\boldsymbol{\omega} \times \boldsymbol{\rho}) = \dot{\boldsymbol{\omega}} \times \boldsymbol{\rho} + \boldsymbol{\omega} \times \dot{\boldsymbol{\rho}}$$

$$= \dot{\boldsymbol{\omega}} \times \boldsymbol{\rho} + \boldsymbol{\omega} \times \left(\dot{\boldsymbol{\rho}}|_{\text{rot}} + \boldsymbol{\omega} \times \boldsymbol{\rho}\right) \qquad (C.15)$$

Combining Eqs. (C.14) and (C.15), we obtain

$$\ddot{\mathbf{r}} = \ddot{\mathbf{r}}_O + \ddot{\boldsymbol{\rho}}|_{\text{rot}} + 2\boldsymbol{\omega} \times \dot{\boldsymbol{\rho}}|_{\text{rot}} + \dot{\boldsymbol{\omega}} \times \boldsymbol{\rho} + \boldsymbol{\omega} \times (\boldsymbol{\omega} \times \boldsymbol{\rho}) \qquad (C.16)$$

Equation (C.16) is the absolute acceleration of particle P. A description of each term is:

$\ddot{\mathbf{r}}_O$ = absolute acceleration of the origin of the moving frame $Oxyz$
$\ddot{\boldsymbol{\rho}}|_{\text{rot}}$ = acceleration of P relative to the moving/rotating frame
$2\boldsymbol{\omega} \times \dot{\boldsymbol{\rho}}|_{\text{rot}}$ = Coriolis acceleration
$\dot{\boldsymbol{\omega}} \times \boldsymbol{\rho}$ = acceleration due to the changing angular rate $\boldsymbol{\omega}$ of frame $Oxyz$
$\boldsymbol{\omega} \times (\boldsymbol{\omega} \times \boldsymbol{\rho})$ = centrifugal acceleration.

We may generalize the results of this section by presenting the Coriolis theorem

$$\left.\frac{d\mathbf{x}}{dt}\right|_{\text{fix}} = \left.\frac{d\mathbf{x}}{dt}\right|_{\text{rot}} + \boldsymbol{\omega} \times \mathbf{x} \qquad (C.17)$$

where $d\mathbf{x}/dt|_{\text{fix}}$ is the time derivative of vector $\mathbf{x}$ with respect to an inertial, non-rotating (fixed) reference frame and $d\mathbf{x}/dt|_{\text{rot}}$ is the time derivative of $\mathbf{x}$ with respect to a rotating frame. The angular velocity of the rotating frame relative to the fixed frame is vector $\boldsymbol{\omega}$.

C.3 Polar Coordinates

Figure C.2 shows particle P moving in the plane XY. Cartesian coordinate frame OXY is a fixed, non-rotating or inertial reference frame. Let $\mathbf{u}_r$ and $\mathbf{u}_\theta$ be unit vectors along the radial and transverse directions, respectively, as shown in Figure C.2. The absolute position vector of particle P is

$$\mathbf{r} = r\mathbf{u}_r \qquad (C.18)$$

where $\mathbf{u}_r$ always points from the origin O to P. The absolute velocity of P is determined by taking the time derivative of all terms in Eq. (C.18)

$$\mathbf{v} = \frac{d\mathbf{r}}{dt} = \dot{\mathbf{r}} = \dot{r}\mathbf{u}_r + r\dot{\mathbf{u}}_r \qquad (C.19)$$

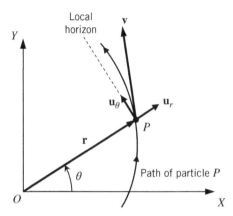

Figure C.2 Planar motion of particle P and polar coordinates.

Note that we must include the time derivative of unit vector $\mathbf{u}_r$ because it changes direction as particle P moves along its path (note also that $\dot{r}$ is the time rate of change of the *magnitude* of vector $\mathbf{r}$). Figure C.2 shows that $\mathbf{u}_r$ rotates with angular velocity $\boldsymbol{\omega} = \dot{\theta}\mathbf{u}_k$ where $\mathbf{u}_k$ is a unit vector normal to the OXY plane with direction out of the page (i.e., $\mathbf{u}_k = \mathbf{u}_r \times \mathbf{u}_\theta$). Hence, the time derivative of $\mathbf{u}_r$ is the cross product of the angular velocity vector $\dot{\theta}\mathbf{u}_k$ and $\mathbf{u}_r$

$$\dot{\mathbf{u}}_r = \boldsymbol{\omega} \times \mathbf{u}_r = \dot{\theta}\mathbf{u}_\theta \tag{C.20}$$

which is a vector in the transverse direction (perpendicular to position vector $\mathbf{r}$). Substituting Eq. (C.20) into Eq. (C.19) yields the absolute velocity of particle P

$$\mathbf{v} = \dot{r}\mathbf{u}_r + r\dot{\theta}\mathbf{u}_\theta = v_r\mathbf{u}_r + v_\theta\mathbf{u}_\theta \tag{C.21}$$

where $v_r = \dot{r}$ and $v_\theta = r\dot{\theta}$ are the radial and transverse velocity components.

The absolute acceleration of P is determined by taking the time derivative of all terms in the velocity equation (C.21):

$$\mathbf{a} = \frac{d\mathbf{v}}{dt} = \dot{\mathbf{v}} = \ddot{r}\mathbf{u}_r + \dot{r}\dot{\mathbf{u}}_r + \dot{r}\dot{\theta}\mathbf{u}_\theta + r\ddot{\theta}\mathbf{u}_\theta + r\dot{\theta}\dot{\mathbf{u}}_\theta \tag{C.22}$$

Substituting the cross products $\dot{\mathbf{u}}_r = \boldsymbol{\omega} \times \mathbf{u}_r = \dot{\theta}\mathbf{u}_\theta$ and $\dot{\mathbf{u}}_\theta = \boldsymbol{\omega} \times \mathbf{u}_\theta = -\dot{\theta}\mathbf{u}_r$ for the time derivatives of the unit vectors, Eq. (C.22) becomes

$$\mathbf{a} = \left(\ddot{r} - r\dot{\theta}^2\right)\mathbf{u}_r + \left(r\ddot{\theta} + 2\dot{r}\dot{\theta}\right)\mathbf{u}_\theta \tag{C.23}$$

Equation (C.23) is the absolute acceleration of particle P where the bracketed terms are the radial and transverse acceleration components. Note that the transverse acceleration term $2\dot{r}\dot{\theta}$ is the Coriolis acceleration.

C.4 Normal-Tangential Coordinates

For launch and entry scenarios, it is usually convenient to express the absolute acceleration in components that are tangent and normal to the flight path. Figure C.3 shows

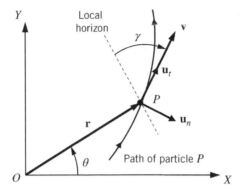

Figure C.3 Planar motion of particle P and normal-tangential coordinates.

particle P moving along its path in the plane XY where OXY is a fixed, non-rotating reference frame. Absolute velocity is

$$\mathbf{v} = v\mathbf{u}_t \qquad \text{(C.24)}$$

where unit vector $\mathbf{u}_t$ moves with P and remains tangent to the path along the direction of motion. The unit vector $\mathbf{u}_n$ is always normal to the path and is chosen so that the cross product $\mathbf{u}_n \times \mathbf{u}_t = \mathbf{u}_k$ where $\mathbf{u}_k$ is perpendicular to OXY and out of the page in Figure C.3. Using Eq. (C.17), the absolute acceleration is

$$\mathbf{a} = \frac{d\mathbf{v}}{dt}\bigg|_{\text{fix}} = \frac{d\mathbf{v}}{dt}\bigg|_{\text{rot}} + \boldsymbol{\omega} \times \mathbf{v} \qquad \text{(C.25)}$$

From Figure C.3, we see that the angular velocity of the rotating coordinates is

$$\boldsymbol{\omega} = \dot{\theta}\mathbf{u}_k = \frac{v\cos\gamma}{r}\mathbf{u}_k \qquad \text{(C.26)}$$

where γ is the angle measured from the local horizon (i.e., the $\mathbf{u}_\theta$ direction; see Figure C.2) to the velocity vector. In flight mechanics problems, angle γ is called the flight-path angle, and it is positive when $\dot{r} > 0$ as shown in Figure C.3. The time derivative of velocity with respect to the rotating normal-tangential frame is

$$\frac{d\mathbf{v}}{dt}\bigg|_{\text{rot}} = \dot{v}\mathbf{u}_t + v\dot{\gamma}\mathbf{u}_n \qquad \text{(C.27)}$$

where $\dot{v}$ is the tangential acceleration due to the change in magnitude of the velocity vector and $v\dot{\gamma}$ is a normal acceleration component due to the rotation of the velocity vector with respect to the local horizon. The cross-product term in Eq. (C.25) is

$$\boldsymbol{\omega} \times \mathbf{v} = \frac{v\cos\gamma}{r}\mathbf{u}_k \times v\mathbf{u}_t = -\frac{v^2\cos\gamma}{r}\mathbf{u}_n \qquad \text{(C.28)}$$

Hence, the absolute acceleration of particle P expressed in normal-tangential coordinates is the summation of Eqs. (C.27) and (C.28):

$$\mathbf{a} = \dot{v}\mathbf{u}_t + \left(v\dot{\gamma} - \frac{v^2\cos\gamma}{r}\right)\mathbf{u}_n \qquad \text{(C.29)}$$

Equation (C.29) can also be obtained from the absolute acceleration in polar coordinates, Eq. (C.23), by using the following substitutions

$$\dot{r} = v\sin\gamma \tag{C.30}$$

$$\ddot{r} = \dot{v}\sin\gamma + v\dot{\gamma}\cos\gamma \tag{C.31}$$

$$r\dot{\theta} = v\cos\gamma \tag{C.32}$$

$$\dot{r}\dot{\theta} + r\ddot{\theta} = \dot{v}\cos\gamma - v\dot{\gamma}\sin\gamma \tag{C.33}$$

and with the polar coordinates expressed in terms of the normal-tangential coordinates:

$$\mathbf{u}_\theta = \cos\gamma\,\mathbf{u}_t - \sin\gamma\,\mathbf{u}_n \tag{C.34}$$

$$\mathbf{u}_r = \sin\gamma\,\mathbf{u}_t + \cos\gamma\,\mathbf{u}_n \tag{C.35}$$

The remainder of this alternative derivation is left to the reader.

Index

Space Flight Dynamics, First Edition. Craig A. Kluever.
© 2018 John Wiley & Sons Ltd. Published 2018 by John Wiley & Sons Ltd.
Companion website: www.wiley.com/go/Kluever/spaceflightmechanics

Printed and bound by CPI Group (UK) Ltd, Croydon, CR0 4YY

27/10/2024

14580309-0002